ALLE ZEIT WACH
1842

K. Y. Kondratyev V. V. Kozoderov O. I. Smokty

Remote Sensing of the Earth from Space: Atmospheric Correction

With 154 Figures and 66 Tables

Springer-Verlag
Berlin Heidelberg New York
London Paris Tokyo
Hong Kong Barcelona
Budapest

Professor Dr. KIRILL Y. KONDRATYEV
Research Centre of Ecological Safety
University emb., 5, 199034 St. Petersburg, Russia

Dr. VLADIMIR V. KOZODEROV
Institute of Computational Mathematics
Academy of Sciences
33, Leninsky pr., Moscow, Russia

Professor Dr. OLEG I. SMOKTY
Institute of Informatics and Automatization
Academy of Sciences
199178, 14th Line, St. Petersburg, Russia

ISBN-13:978-3-642-76749-4 e-ISBN-13:978-3-642-76747-0
DOI: 10.1007/978-3-642-76747-0

Library of Congress Cataloging-in-Publication Data
Kondratyev, K. Y. (Kirill Yakovlevich)
Remote sensing of the earth from space: atmospheric correction / K. Y. Kondratyev, V. V. Kozoderov, O. I. Smokty. p. cm. Includes bibliographical references and index.
ISBN-13:978-3-642-76749-4

1. Aerosols – Mathematical models. 2. Radiative transfer – Mathematical models. 3. Solar radiation – Mathematical models. 4. Remote sensing. I. Kozoderov, V. V. (Vladimir Vasil'evich) II. Smoktiĭ, O. I. (Oleg Ivanovich) III. Title. QC882.42.K66 1992 551.5–dc20

Softcover reprint of the hardcover 1st edition 1992

Production Editor: Isolde Gundermann
Reproduction of the figures: Gustav Dreher GmbH, Stuttgart
Typesetting: K+V Fotosatz GmbH, Beerfelden

32/3145-5 4 3 2 1 0 – Printed on acid-free paper

Preface

The monograph being proposed for the English-speaking research community is concentrated on the atmospheric correction of satellite images as a part of thematic interpretation procedures while processing remote sensing data. For linguistic reasons a large section of the community may have been unaware of the progress made in Russia in this field of science and technology. Meanwhile, Russia was the first country to launch the first artificial satellite in 1957 and to obtain from space for the first time spectra of the Earth's atmosphere in the 1960's. New applications of the radiation transfer theory for the atmosphere-underlying surface system appeared first in Russia in the 1970's. Direct and inverse problems of the atmospheric optics were then formulated giving the scientific basis for studies of natural resources from space. Since that time new mathematical treatments for the atmospheric correction procedures have been widely developed in Russia, including both analytical and numerical techniques to simulate spectral, angular, and spatial distributions of the outgoing radiation in visual and infrared regions. The authors of the book were at the beginning of the scientific approach. A wide range of mathematical improvements to elaborate polinomial approximations for dependencies between atmospheric radiation field and parameters of space surveying was due to the necessity to process satellite images in real time using special software of existed computer means for the studies. Another new scientific approach has included color coding simulation results in terms of spectral reflectivities and optical parameters of the atmosphere. Though the authors are researchers in physics and mathematics, the readers will find in the book some new applications of the techniques under study to optics of natural media and climatological interpretations of temporal data sets for radiation budget components of the Earth. The book summarizes new results concerning models of atmospheric aerosols, radiation and atmospheric constituents interaction, theory and application of satellite data processing. We hope it will be also useful for new ecological research from space.

Authors

Contents

Introduction

The present development of space-based studies of the Earth's environment is characterized by combined use of automated satellites (AS), manned spacecraft (MS), and permanent orbital stations (POS). On the one hand, this is explained by objective necessity to use the space-borne means in studies of global-scale interrelated processes including weather phenomena, climate change, biospheric dynamics, etc. On the other hand, this trend reflects adequately economic need to develop an operational satellite system to observe the atmosphere, land, inland waters and the World ocean to study the environment and Earth's resources. No doubt, in conditions of prolonged periods of orbital space flights automated satellites are the most useful means of solving various problems in remote sensing of the Earth's resources. For the AS to function successfully, there must be a unique automated complex of systems, including (a) on-board scientific instrumentation; (b) a system for recording and preliminary processing of observational data; (c) a system for data transmission, including a sub-system for secondary processing, coding and identifying the data; (d) a system to distribute space-derived information among the users and organizations concerned.

There is, however, a broad class of scientific and technical problems which can only be solved by multipurpose MS and POS. One of the decisive advantages of MS observations is that the cosmonaut can choose observation targets, determine the most favorable conditions for observations, and establish an optimal regime for the on-board instruments in extreme situations. Besides, the cosmonaut, performs an on-board logical analysis of information, carries out its sampled processing, and transmits it to the users at the Flight Control Centers (FCCs) for subsequent correction. There are rapidly improving optical instruments, optimal techniques for data processing, and intensive development of modeling of the processes of radiation transfer, based on the use of analytical, numerical and statistical (Monte-Carlo) techniques. These make it possible to carry out complex experimental and theoretical studies of the laws of spatial, angular and spectral distribution of the fields of the atmosphere-object-background system's radiation to solve the various problems of ecology, atmospheric physics, oceanology, climatology, etc. Independent of the character of problems solved using space-based observations, three kinds of optical information are interesting from the practical point of view.

First, visual observations by the cosmonauts, using the perfect optical device of the human eye, to carry out highly accurate and detailed studies of the environment and outer space.

Second, photographic, TV, lidar, IR and microwave images of the Earth which widen the possibilities for complex analyses of the laws of spatial and temporal variability of the optical and physical parameters of natural formations, backgrounds and natural objects.

Third, multispectral images and spectra of natural formations, whose origin is governed by spectral properties of the incoming solar radiation and by its interaction with natural formations, backgrounds and surface objects.

These three kinds of optical information vary in value and applicability when solving remote sensing problems. A major drawback of space-derived images, despite their high spatial resolution and color characteristics, and those of TV, IR and microwave images, is that their interpretation gives non-unique results. Objectively, this difference is explained because data on the spatial and angular distribution of the brightness field in one spectral interval alone are insufficient to recognize a surface (or above-water) object and to assess its optical and physical parameters. Spectral optical information ensures broader possibilities: it consist of space-recorded spectral brightnesses of the atmosphere, landforms and vegetation, backgrounds and objects in the UV, visible, IR, and microwave spectral regions.

Spectrophotometry of the Earth from space in the wavelength interval 400 – 800 nm was first theoretically proved in the USSR and successfully accomplished in 1968 – 1971 with the MS "Soyuz" and the POS "Salyut-1", using specially developed programs, techniques and instruments.

The obtaining of the first spectra of the twilight halo of the terrestrial atmosphere (MS Soyuz-5), of its day-time horizon (POS Salyut-1), and of basic types of landforms (MS Soyuz-7), successfully achieving the first combined sub-satellite geographical experiment (MS Soyuz-7, 9) have served as the basis for a new perspective for studies of the Earth's resources from space. They have ensured that remote sensing of the environment using optical techniques is a priority of Soviet science and technology.

First experiments in the early 1970s on Earth spectrophotometry from the MS Soyuz and the POS Salyut demonstrated the need to consider atmospheric distortion (atmospheric correction) in practical use of satellite information. Further studies suggested the possibility of atmospheric correction of satellite imagery and spectra, based on the solution of direct and inverse problems of the theory of radiation transfer in the surface-atmosphere system, using specialized computers for information processing.

The physical basis for solving these problems is the recording by on-board optical instruments of the intensities and fluxes of reflected (outgoing) radiation of the atmosphere-object-background system. They also include the degree of its polarization in various intervals of the electromagnetic spectrum (UV, visible, IR, microwave).

The multi-zonal space-derived information obtained on the environment is transmitted, via radio-channels, to Earth where it is recorded on magnetic tapes. Then, it can be either thematically analyzed by digital techniques or, upon its synthesizing as images, visually interpreted. The determination of the shape and state of natural objects is one of the basic objectives of its thematic analysis and interpretation.

The state of natural formations is characterized by a combination of physical and chemical, optical and physical, biometric, and other parameters. The optical-physical parameters of an object form its radiation "image" characterized by a set of constants of the interaction between radiation and the substance of this object. The multidimensional matrix of the optical-physical properties of the object takes into account the major dependence of the outgoing radiation intensity (radiance) on the direction of the object viewing, solar zenith angle and its azimuth, the degree of radiation polarization, etc. The radiation leaving the Earth's surface, upon penetrating the atmosphere, is transformed as a result of interaction with the gas and aerosol components of the atmosphere, being repeatedly re-reflected from the surface. The effect of the atmosphere manifests itself through a complicated transformation of the reflected radiation spectrum, which leads to increased brightness of dark objects and decreased brightness of light objects, as well as to reduced contrasts between objects. The distorting effect of the atmosphere leads to a substantial change of spectral signatures of natural objects for various wavelengths. Thus, the determinantion of the state of natural features by satellite-recorded radiation is connected with the recognition of the radiation "images" (signatures) of these objects transformed by the atmosphere due to multiple scattering, absorption, and refraction.

The complicated character of the dependence of the outgoing short-wave radiation on various parameters can lead to different assessments of the state of natural features from the data of the space-based survey. For instance, considerable difficulties arise in determing the state of vegetation cover in the visible spectrum, since in this region the spectral reflection characteristics of various objects are substantially distorted by the atmosphere. In particular, due to molecular and aerosol scattering of reflected radiation by the atmosphere the vegetation brightness can vary by a factor of two or more.

If the atmosphere-induced distortions are not taken into account, the probability of correct recognition of the signatures of various natural features (especially vegetation) from their space-derived signatures is low. Consideration of atmospheric distortion is particularly urgent because analysis of the space-derived information must be made using an automated processing system of the multi-spectral satellite information requireing precise correction of satellite imagery. Therefore, an assessment of the effect of the real atmosphere on radiation transfer in the atmosphere-object-background system is a necessary stage in studies of the environment and the Earth's resources remote sensing information.

Analysis of the problem connected with consideration of the atmospheric effect on spectral brightnesses of natural phenomena has shown the urgency of the following problems:

1. the determination of angular and spectral dependences of the outgoing radiation intensity in a vertically inhomogeneous atmosphere;

2. the quantitative assessment of the effect of a stratified aerosol atmosphere on the outgoing radiation intensity;
3. the development of algorithms and techniques for practical removal of the distorting effect of the atmosphere on the space-derived surface images using computers.

Practical realization of atmospheric correction of satellite information in real time, requires a presentation of the results of radiation fields modeling as polynomial dependences, which consider basic factors of the radiation field formation. The focus of studies relevant to atmospheric correction of the air- and space-derived data is an approximation of the outgoing radiation intensity, as a function of the angular parameters of the survey and of the surface reflection characteristics by orthogonal polynomials. These make it possible to consider both the results of theoretical modeling and the necessity for a real-time atmospheric correction, in the processing of remote-sensing images and spectra. This setting of the problem of atmospheric correction is scientifically new, compared to the conventional empirical transformations, in which the laws of radiation transformation in the surface-atmosphere system are not considered.

Bearing in mind the high cost of the space rocket complex used to launch satellites, MS, POS, the methodical programme and instrumental provision of the space-based experiments and the interpretation of the data obtained (including visual observations by the cosmonaut), theoretical (simulation) modeling of the radiation fields of the atmosphere-background-object system gains special importance. The general mathematical and physical basis of this modeling is the radiation transfer theory. Using given optical models of the environment and the given power of external (internal) radiation sources it is possible to determine the spatial, angular and spectral values of radiative characteristics of the atmosphere-object-background system (direct problem). Inversely the spatial and spectral distributions of the optical parameters of this system may be determined from the given distribution of the reflected or transmitted radiation (inverse problem).

The atmospheric turbulence and refraction, as a rule, affect the quality of the space-derived images and spectra weakly, a comparative analysis is needed of the applied techniques of the radiation transfer theory and algorithms of atmospheric correction of satellite optical information. This ensures an efficient account of the effect of an aerosol plane-parallel atmosphere during the space-based survey of the Earth on spectral brightnesses, brightness contrasts, albedo, color characteristics and the degree of polarization of natural features in the visible and near IR spectral regions. An advanced technique of spherical harmonics, similarity ratios, and accurate and approximate analytical solutions of the initial equation of radiation transfer, are the mathematical basis for theoretical modeling of radiation fields, in direct and inverse problems, of space-based spectrophotometry.

Numerical and analytical solutions of boundary-value problems of radiation transfer in the atmosphere-object-background system can be solved numerically to give quantitative values for outgoing radiation intensity (radiance). Angular distributions of intensity make it possible to consider correctly the distorting effect of the atmosphere under conditions that, simultaneously with recording outgoing radiation retrieval of the atmospheric parameters affecting the radiation transfer is made. These parameters are: vertical distribution of the concentration of the gaseous and aerosol components of the atmosphere, optically active in the visible and near IR, the composition (complex index of refraction) and size distribution of aerosols. In the first approximation, optical thickness and phase function of the atmosphere can serve as optical parameters.

The concept of the transfer function of the object-background-atmosphere system in the monograph, serves as the basis for atmospheric correction of the space-derived spectrophotometric information. Theoretically, atmospheric correction processing has two stages. Firstly, several inverse problems of radiation transfer theory are solved, associated with retrieval of the optical parameters of the atmosphere and reflection characteristics of the surface. Secondly, the direct problem of transfer theory is solved for the optical characteristics obtained and the given initial boundary conditions. The latter makes it possible to reduce satellite-derived spectrophotometric information to the surface level using the transfer function.

Of principal importance for atmospheric correction are problems concering the non-Lambertian reflection of radiation from the surface and the horizontal diffusion of photons in a spatially inhomogeneous system "atmosphere-Earth-background". All these theoretical problems, together with practical aspects of atmospheric correction of satellite information, made with the use of computers are considered in detail in the monograph. The monograph deals initially with problems of absorption, scattering and attenuation of radiation by individual particles (Mie theory). Then follow problems of radiation transfer in realistically stratified environs. Finally distortions, resulting from the passage of electromagnetic radiation through the atmosphere, are discussed.

Chapter 1 Optical-Physical Models of Atmospheric Aerosol

1.1 Complex Optical Models of Atmospheric Aerosol

The optical parameters of a cloud-free atmosphere as a scattering medium are, firstly, determined by the presence of an aerosol, by its stratification, size distribution, chemical composition and other characteristics. The contribution of an aerosol to the absorption of radiation by the atmosphere in the optical wavelength region 0.2 – 13 μm is also considerable. Besides, at certain sun elevations aerosols affect radiation absorption by the surface. Aerosols are generally understood to be the finest suspended particles, referring to all liquid and solid particles in the troposphere, stratosphere and at higher altitudes, but not cloud particles and ice crystals. An analysis performed recently both in the USSR and in other countries [9, 24, 59, 60, 65, 91, 92, 99, 104, 112, 153, 154, 231, 269] reveals the importance of aerosol radiative effects. However, there is no theory on the formation of the global aerosol field yet which determines the complex problem of the parameterization of its effect on the radiation fields in the atmosphere-surface system.

Techniques for parameterization of the aerosol optical effects can only be developed using more adequate information about the concentration field, microphysical and optical properties of global aerosol [371].

Modeling of the optical state of the atmosphere as a scattering and absorbing natural medium, is reduced, firstly to modeling the optical, chemical and microphysical properties of its aerosol components. Because of varied weather conditions and generation of aerosol particles, this atmospheric material has considerable diurnal and seasonal variations. Analysis of the mechanism for atmospheric aerosol generation shows that various processes of aerosol generation in the atmosphere can be considered independently. For example, formation of soil-erosion and marine aerosols and aerosols which originate during gas condensation. Each process generates particles with a definite chemical composition within a certain size distribution range. Under the influence of diffusion, coagulation and sedimentation processes, the resulting aerosols reach distinctive size distributions which approximate to three modes: nuclear, accumulative, and large-particle mode (large-size aerosol fraction).

In consideration of the aerosol concentration field, one can select the background aerosol that is transported, together with air masses and aerosols from local sources, by air mass motion. Since the concentration and chemical composition of atmospheric aerosols depend strongly on air mass type, the optical properties of the atmosphere and its turbidity also depend on this. Maritime air masses contain large amounts of salt particles – a large-sized aerosol fraction. Arctic air is very pure and the aerosol density in it is lower. Tropical air masses contain many small particles with a high density of the small-sized aerosol fraction (accumulation mode). Various air masses differ not only in particle density but also in the mode structure aerosol concentration and chemical composition.

In motion, air masses may lose one type of aerosol and gain others because of local sources and sinks of aerosols.

Naturally, atmospheric circulation characteristics affect not only the chemical composition and density of aerosols in air masses but also their vertical profile. In mixed air masses atmospheric aerosol can be heavily stratified, with varying concentration and chemical composition. In some cases, the field of aerosol concentration can even have a cluster structure [104], connected with vertical turbulent motion and specific features of condensation processes in the atmosphere. This necessitates the modeling of aerosol as a moving statistical entity, resulting from non-coherent variations of some fractions, that have evolutionary paths of their own and that play different roles in the formation of the atmospheric optical parameters [163]. Modeling of the optical and physical-chemical properties of aerosols [104] is the first and important stage to explain the aerosol effect on the radiation field of the atmosphere-surface system.

A detailed analysis and review of these models is contained in monographs [65, 93, 104]. Most of the proposed models can be divided into: (1) tropospheric, stratospheric, and models of the upper atmosphere; (2) models of various dimensionalities; (3) synthetic models. The simplest models of global aerosol usually estimate optical characteristics, calculated by estimating sphericity and homogeneity of particles using the Mie formulas [26, 43, 162, 206]. In some publications, however, attempts have been made to consider the effects of non-sphericity and inhomogeneity of aerosol particles [235, 270, 322, 407]. One rational approach to calculations of optical characteristics of irregularly-shaped particles is to present them as an equivalent ensemble of spherical particles, randomly oriented in space. This approach to enables use of the Mie theory (developed for spherical, cylindrical and elliptic particles) to calculate these characteristics of polydisperse and irregularly shaped particles. A consideration of irregularly shaped aerosol particles reduces the diffraction maximum and stretches the phase function, together with increasing radiation absorption by particles as opposed to scattering. The latter property of non-spherical particles can be modeled by adding the small-sized, chemically-identical fraction to the large-sized aerosol fraction.

However, a quantitative estimate of possible variations in the model aerosol properties due to these effects and an account of the respective characteristics in the model of global aerosol remain as future problems.

At present, it is impossible to verify an adequate model of global aerosol needed to model the radiation field of the planet Earth. However, a data base sufficient to form a first set of steady-state models has recently been accumulated. These are representative to use as input data in studies of the effect of aerosols on radiation transfer [9, 51, 65, 104, 110, 154].

The problem of aerosol effects in modeling of radiation fields of the atmosphere-surface system can be divided into two stages: (1) the verification of aerosol models which enable determination of a minimum set of independent optical and physical-chemical parameters characterizing the properties of atmospheric aerosols; (2) a detailed parameterization of atmospheric physical processes in the presence of aerosol which corresponds to the complexity of the developed model and to the respective problem.

The developed aerosol models must give a detailed picture of the optical-meteorological conditions observed in the atmosphere. The basic parameters of such models must enable one to introduce information about aerosol in different geographical regions, at various altitudes and at different seasons.

Many physical effects may be included in the developed aerosol models [58, 157, 213, 255, 270, 313, 322, 407].

Summing up, we note that basic directions in studies of radiative aerosol effects at present are as follows: (1) developing realistic (physical-chemical and optical) models of aerosol useful in modeling of the radiation fields of the atmosphere-surface system; (2) comparing available models of radiation transfer in the aerosol atmosphere, to assess the accuracy of the applied algorithms in assessing the effect of input parameters of an aerosol model, initial and boundary conditions on the radiation field of the atmosphere-surface system; (3) revealing basic types of global aerosol needed to model the radiation field of the planet Earth, based on simulation experiments.

It is important that the modeling of the optical characteristics of atmospheric aerosol is a closed one. A closed model is the obtaining of all optical characteristics of atmospheric aerosol (coefficients of absorption, scattering, attenuation, and phase functions) and simultaneous explanation of their variations in connection with the spatial and temporal variations of the aerosol concentration field.

Note, that information about the size distribution of atmospheric aerosol is still inadequate. Most reliable are data on the aerosol optical thickness τ_{aer} at $\lambda = 0.55$ μm and information about vertical profiles of the areosol optical density in the atmosphere. Statistical data on the optical thickness of aerosol τ_{aer}, its diurnal and annual change in different latitudes have been generalized in [104, 268, 279].

Statistical information obtained from regional studies of the spectral-optical characteristics of atmospheric aerosol refers, as a rule, to the attenuation coefficients for the near-surface atmospheric layer [104, 144, 201]. Their extrapolation to other tropospheric layers is incorrect because of vertical variations in the size distribution and chemical composition of atmospheric aerosol. More rational is modeling of the optical characteristics of atmospheric aerosol, based on the use of a priori information on size distribution and chemical composition of aerosols and with explanation of the evolution, source and sink mechanism, small- and large-scale atmospheric circulation. An important circumstance is an explanation of condensation of aerosol particles, strongly affecting the optical properties of atmospheric aerosol in the near surface layer [65, 202, 346, 374].

Many particles of atmospheric aerosol either are hygroscopic or dissolve at a relative humidity over some critical value. The optical properties of particles vary with changing atmospheric humidity, the effective cross-section of attenuation and also the ratio between radiation absorbed and scattered by particle. Variations in the complex refraction index with varying atmospheric humidity can be measured by averaging its real and imaginary parts according to the weight concentration of chemical elements in the particles [65]. An increase of particles' radii with increasing humidity has been considered in [374, 408]. The optical constants for various components of soluble and non-soluble substances have been obtained in [104, 256, 408].

It is important to develop a technique for modeling the vertical distribution of the optical characteristics of atmospheric aerosol. Most acceptable is modeling of the optical characteristics of natural aerosol as a mixture of weighted, by concentration or optical density, optical characteristics of its different fractions. Each aerosol fraction, with its own special chemical composition and size distribution, is given by the respective vertical profile. Thus, the effect of vertical variations in these, on the altitudinal dependences of the optical characteristics of atmospheric aerosol can be taken into account.

It is impossible to construct an adequate model of atmospheric aerosol without taking account of geographical and seasonal factors, meteorological conditions, and the type of surface, either a source or a sink of aerosol formations. Also, it is important to consider the effect of atmospheric pollution and humidity on variations of aerosol optical constants and size distributions. Since there is a large spatial and temporal variability of atmospheric aerosol concentrations and lacking detailed information about its optical properties for a wide range of atmospheric conditions, models of the aerosol optical characteristics must be based on global regional divisions. This approach makes it possible to consider re-distributions of the contributions from independent sources of aerosol formation and mechanisms for generation, in conditions of horizontal homogeneity of aerosols and their optical-physical characteristics.

Models by L. Elterman between 1964 and 1970 [256–260] were the first complex models of atmospheric aerosol. They were based on 79 series of search-light measurements of aerosol scattering at $\lambda = 0.55\ \mu m$ at altitudes 3.7–32 km. These data were included in the atlas of aerosol attenuation values and aerosol optical thickness in the altitude range 0–50 km with a vertical step of $\Delta h = 1$ km for 22 wavelengths in the interval 0.27–4.0 μm. Beyond the range 3.7–32 km for the aerosol attenuation coefficient α_{aer} at $\lambda = 0.5\ \mu m$ the following analytical values were used:

$$\alpha_{aer}(\lambda, z) = \alpha_{aer}(\lambda, z_0) \exp[(z_0 - z)/H]_* \tag{1.1}$$

and for other wavelengths

$$\alpha_{aer}(\lambda, z) = \frac{\alpha_{aer}(\lambda, 0)}{\alpha_{aer}(\lambda_1, 0)} \alpha_{aer}(\lambda_1, z) . \tag{1.2}$$

The scale height $H_* = 1.2$ km in the altitude range $z = 0-3.7$ km and $H_* = 3.75$ km in the range $32-50$ km. The spectral change of aerosol attenuation coefficients $\alpha_{aer}(\lambda, 0)$ and $\alpha_{aer}(\lambda_1, 0)$ in Elterman's model (1968) was obtained by generalizing different experimental data. Such an approach enables one to improve the model some what, with the growing experimental data base on the spectral change of the aerosol attenuation coefficient, depending on the type of aerosol and on atmospheric turbidity. Using this ability of the model, Elterman introduced for the initial models of 1966 and 1968, eight gradations corresponding to 2-to-13-km visibility near the surface [259].

The following development of complex aerosol models concerns the atmospheric aerosol complexity resulting from various mechanisms for the generation and sink of aerosols at several levels. Bearing this in mind, publication [153] suggests a 6-layer model as follows (1) surface-air layer (0 – 1.5 km); (2) troposphere (1.5 – 9.0 km); (3) tropopause (9.0 – 13.0 km); (4) lower stratosphere (13.0 – 22.0 km); (5) upper stratosphere (22.0 – 30 km); (6) high atmospheric layers (30 – 100 km).

Based on analysis of observational data on the size distribution of aerosol particles and calculations of their optical characteristics, a conclusion [374] has been drawn about possibly reducing the number of atmospheric layers: (1) near-surface layer (0 – 1.5 km); (2) troposphere (1.5 – 13 km); (3) stratosphere (13 – 100 km).

The main simplification assumed in this model is the supposition that each particle contains all the uniformly mixed compounds found in chemical analyses of aerosol samples. This enables one to calculate the optical constants of aerosol substances. However, recent studies have shown that the supposed "internal mixing" contradicts the observed properties of aerosol in many cases and distorts its real optical characteristics [65]. Apart from this, this model does not consider organic substances, for instance, any form of carbon is neglected.

This approach just mentioned, in which aerosol is classified by types, sources and sinks, and by vertical startification in the atmosphere, has led to detailed models, for instance, Elterman's model of atmospheric aerosol [259] has been specified [374]. A new aerosol model differentiates the atmosphere into four layers (Fig. 1.1). For the boundary layer (0 – 2 km), ten versions of typical vertical profiles of attenuation coefficients were suggested characterizing both a pure and a polluted maritime atmosphere, with horizontal visibility ranging $S_M = 20-50$ km (figures at the bottom of Fig. 1.1 are visibility values). Pure aerosol is a combination of dust and water-soluble particles (ammonium, calcium sulfate, organic compounds), whose size distribution is approximated by the sum of two log-normal distributions so that number density is governed by the law r^{-4}, and volume density is bimodal with broad maxima in the radius ranges 0.1 – 1 μm and 5 – 100 μm. Urban aerosols are particles of sea salt resulting from evaporation of sea spray, and particles of continental aerosol (small-sized component of background aerosols).

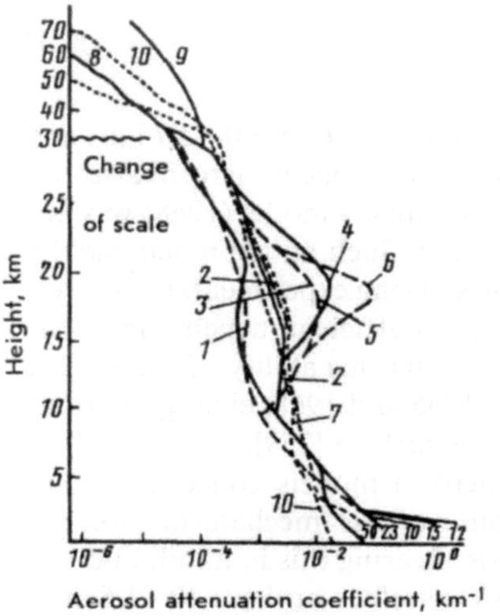

Fig. 1.1. Typical vertical profiles of the aerosol attenuation coefficient at 0.55 μm wavelength: *1* background aerosol (mean global conditions); *2, 3* upon moderate and strong volcanic eruptions, respectively; *4* summer; *5* winter; *6* after powerful volcanic eruptions; *7* the Elterman (1968) model; *8* normal content of aerosol; *9* extremely dust-loaded upper atmosphere; *10* Rayleigh scattering

The upper troposphere aerosol is characterized by two models corresponding to "spring-summer" and "autumn-winter" conditions, in both cases the size distribution being identical to the corresponding continental component of marine aerosol.

For the stratosphere (up to 30 km) two seasonal models (summer and winter) have been suggested, including the effects of volcanic eruptions [228]: i.e. background aerosol and moderate or severe eruptions. The case of the moderate-intensity eruptions is identical to Elterman's model (1968). Background aerosol is thought to consist of droplets of a 75% water solution of H_2SO_4, and its size distribution is determined by Deirmendjian's haze model H [43]. The size distribution of volcanic aerosol (above 30 km) is described by two models: that of fresh volcanic aerosol, and that of volcanic aerosol, in both cases a modified gamma-distribution is applied. Meteoritic dust is a major component of the upper atmosphere aerosol.

Some of the calculations (with Mie formulae) of the optical characteristics for the aerosol models described (coefficients of attenuation, scattering and absorption, phase function of brightness and polarization, etc.) of wavelength 0.2–40 μm have been described in [374]. The spectral changes of the attenuation coefficients for rural and urban aerosol differ little (but dependence of scattering and absorption coefficients on wavelengths are substantially different).

A mean-global model of the size distribution, chemical composition and optical thickness of aerosol (at $\lambda = 0.55$ μm) in the stratosphere and troposphere to calculate the fields of radiation in global mean conditions is suggested in [389]. The data obtained refer to a 3-layer atmosphere (0–3 km, 3–12 km, 12–45 km). However, the vertical profile of the aerosol optical thickness τ_{aer} is described in detail using Elterman's model [259]. Characteristics of this aerosol model are given in Table 1.1.

Table 1.1. A Global model of atmospheric aerosol

H, km	τ_{aer} (0.55 μm)	Composition	Size distribution
0	0.125	15% Sea salt	$dN(r)/dr = C$; $r < r_0$
1	0.083	The same	The same
2	0.058	35% Soil (basalt) 50% Sulfates	$dN(r)/dr = C(r_0k/r)^{\beta+1}$, $r \geqslant r_0$, $r = 0.1$ μm; C – arbitrary constant $0.1 < r < 5$ μm; $\beta = 2$; 5 μm$< r <$50 μm; $\beta = 4$; 50 μm$< r$; $dN/dr = 0$
3	0.042	40% Soil (basalt)	$dN/dr = C$, $r < r_0$
4	0.034		$dN(r)/dr = C(r_0/r)^{\beta+1}$, $r \geqslant r_0$
5	0.028	60% Sulfates	$r = 0.045$ μm; C-arbitrary constant; $0.045 < r < 5$ μm; $\beta = 2.5$
6	0.024	The same	$5 < r < 30$ μm; $\beta = 4.6$; 30 μm$< r$; $dN/dr = 0$
7	0.021	–	–
8	0.017	–	–
9	0.014	–	–
10	0.011	–	–
11	0.008	–	–
12	$5.0 - 10^{-3}$	–	–
15	3.3×10^{-3}	100% Sulfates (a 75% water solution of H_2SO_4)	$dN(r)/dr \sim \exp\left[\frac{-\ln^2(\sqrt{m}/r)}{2\ln^2\sigma}\right]$ $\sigma = 2$; $\sqrt{m} = 0.035$ μm
16	1.8×10^{-3}	–	–
21	8.3×10^{-4}	–	–
24	4.2×10^{-4}	–	–
27	2.0×10^{-4}	–	–
30	8.7×10^{-5}	–	–
33	3.8×10^{-5}	–	–
36	1.6×10^{-5}	–	–
39	6.4×10^{-6}	–	–
42	2.0×10^{-6}	–	–
45	0.0	–	–

The tropospheric aerosol is thought to consist of homogeneous solid spherical particles. The effect of humidity and the possible presence of organic components, combustion products, nitrates and aerosol pollutions are not considered.

Results are also given [389] of calculations of the attenuation coefficient σ, phase function $x(\gamma)$ and its asymmetry factor for different components of aerosol, depending on wavelength λ in the interval $0.1-100$ μm, as well as aerosol optical thicknesses τ_{aer} for all three layers of the atmosphere. The sensitivity of aerosol optical characteristics to variations in the complex refraction index m and size distribution of aerosol was studied. From the data on optical thickness τ_{aer} and attenuation coefficient σ the vertical profile of particles density $N(z)$ was found and compared with the available observational data.

Generally two parameters must be known for an aerosol model: (1) the complex refraction index of particles; (2) the particle size distribution. If we assume particles sphericity and homogeneity, these two characteristics are enough to calculate all optical characteristics of aerosols, using Mie theory [153, 330, 331].

The aerosol vertical profiles are determined by a combining some size distributions and complex indices of refraction, vertically distributed in a different way. The aerosol size distribution is defined using formulas:

$$\frac{dN_i(r)}{dr} = \frac{N_i}{\lg r \sigma_i \sqrt{2\pi}} \exp\left[\frac{-(\lg r - \lg r_i)^2}{2\sigma_i^2}\right] \quad (1.3)$$

$$\frac{dN}{dr} = A r^{\alpha} \exp(-br^{\gamma}) \; . \quad (1.4)$$

The first formula describes a log-normal distribution, and the second formula describes a modified gamma-size distribution of aerosol particles. The size distribution parameters are chosen so that each distribution is normalized to N particles/cm^3. The N value is chosen so that the first coefficient of attenuation calculated at $\lambda = 0.55$ µm be unity. The $N_i(r)$ is the number of particles with a radius below that fixed, and the index refers to various modes constituting the given mixture of aerosol particles. In addition the aerosol model [331] contains tabulated values of the complex index of refraction m, coefficients of attenuation, scattering and absorption, single-scattering albedo, and asymmetry factor in the range $0.2 - 40$ µm for the five types of aerosol (Fig. 1.2): background 1; dry 2; moist 3; inversional 4; hematitic 5.

The model covers the wavelength interval $0.2 - 40$ µm and, additionally, contains information about vertical distributions of temperature, pressure, humidity and optically active gaseous components of the atmosphere in the altitude range $h = 0 - 100$ km. The data on temperature, pressure, humidity, air density and on ozone content are taken for the USA standard atmosphere (1976), for tropical conditions and sub-arctic winter. Also, possible interpolation procedures using the available table have been discussed, to make compatible input data so that a larger vertical resolution of the model can be obtained. An exponential interpolation is recommended for the pressure field, aerosol concentration and atmospheric gases, and a linear interpolation for temperature.

Continuing the idea of regional modeling [104, 139, 374], it was proposed to develop models of the aerosol optical characteristics, taking account of specific conditions, for the following regions:

1. sea basins and oceans, consider wind speeds and dust outbreaks over the ocean;
2. continental mid-latitude regions, including seasonal features;
3. arid and sub-arid regions, taking account of specific features of atmospheric circulation, diurnal change of optical density, and dust outbreaks over the land;
4. heavily populated industrial regions of Western Europe and America;
5. coastal tropical regions;
6. polar latitudes;
7. vast forested and marshy regions.

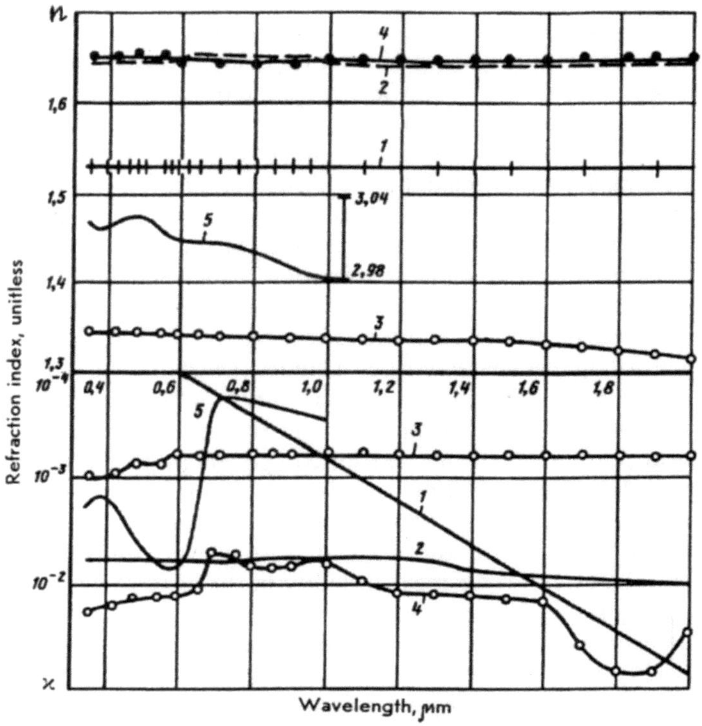

Fig. 1.2. The spectral course of the real and imaginary parts of the complex refraction index for the aerosol models under study

In the proposed [104, 374] model of the vertical structure of atmospheric aerosol, including the surface effects the character of vertical turbulent exchange, atmospheric circulation, type of air masses, and local sources of aerosol, the following zones should be selected:

1. a zone of surface-air haze 0–70 m thick with considerable diurnal variations in aerosol concentration;
2. a surface layer of atmospheric haze $z_1 = 70$ m in which the volume concentration of particles usually decreases with height (except for the atmosphere with a dust outbreak over the ocean for which $z_1 = 0.3-0.5$ km);
3. a layer of active vertical turbulent exchange with vertically constant concentration of aerosol (for arid zones z_2 may be 5–6 km);
4. a layer of the upper troposphere ($z_2 - z_3$) with the aerosol concentration decreasing sharply to the background level (z_3 is the height of the tropopause).

In stratospheric aerosol with general background-level aerosol, a sulfate aerosol layer can be identified ($z_4 - z_5$), in which the maximum of the particle concentration is at heights of 22–26 km in the equatorial region and 17–18 km in polar regions. Above this layer, the aerosol concentration decreases with height (except for, probably, a maximum around 40 km) and is largely determined by the concentration of background aerosol. There is also a zone of global-scale ozone depletion in the stratosphere around a height of 30 km. The heights of maximum density of stratospheric aerosol are not fixed: an optical density maximum is located at 20 km in the near-equatorial region and at 12–15 km in polar regions. With this approach [103, 104] studies have been carried out on dust aerosol, sulfates, global mean fraction of water solutions of the continental salt fraction of aerosol, sea salt particles and their water solutions, industrial aerosols with different values of the real and imaginary parts of the complex refraction index. The developed model of the aerosol optical characteristics includes the coefficients of aerosol attenuation, absorption, scattering and phase function for polarized components. The aerosol models are constructed with the help of computers by fixing the vertical profiles of various aerosol components whose size distribution is determined by superposition of different gamma-distributions (1.3–1.4). The modeling implies the construction of zonal models of the aerosol optical characteristics including diurnal and seasonal variations of atmospheric aerosol and the extent of air mass turbidity in the zone of active turbulent exchange.

Figure 1.3 shows a model of the vertical structure of the atmospheric aerosol optical density for a global mean atmospheric model [104]. In the vertical profile of aerosol there is selected a small-sized aerosol fraction, representing the background aerosol (nuclear mode), which includes 80% of the particles of gas-to-particle conversion origin (sulfates, organic component, soot particles) and 20% of small-sized dust particles.

The proposed [104] global mean model of atmospheric aerosol differs somewhat from the model in [389] in both the chemical composition and size

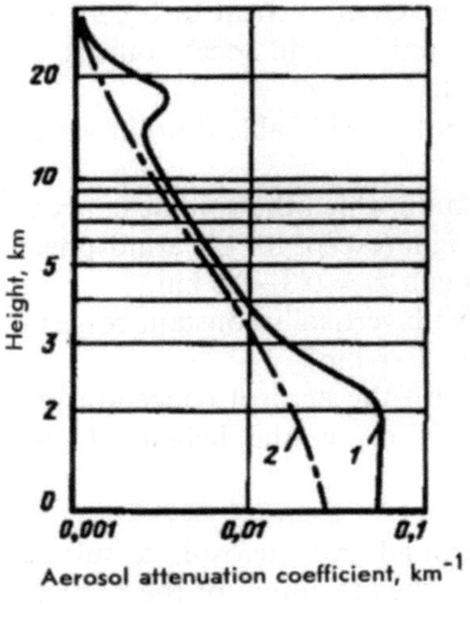

Fig. 1.3. A model *1* of the vertical structure of the optical density of atmospheric aerosol for the global mean atmospheric model; the vertical change of the small-sized aerosol fraction *2*

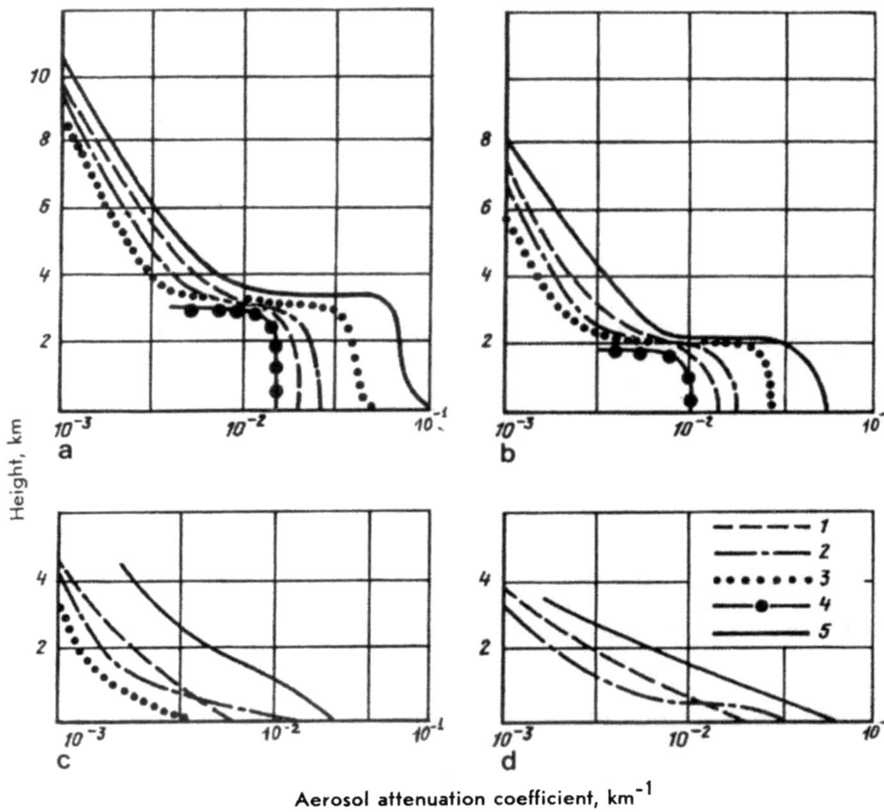

Fig. 1.4a–d. Optical models of the atmospheric aerosol vertical structure in mid-latitudes **a** for summer **b** spring – fall **c** winter **d** the Arctic: *1* small-sized fraction of the background aerosol; *2* soluble fraction of atmospheric aerosol; *3, 4* dust aerosol fraction; *5* large-sized fraction

distribution as well as optical properties. The presence in the model [103] of a small-sized fraction of background- and industrial-aerosol fraction leads to the fact that aerosol absorbs more in the visible and near IR spectral regions [11, 63].

Figure 1.4 shows an optical model of tropospheric aerosol for summer conditions in mid-latitudes. It includes vertical profiles of the optical density of background aerosol, represented by its small-sized fraction, of the soluble salt compounds (mainly, $(NH_4)_2SO_4$) and large-sized and middle-sized fractions of dust aerosol [104].

Figure 1.5 illustrates optical models of atmospheric aerosol over the Atlantic ocean [104]. Curve 2 is the vertical profile of the optical density of the salt aerosol fraction in calm conditions (A), at a wind speed of 5 m/s (B), 10 m/s (C) and 15 m/S (D). Curves 3, 4, 5 are contributions of dust aerosol in conditions of weak, moderate and severe dust outbreaks.

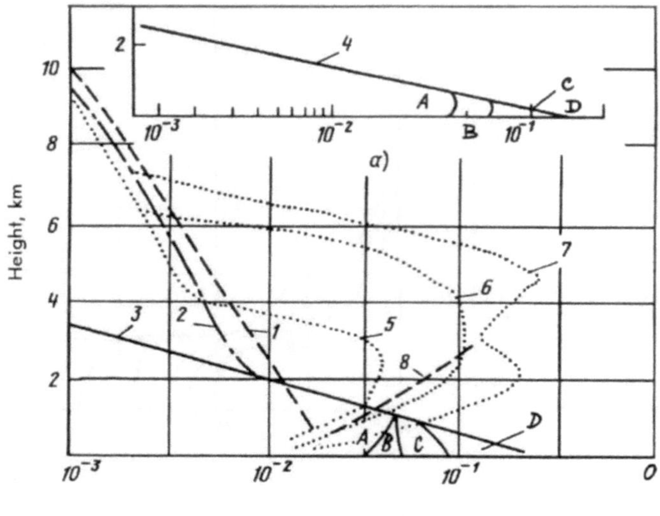

Fig. 1.5. Models of the vertical profiles of optical density for the fractions of *1* small-sized background, *2* global mean salt, *3* marine, *5–8* dust aerosol over the Atlantic ocean with atmospheric humidity $\bar{f} < 75\%$; *4* fall and spring conditions at $\bar{f} > 75\%$ for marine aerosol

The mineral fraction of dust aerosol is characterized by decreasing dust concentration as it sinks onto the water surface. The maximum dust-concentration height depends on the intensity of dust outbreak. For severe dust storms the aerosol optical thickness can reach $\tau_{aer} = 1.2$. Because dust is washed out by cloudiness a minimum of the dust cloud concentration is observed in its vertical profile at a height of 3 km. As the dust cloud is partially washed out, its concentration profile evolves into curve 6.

Figure 1.6 suggests an optical model of atmospheric aerosol for coastal tropical regions in zones of active turbulent exchange [104]. It includes vertical profiles of optical density for small-sized aerosol fractions, sulfate and marine aerosols, and both large- and middle-sized fractions of dust aerosols. If, over the ocean, the concentration of dust in the zone of active exchange decreases with height, so in the coastal zones the concentration of sea salt particles in $z < 1.5$ km decreases at lower elevations.

The latter is determined by the sinking of marine aerosol onto the surface and also by the lifting of sea salt particles to great heights by the upward air fluxes over the continents, due to the growth of the vertical component of turbulent air masses.

Figure 1.7 exemplifies the construction of a microphysical model of atmospheric aerosol, over a sub-arid region, in the zone of active turbulent exchange [104]. It is thought that for the soluble salt aerosol fraction z is most stable and its vertical profile varies weakly both diurnally and seasonally. Nuclear mode 1 is a result of atmospheric photochemical changes in the gas

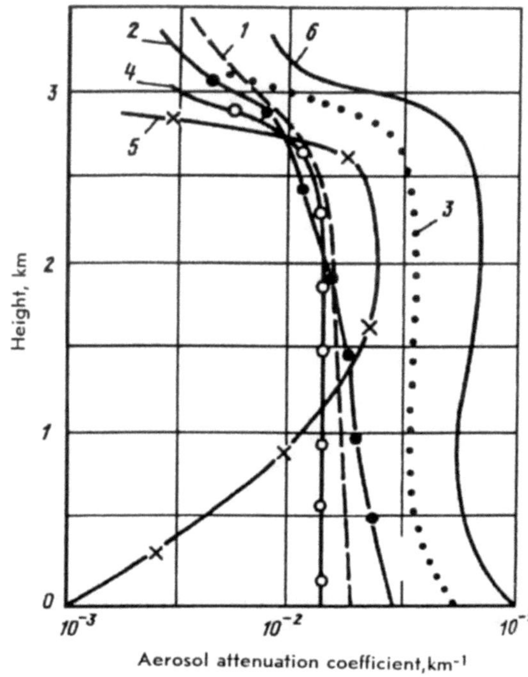

Fig. 1.6. An optical model of the vertical structure of atmospheric aerosol for the tropical coastal regions with onshore winds in the zone of active turbulent exchange: *1* small-sized fraction of background aerosol; *2* salt fraction of background aerosol; *3, 4* dust aerosol fraction; *5* sea salt particles; *6* vertical profile of the aerosol optical density gradient

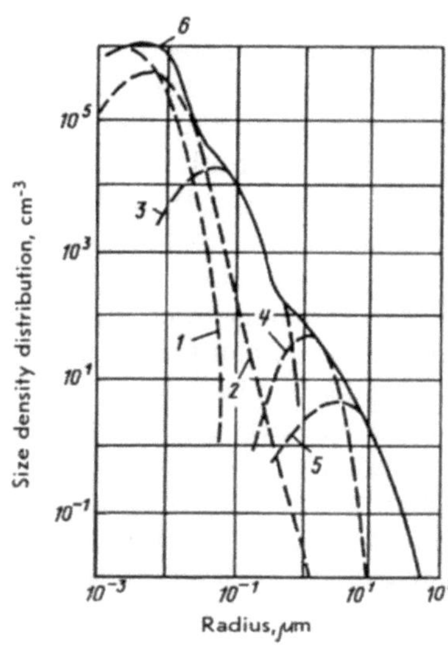

Fig. 1.7. An exemplified model of the microphysical characteristics of atmospheric aerosol for the sub-arid region: *1* the background aerosol fraction; *2* soluble fraction; *3–5* different fractions of dust aerosol; *6* total size distribution of aerosol particles

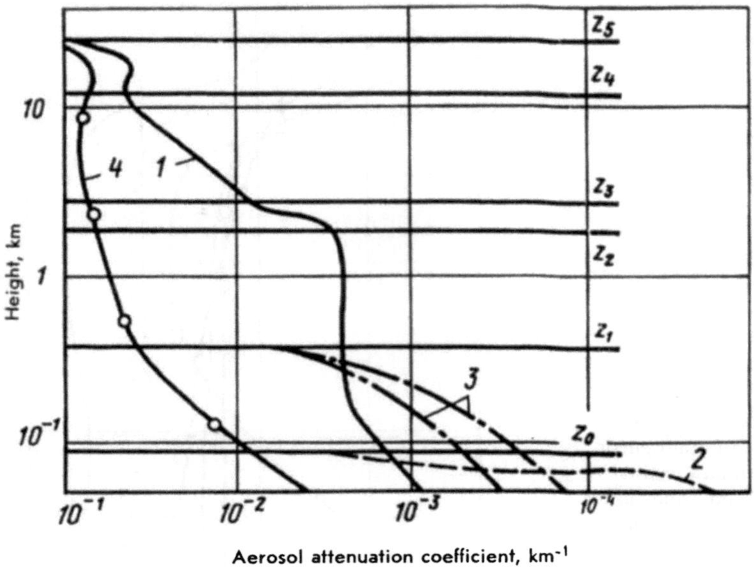

Fig. 1.8. The vertical structure of the optical density of *1* atmospheric aerosol, *2* haze, *3* fog, *4* aerosol over the pole (model)

phase. The size distribution and optical properties of dust aerosol are determined by the surface state and by atmospheric turbidity, in the zone of active turbulent exchange. With a relative humidity below 75%, its effect on the optical characteristics can be neglected.

Figures 1.4 and 1.8 illustrate optical models of the atmosphere for the arctic air masses and mid-latitude winter conditions. These models are characterized by the presence of the small-sized fraction of aerosol, sulfate aerosol, and middle-sized aerosol fraction [104].

The optical models shown in Figs. 1.3 to 1.8 reflect the atmospheric aerosol optical properties at humidity $f<75\%$. The model calculations of the atmospheric aerosol optical characteristics [104, 374] at high humidities ($f>75\%$) show that in the condensation growth of particles, over the continents, small-sized background and global mean salt aerosols play the least roles. Condensation effects occur on larger-sized fractions of dust and salt aerosols, mainly in conditions of calm weather. An increase of atmospheric humidity and continental dust lead to increased atmospheric turbidity, further stretching of phase functions, and a decrease of radiation absorption in the visible and near IR spectral regions.

Figure 1.9 illustrates how marine aerosol size distributions depend on different wind speeds over the ocean, at a relative humidity of $\tilde{f} = 80\%$. With increasing wind speed, the number of particles grows and the number of gigantic particles increases especially. Note, that the vertical profile of the marine aerosol density depends also, on the maritime air turbidity, in the surface-

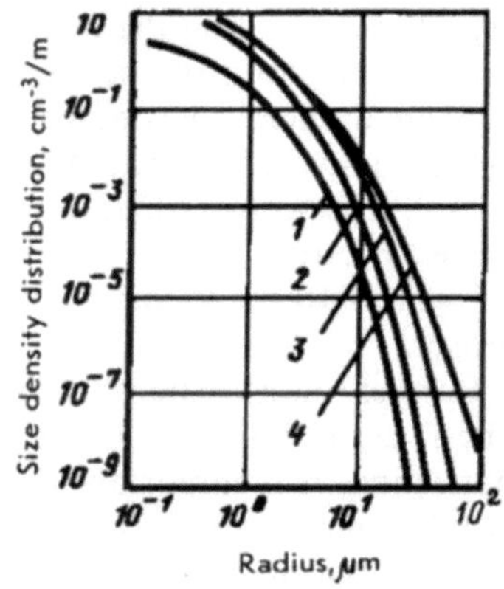

Fig. 1.9. The marine aerosol size distribution for different wind speeds at a relative humidity $\bar{f} = 80\%$: *1* 1 m/s; *2* 5 m/s; *3* 10 m/s; *4* 15 m/s

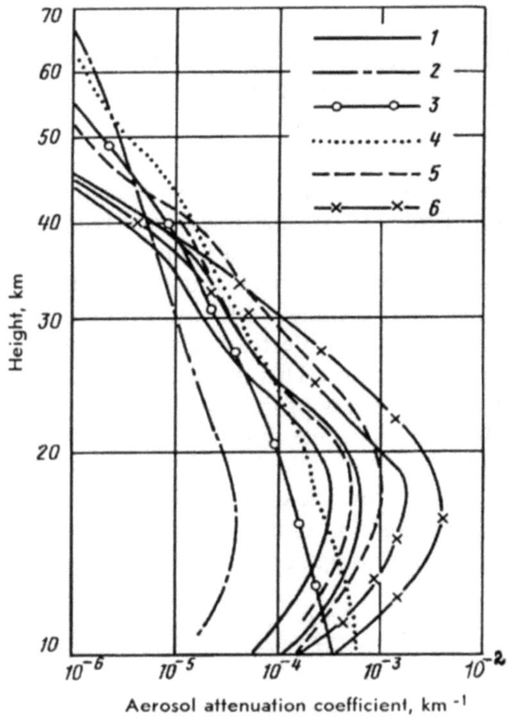

Fig. 1.10. The vertical profiles of optical density at $\lambda = 0.55$ μm for different fractions of stratospheric aerosol (*1*–*6*). Fractions *1*, *5*, *6* are shown for periods of moderate and strong volcanic activity

air layer over the ocean. This determines the generation and sink of aerosol. At an air humidity $\bar{f} < 75\%$, the concentration of marine aerosol and its optical thickness, will be less in cold regions and greater in warm regions. The aerosol in the turbulent exchange zone in cold regions should contain fewer gigantic particles. The stratospheric aerosol has less variable size distribution and chemical composition compared to the tropospheric one, and its optical characteristics depend on the intensity of volcanic activity especially in the upper parts of the stratospheric aerosol layer. This is because volcanic aerosol has different a chemical composition compared to other types of stratospheric aerosol, and hence, different optical characteristics. Compared to sulfate aerosol, it absorbs radiation more strongly in the visible and more weakly in the IR spectrum.

Figure 1.10 shows various models of the vertical structure of stratospheric aerosol.

Further improvements in modeling the optical characteristics of atmospheric aerosol must be made, taking into account the radiative factor. In this connection, it is important to study interactions of radiative, convective and turbulent heat- and mass-exchange with the generation, chemical composition, size distribution and optical properties of atmospheric aerosol.

1.2 Application of Mie Theory in Determining the Optical Parameters of Atmospheric Aerosol

The problem of theoretical assessment of the optical characteristics of such a complicated polydisperse system as real atmospheric aerosol, is, in general, impossible to solve, without some simplifying assumptions. Let an ensemble of homogeneous spherical scatterers, where size distribution follows the law $f(r, z)$ and complex refraction index $m = n + i\kappa$, which depends on the wavelength λ of electromagnetic radiation, be substituted for a set of real aerosol particles. This idealization of atmospheric aerosol, makes it possible to use the results of the theory of scattering of a plane electromagnetic wave, falling in a definite direction onto a homogeneous spherical particle [26, 43, 206].

Monochromatic cross-sections of scattering $k_s(m, \tilde{x})$ and attenuation $k_t(m, \tilde{x})$ calculated for an individual spherical particle of radius r can be written using the following series [43]:

$$k_s(m, \tilde{x}) = \frac{2\pi r^2}{\tilde{x}^2} \sum_{k=1}^{\infty} (2k+1)(|\tilde{a}_k|^2 + |\tilde{b}_k|^2) \; ; \tag{1.5}$$

$$k_t(m, \tilde{x}) = \frac{2\pi r^2}{\tilde{x}^2} \sum_{k=1}^{\infty} (2k+1) \operatorname{Re}(\tilde{a}_k + \tilde{b}_k) \; ; \tag{1.6}$$

$$\tilde{a}_k(m, \tilde{x}) = \frac{\left[\dfrac{A_k(y)}{m} + \dfrac{k}{\tilde{x}}\right] \operatorname{Re}\{\omega_k(\tilde{x})\} - \operatorname{Re}\{\omega_{k-1}(\tilde{x})\}}{\left[\dfrac{A_k(y)}{m} + \dfrac{k}{\tilde{x}}\right] \omega_k(\tilde{x}) - \omega_{k-1}(\tilde{x})} \; ; \tag{1.7}$$

$$\tilde{b}_k(m, \tilde{x}) = \frac{\left[m A_k(y) + \frac{k}{\tilde{x}}\right] \mathrm{Re}\{\omega_k(\tilde{x})\} - \mathrm{Re}\{\omega_{k-1}(\tilde{x})\}}{\left[m A_k(y) - \frac{k}{\tilde{x}}\right] \omega_k(\tilde{x}) - \omega_{k-1}(\tilde{x})} ; \tag{1.8}$$

where $\tilde{x} = \frac{2\pi r}{\lambda}$; $y = m\tilde{x}$.

The Riccati-Bessel function $\omega_k(\tilde{x})$ in (1.7 – 1.8) are expressed through the Bessel cylindric functions of the first kind of half-integer order of complex argument $Y_{k\pm 1/2}(\tilde{x})$:

$$\omega_k(\tilde{x}) = \sqrt{\frac{\pi\tilde{x}}{2}}\, [Y_{k+1/2}(\tilde{x}) + i(-1)^k Y_{-k-1/2}(\tilde{x})] \; . \tag{1.9}$$

Functions $A_k(y)$ meet the following recurrent equations [43]:

$$A_k(y) = -\frac{k}{y} + \left[\frac{k}{y} - A_{k-1}(y)\right]^{-1} , \quad k \geqslant 1 \; ; \tag{1.10}$$

$$A_0(y) = \frac{\sin\tilde{p}\cos\tilde{p} + i\,\mathrm{sh}\,q\,\mathrm{ch}\,q}{\sin^2\tilde{p} + \mathrm{sh}^2 q} , \tag{1.11}$$

where $\tilde{p} = n\tilde{x}$, $q = \kappa\tilde{x}$.

Factors concerning efficiency of scattering $K_{scat}(m, \tilde{x})$ and attenuation $K_{att}(m, \tilde{x})$ are related to cross-sections k_s and k_t in the following way:

$$K_{scat}(m, \tilde{x}) = \frac{k_s(m, \tilde{x})}{\pi r^2} \tag{1.12}$$

$$K_{att}(m, \tilde{x}) = \frac{k_t(m, \tilde{x})}{\pi r^2} \tag{1.13}$$

Note, that with $\tilde{x} > 30$, in some cases calculations with formulas (1.5 – 1.11) cannot be accurate if standard recurrent formulas for Bessel functions are used [53]. This is determined by the growth of these functions at large indices of summing, leading to an overloading of arithmetic computers. Additionally when $\tilde{x}$ values are large the accuracy of calculations must be checked twice, leading to considerable computing time expenditures. Therefore, it is more useful to present formulas (1.5 – 1.11) as some mathematical expressions free of these drawbacks. The same expressions are useful when calculating atmospheric aerosol phase functions. To improve the calculation algorithm, we separate the real and imaginary parts in expression (1.10):

$$\operatorname{Re}A_k(y)=\frac{k\tilde{p}}{\tilde{p}^2+q^2}+\frac{\dfrac{kp}{\tilde{p}^2+q^2}-\operatorname{Re}A_{k-1}(y)}{\left[\dfrac{k\tilde{p}}{\tilde{p}^2+q^2}-\operatorname{Re}A_{k-1}(y)\right]^2+\left[\dfrac{kq}{\tilde{p}^2+q^2}-\operatorname{Im}A_{k-1}(y)\right]^2}\ ; \tag{1.14}$$

$$\operatorname{Im}A_k(y)=-\frac{kq}{\tilde{p}^2+q^2}-\frac{\dfrac{k\tilde{p}}{\tilde{p}^2+q^2}-\operatorname{Im}A_{k-1}(y)}{\left[\dfrac{k\tilde{p}}{\tilde{p}^2+q^2}-\operatorname{Re}A_{k-1}(y)\right]+\left[\dfrac{kq}{\tilde{p}^2+q^2}-\operatorname{Im}A_{k-1}(y)\right]^2}\ ; \tag{1.15}$$

Hence,

$$\operatorname{Re}Q_k=\frac{k}{\tilde{x}}+\frac{\tilde{x}(\tilde{p}\operatorname{Re}A_k-q\operatorname{Im}A_k)}{\tilde{p}^2+q^2}\ ;\qquad \operatorname{Re}P_k=\frac{k+\tilde{p}\operatorname{Re}A_k+q\operatorname{Im}A_k}{\tilde{p}^2+q^2}\ ; \tag{1.16}$$

$$\operatorname{Im}Q_k=\frac{\tilde{x}(\tilde{p}\operatorname{Re}A_k+\tilde{p}\operatorname{Im}A_k)}{\tilde{p}^2+q^2}\ ;\qquad \operatorname{Im}P_k=\frac{\tilde{p}\operatorname{Im}A_k-q\operatorname{Re}A_k}{\tilde{x}}\ . \tag{1.17}$$

In notations of (1.16–1.17) the real and imaginary parts of formula (1.7) become:

$$\operatorname{Re}\tilde{a}_k=1+\frac{R_{1k}-R_{2k}R_{3k}}{1+R_{3k}^2}\ ; \tag{1.18}$$

$$\operatorname{Im}\tilde{a}_k=\frac{R_{1k}R_{3k}+R_{2k}}{1+R_{3k}^2}\ . \tag{1.19}$$

where

$$R_{1k}=\frac{\operatorname{Im}Q_k}{\dfrac{\operatorname{Re}Q_k J_{k+1/2}(\tilde{x})-J_{k-1/2}(\tilde{x})}{J_{-k-1/2}(\tilde{x})}}(-1)^k-\operatorname{Im}Q_k\ ; \tag{1.20}$$

$$R_{2k}=\frac{\operatorname{Re}Q_k+\dfrac{J_{-k+1/2}(\tilde{x})}{J_{-k-1/2}(\tilde{x})}}{\dfrac{\operatorname{Re}Q_k J_{k+1/2}(\tilde{x})-J_{k-1/2}(\tilde{x})}{J_{-k-1/2}(\tilde{x})}(-1)^k-\operatorname{Im}Q_k}\ ; \tag{1.21}$$

$$R_{3k}=\frac{\operatorname{Re}Q_k+\dfrac{J_{-k+1/2}(\tilde{x})}{J_{-k-1/2}(\tilde{x})}+\operatorname{Im}Q_k\dfrac{J_{k+1/2}(\tilde{x})}{J_{-k-1/2}(\tilde{x})}(-1)^k}{\dfrac{\operatorname{Re}Q_k J_{k+1/2}(\tilde{x})-J_{k-1/2}(\tilde{x})}{J_{-k-1/2}(\tilde{x})}(-1)^k-\operatorname{Im}Q_k}\ . \tag{1.22}$$

The real and imaginary parts $\tilde{b}_k(m, \tilde{x})$ can easily be expressed in terms of complexes R_{1k}, R_{2K}, R_{3k}, resulting from the substitution of Q_k by P_k in formulas (1.20–1.22).

If $1 \ll \tilde{x} \ll \lambda' = k + \frac{1}{2}$, then

$$J_{\lambda'}(\tilde{x}) = \frac{1}{2}\sqrt{\frac{2}{\pi\mu'}}\, e^{\mu' - \lambda' \operatorname{arcth}(\mu'/\lambda')} ; \tag{1.23}$$

$$Y_{\lambda'}(\tilde{x}) = -\sqrt{\frac{2}{\pi\mu'}}\, e^{-\mu' + \lambda' \operatorname{arcth}(\mu'/\lambda')} ; \tag{1.24}$$

where $\mu' = \sqrt{\lambda'^2 - \tilde{x}^2}$.

With the help of asymptotic presentations (1.23) and (1.24) the calculation of cylindrical functions in (1.20–1.22) is simplified:

$$\frac{J_{k+1/2}(\tilde{x})}{J_{-k-1/2}(\tilde{x})} = \frac{(-1)^k}{2} \exp\left[(2k+1)\left(\sqrt{1-\left(\frac{\tilde{x}}{k+1/2}\right)^2} - \frac{1}{2} \ln \frac{1+\sqrt{1-\left(\frac{\tilde{x}}{k+1/2}\right)^2}}{1-\sqrt{1-\left(-\frac{\tilde{x}}{k+1/2}\right)^2}}\right)\right] ; \tag{1.25}$$

$$\frac{J_{k-1/2}(\tilde{x})}{J_{-k-1/2}(\tilde{x})} = \frac{(-1)^k}{2} \exp\left\{ \left(k-\frac{1}{2}\right)\sqrt{1-\left(\frac{\tilde{x}}{k-1/2}\right)^2} + \left(k+\frac{1}{2}\right)\sqrt{1-\left(\frac{\tilde{x}}{k+1/2}\right)^2} - \left(k-\frac{1}{2}\right) \ln \frac{1+\sqrt{1-\left(\frac{\tilde{x}}{k-1/2}\right)^2}}{1-\sqrt{1-\left(\frac{\tilde{x}}{k-1/2}\right)^2}} - \left(k+\frac{1}{2}\right) \ln \frac{1+\sqrt{1-\left(\frac{\tilde{x}}{k+1/2}\right)^2}}{1-\sqrt{1-\left(\frac{\tilde{x}}{k+1/2}\right)^2}} \right\} \sqrt{\frac{2k+1}{2k-1}} ; \tag{1.26}$$

$$\frac{J_{-k+1/2}(\tilde{x})}{J_{-k-1/2}(\tilde{x})} = -\sqrt{\frac{2k+1}{2k-1}}\exp\left\{\left(k+\frac{1}{2}\right)\sqrt{1-\left(\frac{\tilde{x}}{k+1/2}\right)^2}-\left(k-\frac{1}{2}\right)\right.$$

$$\times\sqrt{1-\left(\frac{\tilde{x}}{k-1/2}\right)^2}-\left(k+\frac{1}{2}\right)\frac{1}{2}\ln\frac{1+\sqrt{1-\left(\frac{\tilde{x}}{k+1/2}\right)^2}}{1-\sqrt{1-\left(\frac{\tilde{x}}{k+1/2}\right)^2}}$$

$$\left.+\left(k-\frac{1}{2}\right)^{1/2}\ln\frac{1+\sqrt{1-\left(\frac{\tilde{x}}{k-1/2}\right)^2}}{1-\sqrt{1-\left(\frac{\tilde{x}}{k-1/2}\right)^2}}\right\}. \tag{1.27}$$

If $1 \ll \lambda' \ll \tilde{x}$, then, to calculate the Bessel functions of the half-integer index, recurrent formulas can be used [196]:

$$J_{k+1/2}(\tilde{x}) = \frac{(2k-1)}{\tilde{x}}\, J_{k-(1/2)}(\tilde{x}) - J_{k-(3/2)}(\tilde{x})\ ;$$

$$J_{-k-(1/2)}(\tilde{x}) = -\frac{(2k-1)}{\tilde{x}}\, J_{-k+(1/2)}(\tilde{x}) - J_{-k+(3/2)}(\tilde{x})\ ; \tag{1.28}$$

with

$$J_{1/2}(\tilde{x}) = \sqrt{\frac{2}{\pi\tilde{x}}}\sin\tilde{x}\ , \qquad J_{-1/2}(\tilde{x}) = \sqrt{\frac{2}{\pi\tilde{x}}}\cos\tilde{x}\ . \tag{1.29}$$

If $\tilde{x} < 1$, then the ratios of the Bessel functions of half-integer index can be calculated with the formulas:

$$\frac{J_{k+1/2}(\tilde{x})}{J_{-k-1/2}(\tilde{x})} = \left(\frac{\tilde{x}}{2}\right)^{2k+1}\frac{\sum\limits_{j=0}^{\infty} D_{k+i,j}\dfrac{k+1/2}{k+j+1/2}}{\sum\limits_{j=0}^{\infty} C_{kj}}(-1)^k\left(k+\frac{1}{2}\right)$$

$$\times\left[\frac{2}{1}\cdot\frac{2}{3}\cdots\frac{2}{2k+1}\right]^2 ; \tag{1.30}$$

$$\frac{J_{k-1/2}(\tilde{x})}{J_{-k-(1/2)}(\tilde{x})} = \left(\frac{\tilde{x}}{2}\right)^{2k}\frac{\sum\limits_{j=0}^{\infty} D_{k-1,j}}{\sum\limits_{j=0}^{\infty} C_{kj}}(-1)^k\left[\frac{2}{1}\cdot\frac{2}{3}\cdots\frac{2}{2k-1}\right] ; \tag{1.31}$$

$$\frac{J_{-k+1/2}(\tilde{x})}{J_{-k-(1/2)}(\tilde{x})} = \frac{\tilde{x}}{2(-k+(1/2))} \frac{\sum_{j=0}^{\infty} C_{k-1,j}}{\sum_{j=0}^{\infty} C_{kj}} ; \tag{1.32}$$

$$D_{k0} = 1 ; \quad D_{k,j+1} = -\frac{(\tilde{x}/2)^2}{(j+1)(-k+j+3/2)} D_{kj} ;$$

$$C_{k0} = 1 ; \quad C_{k,j+1} = -\frac{(\tilde{x}/2)^2}{(j+1)(-k+j+1/2)} C_{kj} ; \quad j = 0, 1, 2, \ldots$$

To calculate the coefficients of extension of the monochromatic aerosol phase function $x_{aer}(\gamma)$ in series in Legandre polynomials $P_n(\cos\gamma)$, a two-series presentation of these coefficients can be used [234]:

$$B_{n\lambda} = \frac{2\pi r^2(-1)^n}{\tilde{x}^2(2n+1)k_s(m, \tilde{x})} \sum_{j=1}^{\infty} \sum_{k=1}^{j} \left(\frac{2}{1+\delta_{jk}}\right) \times \left\{(2n+1)\left[\frac{j(j+1)+k(k+1)-n(n+1)}{2}\right]^2 W_{jkn}M_{jk} + V_{jkn}N_{jk}\right\} \tag{1.33}$$

where

$$M_{jk} = \frac{(-1)^{j+k}(2j+1)(2k+1)}{j(j+1)k(k+1)} \times [\mathrm{Im}\,\tilde{a}_j \mathrm{Im}\,\tilde{a}_k + \mathrm{Re}\,\tilde{a}_j \mathrm{Re}\,\tilde{a}_k + \mathrm{Im}\,\tilde{b}_j \mathrm{Im}\,\tilde{b}_k + \mathrm{Re}\,\tilde{b}_j \mathrm{Re}\,\tilde{b}_k] ; \tag{1.34}$$

$$N_{jk} = \frac{(-1)^{j+k}(2j+1)(2k+1)}{j(j+1)k(k+1)} \times [\mathrm{Im}\,\tilde{a}_j \mathrm{Im}\,\tilde{b}_j \mathrm{Re}\,\tilde{a}_k + \mathrm{Im}\,\tilde{b}_j \mathrm{Im}\,\tilde{a}_k + \mathrm{Re}\,\tilde{b}_j \mathrm{Re}\,\tilde{a}_k] ; \tag{1.35}$$

δ_{jk} is the Kronecker symbol; $W_{jkn} = 0$; if $(j+k-n)$ is odd; $V_{jkn} = 0$, if $(j+k-n)$ is even.

For W_{jkn} and V_{jkn} not equal to zero the following recurrent formulas are valid:

$$W_{n+1,1,n} = \frac{(n+1)}{(2n+1)(2n+3)} ;$$

$$W_{j+1,k+1,n} = \frac{(j+k-n+1)(j+k+n+2)}{(j+k-n+2)(j+k+n)} W_{jkn} ; \tag{1.36}$$

$$W_{j-2,k,n} = \frac{(j+k+n+1)(j+k-n)(j-k+n)(k-j+n+1)}{(j+k+n)(j+k-n-1)(j-k+n-1)(k+n-j+2)} W_{jkn} ;$$

$$V_{n,1,n} = n(n+1) ;$$

$$V_{j+1,K+1,n} = \frac{(j+k-n+2)(j+k+n+3)}{(j+k-n+1)(j+k+n+2)} V_{jkn} ; \tag{1.37}$$

$$V_{j-2,k,n} = \frac{(j+k+n)(j+k-n-1)(j-k+n-1)(k+n-j+2)}{(j+k+n+1)(j+k-n)(j-k+n)(k-j+n+1)} V_{jkn} \ .$$

A definite calculation of the monochromatic optical characteristics of atmospheric aerosol with these formulas raises difficulties. On one hand, this is explained by considerably forward-stretching the phase function in the given wavelength range, and on the other hand, by a broad spectrum of the aerosol particle sizes. The first circumstance leads to a poor convergence of series (1.33). Because of computer capacity, it is not possible to obtain in one run, the whole set of Bessel functions of half-integer order using standard recurrent formulas due to their fact growth at large $\tilde{x}$ and n values. This necessitates calculations of auxiliary expressions (1.25 – 1.27) and (1.30 – 1.31), using, when n is large, asymptotic presentations of the Bessel functions of half-integer order. The second circumstance explains a substantial difference in the calculation formulas for these expressions depending on the $\tilde{x}$ and n values.

In calculations with formulas (1.7) and (1.8) all partial oscillations were considered, with amplitudes not less than 10^{-19}. In calculations with formulas (1.5) and (1.6) though, as in [53], it is sufficient to consider to the terms $\tilde{Q}_k(m, \tilde{x})$ and $\tilde{P}_k(m, \tilde{x})$, with absolute value is not less than 10^{-7}. The summing in (1.33) was made simultaneously with those of recurrent formulas (1.36), (1.37) due to the great length of 3-D (computer) arrays W_{jkn} and V_{jkn}, which, in calculations, were changed to 1-D ones.

Figure 1.11 shows the course of the real and imaginary parts of the complex amplitudes of electrical partial waves $\tilde{a}_n(m, \tilde{x})$ and monochromatic coefficients $B_{n\lambda}$ depending on the index of summing n for $\tilde{x} = 50$ and $m = 1.53 - 0.01\,i$.

To obtain effective cross-sections and coefficients of extension of the aerosol phase function in any one interval of the solar spectrum, the respective monochromatic parameters should be averaged, not only over the solar spectrum $F(\lambda)$, but also over the size spectrum of particles with the respective distribution function for each of the particles size ranges. Experimental studies of the size distribution functions f(r) of real aerosol particles, show that for individual d-size ranges with the upper limit of radii r_d and lower limit r_{d-1} these distributions are characterized either: (1) by power dependence [208] with the index $\beta_d(z)$, depending on the vertical coordinate z; or (2) by a family of Gauss and log-normal curves [164] where r_0 is the most probable radius, A_0 the width of distribution and N_{r_0} the number of particles with the most probable radius N_{r_0}.

For an adequate account of the aerosol effect on radiation transfer in the atmosphere, vertical profiles of aerosol particles concentration must be known, apart from the spectral change of the complex refraction index m and size distribution functions f(r).

Using (1.5), (1.6) and (1.33), the formulas for calculating effective cross-sections and parameters of the aerosol phase function can be presented as (D is the number of size ranges):

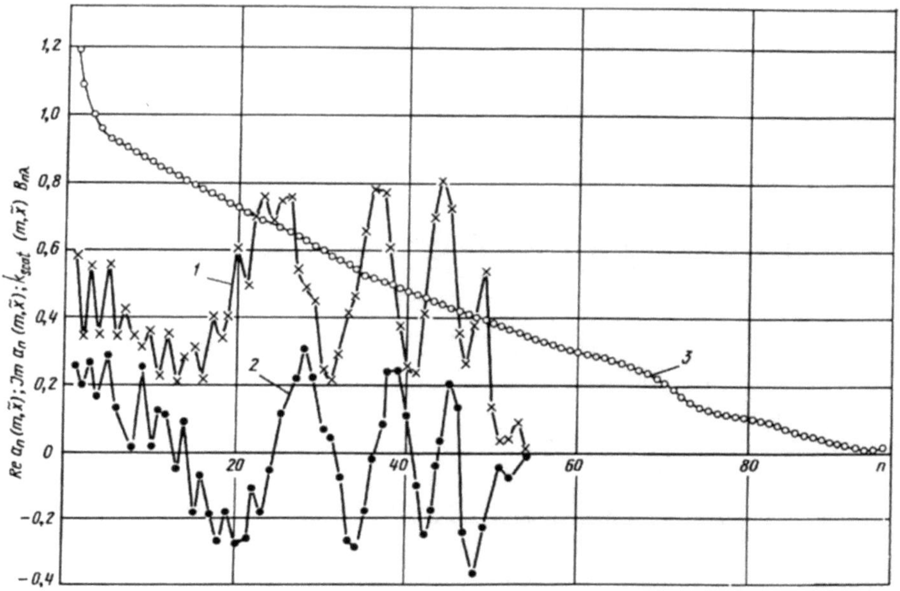

Fig. 1.11. The real (*1*) and imaginary (*2*) parts of the amplitudes of electric partial waves and values of K_{sc} (m, $\tilde{x}$) $B_{n\lambda}$ (*3*) as functions of the index of summation n; x = 50; m = 1.53 − i0.01

$$k_{s(l,l+1)} = \left(\sum_{d=1}^{D} \varrho_d\right)^{-1} \sum_{d=1}^{D} \varrho_d \times \frac{\int_{\lambda_l}^{\lambda_{l+1}} d\lambda F(\lambda) \int_{r_{d-1}}^{r_d} f_d(r, \beta_d) k_s(m, \tilde{x}) dr}{\int_{\lambda_l}^{\lambda_{l+1}} F(\lambda) d\lambda} \tag{1.38}$$

$$k_{t(l,l+1)} = \left(\sum_{d=1}^{D} \varrho_d\right)^{-1} \sum_{d=1}^{D} \varrho_d \times \frac{\int_{\lambda_l}^{\lambda_{l+1}} d\lambda F(\lambda) \int_{r_{d-1}}^{r_d} f_d(r, \beta_d) k_t(m, \tilde{x}) dr}{\int_{\lambda_l}^{\lambda_{l+1}} F(\lambda) d\lambda} ; \tag{1.39}$$

$$B_{n(l,l+1)} = \frac{\sum_{d=1}^{D} \varrho_d \int_{\lambda_l}^{\lambda_{l+1}} d\lambda F(\lambda) \int_{r_{d-1}}^{r_d} f_d(r, \beta_d) k_s(m, \tilde{x}) B_{n\lambda} dr}{\sum_{d=1}^{D} \varrho_d \int_{\lambda_l}^{\lambda_{l+1}} d\lambda F(\lambda) \int_{r_{d-1}}^{r_d} f_d(r, \beta_d) k_s(m, \tilde{x}) dr} . \tag{1.40}$$

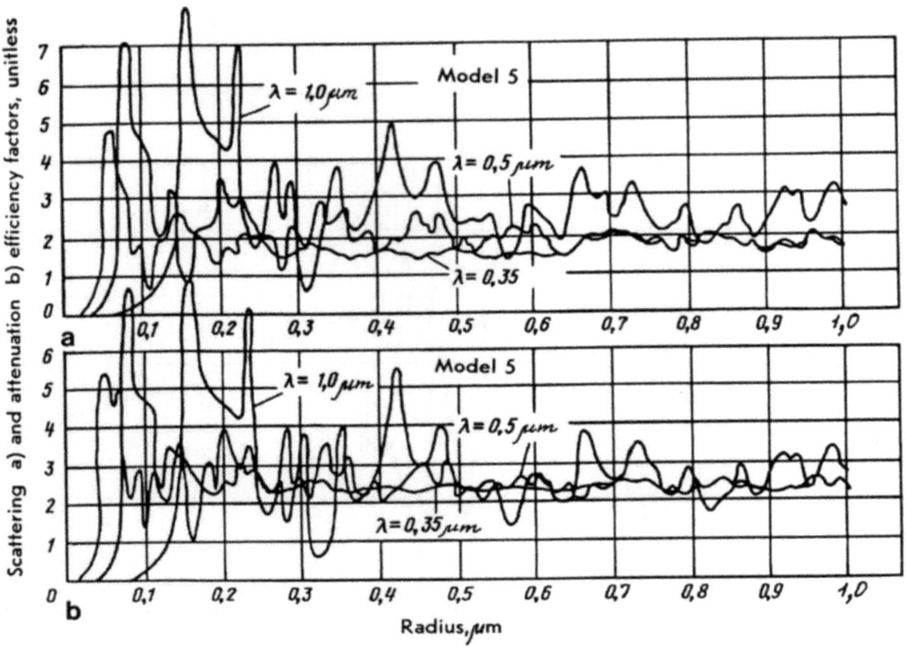

Fig. 1.12 a, b. The fine structure of the cross-sections of scattering (*upper curves*) and attenuation (*lower curves*) for aerosol model 5 (see Fig. 1.2) and three wavelengths λ

Theoretical and experimental studies of the radiation fields of the atmosphere-object-background system have shown that, the radiative regime of this system, is largely determined by the behaviour of polydisperse aerosol particles, in different spectral regions. This problem can successfully be studied by comparing observational data with the respective calculations using Mie theory. To illustrate the spectral properties of atmospheric aerosol needed to solve this problem, models are used of the complex refraction index for aerosol particles shown in Figs. 1.12 and 1.13.

Figures 1.12 and 1.13 present factors of scattering and attenuation efficiency for these aerosol models, calculated using formulas (1.12) and (1.13), depending on particle radius and wavelengths $\lambda = 0.35$, 0.5, and 1.0 μm. The behavior of the curves is complicated, being more regular for model 2 than for model 5.

For models 2 and 5, a comparison was made of theoretical values of phase functions, for the Gauss log-normal size distributions of particles and various wavelengths. Assuming that the aerosol concentration $N = 10^3\ \text{cm}^{-3}$, Figs. 1.14 and 1.15 show the angular change of phase functions for $\lambda = 0.35$ μm, Figs. 1.16 and 1.17 for $\lambda = 1.00$ μm. Figures 1.14 and 1.16 characterize narrow modes of distribution (the width of distribution $A_0 = 400$); those of Figs. 1.15 and 1.17 on the contrary, are broad ($A_0 = 10$).

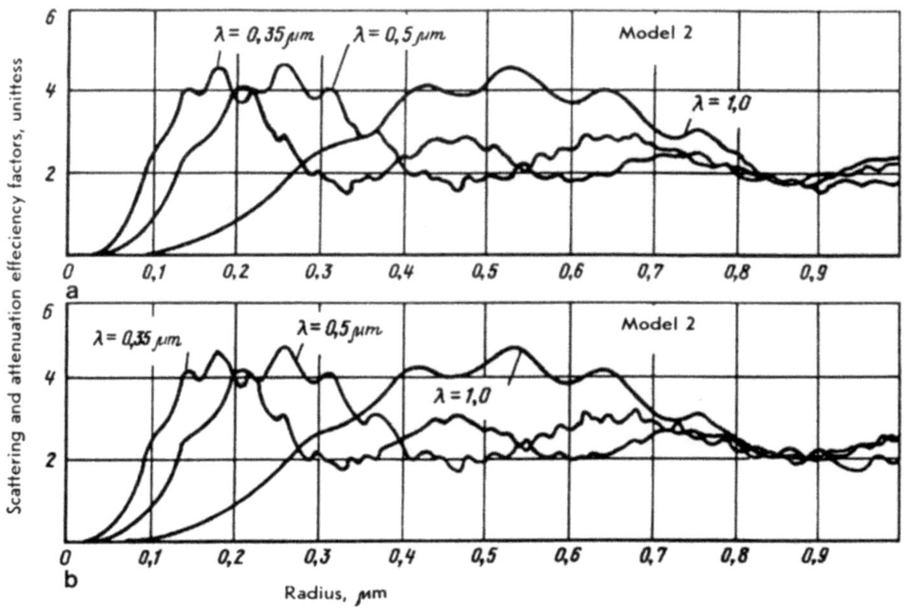

Fig. 1.13 a, b. The fine structure of the cross-sections of scattering (*upper curves*) and attenuation (*lower curves*) for aerosol model 2 (see Fig. 1.2) and three wavelengths λ

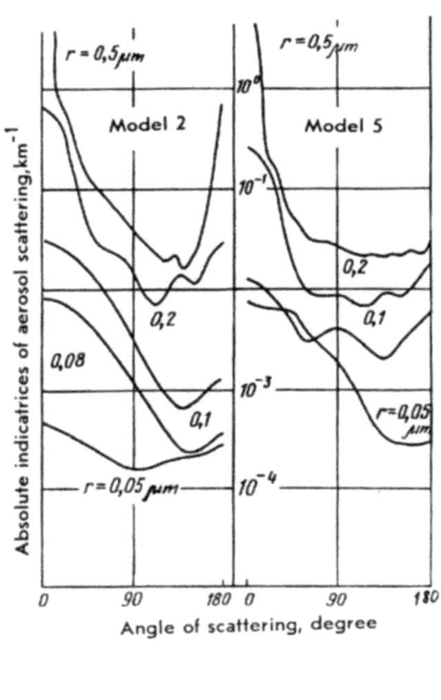

Fig. 1.14. Absolute phase functions for model 2 (*left*) and model 5 (*right*) of atmospheric aerosol for the log-normal particle size distribution for the width $A_0 = 400$ and $\lambda = 0.35$ μm

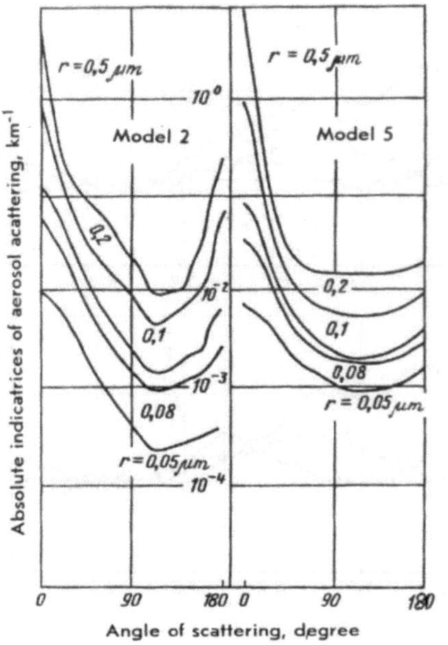

Fig. 1.15. Absolute phase functions for model 2 (*left-hand side*) and model 5 (*right-hand side*) of atmospheric aerosol for the log-normal particles size distribution for the width $A_0 = 10$ and $\lambda = 0.35$ μm

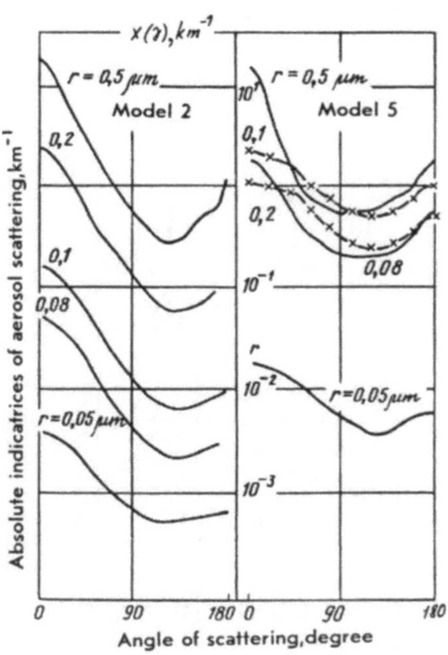

Fig. 1.16. Absolute phase functions for model 2 (*left*) and model 5 (*right*) of atmospheric aerosol for the log-normal particles size distribution for the width $A_0 = 10$ and $\lambda = 1.0$ μm

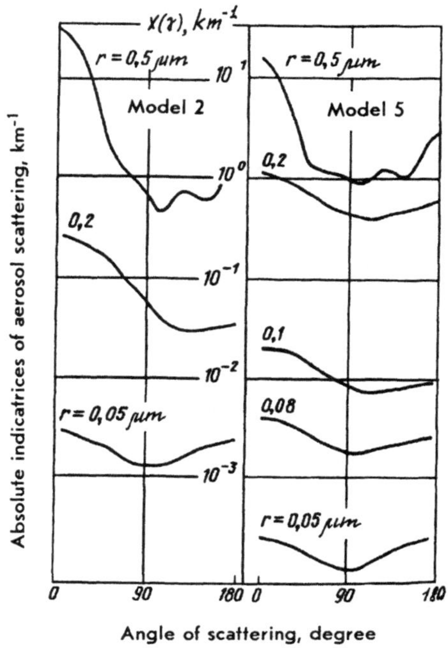

Fig. 1.17. Absolute phase functions for model 2 (*left*) and model 5 (*right*) of atmospheric aerosol for the log-normal particles size distribution for the width $A_0 = 400$ and $\lambda = 0.1$ μm

A comparison of the left-hand and right-hand parts of these graphs, shows substantially different effects using models 2 and 5, on the probability of radiation scattering in different directions. It is seen that, with decreasing r_0, growing A_0 and λ, the phase functions for model 2 approach Rayleigh scattering, and conversely growing r_0 and decreasing A_0 and λ, the forward-scattering grows. The angular course of phase functions for model 5 is more complicated.

By varying r_0, A_0 and N_{r0} values, one can obtain various phase functions for aerosol. Such calculations can be very useful in the interpretation of spectral information obtained with different angles of scattering γ.

Figure 1.18 illustrates the effect of models 2, 3 and 4 on the values of monochromatic cross-sections for each of the five ranges of particle radii: 0.02 – 0.1; 0.1 – 0.5; 0.5 – 1.0; 1.0 – 2.5; 2.5 – 6.0 μm. The integration of cross-sections by sizes was made over each range using the power function of distribution f(r) using the Simpson formula. The refraction index m of aerosol particles was assumed to be independent of their size.

It follows from Fig. 1.18 that cross-sections grow considerably with increasing size of aerosol particles: in the transition from very small particles to very large particles, their cross-sections increase by an order of 4 – 5. However, in a real atmosphere the concentration of the first particles can exceed the concentration of the latter ones by a larger value still. Nevertheless, the contribution of gigantic aerosol particles can be substantial.

Figure 1.18 shows that for very small particles only and for the range r = 0.1 – 0.5 μm partly the scattering cross-sections are inversely proportional

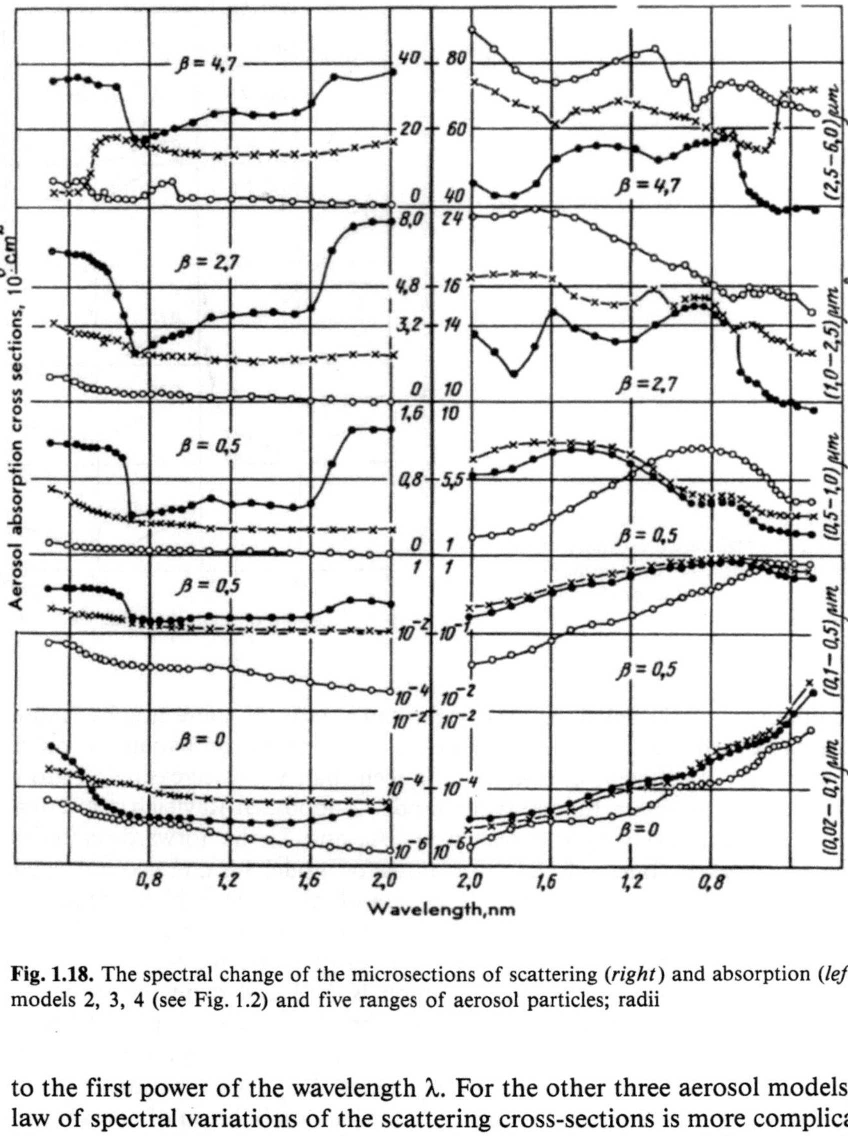

Fig. 1.18. The spectral change of the microsections of scattering (*right*) and absorption (*left*) for models 2, 3, 4 (see Fig. 1.2) and five ranges of aerosol particles; radii

to the first power of the wavelength λ. For the other three aerosol models, the law of spectral variations of the scattering cross-sections is more complicated. There is a sharp decrease in scattering for gigantic particles around the wavelength $\lambda = 0.7$ μm for model 4, coinciding with a sharp increase of the imaginary part of the refraction index.

The scattering exceeds the absorption on aerosols, for all wavelengths λ in all size ranges r, and for all models of their chemical composition, except for a dry aerosol fraction (model 2), for which the absorption exceeds the scattering at $\lambda > 1.5$ μm. This is particularly important for wet aerosols (model 3).

Comparison of calculated values of scattering and absorption for models 2 and 4, in which the real parts of the refraction indices are almost equal at all wavelengths, shows that a change in the imaginary part of the refraction index leads to variations, both in scattering and in absorption. Differences in cross-sections in models 2 and 4 are particularly noticeable for a range of radii 2.5 – 6.0 μm, where the scattering cross-sections differ by a factor of 1.5, at almost all wavelengths. The sum of cross-sections of scattering and absorption for models 2 and 4, varying little in the whole spectral region in question. The attenuation cross-section, under the influence of the imaginary part of the refraction index, changes mostly in the size range 0.02 – 0.1 μm, which is, firstly, connected with a large absorption value in model 4.

The spectral change of absorption (see the left-hand part of Fig. 1.18) is broadly, similar to the spectral change of scattering. The absolute value of the absorption cross-sections is less than for scattering. For model 4, however, these values are comparable, across all particle size-ranges. The transition to more smoothed spectral values is observed with decreasing size of aerosol particles. The data in Fig. 1.18 on atmospheric aerosol absorption show that, the absorption in the visible is much less than the absorption in the IR, only in small-sized aerosol fractions. For large and gigantic particles, the values of absorption in these regions are comparable. In most studies on atmospheric aerosol the absorption in the visible spectrum has been neglected. However, the observed residual absorption of solar radiation in the atmosphere is connected with the aerosol absorption, in this spectral region [92]. The left-hand part of Fig. 1.18 confirms this fact.

Figure 1.19 presents histograms of the spectral change of cross-sections of scattering $k^{(1)}_{s(l,l+1)}$ and attenuation $k^{(1)}_{t(l,l+1)}$ for aerosol model 1, integrated over sizes from 0.02 to 0.1 μ assuming that $\beta_d = 0$. For the same aerosol model 1 in Fig. 1.19b are given the values $k^{(2)}_{s(l,l+1)}$ and $k^{(2)}_{t(l,l+1)}$ obtained in the integration over radii from 2.5 to 6.0 μm at $\beta_d = 5$. Figure 1.19a, b illustrates relationships between scattering and absorption for model 1 between extremely fine and gigantic particles. It is seen from Fig. 1.19a that for fine particles at $\lambda > 1.4$ μm absorption prevails over scattering. On the contrary, at $\lambda < 0.8$ μm scattering exceeds absorption. Figure 1.19 b illustrates a sufficiently neutral course over the spectrum of cross-sections of scattering and attenuation at $\lambda = 1$ μm. The scattering and absorption for gigantic particles are of the same order.

The effect of the size distribution function f(r) for aerosol particles with radii 2.5 – 6.0 μm with the power law of the size distribution is shown in Fig. 1.19c, where the respective values are given for cross-sections $k^{(3)}_{s(l,l+1)}$ and $k^{(3)}_{t(l,l+1)}$ at $\beta_d = 0$. The spectral course of these functions in Fig. 1.19c is similar to that in Fig. 1.19b and, again, close to neutral at $\lambda > 1$ μm.

For radii 0.1 – 1.0 μm of model 1 and $\beta_d = 2$ the spectral change of cross-sections is shown in Fig. 1.19d, and for the range 1.0 – 2.5 μm of the same model and $\beta_d = 3$ in Fig. 1.19e. There is a substantial difference between Fig. 1.19d and Fig. 1.19a – c: the averaged curves of the spectral change (close to hyperbolas in Fig. 1.19a and the curves in Fig. 1.19b, c) at $\lambda > 1$ μm are transformed into the curves of the parabolic type for cross-sections of both scattering and attenuation. Both curves in Fig. 1.19d are approximately

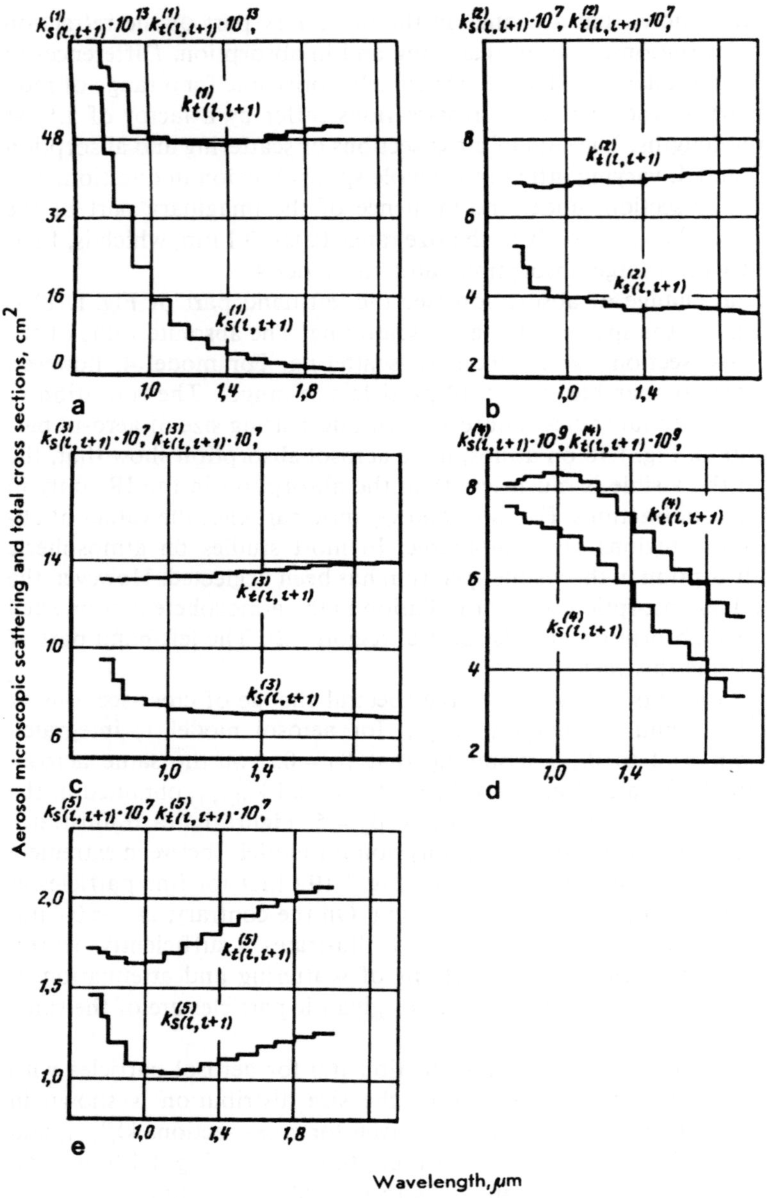

Fig. 1.19 a – e. The spectral change of the microsections of scattering and attenuation for the particles' radii **a** Δr ranging between 0.02 and 0.1 μm and for zero Yunge index β; **b** Δr from 2.5 to 6.0 μm and $\beta = 5$; **c** Δr from 2.5 to 6.0 μm and $\beta = 0$; **d** Δr from 0.1 to 1.0 μm and $\beta = 2$; **e** Δr from 1.0 to 2.5 μm and $\beta = 3$ (e)

equidistant, which illustrates the constancy of absorption in radii 0.1 – 1.0 μm at $\lambda > 1$ μm for a given model of the complex refraction index. At $\lambda < 1$ μm its imaginary part decreases substantially.

An interesting result is shown in Fig. 1.19e, where for model 1 and $\lambda > 1$ μm the cross-section of aerosol scattering increases with growing wavelength, λ. Such an unusual spectral course is connected with the well-known fact of decreasing scattering cross-section with growing parameter $\tilde{x}$, within the interval $10 > \tilde{x} > 5$, when longer waves are scattered more strongly than the short ones. As in Fig. 1.19d, the absorption in radii 1.0 – 2.5 μm is nearly constant for $\lambda > 1$ μm.

The illustrated model values of the spectral change of cross-sections and scattering functions for atmospheric aerosol, show complicated behavior of this atmospheric substance at the present stage of our knowledge. To resolve the parameters of scattering and absorption of radiation by aerosol, which will be used in the atmospheric correction of the space-derived Earth images, it is necessary to perform a closed modeling of the optical characteristics of atmospheric aerosol, based on the available observational data and calculations using Mie theory.

1.3 Closed Modeling of the Optical Characteristics of Atmospheric Aerosol

Simulation modeling of the fields of radiation of the atmosphere-object-background system depends much on the spatial and temporal structure of aerosol formations (see Section 1.1). Of principal importance, is the fact that modeling of the optical characteristics of atmospheric aerosol must be closed. This means that all the optical parameters of aerosol and their variations must be adequate to the spatial and temporal transformations of aerosol concentrations and its physical-chemical properties.

The available statistical information, from regional studies on spectral optical characteristics of atmospheric aerosol refers, as a rule, to the attenuation coefficients of the surface boundary layer [201, 202]. For the whole atmosphere, spectral measurements of all optical parameters of aerosol are difficult to achieve [325 – 329]. Therefore, the vertical profile of aerosol is most rationally expressed through the profile of its optical density. For many geographical regions of the Earth, statistical data have been obtained on the vertical profile of optical thickness of atmospheric aerosol and its annual and diurnal change. These make it possible to judge the validity the modeling concepts of the optical characteristics of atmospheric aerosol. It is known that the optical density of atmospheric haze in the surface air layer suffers considerable temporal variations, whereas the optical thickness of atmospheric aerosol column is less variable. This conclusion confirms the fact that, severe short-term variations of the optical density of atmospheric aerosol, are caused by variations in the structure of the surface aerosol layer. The vertical profile, of the optical density of atmospheric aerosol, was taken as a major characteristic in modeling the optical properties of the Earth's atmosphere, in many studies on atmospheric aerosol [9, 58, 147, 163, 259, 362, 374, 389] (see Sect. 1.1, Figs. 1.3 to 1.10).

Bearing in mind the variability of the microphysical properties of aerosol particles, the modeling of the aerosol effect on the spectral and spatial structures of radiation fields, vertical profiles of spectral and total fluxes, radiation energy balance and input initially should be based on simple models with fixed optical and physical-chemical properties of aerosol particles. However, it is necessary now, to develop models of aerosol formation and transformation, including its space/time variability, the effect of meterorological parameters, and atmospheric dynamics. No doubt, this problem can be solved only by computers. The optical characteristics of aerosol should be programmed by the "aerosol unit", which is a component of a closed system of numerical modeling of radiation processes, in the atmosphere-surface system. Generalizations of all data on global aerosol and it optical properties are necessary. Complex ground- and space-based experiments need to be carried out, with simultaneous measurements of spectral optical characteristics of aerosol, its size distribution, chemical composition, spectral intensities of radiation, spectral and total downward and upward radiation fluxes at different altitudes. With the accomplishment of such combined complex experiments, it is possible not only to test the correctness of modeling the optical characteristics of aerosol and the radiation transfer, but also to specify an application of schemes and methods, and the correctness of the algorithms of this modeling.

The size distribution N(r) of aerosol particles is presented by summing up the size distributions $N_i(r)$ of individual fractions with the formula:

$$N(r) = \sum_{i=1}^{N} N_i(r) = \sum_{i=1}^{N} A_i r^{a_i} \exp[-b_i r^{c_i}] \ , \tag{1.41}$$

where A_i, a_i, b_i, c_i are parameters of the i-th fraction; N the number of fractions. Since the most reliable information on the vertical aerosol profile is obtained from optical density, the spectral coefficients of attenuation $\alpha_{\lambda_{aer}}$, scattering $\sigma_{\lambda_{aer}}$, and absorption $\chi_{\lambda_{aer}}$ are normalized against optical density. Each aerosol fraction is decribed by its vertical profile, which enables one to model vertical variations in its chemical composition, size distribution and optical characteristics. Spectral coefficients of attenuation $\alpha_{\lambda_{aer}}$, scattering $\sigma_{\lambda_{aer}}$, and phase function $\chi_{aer}(\gamma)$ are determined with the formulas:

$$\alpha_{\lambda_{aer}} = \sum_{i=1}^{N} B_i(z)\, \alpha_{i\lambda_{aer}} \ ; \tag{1.42}$$

$$\sigma_{\lambda_{aer}} = \sum_{i=1}^{N} B_i(z)\, \sigma_{i\lambda_{aer}} \ ; \tag{1.43}$$

$$\chi_{\lambda_{aer}} = \sum_{i=1}^{N} B_i(z)\,\chi_{i\lambda_{aer}} \;; \qquad (1.44)$$

$$\chi_{\lambda_{aer}}(z,\gamma) = \sum_{i=1}^{N} B_i(z)\,\chi_{i\lambda_{aer}} \Bigg/ \sum_{i=1}^{N} B_i(z) \;; \qquad (1.45)$$

In formulas (1.42–1.45) the value $B_i(z) = \partial\tau_i/\partial z$ is the vertical profile of optical density; α_i, λ_{aer}, χ_i, λ_{aer}, σ_i, λ_{aer} are normalized values of spectral coefficients of attenuation, scattering and absorption for the i-th fraction of aerosol; $\chi_{\lambda,i}$ is the phase function for the i-th fraction of aerosol.

Figure 1.20 shows how phase function $\chi_{aer}(\gamma)$ is dependent on height over the continents at mid-latitudes [374]. The considerable vertical variations of phase function, due to variations in the size distribution and chemical composition of atmospheric aerosol are interesting. The most symmetric phase function, close to Rayleigh scattering, is observed in the zone of background aerosol. The phase function is steepest in the surface boundary layer of the atmosphere.

Figure 1.21 illustrates the spectral structure of the attenuation coefficients at different heights in the atmosphere for subarid regions. In the surface air layer, the attenuation coefficient α_{aer} reaches a maximum at $\lambda = 1\ \mu m$, which is caused by a considerable contribution to radiation attenuation by the dust fraction. Within the zone of active turbulent exchange, over the continents in the wavelength region 0.3–2 μm, the spectral coefficient of scattering σ_{aer}

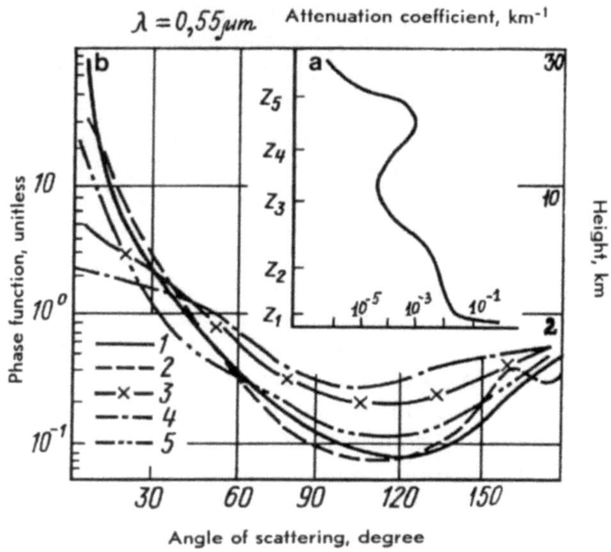

Fig. 1.20. a The vertical profile of optical density and **b** phase function typical of the summer mid-latitudes: *1* boundary-layer model; *2* the layer of active turbulent exchange; *3* the upper troposphere; *4* areas above and below the stratosphere; the stratospheric layer ($Z_6 - Z_5$)

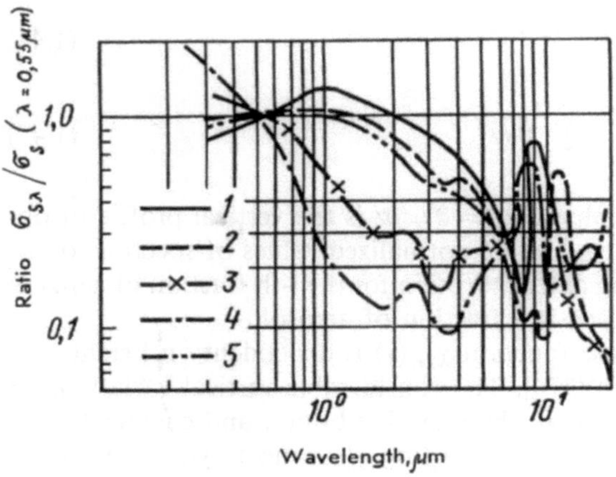

Fig. 1.21. The spectral dependence of the attenuation coefficients for different altitudes (notations see in Fig. 1.20)

varies weakly. In the middle and upper troposphere the coefficient of attenuation α decreases with wavelength, starting from the UV to near IR region. The strongest spectral variations of the attenuation coefficient α_{aer} are observed at altitudes where large aerosol particle concentration is small, and the spectral coefficients of attenuation are determined by the background fraction of aerosol. Variations of the spectral dependences of attenuation coefficients α_{aer} between 0.5 – 4 μm, at different altitudes, are caused largely by variations in the size distribution of atmospheric aerosol and its chemical composition.

Figure 1.1 compares the spectral coefficients of attenuation α_{aer} of global mean background aerosol, salt aerosol, particles of a 75% solution of H_2SO_4 (for 2 size-distributions), volcanic aerosol and cosmic dust particles. The spectral coefficients of attenuation α_{aer} are normalized against optical density $\partial\chi_{aer}/\partial z$ at $\lambda = 0.55$ μm. The optical characteristics of stratospheric aerosol were obtained by a superposition of its various fractions using formulas (1.41 – 1.45). In the absence of volcanic activity, the optical characteristics of stratospheric aerosol are determined by a small-sized fraction of background aerosol, salt sulfate aerosol, and space meteoritic dust.

The latter two aerosol types absorb the shortwave radiation weakly and the longwave radiation strongly [362, 373].

As studies have shown, when closed optical models of the atmosphere are constructed, it is most difficult to achieve an adequate theoretical modeling of the optical properties of the planetary aerosol. There is no general statistical theory how the global field of atmospheric aerosol is formed. Therefore, it is difficult to be confident of any of the former steady-state models of the planetary aerosol, despite their complex character and vertically detailed physico-chemical and optical information. Additionally, due to the vertical mixing of radiation fluxes following multiple scattering in the vertically inhomogeneous atmosphere and re-reflection of photons from the non-Lambertian surface, the need for such detailed information and its effect on the field of diffusively reflected and transmitted radiation, is not clear.

In theoretical modeling of radiation fields in a real terrestrial atmosphere the aerosol models by Elterman and their modifications [256–259, 374] have been widely used during the last decade.

The optical closing of Elterman's aerosol models with respect to polydisperse phase functions $P_{i\,aer}(\gamma)$ and scattering functions $x_{aer}(\gamma)$ can be made using Mie theoretical formulae (see Sect. 1.2). It is necessary here to give additional information about the complex refraction index m and the function of aerosol size distribution f(r). There is no such information in Elterman's models. Therefore, the parameters of the fixed size distribution function f(r) for an adequate refraction index m should be chosen, to provide the calculated aerosol optical characteristics, with the spectral dependence given in initial models. To choose the parameters needed in the wavelength region $\lambda = 400-800$ nm, the spectrum-averaged aerosol refraction index $m = 1.55-0.01i$ and the power function of distribution f(r) of the Yunge type were used:

$$f(r) = \begin{cases} 0 & r > r_{max} \\ c \cdot r^{-\beta} & r_{min} < r < r_{max} \\ 0 & r < r_{min} \end{cases}$$

Calculations of the aerosol spectral coefficient of scattering $\sigma_{aer}(\lambda)$ at the surface, using Mie formulas for a wide range of the values of β, r_{min}, and r_{max} parameters, have shown that best agreements with the respective spectral data of Elterman's model (1968) are obtained with the following values of calculated parameters: [153]: $r_{max} = 10\ \mu m$; $r_{min} = 0.04\ \mu m$; $\beta = 3.5$. Tables 1.2

Table 1.2. The polydisperse phase function of scattering $P_i(\gamma)$ ($i = 1, 2, 3, 4$) for Elterman's aerosol model (1968) ($\lambda = 550$ nm, $m = 1.55-0.01i$; $r_{max} = 10\ \mu m$, $r_{min} = 0.04\ \mu m$, $\beta = 3.5$)

γ, deg	$P_{i,aer}$	$P_{2,aer}$	$P_{3,aer}$	$P_{4,aer}$	γ, deg	$P_{1,aer}$	$P_{2,aer}$	$P_{3,aer}$	$P_{4,aer}$
0	269.8	269.8	269.8	0	95	0.2503	0.1982	0.1194	0.0103
5	26.63	26.64	26.61	−0.7196	100	0.2216	0.1771	0.0907	0.0153
10	11.84	11.83	11.79	−0.5109	105	0.1985	0.1626	0.0676	0.0199
15	7.473	7.456	7.397	−0.4073	110	0.1801	0.1536	0.0487	0.0243
20	5.306	5.285	5.223	−0.3325	115	0.1654	0.1496	0.0334	0.0285
25	3.967	3.940	3.882	−0.2743	120	0.1540	0.1501	0.0208	0.0328
30	3.045	3.011	2.957	−0.2266	125	0.1454	0.1550	0.0103	0.0371
35	2.372	2.333	2.280	−0.1858	130	0.1393	0.1646	0.0014	0.0419
40	1.869	1.823	1.772	−0.1512	135	0.1359	0.1795	−0.0064	0.0469
45	1.487	1.433	1.383	−0.1213	140	0.1351	0.2007	−0.0137	0.0525
50	1.193	1.132	1.084	−0.0956	145	0.1370	0.2299	−0.0217	0.0581
55	0.9652	0.8994	0.8512	−0.0742	150	0.1424	0.2677	−0.0326	0.0627
60	0.7883	0.7184	0.6694	−0.0563	155	0.1516	0.3114	−0.0508	0.0618
65	0.6499	0.5775	0.5269	−0.0412	160	0.1677	0.3491	−0.0831	0.0458
70	0.5409	0.4679	0.4147	−0.0286	165	0.2080	0.3687	−0.1389	0.0127
75	0.4544	0.3827	0.3260	−0.0182	170	0.2998	0.4143	−0.2365	0.0103
80	0.3855	0.3168	0.2556	−0.0098	175	0.4284	0.4923	−0.3982	0.0224
85	0.3303	0.2660	0.1997	−0.0018	180	0.5360	0.6360	−0.5360	0
90	0.2860	0.2272	0.1551	0.0046					

Table 1.3. Atmospheric scattering function x (γ) for Elterman's aerosol model (1968)

γ, deg	λ, nm							
	400	450	500	550	600	650	700	800
1	2	3	4	5	6	7	8	9
0	187.038	196,527	197.337	193.142	185.4	175.2	166.311	147.5
5	12.641	15.328	17.567	19.455	20.94	21.96	22.954	24.30
10	6.053	1.314	8.098	1.268	9.496	9.939	10.368	10.94
15	4.103	1.474	5.291	5.746	6.106	6.367	6.619	6.955
20	3.124	1.354	3.891	4.186	4.4201	4.590	4.754	4.974
25	2.054	2.781	3.017	3.214	3.3708	3.484	3.595	3.743
30	2.066	2.249	2.406	2.538	2.6420	2.718	2.793	2.893
35	1.735	1.853	1.953	2.038	2.1064	2.156	2.204	2.272
40	1.476	1.546	1.607	1.659	1.6977	1.730	1.761	1.801
45	1.269	1.305	1.337	1.364	1.3860	1.403	1.419	1.441
50	1.102	1.113	1.123	1.133	1.1411	1.148	1.154	1.164
55	0.965	0.958	0.954	0.951	0.9488	0.9480	0.947	0.9477
60	0.853	0.843	0.818	0.806	0.7967	0.7907	0.785	0.7786
65	0.761	0.733	0.710	0.690	0.6765	0.6606	0.657	0.6458
70	0.687	0.653	0.623	0.599	0.5813	0.5686	0.556	0.5414
75	0.629	0.589	0.550	0.527	0.5063	0.5281	0.477	0.4599
80	0.584	0.539	0.502	0.471	0.4478	0.4312	0.415	0.3952
85	0.551	0.502	0.462	0.429	0.4030	0.3848	0.367	0.3451
90	0.529	0.477	0.433	0.397	0.3656	0.3498	0.331	0.3066
95	0.518	0.465	0.415	0.367	0.3458	0.3244	0.304	0.2777
100	0.516	0.456	0.405	0.363	0.3306	0.3073	0.285	0.2565
105	0.524	0.458	0.403	0.357	0.3222	0.2971	0.273	0.2420
110	0.538	0.467	0.408	0.358	0.3200	0.2929	0.267	0.2330
115	0.560	0.483	0.418	0.365	0.3234	0.2938	0.265	0.2285
120	0.587	0.504	0.434	0.376	0.3313	0.2944	0.269	0.2287
125	0.619	0.529	0.454	0.392	0.3434	0.3089	0.276	0.2332
130	0.655	0.559	0.478	0.411	0.3591	0.3222	0.287	0.2409
135	0.693	0.591	0.505	0.434	0.3784	0.3386	0.301	0.2525
140	0.734	0.626	0.535	0.460	0.4012	0.3271	0.319	0.2681
145	0.776	0.663	0.568	0.489	0.4278	0.3841	0.342	0.2884
150	0.819	0.701	0.603	0.521	0.4578	0.4128	0.369	0.3138
155	0.860	0.739	0.639	0.556	0.4906	0.4446	0.400	0.3433
160	0.897	0.775	0.673	0.588	0.5219	0.4753	0.430	0.3724
165	0.932	0.809	0.705	0.620	0.5536	0.5065	0.461	0.4033
170	0.977	0.859	0.759	0.677	0.6131	0.5678	0.524	0.4688
175	1.032	0.923	0.831	0.756	0.6971	0.6554	0.615	0.5650
180	1.069	0.967	0.881	0.811	0.7569	0.7184	0.681	0.6348

and 1.3 give values of the polydisperse aerosol phase function $P_{i,aer}$ at $\lambda = 550$ nm and vertically-averaged atmospheric scattering function x(γ). These correspond to Elterman's closed model (1968) for wavelengths $\lambda = 400-800$ nm. The aerosol scattering function $x_{aer}(\gamma)$, suitable for Elterman's model (1968) in this spectral region was obtained in [153].

A given algorithm determines the polydisperse scattering functions for the selected aerosol model. It can be used for atmospheric correction procedures when the complex refraction index m and the type of aerosol size distribution function f(z) are known.

Chapter 2 The Radiation-Environment Interaction

The practical use of the results of the Earth studies from space is largely determined by the possibility of solving two basic problems: developing automated systems to process and interpret video information (ASPIV) and establishing relationships between the parameters characterizing the environmental state and the outgoing radiation field. The development of special processors, the networking and use of computers in series, and the development of computer-processing mathematical-support help in solving the first problem. The solution of the second problem is connected with mathematical description of the environment and of its interaction with electromagnetic radiation.

As applied to studies of the Earth from space, the use of ASPIV has been most successfull in the following directions: (1) video-data compression; (2) retrieval and correction of images; (3) organization, archiving of and search for video-data; (4) analysis of scenes; (5) synthesis of images. There is a persistent trend to change from specialized systems to problem-oriented systems, based on the development of interactive (dialogue) regimes for problem solving. The ASPIVs are based, on special multi-processor configurations to parallel-process multispectral images, and medium and small computers have control functions. Nevertheless, to represent video-information in terms of various brightness matrices requires the development of such methods, that are sensitive to a set of objects with definite position, shape and size, but not to individual matrix elements.

Video-information interpretation must be based on model presentations of atmospheric media and their interaction with radiation. Analysis includes numerical realization of models, enabling the extraction of useful information from a volume of observational data. Brightness must be interpreted as a measure of information about the state characterized by a complicated composition of physical, chemical, biometric and other factors. The conventional treatment of brightness, as a mean energy flux in a certain direction, cannot describe the wave-like character of the outgoing radiation field as a statistical dynamic information system.

Let us dwell upon the conventional definition of the information field. This is a suitable abstraction to describe various means of setting an input information flux, independent of its source: for instance, a signal is an example of a 1-D information field; a spatial field of electromagnetic radiation is a 3-D field; the same field varying in time is a 4-D field, etc. As applied to remote sensing, the information field can be understood as the reflection of a multidimensional discrete space into a final number of values of brightness and/or color, called gradations. Generally, the information field can be considered a vector function. There are various types of information field: a black-white graphic image (two gradations of brightness); a false-color graphic image (an irregular set of gradations of brightness/color); a black-white half-tone image (a certain number of regular gradations of brightness – a black-white photo; an image displayed on the screen); a false-color half-tone image (a quantized half-tone space, for instance, a displayed false-color image, a spectrozonal discretized image, etc.).

We have spoken, so far, about representations of the information field and nothing about the basis of the formation of the outgoing radiation field. The problems of assessing the information content of the measurement data and retrieving from them of the parameters included in models of the radiation-environment interaction are far from being solved. The use of empirical estimates of the respective parameters [173] or an assimilation of theoretical results on the "semi-empirical basis" (as, for instance, in the description of the radiative regime of vegetation cover using the transfer theory [167]) can hardly be considered a cardinal solution for the problem. Theoretical fundamentals of the interaction and the information features of the outgoing radiation field are combined.

2.1 Basic Definitions in Photometry

The outgoing radiation field, containing information on the state of the environment, has statistical-wavelike properties: it is created by multiple uncorrelated elementary sources (atoms), varies as it interacts with the environment, and is a statistical dynamic system. The coherence theory, describing these field transformations is a section of a general theory of information, that studies the ability of the field to transmit information. In the classical theory of the electromagnetic field, the tensions of the electric and magnetic fields are assumed to be measurable functions of coordinates and time: the measured parameters are described by cross-correlation functions (coherence functions) of tension. Actually, the radiation field can oscillate so quickly that real detectors can fail to follow its changes, typically, in the case of optical radiation. Therefore, it is necessary to introduce the respective averaging of quantities received by a radiation sensor.

Some basic definitions are needed to understand the information resources of the outgoing radiation field:

1. Brightness of the sources (Fig. 2.1 a); the quantity of an average flux of energy emitted by the unit area to unit solid angle $d\Omega_{\vec{n}}$ close to $\vec{n}$: $I(\vec{r}, \vec{n})$, where $\vec{r}$ is radius-vector;
2. Average density of energy off the area in the direction $\vec{n}$: $d\vec{L} = d\vec{L}(\vec{r}, \vec{n}) = \vec{n} I(\vec{r}, \vec{n})\, d\Omega_{\vec{n}}$;

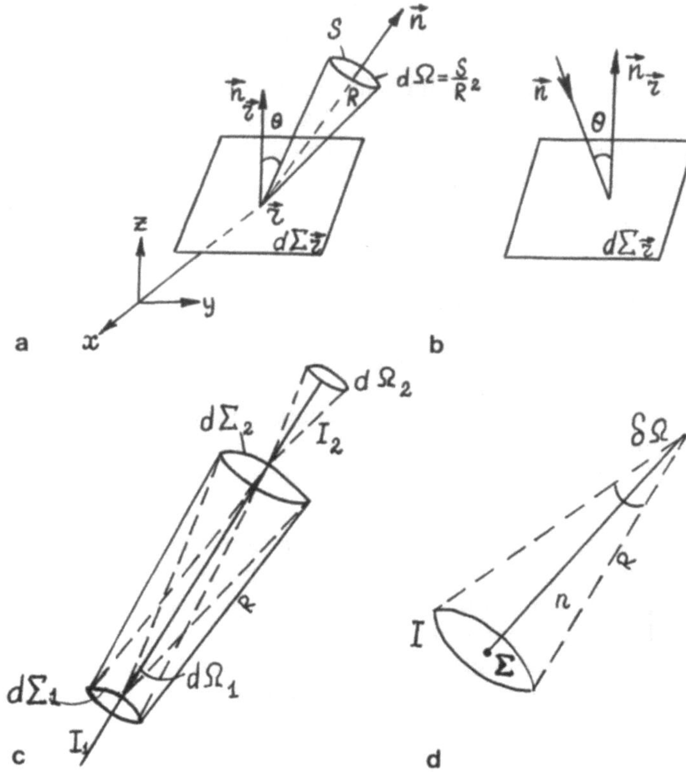

Fig. 2.1 a – d. Explanations to basic definitions of the outgoing radiation field: **a** radiation brightness over an area $d\Sigma_{\vec{r}}$; **b** irradiance of the area; **c** propagation of radiation in a homogeneous medium; **d** emission of a small source Σ

3. Energy flux in the direction $\vec{n}$ across an elementary area $\vec{n}_{\vec{r}} d\Sigma_{\vec{r}}$ with the normal $\vec{n}_r$ is

$$d\vec{L}\vec{n}_{\vec{r}} d\Sigma_{\vec{r}} \quad \text{or} \quad dp = dp(\vec{r}, \vec{n}) = \cos\theta\, I(\vec{r}, \vec{n})\, d\Omega_{\vec{n}}\, d\Sigma_{\vec{r}} \ ,$$

where $\cos\theta = \vec{n}\vec{n}_{\vec{r}}$, i.e., brightness is related to the flux:

$$I(\vec{r}, \vec{n}) = dp(\vec{r}, \vec{n})/[\cos\theta\, d\Omega_{\vec{n}}\, d\Sigma_{\vec{r}}] = [\mathrm{W/(m^2 sr)}] \ ;$$

4. Irradiance F of an elementary area $d\Sigma_{\vec{r}}$ is a total flux of energy incident onto $d\Sigma_{\vec{r}}$ from all directions $\vec{n}$, but from the direction of the normal $\vec{n}_{\vec{r}}$ (Fig. 2.1 b):

$$F = F(\vec{r}) = -\int_{2\pi} \vec{n}_{\vec{r}} d\vec{L} = \int \cos\theta\, I(\vec{r}, \vec{n})\, d\Omega_{\vec{n}} \ ,$$

where the “minus” sign shows that the unit vectors of the direction of propagation of the wave $\vec{n}$ and the normal towards the area $\vec{n}_{\vec{r}}$ form an obtuse angle: $\vec{n}\,\vec{n}_r = -\cos\theta$;

5. Spherical irradiance ϕ is a total flux of energy incident onto $d\Sigma_{\vec{r}}$ from every possible direction: $\phi = \phi(\vec{r}) = \int_{4\pi} I(\vec{r}, \vec{n})\, d\Omega_{\vec{n}}$;
6. If an observation point is far from an emitting plane element of the surface Σ at a distance much larger than the size of the surface, then Σ is usually treated as a point source, introducing the notion "light intensity" as an energy flux per unit solid angle:

$$\tilde{I}(\vec{n}) = \int_{\Sigma} \cos\theta\, I(\vec{r}, \vec{n})\, d\Sigma_{\vec{r}} \;;$$

7. Volume density of energy is determined by an expression:

$$W(\vec{r}) = c^{-1} \int I(\vec{r}, \vec{n})\, d\Omega_{\vec{n}} \;;$$

where c is the velocity of radiation propagation (light velocity), is a volume filled by radiation across a unit section $\Sigma = 1$ per unit time, where $I(\vec{r}, \vec{n})\, d\Omega_{\vec{n}}$ is energy per unit solid angle.

In a free space (a homogeneous non-absorbing medium) radiation brightness, introduced for an assumed area, penetrated by radiation beams, is the same along the beam tube (Fig. 2.1 c). It is easily seen, if we write an energy flux $I_1 d\Sigma_1 d\Omega_1$ incident from $d\Sigma_1$ onto a section $d\Sigma_2$ and set it equal to the flux $I_2 d\Sigma_2 d\Omega_2$ outgoing from $d\Sigma_2$, which gives $I_1 = I_2$. However, energy density associated with the notion "radiation intensity" usually used in calculations, decreases at large distances as R^{-2} (Fig. 2.1 d):

$$W = c^{-1} I\, \delta\Omega = c^{-1} I \Sigma / R^2 \;.$$

Thus radiation brightness is not equal to intensity.

Now let us move on to the interaction between radiation and the environment, which transforms the ray characteristics of the radiation field described above.

2.2 Equations of Interaction Between Radiation and Randomly Inhomogeneous Media

The outgoing radiation field is formed by its interaction with the environment of a randomly inhomogeneous character. The vector electromagnetic field is described by tensions $\vec{E}(\vec{r}, t)$ and $\vec{H}(\vec{r}, t)$ (t, time). The transverse position of this field (the electric field vector $\vec{E}$ is perpendicular to the magnetic field vector $\vec{H}$, and both are normal to the vector of energy density $\vec{L}$) explains a characteristic phenomenon of radiation polarization, i.e., non-chaotic (oriented) spatial distribution of these vectors or their rotation in plane ($\vec{E}$, $\vec{H}$). Due to the correlation between $\vec{E}$ and $\vec{H}$, equations for $\vec{E}$ only will be considered below. The following equation[1] holds for the components of the E random field upon its interaction with the environment:

$$\hat{L}_{\alpha\beta}E_{\beta} \equiv (\hat{L}^{(0)}_{\alpha\beta} - \hat{V}_{\alpha\beta})E_{\beta} = q_{\alpha} \ , \tag{2.1}$$

where α, $\beta = (x, y, z)$, $\hat{L}$ is the sought-for wave operator of the scattering medium; $L^{(0)}$ is the determined operator for which an inverse (integral) operator – Green's function $\hat{G}_0$ is known; $\hat{V}$ is a random operator describing the statistical variations of the medium; q is a function of the sources which can be both determined and random. The statistical characteristics of the medium (operator $\hat{V}$) and sources (q) should be known; an unknown random Green's operator $\hat{G} = \hat{L}^{-1}$ is to be determined, which characterizes the process of interaction between radiation and the environment.

This process of interaction at a microlevel (on the characteristic scale of radiation wavelength) is described by Maxwell's equations. A major difficulty arising here is the transition from microcharacteristics of interaction to macrocharacteristics. In remote sensing data interpretation, the latter are of particular interest: this is the level of individual elements of the vegetation structure in the optical range, of individual droplets in clouds and precipitation in the microwave interval, etc. Here the validity of Maxwell's equations for these environmental elements can be assumed. Thereby, assuming the statistical homogeneity of the environment on smaller scales, taking the scales of elements for the microlevel of the interaction and introducing the notion "brightness" for this level. Then, the problem is reduced to the averaging of the brigthness of natural objects by the orientation and viewing angle of the measuring instruments – this is the macrolevel of interaction (the brightness field), where results must be presented in terms of non-photometric (physical) parameters [78].

Bearing in mind these considerations, we shall examine Maxwell's equations, on the assumption that the medium is not magnetic:

$$\left.\begin{aligned}
&\operatorname{rot}\vec{H}(\vec{r}, t) = \frac{1}{c}\frac{\partial\vec{D}(\vec{r}, t)}{\partial t} + \frac{4\pi}{c}\vec{j}(\vec{r}, t) \ , \\
&\operatorname{rot}\vec{E}(\vec{r}, t) = -\frac{1}{c}\frac{\partial\vec{H}(\vec{r}, t)}{\partial t} \ , \\
&\operatorname{div}\vec{D}(\vec{r}, t) = 4\pi\varrho(\vec{r}, t) \ , \\
&\operatorname{div}\vec{H}(\vec{r}, t) = 0 \ ,
\end{aligned}\right\} \tag{2.2}$$

where $\vec{D}$ is the electrical induction vector; $\vec{j}$ is the volume density of conduction currents; ϱ is the respective volume density of charges; the latter three characteristics determine the secondary (scattered) radiation field.

If we time-extend the vectors of the field and medium in Fourier integrals and introduce the material bonds for the Fourier components:

[1] The summation is made over the repeated Greek indices.

$$\vec{D}(\vec{r}, \omega) = \varepsilon'(\vec{r}, \omega)\vec{E}(\vec{r}, \omega) ,$$

$$j(\vec{r}, \omega) = \sigma(\vec{r}, \omega)\vec{E}(\vec{r}, \omega) ,$$

where ε' is the real part of the di-electrical penetrability of the medium; σ is its conductivity; ω is the radiation frequency, the following equation for the electrical field can easily be obtained:

$$\Delta\vec{E}(\vec{r}, \omega)+\frac{\omega^2}{c^2}\varepsilon(\vec{r}, \omega)\vec{E}(\vec{r}, \omega)-\text{grad div}\,\vec{E}(\vec{r}, \omega) = 0 \tag{2.3}$$

where Δ is the Laplace operator,

$$\varepsilon(\vec{r}, \omega) = \varepsilon'(\vec{r}, \omega)+\frac{4\pi i}{\omega}\sigma(\vec{r}, \omega)$$

is the complex di-electrical penetrability which is a tensor for an anisotropic medium. The $\vec{D}$ and $\vec{E}$ vectors are interrelated by the $\vec{P}$ polarization vector:

$$\vec{D} = \vec{E}+4\pi\vec{P} ,$$

which, in the absence of the dispersion of the medium ($\vec{D} = \varepsilon\vec{E}$) is

$$\vec{P} = \frac{\varepsilon-1}{4\pi}\vec{E} ,$$

and when the $\vec{D}-\vec{E}$ correlation is of non-local character,

$$\vec{P} = \int\chi(\vec{r}, \vec{r}')\vec{E}(\vec{r}')d\vec{r}' ,$$

where χ is the polarizability of a medium with spatial dispersion.

We re-write Eq. (2.3) as

$$\left(\Delta+\frac{\omega^2}{c^2}-\text{grad div}\right)\vec{E}(\vec{r}, \omega) = \frac{\omega^2}{c^2}V(\vec{r}, \omega)\vec{E}(\vec{r}, \omega) , \tag{2.4}$$

where $V(\vec{r}, \omega) = \varepsilon(\vec{r}, \omega)-1$. We presented Maxwell's equations similarly to (2.1): the operator in the left-hand part of (2.4) corresponds to the operator $\hat{L}^{(0)}$ in (2.1), if the problem is solved with respect to an average parameter $\varepsilon(\vec{r}, \omega)$; the V operator in the right-hand part of (2.4) corresponds to the $\hat{V}$ operator in (2.1), if the problem is solved with respect to the fluctuation component of di-electrical penetrability $\tilde{\varepsilon}(\vec{r}, \omega) = V(\vec{r}, \omega)$, characteristic of the random fluctuations of the environment.

In the simplest case of a semi-finite medium, the solution of (2.4) is performed through the Green function $G_0^{\varrho\nu}(\vec{r}, \omega)$, which satisfies the requirements of radiation at infinity [197]:

$$G_0^{\mu\nu}(\vec{r}, \omega) = \frac{1}{4\pi k_0^2}\Lambda^{\mu\nu}(\vec{r}, \omega)+\frac{1}{3k_0^2}\delta^{\mu\nu}\delta(\vec{r}) ,$$

$$\Lambda^{\mu\nu}(\vec{r}) = k_0^2\,\delta^{\mu\nu}+\frac{\partial^2}{\partial x^\mu\partial x^\nu}\,\frac{\exp(ik_0|\vec{r}|)}{|\vec{r}|} ,$$

$$k_0 = \frac{\omega}{c} ,$$

$$\delta^{\mu\nu} = \begin{cases} 1 \text{ at } \mu = \nu \\ 0 \text{ at } \mu \neq \nu \end{cases} .$$

Having written (2.4) as an integral equation

$$E^{\mu}(\vec{r}, \omega) = E_0^{\mu}(\vec{r}, \omega) + k_0^2 \int_{V'} G_0^{\mu\nu}(\vec{r}-\vec{r}')V(\vec{r}, \omega)E^{\nu}(\vec{r}', \omega)d\vec{r}' , \tag{2.5}$$

where $E_0^{\mu}(\vec{r}, \omega)$ is the field in a free space, V' is the integration volume, we obtain for the Fourier components of the electrical field:

$$E^{\mu}(\vec{P}, \omega) = E_0^{\mu}(\vec{P}, \omega) + \frac{k_0^2}{(2\pi)^3} \int G_0^{\mu\nu}(\vec{P})V(\vec{P}-\vec{q}, \omega)E^{\nu}(\vec{q}, \omega)d\vec{q} . \tag{2.6}$$

In (2.5) and (2.6) the first right-hand item is the incident radiation field, the second item is the field induced by the medium; if the medium is anisotropic, then beneath the integrals in (2.5) and (2.6) expressions of $G_0^{\mu\nu}V^{\nu\varrho}E^{\varrho}$ type will be substituted for those written earlier.

Then we introduce the scattering operator $T^{\nu\varrho}(\vec{r}'', \vec{r}')$:

$$E^{\mu}(\vec{r}, \omega) = E_0^{\mu}(\vec{r}, \omega) + \int G_0^{\mu\nu}(\vec{r}-\vec{r}'')T^{\nu\varrho}(\vec{r}'', \vec{r}')E_0^{\varrho}(\vec{r}')d\vec{r}''d\vec{r}' , \tag{2.7}$$

which, in contrast to the well-known scattering amplitude, determines the scattering field not only far from the scattering medium, but also near by and within it.

Equation (2.7) for the Fourier components becomes

$$E^{\mu}(\vec{P}, \omega) = E_0^{\mu}(\vec{P}, \omega) + \frac{1}{(2\pi)^3} G_0^{\mu\nu}(\vec{P}) \int_{-\infty}^{\infty} T^{\nu\varrho}(\vec{P}, \vec{q})E_0^{\varrho}(\vec{q})d\vec{q} . \tag{2.8}$$

If the incident field is a plane wave, then

$$\vec{E}_0(\vec{r}, \omega) = \vec{A}(\vec{k}_0, \omega)e^{i\vec{k}_0\vec{r}} ,$$

where the $\vec{k}_0$ vector determines the direction of the incident radiation. Then

$$\vec{E}_0(\vec{P}, \omega) = (2\pi)^3\vec{A}(\vec{P}, \omega)\delta(\vec{k}_0 - \vec{P}) . \tag{2.9}$$

Using (2.9), (2.6) and (2.8), we obtain the following equation for the Fourier transformation of the scattering operator:

$$\begin{aligned} T^{\nu\varrho}(\vec{P}, \vec{k}_0)A^{\varrho}(\vec{k}_0, \omega) = {} & k_0^2 V(\vec{P}-\vec{k}_0, \omega)A^{\nu}(\vec{k}_0, \omega) \\ & + \frac{k_0^2}{(2\pi)^3} \int V(\vec{P}-\vec{q}, \omega)G_0^{\nu\varrho}(\vec{q}) \\ & \times T^{\varrho\sigma}(\vec{q}, \vec{k}_0)A^{\sigma}(\vec{k}_0, \omega)d\vec{q} . \end{aligned} \tag{2.10}$$

If variations in polarization in the scattered radiation can be neglected, then the $T^{\nu\varrho}$ and $G_0^{\nu\varrho}$ tensors are reduced to scalars and Eq. (2.10) is transformed:

$$T(\vec{P}, \vec{k}_0) = k_0^2 V(\vec{P}-\vec{k}_0, \omega) + \frac{k_0^2}{(2\pi)^3} \int V(\vec{P}-\vec{q}, \omega) G_0(\vec{q}) T(\vec{q}, \vec{k}_0) d\vec{q} \ , \qquad (2.11)$$

where

$$G_0(\vec{q}) = \frac{1}{k_0^2 - q^2} \ .$$

Equation (2.1) relates the scattering properties of individual elements of the environment (the scattering operator) and their physical-chemical properties (complex di-electrical penetrability). This equation can be solved with respect to either $T(\vec{p}, \vec{k}_0)$ or $V(\vec{p}-\vec{q})$. In the first case, the scattering properties of the environment can be calculated from the known physical-chemical characteristics; in the second case – vice versa. In case of scattering on the ensemble of natural elements ("a random cloud of scatterers" using the terminology in [67, 78, 168]) we write (2.11) in terms of operators

$$\hat{\hat{T}} = \hat{V} + \hat{V}\hat{G}_0\hat{\hat{T}}$$

where the operator

$$\hat{V} = \sum_{l=1}^{N} \hat{V}^{(l)}$$

corresponds to the whole medium with the scattering inhomogeneity $\hat{V}^{(l)}$ to a single l-element, we obtain a system:

$$\hat{\hat{T}}^{(N,l)} = \hat{T}^{(l)} + \sum_{m \neq l} \hat{T}^{(l)} \hat{G}_0 \hat{\hat{T}}^{(N,m)} \qquad (2.12)$$

where the operator $\hat{\hat{T}}^{(N,l)} = \hat{V}^{(l)}(1+\hat{G}_0\hat{\hat{T}})$ corresponds to the scattering of a l-element, with the mutual effect of all N elements of the medium taken into account: $\sum_l \hat{\hat{T}}^{(N,l)} \equiv \sum_l V^{(l)}(1+\hat{G}_0\hat{\hat{T}})$, the operator $\hat{T}^{(l)} = \hat{V}^{(l)}(1+\hat{G}_0\hat{T}^{(l)})$ is known from the solution for (2.11) for an individual element.

However, practically, the radiation field itself is not measured but the square values for this field, which necessitates an analysis of radiation transformation in the system "natural object – instruments".

2.3 The Filtering Role of the Optical Radiation Sensors

The average energy flux density just introduced, is a square function of the electromagnetic field tension, and for instant values of the complex field it is:

$$\vec{L} = \mathrm{Re}\, \frac{c}{4\pi} [\vec{E}\vec{H}^*] \ . \qquad (2.13)$$

However, a consideration of the field's effect on a radiation sensor implies that certain conditions, i.e. the processes of propagation and scattering of radiation in a medium, have been fulfilled. When measured by a sensor, the radiation field splits into a set of incoherent (statistically independent) wave packets. The latter are a modified notion of plane monochromatic waves, in which the direction of energy propagation coincides with that of the wave front. In inhomogeneous media, the idea of a wave packet has no meaning and hence, the already mentioned idea of brightness cannot be used according to definitions introduced earlier as spherical waves and partially coherent beams in different directions appear. Thus, the quasi-homogeneity of the medium is a requirement when using beam representations and the related theory of radiation transfer (see later). When there is a severe statistical inhomogeneity of the medium, the beam tubes become twisted and it becomes impossible to introduce the sections of interaction typical of the transfer theory. The quasi-homogeneity of the medium is necessary but insufficient for the idea of wave packets, carrying the energy of the field, as an information system and being a prototype of the beam characteristics of the phenomenological transfer theory: it is necessary that there should be no reduction of the Fourier components $\vec{E}$ and $\vec{H}$ within the coherence area:

$$\varrho_K \approx \frac{4\pi^2}{(k_0 \varrho_0)^2} = \frac{\lambda^2}{\upsilon^2}$$

during the time of coherence $t_K \approx \dfrac{1}{2\omega_0}$, where $\lambda = \lambda_0/n$, n is the real part of the refraction index of the medium; $\upsilon = \varrho_0$ is the angle of flare of the beam's rays (Fig. 2.2a).

The tensions of the electric and magnetic fields in such a wave formation (wave packet) containing a set of plane monochromatic waves, are:

$$\left.\begin{aligned} \vec{E}(\vec{r}, t) &= \vec{E}_0 \exp\left\{i\bar{\omega}\left(t - m\frac{\vec{l}\vec{r}}{c}\right)\right\} g(\vec{r}, t) \\ \vec{H}(\vec{r}, t) &= m[\vec{l}\vec{E}_0] \exp\left\{i\bar{\omega}\left(t - m\frac{\vec{l}\vec{r}}{c}\right)\right\} g(\vec{r}, t) \end{aligned}\right\}, \tag{2.14}$$

where the amplitude $\vec{E}_0$ and a function

$$g(\vec{r}, t) = \int_{\omega_0} \int_{|\vec{\varrho}|} E_0(\omega', \vec{\varrho}) \exp\left\{i\omega'\left(t - m\frac{\vec{l}\vec{r}}{c}\right) - i\frac{\omega}{c} m \vec{\varrho}\vec{r}\right\} d\omega' d\vec{\varrho} \tag{2.15}$$

depends weakly on $\vec{r}$ and t; $\bar{\omega}$ is an average frequency of radiation; $|\omega'| \leqslant \omega_0 \ll \bar{\omega}$: $\omega = \bar{\omega} - \omega'$; $k = (\omega/c)m$ is a wave number; $m = n - i\kappa$ is the complex refraction index of the medium; $\vec{l}$ is the wave normal along the beam's axis; $\vec{\varrho}$ is the normal in perpendicular direction: $\vec{l}\vec{\varrho} = 0, |\vec{\varrho}| \leqslant \varrho_0 \ll 1$; $E_0F_0(\omega, \vec{\varrho})$ is the smoothly varying density, becoming zero outside the interval $|\omega'| \leqslant \omega_0 \ll \bar{\omega}$.

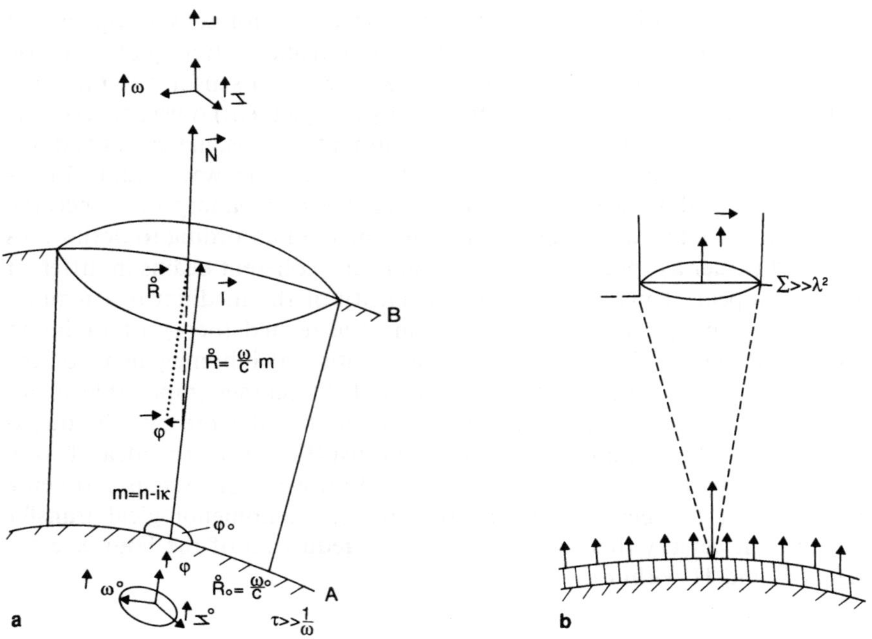

Fig. 2.2 a, b. A scheme of a wave packet **a** before the interaction with the medium (*A*) and after it (*B*); **b** the effect of a wave packet on the radiation sensor

Introducing the wave normal vector

$$\vec{N} = \mathrm{Re}\frac{\vec{k}}{\vec{k}_0} \ ,$$

and the attenuation vector

$$\vec{K} = -\mathrm{Im}\frac{\vec{k}}{\vec{k}_0} \ ,$$

where

$$\vec{k} = k_0 m(\vec{l} + \vec{\varrho}) \ ,$$

we can obtain, proceeding from (2.13):

$$\vec{L} = \frac{c}{4\pi} e^{-2k_0\vec{K}\vec{r}} \{(\vec{E}\vec{E}^*)\vec{N} + i[\vec{K}[\vec{E}\vec{E}^*]]\} \ . \tag{2.16}$$

The requirement of the quasi-homogeneity of the wave packet along the ray (a small attenuation of the field) and across it (a small angular divergence) leads, respectively, to the following expressions:

$$2\kappa\omega t_K \ll 1 \, (\text{or } \kappa \ll \omega/\bar{\omega}) \ .$$

Using a presentation of the complex normal along the axis

$$\vec{l} = \vec{l}_R + i\vec{l}_I \ ,$$

$$|\vec{l}_I| \ll \frac{1}{4\pi k_0 \sqrt{\varrho_K}} \ll 1$$

and

$$\vec{N} = n\vec{a} + x\vec{b} \ ; \quad \vec{K} = x\vec{a} - n\vec{b} \ ,$$

where

$$\vec{a} = \mathrm{Re}(\vec{l} + \vec{\varrho}) \ , \quad \vec{b} = \mathrm{Im}(\vec{l} + \vec{\varrho}) \ ,$$

we can see that the conditions needed to present the beam characteristics in terms of the wave properties of the field are as follows:

$$\vec{N} \approx n\vec{l}_R \ ; \quad \vec{K} \approx x\vec{l}_R - n\vec{l}_I \ . \tag{2.17}$$

With the limitations ensuring the consideration of the dynamic characteristics of the wave packets, as an energy basis for beam presentations, the $[\vec{E}\vec{E}^*]$ vector must be oriented towards $\vec{l}_k$, which gives

$$[\vec{E}\vec{E}^*] \approx i[\vec{E}\vec{E}^*]\tilde{q}\vec{l}_R \ , \tag{2.18}$$

where $\tilde{q}$ is the degree of ellipticity of polarization. Then

$$\vec{L} \approx \frac{cn}{4\pi}(\vec{E}_0\vec{E}_0^*)\, e^{-2k_0\vec{K}\vec{r}}(\vec{l}_R + \tilde{q}\,[\vec{l}_R\vec{l}_I])\,\tilde{q}(\omega_0, \varrho_0) \ , \tag{2.19}$$

where $\tilde{q}(\omega_0, \varrho_0) = \langle g(\vec{r}, t)\, g^*(\vec{r}, t)\rangle|_{t_K, \varrho_K}$ is a value averaged in time and over the coherence area, and the energy flux is transferred in the direction $\vec{l}_R + \tilde{q}\,[\vec{l}_R\vec{l}_I]$, differing from the direction of the wave normal $\vec{N}$ at a group speed:

$$U \approx \frac{c}{n + \omega(\partial n/\partial\omega)} \ .$$

In the sensors measuring the fluxes of any dynamic quantity (power, pulse, pulse momentum, etc.), a filtration of the $\vec{L}$ vector takes place, so that the field brightness is not measured but energetic irradiance:

$$F = \langle \vec{\tilde{L}}\vec{f}\rangle|_{\Sigma,\tau} \ ,$$

where $\vec{f}$ is the internal normal to the receiving surface (Fig. 2.2b); $\vec{\tilde{L}}$ is the vector of the energy flux density filtered by the receiver:

$$\vec{\tilde{L}} = \frac{c}{4\pi} e^{-2k_0\vec{K}\vec{r}} \int f(\vec{l})\, f(\vec{l}')\, [\vec{E}(\omega, \vec{l})\, \vec{H}^*(\omega', l')]$$
$$\times e^{i(\omega-\omega')t} e^{ik_0(\vec{N}-\vec{N}')\vec{r}}\, d\omega\, d\omega'\, d\Omega\, d\Omega' \ , \tag{2.20}$$

$f(\vec{l})$ is the angular filter depending on the design of the instrument. The sensor of the type considered, averages the $\vec{L}\vec{f}$ value over both the area of the entrance window and the period equal to its time constant, τ. Bearing in mind small values of ω_0/ω and ϱ_0, practical stability of $\exp(-2k_0\vec{K}\vec{r})$ within the entrance window, characteristic, for the optical range firstly and the quasi-homogeneity of all the Fourier components, using (2.19), we obtain

$$F \approx \frac{cn}{4\pi} e^{-2k_0\vec{K}\vec{r}} (\vec{E}_0\vec{E}_0^*)(\vec{l}\vec{f})\tilde{q}(\omega_0, \varrho_0) \ . \tag{2.21}$$

If $1/\omega \ll \tau \leqslant t_K$ and $\lambda^2 \ll \Sigma \ll \varrho_K$, then the instrument perceives a wave packet as a single coherent formation (a plane monochromatic wave). However, if $\tau \gg t_K$ or $\Sigma \gg \varrho_K$, then the instrument splits the wave packet into non-coherent parts, affecting the instrument independently.

With a quasi-homogeneous and quasi-stationary radiation field, the angular filtration takes place due to scalar product $\cos\theta = \vec{l}\vec{f}$. Here an averaged energetic irradiance measured in the direction $\vec{l} = \vec{f}$ will be

$$F(\vec{l}) = \int F(\omega, \vec{l})\, d\omega \ , \tag{2.22}$$

where

$$F(\omega, \vec{l}) = \int I(\omega, \vec{l}) \cos\theta\, d\Omega \ , \tag{2.23}$$

$$I(\omega, \vec{l}) = \frac{cn}{4\pi} e^{-2k_0\vec{K}\vec{r}} \vec{E}(\omega, \vec{l}) \vec{E}^*(\omega, \vec{l}) \tag{2.24}$$

is the spectral brightness density of a wave packet propagating in the direction $\vec{l}$. Here, it is assumed, that the angular size of an object is much larger than the size of the wave packet cut off by the instrument (Fig. 2.2b).

In photo-energetic sensors based on the principle of photon counting for period of a time $\tau \gg t/\omega$ in some volume $\upsilon \gg \lambda^3$, instead of (2.23) the spectral density of spherical irradiance is:

$$\phi(\omega) = \oint I(\omega, \vec{l})\, d\Omega \ , \tag{2.25}$$

instead of (2.22), the total spherical irradiance of volume, υ:

$$\phi = \int \phi(\omega)\, d\omega = \int I(\vec{l})\, d\Omega \ , \tag{2.26}$$

equal to

$$\phi = \frac{cn}{4\pi} \langle \vec{E}(\vec{r}, t) \vec{E}^*(\vec{r}, t) \rangle |_{\tau, \upsilon} \ , \tag{2.27}$$

$I(\vec{l}) = I(\omega, \vec{l})\, d\omega$ is the total brightness. So, in both cases considered it is necessary to have a theoretical scalar product $(\vec{E}\vec{E}^*)$.

Confining oneself to the case of recording the energetic irradiance we relate the focal plane field of the instrument to the field incident onto the entrance aperture [44]:

$$\vec{E}(x', y', z, t) = \frac{i e^{-ik_0\tilde{F}}}{\lambda\tilde{F}} \int_{-\infty}^{\infty} \vec{E}(x, y, z, t) e^{i(k_0/\tilde{F})(yy' + xx')} M(x, y)\, dx\, dy \ , \tag{2.28}$$

where $\tilde{F}$ is the focal distance; (x, y, z) are the coordinates of an object; (x′, y′, z′) are the coordinates of its image; $\vec{E}(x, y, z) M(x, y)$ is the distribution of the field over the area of the entrance aperture, where the function $M(x, y) = 1$. The introduction of the function $M(x, y)$ makes it possible to extrapolate the limits of integration in (2.28) onto infinity. The square of the field module in the focal plane is

$$|\vec{E}(x', y', z, t)|^2 = \frac{1}{\lambda^2 \tilde{F}^2} \int_{-\infty}^{\infty} \vec{E}(x, y_1, z, t)\vec{E}^*(x_2, y_2, z, t) \times M(x_1, y_1)M(x_2, y_2) \times e^{i(k_0/\tilde{F})[(x_1-x_2)x'+(y_1-y_2)y']} dx_1\,dx_2\,dy_1\,dy_2 \ . \tag{2.29}$$

It is seen that the sub-integral expression is proportional to the value $I(\omega, \vec{I})$ from (2.24). Using (2.11), and (2.12), the scattered field is connected with the incident field, with the limitations mentioned, in the following way:

$$E^{\mu}_{scat} = G_0^{\mu\nu}(\vec{P}, \omega)T^{\nu\varrho}(\vec{P}-\vec{k}_0, \omega)A^{\varrho}(\omega) \ , \tag{2.30}$$

then, when variations in polarization in scattering from (2.29), (2.30) and (2.9) are ignored, upon substituting the variables

$$x_1 - x_2 = \alpha_x \ , \quad \frac{x_1+x_2}{2} = \beta_x \ , \quad y_1 - y_2 = \alpha_y \ , \quad \frac{y_1+y_2}{2} = \beta_y \ ,$$

we obtain

$$\begin{aligned} |\vec{E}(x', y', z, t)|^2 = {} & \frac{|A(\omega)|^2}{(2\pi)^6\lambda^2\tilde{F}^2} \int G_0(\vec{P})G_0^*(\vec{q})T(\vec{P}, \vec{k}_0)T^*(\vec{q}, \vec{k}_0) \\ & \times M\left(\beta_x + \frac{\alpha_x}{2}, \ \beta_y + \frac{\alpha_y}{2}\right)M\left(\beta_x - \frac{\alpha_x}{2}, \ \beta_y - \frac{\alpha_y}{2}\right) \\ & \times e^{i[\beta_x(P_x-q_x)+\beta_y(P_y-q_y)+z(P_z+q_z)]} \\ & \times e^{i[\alpha_x((P_x+q_x/2)+(k_0x'/\tilde{F}))]}e^{i[\alpha_y(P_y+q_y/2)+(k_0y')/\tilde{F}))]} \\ & \times d\alpha_x\,d\alpha_y\,d\beta_x\,d\beta_y\,d\vec{P}\,d\vec{q} \end{aligned} \tag{2.31}$$

Equation (2.31) makes it possible to determine, from the measured values (the left-hand part of the equation), the parameters of the environmental state contained in the operator T, with Green's function G_0 and if the characteristics of the measuring instruments and incident radiation are known. The theoretical solution of the problem of remote sensing data interpretation is, thus, reduced to modeling using the operator T from (2.11), and mindful of (2.12), with given parameters of the state, and from the right-hand of Eq. (2.31) to retrieving the resultant measurements. However, the problem is complicated by the necessity to account for the multiple scattering of radiation.

2.4 Coherence Functions

Returning to the initial problem (2.1), we shall consider the equations for the Fourier components of the field, which enable consideration of the multiple scattering in the medium and confirm the phenomenological theory of radiation transfer [12]. First, we shall consider a scalar model for individual components $\vec{E}$ and $\vec{j}$, satisfying the conditions of transversity. Then we shall generalize the final expressions for electromagnetic fields, including polarization.

The initial equations are:

the Dyson equation

$$\hat{D}\langle U\rangle \equiv (\hat{L}_0 - \hat{V}^e)\langle U\rangle = \langle q\rangle \tag{2.32}$$

and the equation of Boethe-Solpeter

$$(\hat{D}_1 \hat{D}_2^* - \hat{K}_{12})\langle U_1 U_2^*\rangle = \langle q_1 q_2^*\rangle \tag{2.33}$$

or, in terms of the Green function,

$$\langle G\rangle = \hat{D}^{-1} \ , \quad \langle G_1 G_2^*\rangle = \langle G_1\rangle\langle G_2^*\rangle + \langle G_1\rangle\langle G_2^*\rangle K_{12}\langle G_1 G_2^*\rangle \ ,$$

where $U = U(x)$; $x = (\vec{r}, t)$; $\hat{D}$ is the Dyson operator describing the distribution of an averaged field; $\hat{V}^e$ is the mass operator of the effective inhomogeneity of the medium in an averaged field; the real operator $\hat{L}_0$, we have obtained already from Maxwell's equations, the sign $\langle\rangle$ indicates an averaging over the ensemble of realizations; * is the complex conjunction;

$$\hat{K}_{12} f(x_1, x_2) = \int \hat{K}\begin{pmatrix} x_1, & x_1' \\ x_2, & x_2' \end{pmatrix} f(x_1', x_2')\, dx_1'\, dx_2' \ , \tag{2.34}$$

$\hat{K}$ is the intensity operator (the operator's core $\hat{K}_{12}$), characterizing the effect of effective medium inhomogeneities on the radiation field correlation. The presentation of the radiation field as a sum of an average $\langle U\rangle$ and fluctuating U' of the fields and a coherence function

$$\Gamma_{12} = \langle U_1\rangle\langle U_2^*\rangle + \langle \tilde{U}_1 \tilde{U}_2^*\rangle \ , \tag{2.35}$$

characterizing the statistical connection in 2 spatial-temporal points, enables (2.33) to be written, averaging source fluctuations and an assumption that $\langle U\rangle = 0$, as

$$\hat{D}_1 \hat{D}_2^* \Psi_{12} = \hat{K}_{12}\Psi_{12} + \langle q_1 q_2^*\rangle \ , \tag{2.36}$$

where

$$\Psi_{12} = \langle \tilde{U}_1 \tilde{U}_2^*\rangle \ .$$

The real operators $\hat{V}^e$ and $\hat{K}_{12}$ generally, can only be obtained as expansions in infinite series by the powers of momenta and cumulants. The latter are a measure of the statistical relationships of a certain number of factors in a group. In a one-group approximation, i.e. in keeping with cumulant expansions of the operators $\hat{V}^e$ and $\hat{K}_{12}$, only the first cumulants $\hat{V}^e \approx \hat{V}^{(1gr)}$, $\hat{K}_{12} \approx \hat{K}_{12}^{(1gr)}$, is it possible to obtain the expressions relevant for the weakly non-local operators (compact) $\hat{V}^e$ and $\hat{K}_{12}$, with the non-locality radius of an order of correlation radii in a non-homogeneous medium. This assumption, makes it possible to introduce the scattering section per unit volume, as some local characteristic of the scattering medium. Due to the compact nature of $\hat{K}_{12}$ in the description of scattering in a one-group approximation, it is enough to consider the Ψ_{12} values at small divergences of an order of correlation radius for the fluctuations of the medium. Assuming that fluctuations of the scattered field are quasi-homogeneous, where the points diverge slightly the Ψ_{12} function varies rapidly by the difference coordinate $\varrho = x_1 - x_2$. However, the coordinate of the "gravity center" $R = (x_1 + x_2)/2$ produces only small variations. This circumstance enables the introduction of a small dimensionless parameter, equal to the ratio of the efficient wavelength to the characteristic length of "free run". The subsequent supposition, that the attenuation of an average field due to its scattering and absorption, is much less than an average field in the medium

$$|\hat{D}^{Im}(K) \equiv \operatorname{Im}\hat{D}(K)| \ll |\hat{D}^{Re}(K) \equiv \operatorname{Re}\hat{D}(K)| \ , \tag{2.37}$$

where

$$K = (\vec{k}, \omega) \ , \quad Kx = \vec{k}\vec{r} - \omega t \ , \quad \hat{D}(K) = \int \hat{D}(x') e^{iKx'} d^4x' = \hat{L}_0(K) - \hat{V}^e(K) \ ,$$

and the expansion of the spectrum of fluctuations of the scattered field

$$J_K(R) = \Psi_{12}(R, K) \equiv (2\pi)^{-4} \int \Psi_{12}(R, \varrho') e^{-iK\varrho'} d^4\varrho' \tag{2.38}$$

by the small parameter in a series leads to the following equations of the zero and first order expansion [8]:

$$\hat{D}^{Re}(K) J_K^{(0)}(R) = 0 \ , \tag{2.39}$$

$$\begin{aligned} \hat{D}^{Re}(K) J_K^{(1)}(R) = & -[i\hat{D}^{Im}(K) + (2l)^{-1} (\partial_K \hat{D}^{Re}(K)) \partial_R] J_K^{(0)}(R) \\ & + (\hat{D}^*(K))^{-1} \{\hat{K}^{(1gz)} J_K^{(0)}(R)\} \ . \end{aligned} \tag{2.40}$$

It follows from (2.39) that a major term of expansion of the spectral density J_K should be localized on the so-called disperse surface

$$\hat{D}^{Re}(K) = 0 \ . \tag{2.41}$$

Mathematically, it means that, in an assumed weakly conservative medium the respective matrix is Hermitian, it can be expanded in eigen vectors, and the respective determinant is equal to the product of eigen values. From the physical point of view, it means that in the medium, due to its interaction with radiation, there appears the whole spectrum of frequences values ω and directions $\vec{n}$. However, the wavenumber $k_1(\omega)$ connected with the solution of disperse equation (2.41), is approximately equal to the real part of the effective wavenumber $k^e = 2\pi/\lambda^e$:

$$k_1(\omega) \approx \operatorname{Re} k^e \ .$$

Thus,

$$J_K^{(0)}(R) = I_{\omega\vec{n}}(R) k^{-2} \delta(k - k_1(\omega)) \ , \tag{2.42}$$

where the frequency-angular spectrum $I_{\omega\vec{n}}$ in accordance with (2.40–2.42), satisfies the transfer equation:

$$[(\partial_K \hat{D}^{Re})\partial_R - 2\hat{D}^{Im}] I_{\omega\vec{n}}(R) = - |\partial_k \hat{D}^{Re}|^{-1} 2\pi k^2 \{\hat{K}^{(1gz)} I_{\omega\vec{n}}(R) k^{-2} \delta(k-k_1)\} , \tag{2.43}$$

which can be written conventionally

$$(d_s + \alpha_t) I_{\omega\vec{n}}(R) = \int \sigma'(\omega', \vec{n}' \to \omega, \vec{n}) I_{\omega'\vec{n}'}(R) d\omega' d\Omega + \varepsilon'_q ,$$

where $d_s = \dfrac{\partial_K \hat{D}^{Re}}{|\partial_k \hat{D}^{Re}|} \partial_R$ characterizes the change of field along the beam parameter, s (this change is connected with the group velocity and radiation transformation along the chosen direction $\vec{n}$),

$$\alpha_t = 2\hat{D}^{Im}/|\partial_k \hat{D}^{Re}| = 2\mathrm{Im}(\hat{L}_0(K) - \hat{V}^{(1gr)}(K)/|\partial_k \hat{D}^{Re}|$$

is the total section of interaction equal to the sum of sections of absorption (first item) and scattering (second item),

$$\sigma'(\omega', \vec{n}' \to \omega, \vec{n}) = C_1 \hat{K}^{(1gr)}(K' \to K)$$

is the phase function, $C_1 = \dfrac{2\pi k^2}{|\partial_k \hat{D}^{Re}|^2}$, and ε'_q is a function of the sources.

The enumerated assumptions result in the following relationship between radiation intensity I and the correlation function of the field Ψ_{12}:

$$\Psi_{12}(R, \varrho) \equiv \langle \tilde{U}(R+\varrho/2) \tilde{U}^*(R-\varrho/2) \rangle = \int J_K^{(0)}(R) \exp(ik\varrho) d^4K = \int I_{\omega\vec{n}}(R) \exp[i(\vec{n}\vec{\varrho} \mathrm{Re}\, k^e - \omega\tau)] d\omega d\Omega_{\vec{n}} . \tag{2.45}$$

A generalization of (2.45) for the case of a multi-component electromagnetic field can be written as

$$\Psi_{\alpha\beta}^{(E)}(\vec{r}, t) = \int_0^\infty d\omega \oint I_{\alpha\beta}(\omega, \vec{n}) e^{i(k^e \vec{n}\vec{r} - \omega t)} d\Omega_{\vec{n}} = \langle \tilde{E}_\alpha(\vec{r}_1, t_1) \tilde{E}_\beta(\vec{r}_2, t_2) \rangle . \tag{2.46}$$

The coherence function, averaged over the parameters of the radiation sensor, is

$$\Gamma_{\alpha\beta}(\gamma, \vec{\varrho}) = \int_0^\infty \oint m_{\alpha\beta}^{ik}(\vec{l}) J_{jk}(\omega, \vec{l}) e^{-i\omega\gamma + i\vec{k}\vec{\varrho}} d\omega d\Omega , \tag{2.47}$$

where

$$m_{\alpha\beta}^{ik}(\vec{l}) = C_{\alpha j}(\vec{l}) C_{\varrho k}(\vec{l}) ,$$

$C_{\alpha j}(\vec{l}) = (\vec{e}_j)_\alpha$ is complex, directing cosines of the j-th basis vector $\vec{e}_j$ in the direction $\vec{l}$ (j, k = 1, 2).

$$J_{jk}(\omega, \vec{l}) = \frac{cn}{4\pi} E_j(\omega, \vec{l}) E_k^*(\omega, \vec{l}) \tag{2.48}$$

are the components of the density matrix for a wave packet with the frequency ω and direction of propagation $\vec{l}$, the density matrix being connected with the known Stocks parameter S_i describing the polarization, with the relationships

$$S_i(\omega, \vec{l}) = S_P(J\sigma^i) \ , \quad J(\omega, \vec{l}) = \frac{1}{2}S_i(\omega, \vec{l})\sigma^i \ , \tag{2.49}$$

where S_P denotes the trace of the matrix, σ^i the Pauli matrices:

$$\sigma^1 = \begin{pmatrix} 1 & 0 \\ 0 & 1 \end{pmatrix} , \quad \sigma^2 = \begin{pmatrix} 1 & 0 \\ 0 & -1 \end{pmatrix} , \quad \sigma^3 = \begin{pmatrix} 0 & 1 \\ 1 & 0 \end{pmatrix} , \quad \sigma^4 = \begin{pmatrix} 0 & -i \\ i & 0 \end{pmatrix} .$$

Thus, the basic photometric values $I(\omega, \vec{l})$, $J_{jk}(\omega, \vec{l})$, $S_i(\omega, \vec{l})$ are coefficients of the spatial-temporal Fourier expansion of correlation functions of the tension of the electrical part of the radiation field.

Taking $\vec{\varrho} = 0$ after integrating (2.25) along the direction $\vec{l}$, we obtain

$$\Gamma(\gamma, \vec{\varrho} = 0) = \int_0^\infty \phi(\omega) e^{-i\omega\gamma} d\omega \ , \tag{2.50}$$

i.e., the spectral density of spherical irradiance coincides with the spectral density of expansion $\Gamma(\gamma, \vec{\varrho} = 0)$ in a time spectrum. Similarly, after integrating the right-hand part of (2.49) over ω at $j = 0$, using (2.23),

$$\Gamma(\gamma = 0, \vec{\varrho}) = \oint I(\vec{l}) e^{i\vec{k}\vec{\varrho}} d\Omega \ ,$$

i.e., the brightness of a wave packet is equal to the spectral density of expansion $\Gamma(\gamma = 0, \vec{\varrho})$ in the space spectrum.

2.5 The Structure of the Brightness Field and the Structuring of Computation Programs

The theoretical results obtained suggest the following succession of transformations of information fields on aircraft and satellite images: (a) brightness field (b) coherence function (c) spatial structure of the parameters of the physical state of objects.

The solution for the basic problem of interactions of the general type (2.33) seems promising in the light of ideas about structural programming [203]. Really, the formal technique of the Feinman diagrams [8, 168] is nothing but a synthetic (or structural) description close to that used in the program of structural recognition of images [203].

There are three approaches to presenting video-information: direct element-by-element presentation (as if the machine is overwhelmed by the details of an image); presentation based on the orthogonal transformation by Fourier, Adamar, etc. (the machine does not "see" details); synthetic (structural) presentation – a description of the image with a set of standard elements (for instance, angle, stretch, graphs, "tree", etc.).

The constructing of a pyramidal structure of interaction [5] based on the presentation of the information field just described is an elementary rule of transformation of a group of elements with its propagation "in breadth" – onto other elements, and "deep into" – to other levels. In particular, recursion is one of the ways to describe the structure of relationships between natural elements and their brightnesses [5].

In recursive description, mass operations with elements (the group presentation of the mass operator $\hat{V}$ was mentioned earlier) can be made by describing the operation for an elementary cell of the structure only, with subsequent application to the whole information field. The Feinman diagrams for the mass operator and intensity operator, are an intuitive approach to recursion and to the respective description of the "tree" of connections. The regulation of the structural elements is actually equivalent to the evolution of a multi-dimensional array into a 1-D successiveness – here is the possibility to use multi-group approximations. The evolving of the structure makes it easier to realize the problem considered on parallel devices.

Figure 2.3 exemplifies a scheme of structuring of calculations of the object's brightness at the level of an farmer's field: the transition from microlevel – individual elements of the vegetation structure (a) to a description of a plant (b) and then the structure of a crop (c). Here we use the structure of an individual object as a graph $\tilde{G}(\tilde{S}, \tilde{R})$, where $\tilde{S}$ is a multitude of apexes, $\tilde{R}$ is a multitude of arcs. The arc of the graph corresponds to the use of the operator $\hat{Z}$, the apex indicates one of the B multitude of possible versions of the substitution of concrete objects. If the B multitude is finite, then the apex of the structure of the object splits into B apexes, each being marked and corresponding to one version of B. A complicated object in Fig. 2.3 is presented as a hierarchic (tree-shaped) structure. The graph is a reference or elementary cell of the structure, and the operator $\hat{Z}$ describes the law of the development of the structure. Figure 2.3 corresponds to a 2-group approximation describing at the initial stage (a) with the help of (2.33), the dynamics of scattering on a 2-element group, with the subsequent presentation of results using the k-th tree with the (b) graph, and at the final stage (c) the graph can be presented as a tree with marked apexes from $1'$ to k' or as a uniform rectangular grid.

The development of the means of analysis of video-information, parallel computations, personal computers, structural methods for programming, and other scientific and technical achievements makes it possible to look at the problem of the practical use of the remote sensing data from other points of view. Studies and establishing of connections between natural parameters and the characteristics of the outgoing radiation field become possible. This problem can be posed as that of studying the interaction of radiation with randomly inhomogeneous media, characterized by the distribution of physical-chemical, biometric, etc., parameters of the state. A proposed model links processes of interaction with analysis of the information content of the brightness fields recorded by instruments. Based on the structure of the brightness field, it is proposed to retrieve the structure of the respective operators radiation interaction with the environment and the structure of the parameters characterizing the pyhsical state of natural objects. There are certain possibilities to solve

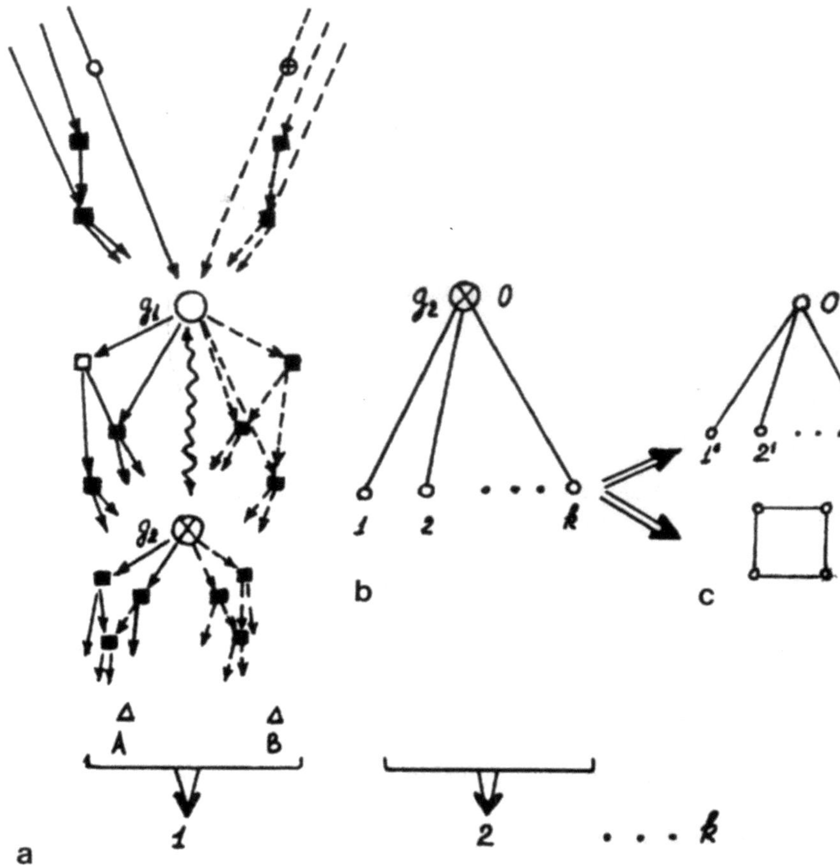

Fig. 2.3a–c. Processes of scattering, corresponding to a 2-group approximation **a**: *A* a wave packet resulting from scattering of the coherent field on element g_2, involving the radiative connection of elements g_2 and g_1; *B* a wave packet for a complex-conjugated field; **b** a graph of the "tree" connections of element g_2 with each of the remaining elements numbered 1 to k, with the group of connection fixed at the preceding stage (g_1, g_2); **c** a generalization of the process for macroobjects with the marked apexes of the graph as the *k*-th "tree" or a rectangular grid

this problem in the light of "old" physical methods of statistical optics/radiophysics and "new" achievements of calculation techniques.

2.6 Partially Coherent Waves Coherence Matrices

1. The parameter observed in the optical wave range is known not to be the complex amplitude of the field $U(\vec{\varrho})$ but its intensity $I(\vec{\varrho})$. In the extreme case

of coherent waves the intensity is related to the amplitude by a simple relationship:

$$I(\vec{\varrho}) = U(\vec{\varrho})U^*(\vec{\varrho}) \ . \tag{2.51}$$

When waves are partially coherent, the situation is different.

The intensity of a partially coherent field is expressed, not through the amplitude of the field $U(\vec{\varrho})$, but through its mutual coherence function

$$\Gamma_{12}(\tau) = \langle U_1(t+\tau)U_2^*(t)\rangle \ .$$

The latter determines, in the second order, the correlation between complex fields in two spatial points, P_1 and P_2, for moments of time separated by a period τ. Here

$$U_1(t) \equiv U(\vec{\varrho}_1, t) \ ,$$

$$U_2(t) \equiv U(\vec{\varrho}_2, t) \ .$$

The brackets $\langle \ldots \rangle$ indicate time averaging over the period T, corresponding to the inertia of the measuring instrument:

$$\langle U_1(t+\tau)U_2^*(t)\rangle \equiv \frac{1}{T}\int\limits_{t-T/2}^{t+T/2} U_1(t+\tau)U_2^*(t)\,dt. \tag{2.52}$$

In a general case $\Gamma_{12}(\tau)$ depends also on t. In the case of stationary fields $\Gamma_{12}(\tau)$ does not depend on t.

In a more precise determination of the mutual coherence function a time average is substituted by an average over the ensemble. In this case, $\Gamma_{12}(\tau)$ is presented by a 2-order moment function of the 2-D distribution connected with a random function $U(\vec{\varrho}, t)$.

The normalized mutual coherence function

$$\gamma_{12}(\tau) = \Gamma_{12}(\tau)/[\Gamma_{11}(0)\Gamma_{22}(0)]^{1/2} \tag{2.53}$$

is called the complex order of coherence. Apparently,

$$0 \leqslant |\gamma_{12}(\tau)| \leqslant 1 \ . \tag{2.54}$$

The field is completely coherent, if $|\gamma_{12}(\tau)| = 1$ for all pairs of points P_1 and P_2 and all values of the time delay parameter τ.

In a practical sense, the field is considered coherent if an approximate equation $|\gamma_{12}(\tau)| \approx 1$ holds for all pairs of points within a sufficiently large space, and for all values of the τ parameter within the large time interval under consideration. A laser characterized by a good radiation coherence and its high intensity, is a major source of such fields in optics. When conventional, "thermal" radiation sources are used, good coherence can only be reached with the help of spectral (increased monochromaticity) and spatial (decreased size of the source) filtration, leading to considerable reduction of light intensity.

By definition,

$$I(P_1) = \Gamma_{11}(0) = \langle U_1(t)U_1^*(t)\rangle \ . \tag{2.55}$$

The intensity of the field I(P) is a parameter observed experimentally. However, simple laws of propagation exist not for the intensity of the field but for its mutual coherence function. It is with respect to the latter that the space can be considered a linear element of the optical scheme.

The function $\Gamma_{12}(\tau)$ satisfies wave equation (see [180]):

$$\square_s \Gamma_{12}(\tau) = 0 \ , \quad s = 1, 2 \tag{2.56}$$

in each of its variables $\vec{\varrho}_1$ and $\vec{\varrho}_2$. The s index refers here to the spatial points P_1 and P_2; the wave operator in (2.56) is

$$\square_s \equiv \nabla^2 - \frac{1}{c^2} \frac{\partial^2}{\partial \tau^2} \ ,$$

where c is the light velocity in a free space. In an other way, it can be written as

$$\nabla_1^2 \nabla_2^2 \Gamma_{12}(\tau) = \frac{1}{c^4} \frac{\partial^4 \Gamma_{12}(\tau)}{\partial \tau^4} \ . \tag{2.57}$$

Equations (2.56) and (2.57) immediately follow from (2.52), bearing in mind the commutativity relationship between a linear transformation of a random function and its averaging, and in accordance with the laws of transformation of moment functions [180].

2. A quasi-monochromatic field in the theory of partial coherence serves as an analog to the monochromatic field considered above, for which the spatial and time variables are separated. The field is quasi-monochromatic, if its spectral width $\Delta\nu$ is much less than an average frequency of oscillations ν_0:

$$\Delta\nu \ll \nu_0 \ . \tag{2.58}$$

In many experimental situations, the conditions are such that

$$\tau \ll 1/\Delta\nu \ . \tag{2.59}$$

In these conditions, which are substantial for the quasi-monochromatic approximation, the spatial and temporal variables of the mutual coherence function are separated, and $\Gamma_{12}(\tau)$ can be presented as

$$\Gamma_{12}(\tau) \approx \Gamma_{12}(0) \exp(-i2\pi\nu_0\tau) \ , \tag{2.60}$$

which reminds us of the equation for the monochromatic field.

The function of spatial variables

$$\Gamma_{12}(0) \equiv \Gamma_{12}$$

is called the mutual intensity function, and the parameters $\Delta t = 1/\Delta\nu$ and $\Delta l = c\Delta t$ are the time and length of coherence; the distance between P_1 and P_2, for which $|\gamma_{12}(0)|$ is equal to a given quantity, for instance, $|\gamma_{12}(0)| = 0.88$, is the interval of spatial coherence. For a quasi-monochromatic field the module of the complex coefficient of coherence corresponds to the visibility parameter of the interference bands (see [180]), introduced in classical optics:

$$|\gamma_{12}(\tau)| = \frac{I_{max} - I_{min}}{I_{max} + I_{min}} . \tag{2.61}$$

Equation (2.60), under conditions of (2.59), follows from the Fourier transformation formula:

$$\Gamma_{12}(\tau) = \int_0^{\infty} \bar{\Gamma}_{12}(\nu) \exp(-i2\pi\nu\tau) d\nu , \tag{2.62}$$

where

$$\bar{\Gamma}_{12}(\nu) = \begin{cases} \int_{-\infty}^{\infty} \Gamma_{12}(\tau) \exp(i2\pi\nu\tau) d\tau , & \nu > 0 \\ 0 , & \nu < 0 . \end{cases} \tag{2.63}$$

In writing these formulas the authors considered the analytical character of the function $\Gamma_{12}(\tau)$. It is clear that

$$\Gamma_{12}(\tau) = \exp(-i2\pi\nu_0\tau) \int_0^{\infty} \bar{\Gamma}_{12}(\nu) \exp[-i2\pi(\nu - \nu_0)\tau] d\nu ,$$

and if

$$\Gamma_{12}(\tau_0 + \tau') = \exp(-i2\pi\nu_0\tau') ,$$

$$\int_0^{\infty} \bar{\Gamma}_{12}(\nu) \exp\{-i2\pi[\nu\tau_0 + (\nu - \nu_0)\tau']\} d\nu , \tag{2.64}$$

then

$$\tau' \ll 1/\Delta\nu ,$$

$$\Gamma_{12}(\tau_0 + \tau') \approx \Gamma_{12}(\tau_0) \exp(-i2\pi\nu_0\tau') . \tag{2.65}$$

3. The mutual intensity function Γ_{12} which we shall also denote as $\Gamma(P_1, P_2)$, in a free space satisfies a pair of Helmholtz equations

$$\nabla_s^2 \Gamma_{12} + k^2 \Gamma_{12} = 0 , \quad s = 1, 2 \tag{2.66}$$

where $k = 2\pi/\lambda$ is the wave number. The solution for these equations is written as

$$\Gamma(P_1, P_2) = \iiiint_{-\infty}^{\infty} \Gamma(Q_1, Q_2) \frac{\exp[ik(\varrho_3 - \varrho_4)]}{\varrho_3 \varrho_4} d\xi_1 d\xi_2 d\eta_1 d\eta_2 , \tag{2.67}$$

which expresses a generalized Huyghense principle.

An equation of propagation of the mutual coherence function through a random linear system is written as

$$\Gamma(\vec{r}_1, \vec{r}_2) = c \iint \Gamma(\vec{\xi}_1, \vec{\xi}_2) K(\vec{\xi}_1, \vec{r}_1) K^*(\vec{\xi}_2, \vec{r}_2) d\vec{\xi}_1 d\vec{\xi}_2 , \tag{2.68}$$

where $K(\vec{\xi}, \vec{r})$ is the pulse response of the system; $\vec{\xi}$ and $\vec{r}$ are radius-vectors of the point in the input and output planes of the system, respectively. In the case of an image-forming system

$$\Gamma_u(\vec{r}_1, \vec{r}_2) = D \iint \Gamma_0(\vec{\xi}_1, \xi_2) \exp\left\{\frac{ik(\vec{\xi}_1^2 - \vec{\xi}_2^2)}{2z_1}\right\} K\left(\frac{\vec{r}_1}{z_2} + \frac{\vec{\xi}_1}{z_1}\right) K^*\left(\frac{\vec{r}_2}{z_2} + \frac{\vec{\xi}_2}{z_1}\right) d\vec{\xi}_1 d\vec{\xi}_2 \ , \tag{2.69}$$

where

$$D = B \exp\left\{\frac{ik(\vec{r}_1^2 - \vec{r}_2^2)}{2z_2}\right\} \ ,$$

$$B(\vec{\alpha}/a) = \begin{cases} 1 \text{ for } |\alpha| < a \ , \\ 0 \text{ for } |\alpha| > a \ , \end{cases} \tag{2.70}$$

In the extreme case of a completely incoherent source, for which

$$\Gamma(Q_1, Q_2) = I(Q_1)\delta(\xi_2 - \xi_1, \eta_2 - \eta_1) \ , \tag{2.71}$$

Eq. (2.67) becomes

$$\Gamma(P_1, P_2) = \iint_{-\infty}^{\infty} I(\xi, \eta) \frac{\exp[ik(\varrho_1 - \varrho_2)]}{\varrho_1 \varrho_2} d\xi d\eta \ . \tag{2.72}$$

This is the mathematical expression of a theorem known in classical optics which states that the light from a finite-size incoherent source becomes more and more coherent with spreading. The temporal coherence does not change, with the propagation of the light, in a medium without dispersion. In another extreme case, when the source is totally coherent, and

$$\Gamma(Q_1, Q_2) = U(Q_1)U^*(Q_2) \ , \tag{2.73}$$

we come to the conventional diffractive integral

$$U(P) = \int_{-\infty}^{\infty} U(Q) \frac{\exp(ik\varrho)}{\varrho} d\xi d\eta \ . \tag{2.74}$$

In the Fresnel approximation, when

$$\varrho \approx z + \frac{(x-\xi)^2}{2z} + \frac{(y-\eta)^2}{2z} \ ,$$

Eq. (2.67) becomes

$$\begin{aligned} \Gamma(P_1, P_2) = \exp\left[ik\frac{(x_1^2 - x_2^2) + (y_1^2 - y_2^2)}{2z}\right] \iiiint_{-\infty}^{\infty} \Gamma(Q_1, Q_2) \exp\left[\frac{ik(\xi_1^2 + \eta_1^2)}{2z}\right] \\ \times \exp\left[\frac{-ik(\xi_2^2 + \eta_2^2)}{2z}\right] \exp\left[-\frac{ik}{z}(x_1\xi_1 + y_1\eta_1)\right] \\ \times \exp\left[\frac{ik}{z}(x_2\xi_2 + y_2\eta_2)\right] d\xi_1 d\xi_2 d\eta_1 d\eta_2 \ . \end{aligned} \tag{2.75}$$

In the Fraunhofer approximation the first two exponential factors in the subintegral expression are omitted and the transformation of function Γ_{12} described by (2.75) becomes the Fourier transformation.
The distribution of intensity in a distant zone is expressed by

$$I(x, y) = \iiiint_{-\infty}^{\infty} \Gamma(Q_1, Q_2) \exp\left\{-\frac{ik}{z}[x(\xi_1-\xi_2)+y(\eta_1-\eta_2)]\right\} \times d\xi_1 d\xi_2 d\eta_1 d\eta_2 \;. \tag{2.76}$$

There are certain relationships of mutual substitution, between the order of coherence and amplitude (or intensity) of the fields produced, by homogeneous coherent and non-coherent sources of equal sizes. For a coherent source, the order of radiation coherence is constant by module (is unity), and the propagation of the complex amplitude in a distant zone is the Fourier transformation of the aperture function. On the other hand, for a non-coherent source the intensity of the resulting field is constant (when the angles are not too large), and the order of coherence is distributed, using the same law as the amplitude for a coherence case.

4. The scalar wave theory just considered is adequate for a large class of optical phenomena, in which the polarization of light is negligible. In a general case, however, light as an electromagnetic field requires a vector-field formalism to describe the processes of its propagation, diffraction and interaction with substance.

To study the polarization properties of light, a mutual coherence matrix is introduced, determined by the relationship for the two stationary field points P_1 and P_2 [180]:

$$J^{(1,2)}(\tau) \equiv J(P_1, P_2, \tau) = \langle \varepsilon(\vec{\varrho}_1, t)\varepsilon^+(\vec{\varrho}_2, t+\tau)\rangle \;. \tag{2.77}$$

The symbol of averaging $\langle \ldots \rangle$ in (2.77) includes a direct product of the vector-column of the electromagnetic field

$$\varepsilon(\vec{\varrho}, t) = \begin{pmatrix} Ex(\vec{\varrho}, t) \\ Ey(\vec{\varrho}, t) \\ Ez(\vec{\varrho}, t) \end{pmatrix} \tag{2.78}$$

and its Hermitian conjugated vector-column

$$\varepsilon^+(\vec{\varrho}, t) = (E_x^*(\vec{\varrho}, t) \;,\quad E_y^*(\vec{\varrho}, t) \;,\quad E_z^*(\vec{\varrho}, t)) \;. \tag{2.79}$$

The mutual coherence matrix gives a total vector description of the physical process, and in a general sense it must be identified with an optical signal. A conventional determination of the coherence matrix can be made from (2.77) at $\vec{\varrho}_1 = \vec{\varrho}_2$ and $\tau = 0$.

Each component of the matrix $J^{(1,2)}$ propagates in space in accordance with the wave equation

$$\Box_s J_{ij}^{(1,2)}(\tau) = 0 \qquad \begin{pmatrix} s = 1, 2 \\ ij = x, y, z \end{pmatrix} . \tag{2.80}$$

In a free space, and in an isotropic di-electric medium, the process of propagation does not mix either the components of the coherence matrix or the Dekartian components of the field $\vec{E}$. All the components of the matrix $J^{(1,2)}$, can be measured experimentally with two point apertures, following a procedure similar to that used to measure four independent substantial parameters, of the conventional (not mutual) coherence matrix [180].

The basic parameter of the scalar theory of light – the mutual coherence function $\Gamma_{12}(\tau)$ – has all the properties of the trace matrix $J^{(1,2)}$ and can be identified with the latter:

$$\Gamma_{12}(\tau) = \mathrm{Tr}\,[J^{(1,2)}(\tau)] \equiv J_{xx}^{(1,2)}(\tau) + J_{yy}^{(1,2)}(\tau) + J_{zz}^{(1,2)}(\tau) \ . \tag{2.81}$$

The processes of propagation of the two parameters $\Gamma_{12}(\tau)$ and $\mathrm{Tr}\,[J^{(1,2)}(\tau)]$ are strictly similar. Both parameters at $P_1 = P_2$ and $\tau = 0$ give a correct value of the field intensity in the point P_1. Generally, there is no simple relationship between the parameter $U(\vec{\varrho}, t)$ of the scalar theory of diffraction and the Dekartian components of the vector field $\vec{E}(\vec{\varrho}, t)$. However, this relationship exists between the respective parameters – variable quadratic fields. The results from the extensive experiments connected with the measurement of light intensity, can be described using the scalar theory based on the mutual coherence function $\Gamma_{12}(\tau)$, identified with a trace of the mutual coherence matrix. The scalar theory can always be applied, when the interactions do not cause the mixing of the components of the vector field, and the information on the light polarization is of secondary value [180].

Chapter 3 Theoretical Models of the Non-Polarized Optical Radiation in the Atmosphere-Surface System

3.1 Justification of the Necessity to Consider the Distorting Effect of the Atmosphere when Surveying the Earth from Space

In studies of the Earth from space, the spectrophotometry of natural formations, surface and above-water objects on the day-side of the Earth, seen from MS and POS, are of great interest [1, 25, 141, 143, 191].

First spectra of the basic surface types obtained from the MS Soyuz-7 [141], and the successful accomplishment of the first combined optical experiment on Soyuz-9 [142] emphasized space-based spectrophotometry as a major study of the Earth from space, using optical remote sensing techniques [14, 202, 72, 133, 169]. Because of the high cost of applied programs of remote sensing of Earth from space, concern has developed for theoretical radiation field, modelling various problems of space-based spectrophotometry of natural formations.

Analysis of the first space-derived spectra of natural formations has shown that, spectrophotometry of the day-side of the Earth from MS and POS, is most efficient at high sun elevations and with viewing angles within 20–30° from the nadir [141, 143, 191]. Here, when modeling the radiation fields of the atmosphere-background-object system and interpreting experimental data, it is sufficient just to consider models of a plane-parallel planetary atmosphere and surface. The mathematical apparatus of the theory of multiple anisotropic scattering of light, widely used in some spheres of current astrophysics, geophysics, oceanology and atmospheric optics serves as the formal basis for modeling the radiation fields in direct and inverse problems of the space-based spectrophotometry [7, 8, 29, 30, 35, 41, 48, 50, 55–57, 62, 67, 71, 118, 121, 122, 161, 184, 187, 194, 204, 205, 277, 312, 319, 405].

When passing through the atmospheric, electromagnetic radiation reflected from the Earth's surface gets disturbed, due to multiple scattering and absorption by gaseous and aerosol atmospheric substances. Thus, an interpretation of the space-derived information on natural resources, is closely connected with an account of the radiative effect of the atmosphere, which distorts information on the surface state [68, 100, 115, 288].

Processes of scattering and absorption of electromagnetic radiation in the atmosphere and reflection of radiation from the surface, with details of the phenomena under consideration, are described by mathematical equations and energy balance relationships of the radiation transfer theory. These contain the radiative parameters of the atmosphere, surface (above-water) objects and the environment.

Among many distortions, to which the information on the state of the Earth's natural formations is subject, the basic[1] ones are those introduced by scattering and absorption of radiation reflected from these formations, and determined by the dependence of reflection characteristics on sun elevation. An account of these distortions and corrections introduced to the space videoinformation obtained necessary procedures in automated computer-processing of this information.

The correction of satellite images is divided into two successive stages: (1) radiometric correction – involving instrument distortions of satellite informations; and (2) radiative correction – involving atmospheric effects and differences in the reflective characteristics of natural formations due to changing survey conditions. The first distortions can be eliminated by technical means (by improving satellite equipment). The second, however, requires scientifically proven techniques to consider atmospheric radiation transformation resulting from powerful space/time variations of atmospheric parameters affecting transmission of the optical images through a medium [115, 116, 127, 349, 392]. The atmosphere-induced distortions generally exceed the distortions introduced by satellite instruments. Thus, the solution of radiometric correction problems of satellite information consists in improving the respective instruments with optimal intervals of space survey [102]. As for radiative correction, an elimination of distortions is determined by the status of the theory of the radiation-atmosphere-surface interaction and by the available reliable experimental data on radiation scattering and absorption [293].

A conclusion can thus be drawn that problems associated with satellite information correction not only results from imperfect instruments and means of receiving and processing satellite images but largely by insufficient resolution of the phenomena causing distortion of the radiative characteristics of natural formations [146]. The problem of correcting satellite information is reduced, thereby, to the problem of radiation interaction with the atmosphere and the object-background system. The final result of this interaction being determined not only by the type of inhomogeneous surface, the optical state of the atmosphere and spectral interval but also by irradiance conditions and viewing direction. The many parameters involved in this radiation field complicates considerably the unique solution of inverse problems to determine the optical parameters of this system in the visible.

An account of atmospheric distortion of the atmosphere is based on certain model ideas of radiative processes of interaction between radiation and the atmosphere-object-background system. Numerous studies have shown that the optical characteristics of the atmosphere, that affect the distortion of satellite

[1] For conditions of space-based survey and for parameters of the on-board optical instruments with viewing angles 2°, the effect on the quality of the space-derived images due to atmospheric turbulence and refraction can be neglected.

information, are largely determined by atmospheric aerosol (see Chapter 1). This suffers substantial spatial and temporal variations in its microphysical and optical properties.

Based on the data of Landsat-1 for different types of natural surface features (plow-land, forests, drainage basins), the effect of an intermediate atmospheric thickness on the recognition of objects by their spectral brightnesses has been considered in [261]. The effect of atmospheric haze, on the brightness of the surface-atmosphere system, was assessed from measurements over the ocean, where it reached 75% with respect to total brightness of the system. Recognition taking into account and ignoring the atmospheric transfer function has given substantially different results. The greatest difference was observed in recognition of forests. In 62% of cases, the coniferous forests were classified as broad-leaved, because the effect of atmospheric haze was ignored.

To compare theoretical calculations of radiation fields with Nimbus-6 data, a numerical model was made [384] of the shortwave (solar) radiation (wavelength interval 0.2 – 5.0 μm) in the Earth's atmosphere. The phase function is presented as an expansion in Legandre polynomials, and the sought-for radiation intensity as Fourier series in cosines (with use of the discrete ordinates technique). Calculations for 96 wavelenghts have been made for various types of surface (ocean, desert, snow cover) in a cloud-free atmosphere or with isolated (outside the atmosphere) stratus clouds. The results obtained have shown a good agreement with the August 1975 Nimbus-6 measurements of reflected radiation.

One project to improve the quality of images distorted by intermediate atmospheric layers consists of using analogue and digital intensification of contrasts and of adaptive systems of sensors. Another possibility is to obtain images on wavelengths where the distorting effect of the atmosphere is minimized, emphasizing the urgency of studies of how the quality of images depends on wavelength. Such studies are also interesting because they reveal the relative role of various factors that determine the parameters of the atmospheric transfer function (turbulence, scattering, wind speed, humidity). Although the deterioration of images is largely ascribed to the effect of turbulence, the contribution of other factors (for instance, aerosol scattering) can be substantial, too. In arid and semi-arid regions, a consideration of the effect of winds and air humidity in determining the concentration and size distribution of aerosol particles becomes particularly important. In this connection, observations were made [315] to assess the effect of meteorological conditions and aerosol properties on the quality of images in conditions of the northern part of the Negev desert (40 km from the Mediterranean coastline).

Analysis was made of the visibility conditions of a black-white grid observed along a horizontal air-surface route 4.1 km long at different wavelengths (400, 450, 500, 550, 600, 650, 700, 750, and 800 nm). It showed that, at relatively strong winds, the concentration of soil aerosol grows, determining an attenuation of low-frequency spatial contrasts in the near IR. In periods of minimum intensity of turbulence, the aerosol causes still stronger attenuation of high-frequency contrasts, confirming the previous conclusions about the spectral dependence of the aerosol transfer function. With a reduced aerosol concentration the images are the best not in the visible but in the near IR. Apparently, these laws are universal and independent of geographical conditions. Since the size distribution of the soil aerosol particles remains relatively constant at heights up to 5 km, its effect must be equivalent for conditions of horizontal and oblique visibility. Three types of aerosol size distribution have been considered: (1) mode A (soil aerosol with radii 1 – 10 μm); (2) mode B (prevailing radius of particles about 30 μm); (3) mode C (background aerosol, 0.02 – 0.5 μm). Aerosols of the first type reduce contrasts in the IR, both at low and high spatial frequences. Aerosols of the third type have a similar effect in the visible.

An interpretation of measurements of ocean brightness made with the Nimbus-7 5-channel (0.44 – 0.75 μm) CZCS requires correction of the distorting effect of atmospheric thickness (atmospheric transfer function). In this case, a relative contribution of the ocean to the brightness of the ocean-atmosphere system constitutes about 5%, whereas the share of the atmospheric haze brightness reaches 95%. Therefore, calculations were made [294] of the brightness of the system, to analyze the effect of the variability of various factors, on the formation of outgoing shortwave radiation. Analysis was made of molecular absorption effects, molecular and aerosol scattering in the atmosphere, of similar factors of radiation transfer in the ocean, the effect of foam on the ocean surface and absorption of chlorophyll by phytoplankton in the ocean (the contribution of yellow substance is ignored, since it can be substantial only at 0.55 μm). Calculations of the system's brightness in nadir were made for sun height 32.5° and wavelengths 0.44, 0.52, 0.55, 0.67, and 0.75 μm, corresponding to the channels of the colour scanner. Variations in the output parameters, characteristic of the real ocean variability and a cloud-free atmosphere were prescribed. The results obtained show that, maximum variations in the brightness of the system ranging between 1% and 10% are caused by variations in atmospheric transparency and by the foam-covered ocean surface. Brightness variations influenced by other factors do not exceed 1%. These results impose strict constraints on the accuracy of the determination of the atmospheric transfer function and on an analysis of the ocean surface state, in solving the inverse problems of retrieving, for instance, chlorophyll concentration in the ocean.

The problem of atmospheric effects and the scheme of scanner image recording, on the spatial resolution of space-derived images of different types of surface has been considered in [379]. The scanner survey system has a finite instant FOV of a definite size. In recording an image pixel, the sensor averages the intensity of radiation reflected from surface elements which reached the scanner's instant FOV. Due to the scattering of radiation by atmospheric substances, the FOV catches also the radiation reflected from nearby surface elements. Therefore, the resolution of space-derived image deteriorates. Calculations have shown that the 30-m resolution of the Landsat thematic mapper may decrease to 100 m, when the atmosphere is turbid. Comparisons of these two effects have shown that, for fields larger than 200 m, the atmospheric effect on the quality of image is greater than the effect of intensity averaging over the instant FOV of the scanner. Hence, to improve the qualitative classification of images containing these fields, further enhancement of the scanner's resolution is not needed, until respective algorithms of atmospheric correction involving surface inhomogeneities, are developed.

To correct for the atmospheric effect, based on theoretical modeling of the processes of radiation interaction with the surface and atmosphere, knowledge is needed of atmospheric characteristics at the moment of image recording. Often these data are absent. Therefore, methods of correction based on the use of the properties of the objects under study only, are interesting for certain remote sensing problems. For instance, to process images containing farmland, a morphological smoothing operator has been considered [16] that physically raises the contrasts of the field boundaries. However, it smoothes the intra-field fluctuations of brightness and distorts the angles of the fields.

An operator of weak contrast has been suggested in [17], which retrieves the smoothed boundaries of the fields, improves the contrast of the intra-field brightness fluctuations and does not distort the field geometry. The problems of the size of window used in processing with these operators and of the number of operations have been considered. These retrieval techniques are based on the fact that due to atmospheric and instrumental effects, the function of brightness at the distance to the interface between the fields is convex-concave. The retrieval of the field boundaries fields by the weak contrasting operator improves the accuracy of classification, and by contrasting intra-field fluctuations of brightness it is possible to judge the state of an object. However, such processing of images only partially removes the background effect, since these operators are functioning in local areas within 3×3, 5×5 pixels. It should be noted that, the given operators do not cause an anomalous sharpness of the field boundaries, in contrast to techniques described briefly in [17].

An approximate solution obtained by Sobolev [184] is used for the atmospheric correction of remote sensing data of sea surface made at different heights in the visible and near IR.

Radiation intensity, recorded at the height h_0, is represented by the following terms: atmospheric haze intensity (no surface); intensity of solar radiation diffused by the atmosphere and mirrored by water surface in the viewing direction; "light spot" intensity (direct solar radiation mirrored by water surface in the viewing direction); intensity of radiation scattered through water thickness. The dependence of atmospheric haze on the height of recording h_0 is obtained in the following way. The atmosphere is represented as two layers, one lying below h_0, the other above h_0. The atmospheric haze intensity at the height h_0 is expressed by the formula by Sobolev for the lower layer, where, at the upper boundary an irradiance appears, determined by scattering of solar radiation there. To calculate the lower layer's coefficient of diffused reflection, the phase function is used vertically averaged from zero to h_0. To calculate the irradiance of the upper boundary of the lower layer, the phase function is vertically averaged from h_0 to infinity. The intensity of solar radiation diffusively transmitted by the atmosphere and mirrored by a water surface is calculated using Sobolev's formula, too. For diffusively transmitted radiation, the effect of the background spectral albedo is considered using an approximate formula. It expresses the sea surface albedo through the ratio of diffusive irradiance to total irradiance at the surface level. The Fresnel coefficient of reflection for a flat surface is used to measure specular reflection. Direct solar radiation mirrored in the direction of viewing is modelled taking wind speed into account.

When performing atmospheric correction to estimate the intensity of radiation scattered in a water thickness, it is necessary to determine the aerosol optical thickness of a layer below h_0. For this purpose, near IR region data were used in which the volume emission of water is negligible. The aerosol optical thickness of the whole atmospheric layer is assumed (either measured or calculated for standard models), and by using the iterative procedure from near IR region data, the aerosol optical thickness of the layer below h_0 is estimated. There is a good consistency between theoretically calculated intensity of radiation coming out of the sea water, carrying information about suspended and dissolved substances and experimental data obtained for different heights and a wide range of viewing angles.

Within the model in [184] it is assumed that atmospheric haze (no reflecting surface) has a strong anisotropy and is described with beta-functions, and background haze (i.e., radiation reflected at least once, from the surface) is considered isotropic. Due to this simplification, a system of differential equations is obtained for the upward and downward fluxes. Intensities expressed through fluxes are introduced to the formal solution of the radiation transfer equation and analytical expressions are obtained for the intensities of the atmospheric and background hazes. The surface is assumed to orthotropically reflect the incident radiation.

This simplified model was used to assess the atmospheric effect on remote sensing data with the help of water bodies – unpolluted lakes [245]. The coefficients of the surface and volume reflection for water are considered known and constant. Using this model, the coefficients of reflection are related to radiation intensity of the under-water object, at the top of the atmosphere, and the atmospheric optical thickness is estimated for a given type of the aerosol phase function. As the experimental results show, this approach requires a very pure lake. In addition, both experimental and calculated data are statistically different, which can be caused by systematic calibration error or by an inadequate atmospheric model.

3.2 Calculating Facilities of the Analytical and Numerical Methods of the Radiation Transfer Theory

A detailed study of the problem of the non-polarized shortwave radiation transfer taking account of multiple scattering, stretching of atmospheric phase function, and diffused reflection from an inhomogeneous surface is contained in monographs [114, 152, 184, 187, 405]. Numerous theoretical studies of radiation transfer in the atmosphere are based on modifications of classical expressions obtained analytically [187, 205]. Publications [94–100] use an advanced method by Sobolev to resolve the transfer function of the atmosphere. Calculations [156] of the brightness of atmospheric light at anisotropic scattering were made using the successive approximations technique. The dependence of multiple scattering of the observed sky radiation on wavelength, solar zenith angle, atmospheric turbidity, surface albedo and aerosol absorption has been studied in [218, 219].

The use of large computing capacity involving spherical harmonics techniques [27] has only recently come into practice to study the radiation transfer in the atmosphere [11, 74–77, 80, 81, 182, 242–244, 360].

Mathematical aspects of the solution of the transfer equation using the spherical harmonics technique have been described in [242]. By expanding the radiation intensity and phase function in series by Legendre polynomials $P_n \cos(\gamma)$, the integro-differential equation is reduced to a system of algebraic equations, which is solved numerically for both a homogeneous and inhomogeneous atmosphere. Depending on the aerosol size distribution and radiation wavelength, the number of terms of expansion of phase function varies between 50 and 150 and for clouds in the visible spectrum it reaches 300–500. Results of an application of this technique for Reyleigh aerosol and cloudy atmosphere with optical thicknesses below 10 have been discussed. For a nonabsorbing Rayleigh atmosphere first, three terms of expansion of phase function ensure calculations of the angular distribution of radiation with an accuracy to the fourth significant figure.

For two other atmospheric models, it is necessary to take into account 69 and 169 terms of expansion, respectively. Calculations were made of the intensity of the upward and downward radiation for each of the three atmospheric models with optical thicknesses of 0.1, 1.0, and 10.0 and solar zenith angles of 0, 40°, 80°. The order of calculation approximation is less than the number of terms of expansion of phase function, where accuracy of presentation is 0.1% in all cases apart from a Rayleigh atmosphere.

Results have been given [244] of calculations of the upward and downward radiation fluxes for 8 types of atmosphere based on the spherical harmonics technique. Using this technique, the authors of [243] calculated the vertical profiles of the shortwave radiation fluxes for altitudes up to 50 km with 1 km steps. Eighty three spectral intervals from 0.285 to 2.5 μm were studied with solar zenith angles 0°–80°, with 15° steps. Calculations were made for models of the mid-latitude summertime atmosphere with the complex refractive index of aerosol $m = 1.5 - i0.01$, and with only its real part, the spectral surface albedos being equal 0, 0.05, 0.1, 0.2, 0.3, 0.4, 0.6, and 0.8. The aerosol size distribution was described with the L-Deirmendjian haze model. If the optical thickness of major atmospheric layers exceeded 0.2, then these layers were divided into sub-layers. The technique turned out to be useful for optically thick atmospheric layers, too [230].

The spherical harmonics technique is very efficient in calculating the shortwave radiation transfer in a real atmosphere and has some advantages compared to the numerical methods used most widely in atmospheric optics. In particular, this technique makes it posssible to consider the effect of individual radiation absorbers and scatterers, high-order multiple radiation scattering, strongly stretched phase functions, etc. However, classically, the spherical harmonics technique, as applied to a real atmosphere, requires solution of systems of high-order algebraic equations. Small computers are not powerful enough to realize this technique without double checking calculation accuracy, which raises doubts about its applicability. The suggested improvement of the technique consists of the following: (1) a division of total radiation intensity into direct solar and diffused radiation; (2) a presentation of the radiation equation for diffused radiation as three-diagonal systems of linear algebraic equations with respect to the moments of the angular distribution of intensity; (3) solution of these systems by an efficient algorithm and establishing relationships between the moments of angular distribution, at external and internal boundaries of individual atmospheric layers. As a result of these improvements, a technique has been developed to calculate the radiation transfer in the atmosphere which combines a high accuracy of presentation of the outgoing radiation intensity at real viewing angles of satellite-borne instruments, with low computing time expenditures [85]. The procedure is similar to that described in [395–399].

Considerable progress has now been achieved in calculating radiation transfer with Monte-Carlo simulations, the calculating possibilities of which have been described in detail in [52, 128, 161, 361]. Results are particularly good determining the order of polarization and direction of polarization of multiple radiation scattering in an aerosol atmosphere [153, 361]. The Monte-

Table 3.1. A comparison of the basic, most widely used techniques to calculate the shortwave radiation transfer in the atmosphere

Calculation technique	Accuracy (%)	Computing time	Drawbacks
Discrete ordinates	1	–	Fairly idealized real atmosphere
Monte Carlo simulation	2–4	1.5 hour	Much computing time
Successive approximations	10–15	–	Real stratification of the atmosphere is neglected
Spherical harmonics	0.1	10–20 min [a]	Plane-parallel geometry of the problem, much computing time
Advanced method of spherical harmonics	10, viewing angle 60°	1.5 min	Plane-parallel geometry of the problem

[a] For IBM-360 computers

Carlo method requires, however, considerable computer time expenditures [3, 153, 361]. Using this method, important application outputs have been obtained, for instance, on the radiative regime of the ocean-atmosphere system [361].

Recently characteristic techniques [35, 36] has become widely used to solve the problems of atmospheric optics, as well as numerical methods based on invariance principles [393]. A "doubling" method has turned out to be rather efficient in calculations, too [275–277, 405]. This method requires knowledge of matrices of transmission and reflection of layers, which generally are known with an inadequate accuracy however.

Table 3.1 compares the potential of some calculating techniques for short-wave radiation transfer in the atmosphere. A comparison of the potential of the successive approximations technique with other techniques has been made in [114], and of the Monte-Carlo method in [128].

3.3 Setting the Problem of Theoretical Determination of Scalar Asymptotic Transfer Functions of the Atmosphere

Experience of interpreting first spectra of natural formations obtained from Soyuz-7 has shown that explanation of the transforming and attenuating effect of the atmosphere [66], is the most important aspect of initial processing and analysis of spectrophotometric information from a space-based platform. An atmospheric layer, between a spectrophotometered object and the on-board recording instrument, distorts the spatial and angular structure of reflection spectra. This is because it is transparent to low frequencies and strongly distorts high frequencies of space-derived images due to processes of atmospheric turbulence, refraction, multiple scattering, and real absorption of solar radiation [57, 69]. Where the orbital motion of the space-based platform is ($H_{orb} = 250$ to 400 km, $V_{orb} = 8$ km/s) and at 2° viewing angles for on-board optical instruments, the first two effects negligibly affect the spectral brightnesses of the Earth recorded from space. The most substantial effect of space-derived spectrophotometric information is exhibited by the processes of multiple scattering and real absorption of solar radiation, which reduce the apparent contrast of brightnesses of the object-background system and "blur" the image of an object in the focal plane of the optical sensor's objective. Really, only by correcting the attenuating affect of the atmosphere and reducting the space-derived image spectra to surface level, is it possible to solve basic problems of remote sensing of the environment successfully. These include recognising the objects of survey, the choice of optimal spectral zones, construction of automated systems of processing, coding and classifying the spectra of reflection. Here the determination of optimal technical, natural and physical conditions of survey at the stage of planning the respective aerospace experiments must answer several questions. Namely, to what extent does the atmosphere transform the brightness contrasts between a surface or above-water object and the environment background, in terms of a threshold value of the identified contrast $k_{thr} \simeq 0.1$, which depends on the sensitivity and frequency-contrast characteristics (FCCh) of the on-board spectrophotometers?

Similar problems occur in the 2-D and 3-D theory of radiation transfer: when assessing the frequency filtering ability of an aerosol atmosphere, limited from beneath by the horizontally inhomogeneous reflecting base [18, 81, 82, 134–136, 174, 192, 212, 316]: when calculating the FCCh of spectral-zonal films, objectives of radiation receivers, and tracks of automated optical information processing systems [10, 61], and when correcting the images of natural formations and objects ($\lambda = 400$ to 800 nm) [31, 77, 80, 84].

To take account of the transforming and attenuating effect of atmospheric haze on spectral brightness and brightness contrasts of natural features and backgrounds, either photographed or spectrophotometered from space in the wavelength interval 400 to 800 nm, Kondratyev and Smokty [94–97] suggested a concept of the transfer operator for the system "atmosphere-object-background". This concept implies a reduction of satellite spectrophotometric data to the surface level in two interrelated stages. At the first stage, based on the angular, spatial and spectral distribution of the reflected radiation intensity measured from a space platform, some inverse problems of atmospheric optics are solved connected with the phase function retrieval $x(\gamma)$, quantum survival probability, optical thickness τ_0 and surface albedo A [178, 312, 314]. At the second stage, with the solution of the direct problem of radiation transfer theory, space-derived spectrophotometric information is reduced to surface level [31, 32, 101]. In this case, it is important to determine the volume of spectral transfer function values, probable limits of its variations depending on the optical characteristics of an aerosol atmosphere, the reflecting properties of surface, viewing direction, solar irradiance conditions, and meteorological situation in the sub-satellite zone. Also, for inverse problems, it is necessary to choose, from the spectral brightnesses of natural formations measured from space (including light transmission through the portholes of MS and POS), such parameters as the brightness of atmospheric haze and atmospheric transparency over the geographical regions under study [179].

To solve these problems, apart from an advanced technique of spherical harmonics, accurate expressions of the brightness coefficients of a plane-parallel planetary atmosphere of a finite optical thickness τ_0 will be used, obtained for an arbitrary three-term phase function $\bar{x}(\gamma)$. Similarity ratios [177, 405, 216] make it possible to move, from the field of multiple radiation scattering for an approximating model phase function, to the field of diffusively reflected and transmitted radiation for a strongly stretched phase function $x(\gamma)$. Based on this, constituents of spectral transfer functions of the atmosphere-background-object system were determined for wavelength interval 400 to 800 nm for a closed aerosol model by Elterman (1968), for different types of the orthotropic and non-orthotropic reflecting base, different directions of viewing θ and irradiance θ_0 of the atmospheric top. Investigation of horizontal diffusion of photons due to horizontal inhomogeneity of a real surface, has shown that a major contribution to the field of reflected radiation of a 2-D medium, is made by photons, singly scattered, from the surface with real albedo A, and repeatedly reflected by it at an average albedo $\bar{A}$ [174, 178]. This conclusion made it possible, when introducing an atmospheric radiative correction to space images and solving the related inverse problems of atmospheric optics, to use a simpler 1-D theory of radiation transfer [101, 132, 175, 312], on the assumption that, in a wide range of spatial frequencies, the atmosphere uniformly reduces radiation reflected by a horizontally inhomogeneous underlying surface.

Analysis of theoretical and experimental data on the brightness field of the Earth has shown that radiation reflection from the surface contributes most to photometric quantities measured from space, with maximum sun elevations and nadir viewing [68]. Therefore, during the Soyuz and Salyut missions the Earth was spectrophotometered in nadir ($\theta = 0°$) at small solar zenith angles. Thus, in theoretical modelling of the effect of the atmospheric radiative factors on spectrophotometric information obtained from space, to use the environmental model. This consists of plane-parallel layers illuminated by down-falling parallel solar beams.

Theoretical determination of the scalar transfer function for given optical parameters of the environment, includes a solution of basic problems of the radiation transfer theory, connected with multiple anisotropic scattering of light in a plane absorbing layer of a finite optical thickness τ_0 with the reflecting base. These problems can be solved by using either precise or approximate analytical methods, numerical methods, or Monte-Carlo simulations. Here the transfer function of the atmosphere-object-background system is estimated by using similarity ratios, precise or approximate solutions from the transfer theory obtained for an approximating 3-term phase function [177], in two most important asymptotic cases: (1) low frequencies – a homogeneous background plane, and (2) high frequencies – small-size surface object.

Let a beam of parallel sun rays, creating a spectral irradiance πS_λ for a site normal to incident radiation, fall onto the upper boundary of a plane-parallel planetary atmosphere at an angle $\theta_0 = \arccos \zeta$ to the normal. Suppose, the atmosphere is limited from beneath by a set of reflecting surfaces forming a system "object-background". In an atmosphere, supposedly horizontally homogeneous, processes of multiple scattering and real absorption of light take place at an arbitrary spectral phase function $x(\gamma)$ and quantum survival probability Λ. At first, a polarization of scattered radiation will be ignored, as well as radiation refraction [131]. We are to assess the effect of the scattering and absorbing atmospheric layer on the spectral brightness of the background $I_{b\lambda}$ and surface (above-water) objects $I_{ob\lambda}$ measured from space.

Let the spectral brightnesses of the background $I_{b\lambda}$ and object $I_{ob\lambda}$ be:

$$I_{b\lambda} = T_\lambda I^0_{b\lambda} + I^b_{h\lambda}\,, \qquad I_{ob\lambda} = T_\lambda I^0_{ob\lambda} + I^{ob}_{h\lambda}\,, \tag{3.1}$$

where $I^{ob}_{h\lambda}$ and $I^b_{h\lambda}$ are the spectral brightness of atmospheric haze over an object and its background, respectively; $I^0_{b\lambda}$ and $I^0_{ob\lambda}$ the spectral brightness of the background and object at the surface level; $T_\lambda = \exp(-\tau_{o\lambda}/\eta)$ – the spectral factor of radiation attenuation in the direction arc cos η (atmospheric transparencies over an object and over the background are taken to be similar). Linear sizes of the surface (above-water) object and its background as well as their relationships are assumed to be arbitrary.

The spectral transfer functions for the brightness of the background $\Pi_{b\lambda}$ and of the object $\Pi_{ob\lambda}$ are assessed using the formulas:

$$I^0_{b\lambda} = \Pi_{b\lambda} I_{b\lambda} , \quad I^0_{o\beta\lambda} = \Pi_{ob\lambda} I_{ob\lambda} . \tag{3.2}$$

The wavelength index λ will now be omitted for simplicity. Substituting (3.1) and (3.2), we obtain

$$\Pi_b = (1/p_b)T , \quad \Pi_{ob} = (1/p_{ob})T , \tag{3.3}$$

where functions

$$p_b = 1 + I^b_h/(T I^0_b) , \tag{3.4}$$

$$p_{ob} = 1 + I^{ob}_h/(T I^0_{ob}) . \tag{3.5}$$

Knowledge of transfer functions Π_b and Π_{ob} makes it possible, not only to reduce satellite optical data to the surface level, but also to select the effect of atmospheric haze from them. Substituting (3.1) – (3.2) to (3.3) – (3.5), we obtain relationships important for setting and solving some inverse problems of the space atmospheric optics:

$$I_{hb} = I_b(1 - p_b^{-1}) = I_b(1 - \Pi_b T) ; \tag{3.6}$$

$$I_{hob} = I_{ob}(1 - p_{ob}^{-1}) = I_{ob}(1 - \Pi_{ob} T) . \tag{3.7}$$

By determining the spectral contrast ($k \geqslant 0$) between the brightness of the background I_b and object I_{ob}, we have

$$k = \frac{I_{ob} - I_b}{I_{ob}} \geqslant 0 , \tag{3.8}$$

if $I_b \leqslant I_{ob}$. At $I_b \geqslant I_{ob}$ formula (3.8) changes, respectively. Below we shall consider (3.8) to be valid for brightness contrasts $k \geqslant 0$. Substituting (3.1) into (3.8), we obtain

$$k_0 = P_{ob} \cdot k , \tag{3.9}$$

where k_0 is the spectral contrast of the brightness of the background and object at the surface level

$$k_0 = \frac{I^0_{ob} - I^0_b}{I^0_{ob}} \geqslant 0 . \tag{3.10}$$

The P_{ob} function is determined from the formula

$$P_{ob} = \frac{k_h}{k} + p_{ob}\left(1 - \frac{k_h}{k}\right) = p_{ob} + \frac{k_h}{k}(1 - p_{ob}) \tag{3.11}$$

where k_h is the spectral contrast of the brightnesses of atmospheric haze over the object and over the background

$$k_h = \frac{I_h^{ob} - I_h^b}{I_h^{ob}} \geqslant 0 \; . \tag{3.12}$$

Using $k_h \sim 0$ (3.11), it follows that $P_{ob} \approx p_{ob}$. We shall call the P_{ob} function the spectral transfer function of the atmosphere for contrasts of brightness of the object and its background, when $I_{ob} \geqslant I_b$. A similar determination is easily obtained at $I_b \geqslant I_{ob}$ too.

It must be emphasized that a reduction of the satellite spectrophotometric information to surface level must be made strictly within a satellite experiment (without using the spatially combined synchronous data of aircraft or surface optical measurements). Therefore, for correct assessment of Π and P, transfer functions from satellite monitoring data, it is necessary first to solve some inverse problems of atmospheric optics. From the brightness of a medium measured from space, its optical characteristics must be determined at different levels first, down to the planetary surface, and then the reduction of spectrophotometric data must be made for surface albedo values. Another method of atmospheric correction of space images and spectra consists of writing and solving either accurate or approximate integral equations directly for Π_{ob} and Π_b scalar transfer functions. For instance, in the simplest case of an infinitely extended background with the albedo $A = 0$ (sea), introducing the Π_b^m transfer function as a ratio of azimuthal harmonics of transmission coefficients σ^m and reflection coefficients ϱ^m, and then using linear integral equations for azimuthal constituents ϱ^m and σ^m obtained by Sobolev [187], we obtain, by simple transformations:

$$\begin{aligned} \varrho_{ex}^m(\eta,\zeta)\Pi_b^m(\eta,\zeta,\tau_0)T^m(\eta) &= \frac{\Lambda}{2}\int_0^1 \varrho_{ex}^m(\eta,\zeta)\Pi_b^m(\eta',\zeta,\tau_0)A^m(\eta,\zeta') \\ &\times \frac{\eta' d\eta}{\eta'-\eta} - \frac{\Lambda}{2}\, e^{-\tau_0/\eta}\int_0^1 \varrho_{ex}^m(\eta',\zeta)A^m(\eta,-\eta')\,\frac{\eta' d\eta'}{\eta+\eta'} \\ &+ A^m(\eta,\zeta)\,\frac{\Lambda}{4}\,\frac{e^{-\tau_0/\eta} - e^{-\tau_0/\zeta}}{\eta-\zeta} \end{aligned} \tag{3.13}$$

where the $\varrho_{ex}^m(\eta,\zeta)$ values are measured at the top of atmospheric by satellite, and functions T^m and $A^m(\eta,\zeta)$ are estimated from relationships given in [187].

The problem of determination of transfer functions and their constituents, is easily simplified with the help of combined subsatellite geophysical experiments [142, 170, 240]. Then the Π and P transfer functions can be found from a simple comparison of reflected radiation intensities measured from space and from a flying laboratory (or a ship).

Another possibility consists of the preliminary determination of their constituents (haze brightness I_h and atmospheric transparency T) from relationships (3.1) considered as a system of linear algebraic equations. Formally, the system of algebraic equations (3.1) is uncertain, since the atmospheric haze brightness over an object and over the background are not equal: $I_h^b \neq I_h^{ob}$. Hence, a unique solution of this system is only possible when intensities I_h^b and I_h^{ob} either coincide or are close (i.e., the interface of two homogeneous media, a small object against an extended background). The technique of "smoothing" the brightness of atmospheric hazes over the background I_h^b and over an object I_h^{ob} has been described in [31, 68, 97].

3.4 An Accurate Solution for the Problem of Diffused Reflection and Transmission of Light by the Terrestrial Atmosphere for Model Phase Functions

Theoretical studies of radiation transfer with non-isotropic scattering have been started by Ambartsumyan [7], who, with the use of his invariance principle, obtained a correct solution for the problem of diffused reflection of light by a semi-infinite medium ($\tau_0 = \infty$) with phase function $x(\gamma)$ expanded in Legendre polynomials $P_n(\cos\gamma)$. Then, by the same method Chandrasekhar [205], solved the problem of diffused reflection and light transmission by a plane layer of a finite optical thickness ($\tau_0 < \infty$). Another way to solve this classical problem has been proposed in [184, 187].

During the last decade, Sobolev has substantially advanced the theory of multiple anisotropic scattering of light, developing a number of common techniques to solve the problems of radiation transfer in scattering and absorbing plane-parallel media with arbitrary phase functions $x(\gamma)$ and quantum survival probabilities Λ [187]. The importance of these techniques consists in that they enable strict assessment of the accuracy of calculations by numerical methods, by Monte-Carlo simulations and from approximate formulas of radiation transfer theory. Considering accurate solutions of some model problems to be reference ones, it is possible to study the effect of anisotropy of atmospheric phase functions and of variations of other optical parameters on the field of diffused radiation depending on spectral interval, irradiance conditions and viewing geometry [189]. In addition, based on accurate solution of individual classical problems of the transfer theory, it is sometimes possible to construct new approximate solutions for interesting practical problems in the optics of natural features [209, 312].

An accurate solution for the classical problem of diffused reflection and transmission of light by a plane-parallel planetary atmosphere of a finite optical thickness ($\tau_0 < \infty$), with a 2-term phase function $x(\gamma) = 1 + x_1 \cos\gamma$ was obtained first by Chandrasekhar [205]. Then Sobolev [184] obtained it in a simpler way. An accurate solution of a similar problem in the case of a 3-term phase function $x(\gamma) = 1 + x_1 \cos\gamma + x_2 P_2(\cos\gamma)$ has been given in [88]. A partial case of an azimuthally (φ) averaged three-term phase function is considered in [88]. Approximate asymptotic formulae for brightness coefficients in the case of large optical thicknesses ($\tau_0 \gg 1$), pure scattering ($\Lambda = 1$), two-term and three-term phase functions are given in [64, 226, 348, 360]. With real absorption in a medium ($\Lambda < 1$), asymptotic formulas for brightness coefficients in such phase functions are obtained in [187].

The classical problem of diffused reflection of light by a semi-infinite plane-parallel planetary atmosphere ($\tau_0 = \infty$) with a three-term phase function was solved accurately first by Chandrasekhar [205, 281], too. It has been considered in detail in a number of publications [90, 113, 281 – 283]. In particular, an azimuthal dependence of the diffusively reflected radiation field was considered and some calculations were made [113]. A correct solution of this problem for a four-term phase function was obtained in [176]. This problem for an azimuthally averaged four-term phase function is partly considered in [89]. Some model calculations of brightness coefficients for this phase function are given in [227].

For the radiation transfer theory, approximate analytical and numerical methods have always been important allowing us, with reasonable accuracy, to calculate the spatial, angular and spectral distributions of the brightness of a plane-parallel atmosphere for real phase functions.

Recently a method after Sobolev has been widely used to solve problems of environmental optics [184]. The accuracy of this method is determined by the stretching of phase function $x(\gamma)$, by optical thickness τ_0, solar zenith angle θ_0, viewing angle θ and azimuth φ [2]. For the terrestrial atmosphere in the visible spectrum ($\lambda = 400$ to 800 nm) at not very low sun elevations $\theta_0 < 60°$ above horizon and viewing near nadir ($\theta < 10°$) Sobolev's method averages 10 – 15% accuracy [2]. However, in order to model theoretically radiation fields, involving the solution of inverse problems of atmospheric optics; to construct the theory of atmospheric transfer function; and to introduce atmospheric radiative correction of space images and spectra of the Earth, it is necessary to use optical information obtained with a higher accuracy. Here, the uncertainty of Sobolev's method in its classical form can be reduced by more accurately modeling the diffused part of the total radiation field [20, 176, 177].

Let a beam of parallel solar rays creating at the upper boundary of the planetary atmosphere the spectral irradiance of the site normal to solar radiation πS_λ, fall onto the upper boundary of a plane-parallel planetary atmosphere of a finite optical thickness $\tau_0 < \infty$ at an angle arc cos $\zeta = \theta_0$ to the normal and azimuth φ_0. The wavelength index λ will be omitted. In a medium limited from beneath by a horizontally homogeneous Lambertian base with albedo A_λ, processes of multiple scattering and real absorption take place with an arbitrary normalized phase function $x(\gamma)$ and quantum survival probability, Λ, independent of height. The direction of scattered radiation at an optical depth τ will be determined by an angle arc cos $\eta = \theta$ between scattered radiation and the normal to the layer as well as azimuth φ calculated from the sun vertical $\varphi_0 = 0$ (Fig. 3.1). The effects of refraction and polarization at in scattering solar photons in the atmosphere are ignored. Then, the radiation transfer equation becomes:

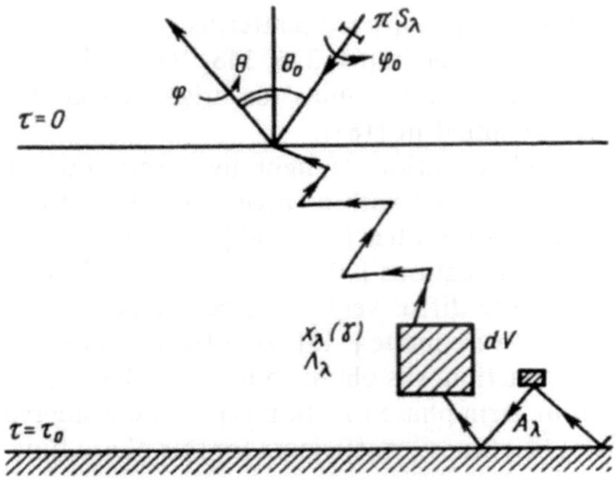

Fig. 3.1. Geometry of sunlight propagation in a plane-parallel planetary atmosphere

$$\eta \frac{dI}{d\tau} = -I + \frac{\Lambda}{4\pi} \int x(\gamma) d\omega + \frac{\Lambda S}{4} x(\gamma_1) e^{-\tau/\zeta} \tag{3.14}$$

with boundary conditions

$$I(0, \eta, \zeta, \varphi) = 0 \quad \text{for } \eta > 0 \tag{3.15}$$

$$I(\tau_0, \zeta) = \frac{A}{\pi} \int_0^{2\pi} d\varphi \int_0^1 I(\tau_0, -\eta, \zeta, \varphi) \eta \, d\eta \tag{3.16}$$

where $d\omega = \sin\theta \, d\theta \, d\varphi$ is an element of solid angle Ω, and the angle of direct solar radiation scattering γ_1 is determined from the relationship $\cos\gamma_1 = -\cos\theta\cos\theta' + \sin\theta\sin\theta'\cos\varphi$. Presenting an arbitrarily taken atmospheric phase function $x(\gamma)$ as [216] $L\delta(\gamma-\gamma_0) + \tilde{x}(\gamma)$, normalizing the remaining phase function $\tilde{x}(\gamma)$ instead of initial transfer equation (3.14), we obtain the following equation:

$$\eta \frac{dI}{d\tau_*} = -I + \frac{\Lambda_*}{4\pi} \int_\Omega I\tilde{x}(\gamma) d\omega + \frac{\Lambda_* S}{4} \tilde{x}(\gamma_1) e^{-\tau_*/\zeta} \tag{3.17}$$

for boundary conditions (3.15)–(3.16). Note, that for the planetary atmosphere of an infinitely large optical thickness ($\tau_0 \gg 1$), the lower boundary condition (3.16) transforms into zero condition:

$$I(\tau_0 \gg 1, \zeta) = 0 \ .$$

Optical parameters τ_* and Λ_* in (3.17) are changed with respect to their initial values τ and Λ based on similarity ratios given in [216]:

$$\tau_* = \left(1 - \Lambda \frac{L}{2}\right)\tau \ , \qquad \Lambda_* = \frac{1-0.5\,L}{1-0.5\,L\Lambda} \Lambda \ . \tag{3.18}$$

Interpolation coefficients $|x_n| < 2n+1$ in the 3-term and 4-term remaining phase functions $\tilde{x}(\gamma)$:

$$\bar{x}(\gamma) = 1 + x_1 P_1(\cos\gamma) + x_2 P_2(\cos\gamma) \ , \tag{3.19}$$

$$\bar{x}(\gamma) = 1 + x_1 P_1(\cos\gamma) + x_2 P_2(\cos\gamma) + x_3 P_3(\cos\gamma) \ , \tag{3.20}$$

are determined, either from a standard expansion of initial phase function $x(\gamma)$ in total orthonormalized system of Legendre polynomials $P_n(\cos\gamma)$, or from the condition of best approximation, on an average, of function $\bar{x}(\gamma)$ by 3 or 4 first polynomials $P_n(\cos\gamma)$ [176, 177]. Thus, the solution for initial problem (3.14) – (3.16) with an arbitrary atmospheric phase function $x(\gamma)$, with a maximum in the interval of small angles $\gamma_0 \sim 0°$, is reduced to the solution of a classical problem for the approximating 3- and 4-term phase function (3.19) – (3.20). An approximation of model phase functions by 3- and 4-term interpolation phase functions without using similarity ratios has been made in [88, 89]. This problem is solved using similarity ratio (3.18) and choosing a maximum at $\gamma \sim 0°$ of real atmospheric phase functions, corresponding in the visible and near IR to Elterman's aerosol model (1968) [153].

Practical value of accurate solutions of transfer equation (3.14) based on these approximations is determined by how well the phase functions of real media are modeled by expressions of the (3.19) or (3.20) type. Since the number of terms of expansion of real phase functions in Legendre polynomials $P_n(\cos\gamma)$ rapidly grows with their stretching, it is important to know for which values of physical parameters of scattering polydisperse media a function $x(\gamma)$ can be approximated with a high accuracy using expressions (3.19) – (3.20).

Results of approximation of these phase functions dissipated with the help of 3- and 4-term interpolation functions (3.19) – (3.20) are shown in Table 3.2.

Table 3.2. An estimation of function $x(\gamma)$ by four- and three-term interpolation phase functions

Parameters of distribution function			Four-term phase function			Three-term phase function	
m	r	n	x_1	x_2	x_3	x_1	x_2
1.50	0.2	0	1.006	0.795	0.215	1.076	0.795
		2	0.208	0.498	0.098	0.240	0.498
		6	0.083	0.497	0.028	0.092	0.297
1.50	0.5	0	1.972	2.470	1.635	2.670	2.470
		2	1.180	0.909	0.269	1.269	0.909
		6	0.529	0.566	0.113	0.566	0.566
1.38	0.5	0	2.079	2.740	1.875	2.879	0.740
		2	1.110	0.869	0.266	1.198	0.869
		6	0.510	0.568	0.092	0.540	0.568
1.50	1.0	2	1.948	2.285	1.432	2.560	2.285
		6	1.615	1.266	0.432	1.789	1.265
		2	2.028	2.450	1.569	2.698	2.459
1.38	1.0	6	1.560	1.283	1.494	1.759	1.283

Table 3.3. Comparison between real phase functions and four-term phase functions

degree	Real phase function	Four-term phase function	Real phase function	Four-term phase function	Real phase function	Four-term phase function
0	3.40	3.36	3.20	2.21	4.42	4.31
10	3.29	3.28	2.17	2.16	4.26	4.19
20	3.02	3.04	2.04	2.04	3.82	3.87
30	2.62	2.68	1.85	1.85	3.20	3.33
40	2.16	2.24	1.62	1.62	2.53	2.70
50	1.71	1.79	1.37	1.35	1.89	2.00
60	1.32	1.36	1.15	1.14	1.36	1.46
70	1.00	1.00	0.96	0.95	0.96	0.96
80	0.76	0.72	0.81	0.81	0.67	0.60
90	0.60	0.55	0.72	0.72	0.48	0.37
100	0.50	0.45	0.68	0.68	0.36	0.25
110	0.48	0.41	0.87	0.68	0.29	0.22
120	0.42	0.41	0.70	0.71	0.25	0.22
130	0.43	0.43	0.75	0.76	0.24	0.24
140	0.44	0.45	0.81	0.81	0.24	0.26
150	0.47	0.46	0.87	0.86	0.25	0.25
160	0.49	0.46	0.91	0.89	0.26	0.22
170	0.50	0.46	0.94	0.92	0.28	0.22
180	0.51	0.46	0.95	0.92	0.28	0.22

Values of x_i coefficients were chosen using the least square technique [375]. In Table 3.3, some real phase functions are compared to those of the four-term interpolation in (3.20) giving approximate values. As was expected, the order or approximation decreases with growing n parameter and decreasing average radius of particles $\bar{r}$, i.e., with decreased stretching of an initial phase function. Table 3.4 exemplifies an approximation of real cloud phase functions with the help of a 4-term interpolation in (3.20) for particles distributed after the generalized gamma-law (1.4). Real cloud phase functions $x(\gamma)$ were calculated using Mie theory (Table 3.5) for the optical parameters corresponding, in the visible spectrum, to the atmosphere of Venus [64, 216, 276].

The estimation of cloud phase functions was made without their maximum values at scattering angles $\gamma_{cut} = 5$, 10, 15, and 20°. In Table 3.4, along with the L parameter of a small-angle estimation and approximating coefficients $|x_1| < 3$, $|x_2| < 5$, $|x_3| < 7$, values are given of the first four coefficients of expansion x_i^* $(i = 0, 1, 2, 3)$ of a given phase function in Legendre polynomials $P_n(\cos\gamma)$. For comparison with the complex refractive index m, Table 3.5 shows values of the cloud phase function $x(\gamma)$ and its parameters calculated with the larger part of the m index.

For the problems of space-based spectrophotometry and atmospheric optics it is important to consider the whole atmospheric thickness, so we introduce, at $\tau_0 < \infty$, the coefficients of reflection $\varrho(\eta, \zeta, \varphi, \tau_0)$ and transmission $\sigma(\eta, \zeta, \varphi, \tau_0)$, instead of radiation intensities at the atmospheric top $(\tau = 0)$ and base $(\tau = \tau_0)$ [23]:

Table 3.4. An estimation of Venusian cloud phase functions $x(\gamma)$ by four-term interpolation phase functions $x(\gamma) = x_1, P_1 + x_2 P_2 + x_3 P_3$

λ, μm	m	γ_{cut} degree	A	x_1^*	x_2^*	x_3^*	x_1	x_2	x_3
0.365	1.46 − 0.01 i	5	0	0.6552	0.7057	0.4958	1.97	4.99	6.47
		10	0.3367				0.74	2.05	1.51
		15	0.4090				0.48	1.18	0.56
		20	0.4454				0.40	0.96	0.32
	1.46	5	0	0.5985	0.6503	0.4093	1.66	5.00	5.10
		10	0.2584				0.72	1.90	1.25
		15	0.3164				0.52	1.24	0.53
		20	0.3501				0.45	1.05	0.31
0.55	1.44 − 0.01 i	5	0	1.1149	1.0319	0.7825	2.99	5.00	7.00
		10	0.3400				1.97	4.89	4.49
		15	0.5791				1.20	2.33	1.57
		20	0.6990				0.97	1.56	0.69
	1.44	5	0	1.0327	0.9685	0.6782	2.89	5.00	6.99
		10	0.2805				1.76	4.45	3.81
		15	0.4801				1.15	2.28	1.38
		20	0.5876				0.96	1.64	0.63
0.99	1.43 − 0.01 i	5	0	1.4366	1.1853	1.8768	2.81	5.00	6.40
		10	0.870				2.57	5.00	5.28
		15	0.2510				2.22	4.29	3.67
		20	0.5145				1.95	3.15	2.35
	1.43	5	0	1.3857	1.1498	0.8246	2.67	5.00	5.93
		10	0.0788				2.45	5.00	4.90
		15	0.2293				2.12	4.08	3.41
		20	0.4717				1.87	3.04	2.18

$$\begin{aligned} S\zeta\varrho &= I(0, -\eta, \zeta, \varphi) \ , \\ S\zeta\sigma &= I(\tau_0, \eta, \zeta, \varphi) \ , \quad \eta > 0 \ . \end{aligned} \tag{3.21}$$

As shown independently by Sobolev [183] and van de Hulst [404], the values of brightness coefficients ϱ and σ, with surface albedo A are easily related to ϱ and σ values at $A = 0$. Therefore, when considering the theory of the multiple scattering of light in a plane-parallel planetary atmosphere of a finite optical thickness ($\tau_0 < \infty$) further, we shall confine ourselves to $A = 0$ using a 3-term phase approximating function (3.9).

We write total brightness coefficients ϱ and σ as parts of a Fourier series with respect to azimuth φ:

$$\varrho(\eta, \zeta, \varphi, \tau_0) = \varrho^0(\eta, \zeta, \tau_0) + 2 \sum_{m=1}^{2} \varrho^m(\eta, \zeta, \tau_0) \cos m\varphi \ , \tag{3.22}$$

$$\sigma(\eta, \zeta, \varphi, \tau_0) = \sigma^0(\eta, \zeta, \tau_0) + 2 \sum_{m=1}^{2} \sigma^m(\eta, \zeta, \tau_0) \cos m\varphi \ . \tag{3.23}$$

Table 3.5. Cloud phase function $x(\gamma)$ of the atmosphere of venus using parameters of a generalized gamma distribution $r = 1.05\ \mu m$, $\alpha = 11$, $b = 14$

γ degree	λ = 0.365 μm		λ = 0.550 μm		λ = 0.990 μm	
	m = 1.46 − 0.01 i	m = 1.46	m = 1.44 − 0.01 i	m = 1.44	m = 1.43 − 0.01 i	m = 1.43
0	243.2672	188.2867	106.4405	89.0614	33.8935	31.4430
5	99.8591	78.3958	70.3470	59.2426	29.9154	27.8146
10	14.1795	12.7902	23.2711	20.4926	20.9223	19.6602
15	5.0044	5.4608	6.6989	6.6868	12.4295	11.8566
20	3.1969	3.7974	3.3717	3.7123	6.9447	6.8076
30	1.7300	2.1920	1.8889	2.1529	2.7576	2.8391
40	1.0436	1.3255	1.2150	1.3718	1.5001	1.5643
50	0.6507	0.8159	0.7934	0.8835	0.8738	0.9133
60	0.4136	0.5082	0.5190	0.5734	0.5278	0.5532
70	0.2532	0.3254	0.3440	0.3810	0.3353	0.3531
80	0.1783	0.2210	0.2341	0.2636	0.2261	0.2399
90	0.1216	0.1556	0.1649	0.1898	0.1633	0.1747
100	0.0853	0.1104	0.1213	0.1418	0.1274	0.1379
120	0.0491	0.0637	0.0821	0.0978	0.1044	0.1173
140	0.0587	0.0909	0.1208	0.1614	0.1444	0.1737
160	0.2647	0.5406	0.4424	0.6508	0.2983	0.3921
180	0.2982	0.7741	0.3924	0.6947	0.2398	0.3348

The azimuthal harmonics ϱ^m and σ^m are routinely expressed in terms of φ_i^m and ψ_i^m – the Ambartsumyan functions [187]:

$$\varrho^m(\eta,\zeta,\tau_0) = \frac{\Lambda}{4}\sum_{i=m}^{2} c_i^m(-1)^{i+m}\,\frac{\varphi_i^m(\eta,\tau_0)\,\varphi_i^m(\zeta,\tau_0) - \psi_i^m(\eta,\tau_0)\,\psi_i^m(\zeta,\tau_0)}{\eta+\zeta}\ ,$$

$$\sigma^m(\eta,\zeta,\tau_0) = \frac{\Lambda}{4}\sum_{i=m}^{2} c_i^m\,\frac{\psi_i^m(\zeta,\tau_0)\,\Psi_i^m(\eta,\tau_0) - \psi_i^m(\eta,\tau_0)\,\Psi_i^m(\zeta,\tau_0)}{\eta-\zeta} \tag{3.24}$$

where c_i^m coefficients are

$$c_i^m = x_i\ , \qquad c_i^m = x_i\,\frac{(i-m)!}{(i+m)!} \tag{3.25}$$

As Sobolev has shown [187], the Ambartsumyan functions φ_i^m and ψ_i^m in an arbitrary phase function $x(\gamma)$ are expressed through the Chandrasekhar functions X^m and Y^m and some polynomials q_i^m and s_i^m to the power $(N-m)$ with regard to η:

$$\varphi_i^m(\eta) = [X^m(\eta)\,q_i^m(\eta) + (-1)^{i+m}\,Y^m(\eta)\,s_i^m(-\eta)]\,P_m^m(\eta)\ , \tag{3.26}$$

$$\psi_i^m(\eta) = [X^m(\eta)\,s_i^m(\eta) + (-1)^{i+m}\,Y^m(\eta)\,q_i^m(-\eta)]\,P_m^m(\eta)\ . \tag{3.27}$$

The q_i^m and s_i^m polynomials are determined from a system of integral equations obtained in [187]. The Chandrasekhar functions $X^m(\zeta,\tau_0)$ and $Y^m(\zeta,\tau_0)$ are expressed through the Sobolev function $\phi^m(\tau,\tau_0)$ in the following way [187]:

$$X^m(\zeta,\tau_0) = 1 + \int_0^{\tau_0} \phi^m(\tau,\tau_0)e^{-\tau/\zeta}d\tau$$

$$Y^m(\zeta,\tau_0) = e^{-\tau_0/\zeta} + \int_0^{\tau_0} \phi^m(\tau_0-\tau,\tau_0)e^{-\tau/\zeta}d\tau \ , \qquad m = 0,1,\ldots,N \ . \tag{3.28}$$

To find the Sobolev function ϕ^m, the following integral equation must be solved [187]:

$$\phi^m(\tau,\tau_0) = K^m(\tau) + \int_0^{\tau_0} K^m(|\tau-\tau'|)\phi^m(\tau',\tau_0)d\tau' \ (m = 0,1,\ldots,N) \tag{3.29}$$

where the core $K^m(\tau)$ is

$$K^m(\tau) = \int_0^1 \psi^m(\eta)e^{-\tau/\eta}\frac{d\eta}{\eta} \ . \tag{3.30}$$

The $\psi(\eta)$ functions are:

$$\psi^m(\eta) = \frac{\Lambda}{2} P_m^m(\eta)A_m^m(\eta,\eta) \tag{3.31}$$

where

$$A_m^m(\eta,\eta') = \sum_{i=m}^{N} c_i^m R_{im}^m(\eta)P_i^m(\eta')$$

$$m = 0,1,2,\ldots,N$$

$$i = m, m+1, m+2,\ldots,N \ . \tag{3.32}$$

Polynomials $R_{im}^m(\eta)$ are of the power $(i-m)$ and determined from recurrent relationships [187]:

$$(i-m+1)R_{i+1}^m, m(\eta) + (i+m)R_{i-1}^m, m(\eta) = (2i+1-\Lambda x_i)\eta_{im}^m(\eta)$$

$$R_{m}^m(\eta) = 1 \ , \qquad R_{im}^m(\eta) = 0 \quad \text{at } m>1 \ , \qquad m = 0,1,2,\ldots,N \ ;$$

$$i = m, m+1 \ , \quad m+2,\ldots,N \ . \tag{3.33}$$

According to the classical scheme [187], to solve this basic problem in the case of an approximating 3-term phase function (3.19), it is necessary first, to find polynomials q_i^m and s_i^m $(m = 0,1,2;\ i = m, m+1,\ldots,2)$. Then integral Eq. (3.29) for the Sobolev functions ϕ^m $(m = 0,1,2)$ can be solved, and using formula (3.28), X^m and Y^m functions after Chandrasekhar $(m = 0,1,2)$ can be determined. Then, using formulas (3.26–3.27), φ_i^m and ψ_i^m Ambartsumyan functions must be found $(m = 0,1,2;\ i = m, = 1,\ldots,2)$, and finally, total coefficients of reflection $\varrho(\eta,\zeta,\varphi,\tau_0)$ and transmission $\sigma(\eta,\zeta,\varphi,\tau_0)$ must be determined from formula (3.24). However, generally for an arbitrary height-independent phase function $x(\gamma)$, this scheme can be simplified substantially, if, instead of the theory of multiple light scattering, we consider their linear algebraic combinations [177].

3.5 Algorithm to Determine the Azimuthal Harmonics of ϱ^m and σ^m Brightness Coefficients

The practical use of q_i^m and s_i^m polynomials to determine the azimuthal harmonics of ϱ^m and σ^m brightness coefficients is difficult because these polynomials must be found simultaneously from a system of integral equations after Sobolev [187].

To simplify the solution for this system for a general arbitrary phase function $x(\gamma)$, we introduce new polynomials Q_i^m and S_i^m $(m = 0, 1, 2, \ldots, N)$ which are equal to linear algebraic combinations of the original polynomials q_i^m and s_i^m [177]:

$$Q_i^m(\eta) = q_i^m(\eta) + (-1)^{i+m} s_i^m(\eta) \ , \quad S_i^m(\eta) = q_i^m(\eta) - (-1)^{i+m} s_i^m(\eta) \quad (3.34)$$

Each of Q_i^m and S_i^m polynomials is now determined individually from independent integral equations easily obtained from a traditional system of equations for q_i^m and s_i^m:

$$\begin{aligned} Q_i^m(\eta) = {} & \frac{\Lambda}{2}\eta \int_0^1 P_m^m(\eta') A_m^m(\eta', \eta') X^m(\eta') \frac{Q_i^m(\eta') - Q_i^m(\eta)}{\eta' - \eta} d\eta' \\ & - \frac{\Lambda}{2}\eta \int_0^1 P_m^m(\eta') X^m(\eta') Q_i^m(\eta') \frac{A_m^m(\eta', \eta') - A_m^m(\eta, \eta')}{\eta' - \eta} d\eta' \\ & + \frac{\Lambda}{2}\eta \int_0^1 P_m^m(\eta') Y^m(\eta') A_m^m(\eta', \eta') \frac{Q_i^m(-\eta') - Q_i^m(-\eta)}{\eta' - \eta} d\eta' \\ & - \frac{\Lambda}{2}\eta \int_0^1 P_m^m(\eta') Y^m(\eta') Q_i^m(-\eta') \frac{A_m^m(\eta', \eta') - A_m^m(\eta, \eta')}{\eta' - \eta} d\eta' \\ & + R_{im}^m(\eta) \ , \end{aligned} \quad (3.35)$$

$$\begin{aligned} S_i^m(\eta) = {} & \frac{\Lambda}{2}\eta \int_0^1 P_m^m(\eta') A_m^m(\eta', \eta') X^m(\eta') \frac{S_i^m(\eta') - S_i^m(\eta)}{\eta' - \eta} d\eta' \\ & - \frac{\Lambda}{2}\eta \int_0^1 P_m^m(\eta') X^m(\eta') S_i^m(\eta') \frac{A_m^m(\eta', \eta') - A_m^m(\eta, \eta')}{\eta' - \eta} d\eta' \\ & - \frac{\Lambda}{2}\eta \int_0^1 P_m^m(\eta') Y^m(\eta') A_m^m(\eta', \eta') \frac{S_i^m(-\eta') - S_i^m(-\eta)}{\eta' - \eta} \delta\eta' \\ & + \frac{\Lambda}{2}\eta \int_0^1 P_m^m(\eta') Y^m(\eta') S_i^m(-\eta') \frac{A_m^m(\eta', \eta') - A_m^m(\eta, \eta'))}{\eta' - \eta} d\eta' \\ & + R_{im}^m(\eta) \quad (m = 0, 1, 2, \ldots N) \ . \end{aligned} \quad (3.36)$$

Similarly to (3.34), we introduce new functions Ω_i^m and Θ_i^m $(m = 0, 1, 2, \ldots, N)$ equal to linear algebraic combinations of classical φ^m, ψ_i^m functions by Ambartsumyan [177]:

$$\Omega_i^m(\eta) = \varphi_i^m(\eta) + (-1)^{i+m}\psi_i^m(\eta) , \qquad \Theta_i^m(\eta) = \varphi_i^m(\eta) - (-1)^{i+m}\psi_i^m(\eta) . \tag{3.37}$$

Using (3.37) and the known system of linear integral equations for φ_i^m and ψ_i^m functions [184], we obtain independent linear integral equations for each of Ω_i^m and Θ_i^m functions. Using definitions from (3.34), (3.37) and relationships (3.26 – 3.27), we can easily show that, in contrast to the standard functions φ_i^m and ψ_i^m, the new functions Ω_i^m and Θ_i^m are related to functions X^m and Y^m by only one of the polynomials: either Q_i^m or S_i^m [177]

$$\Omega_i^m(\eta) = P_m^m(\eta)[X^m(\eta)Q_i^m(\eta) + Y^m(\eta)Q_i^m(-\eta)] , \tag{3.38}$$

$$\Theta_i^m(\eta) = P_m^m(\eta)[X^m(\eta)S_i^m(\eta) - Y^m(\eta)S_i^m(-\eta)] . \tag{3.39}$$

Based on relationships (3.24 – 3.25) and (3.37), it is easy to express the azimuthal harmonics of brightness coefficients ϱ^m and σ^m through new functions Ω_i^m and Θ_i^m [177]:

$$\varrho^m(\eta,\zeta) = \frac{\Lambda}{8}\sum_{i=m}^{N}(-1)^{i+m}C_i^m\frac{\Omega_i^m(\eta)\Theta_i^m(\zeta) + \Theta_i^m(\eta)\Omega_i^m(\zeta)}{\eta+\zeta} , \tag{3.40}$$

$$\sigma^m(\eta,\zeta) = \frac{\Lambda}{8}\sum_{i=m}^{N}(-1)^{i+m}C_i^m\frac{\Omega_i^m(\eta)\Theta_i^m(\zeta) - \Theta_i^m(\eta)\Omega_i^m(\zeta)}{\eta-\zeta} . \tag{3.41}$$

By analogy with (3.37), we now introduce both the sum and the difference of the azimuthal harmonics of the total coefficients of diffused reflection $\varrho(\eta,\zeta,\varphi,\tau_0)$ and transmission $\sigma(\eta,\zeta,\varphi,\tau_0)$:

$$\varrho^m(\eta,\zeta,\tau_0) + \sigma^m(\eta,\zeta,\tau_0) = R^m(\eta,\zeta,\tau_0) , \tag{3.42}$$

$$\varrho^m(\eta,\zeta,\tau_0) - \sigma^m(\eta,\zeta,\tau_0) = r^m(\eta,\zeta,\tau_0) . \tag{3.43}$$

Using an integral relationship of the Ambartsumyan functions φ_i^m and ψ_i^m with the azimuthal harmonics ϱ^m and σ^m [187], we obtain:

$$\begin{aligned}\varphi_i^m(\eta) &= P_i^m(\eta) + 2\eta\int_0^1 P_i^m(-\zeta)\varrho^m(\eta,\zeta)d\zeta ,\\ \psi_i^m(\eta) &= P_i^m(\eta)e^{-\tau_0/\eta} + 2\eta\int_0^1 P_i^m(\zeta)\sigma^m(\eta,\zeta)dz .\end{aligned} \tag{3.44}$$

Then, based on definitions in (3.37), (3.42 – 3.43) and the relationship

$$P_i^m(\zeta) = (-1)^{i+m}P_i^m(-\zeta) \tag{3.45}$$

we obtain integral expressions for Ω_i^m and Θ_i^m functions through brightness coefficients R^m and r^m:

$$\Omega_i^m(\eta) = P_i^m(\eta)[1 + (-1)^{i+m}e^{-\tau_0/\eta}] + 2\eta\int_0^1 P_i^m(-\zeta)R^m(\eta,\zeta)d\zeta \tag{3.46}$$

$$\Theta_i^m(\eta) = P_i^m(\eta)[1 - (-1)^{i+m}e^{-\tau_0/\eta}] + 2\eta\int_0^1 P_i^m(-\zeta)r^m(\eta,\zeta)d\zeta . \tag{3.47}$$

Summing and extracting (3.40 – 3.41), in accordance with (3.42 – 3.43), we have

$$R^m(\eta,\zeta) = \frac{\Lambda}{4} \sum_{i=m}^{N} (-1)^{i+m} c_i^m \frac{\eta\,\Omega_i(\eta)\Theta_i^m(\zeta) - \zeta\,\Theta_i^m(\eta)\Omega_i^m(\zeta)}{\eta^2-\zeta^2} , \tag{3.48}$$

$$r^m(\eta,\zeta) = \frac{\Lambda}{4} \sum_{i=m}^{N} (-1)^{i+m} c_i^m \frac{\eta\,\Omega_i^m(\zeta)\Theta_i^m(\eta) - \zeta\,\Theta_i^m(\zeta)\Omega_i^m(\eta)}{\eta^2-\zeta^2} . \tag{3.49}$$

Substituting (3.48) and (3.49), respectively, into (3.46) and (3.47), we can obtain a system of non-linear equations to find Ω_i^m and Θ_i^m functions. However, independent linear integral equations are theoretically of great importance for the determination of these functions.

Note, that from (3.40 – 3.41) or (3.48 – 3.49), the following relationship is easily obtained for a new function $E^m(\eta,\zeta)$:

$$E^m(\eta,\zeta) = \eta R^m(\eta,\zeta) + \zeta r^m(\eta,\zeta) = \frac{\Lambda}{4} \sum_{i=m}^{N} (-1)^{i+m} c_i^m \Omega_i^m(\eta)\Theta_i^m(\zeta)$$

$$= (\eta+\zeta)\varrho^m(\eta,\zeta) + (\eta-\zeta)\sigma^m(\eta,\zeta) . \tag{3.50}$$

It follows from (3.50) that a consideration of the $E^m(\eta,\zeta)$ function is equivalent to a consideration of the $R^m(\eta,\zeta)$ and $r^m(\eta,\zeta)$ functions, the $\varrho^m(\eta,\zeta)$, $\sigma^m(\eta,\zeta)$, and $R^m(\eta,\zeta)$, $r^m(\eta,\zeta)$ azimuthal harmonics being expressed in terms of an asymmetric function $E^m(\eta,\zeta)$

$$\varrho^m(\eta,\zeta) = \frac{E^m(\eta,\zeta)+E^m(\zeta,\eta)}{2(\eta+\zeta)} , \qquad \sigma^m(\eta,\zeta) = \frac{E^m(\eta,\zeta)-E^m(\zeta,\eta)}{2(\eta-\zeta)} ,$$

$$R^m(\eta,\zeta) = \frac{\eta E^m(\eta,\zeta) - \zeta E^m(\zeta,\eta)}{v^2-\zeta^2} , \quad r^m(\eta,\zeta) = \frac{\eta E^m(\zeta,\eta) - \zeta E^m(\eta,\zeta)}{\eta^2-\zeta^2} . \tag{3.51}$$

Symmetric functions $R^m(\eta,\zeta)$ and $r^m(\eta,\zeta)$, in contrast to $\varrho^m(\eta,\zeta)$ and $\sigma^m(\eta,\zeta)$, satisfy independent linear integral equations which, along with a similar equation for the $E^m(\eta,\zeta)$ function, will be considered later. Note also, that the introduction of $E^m(\eta,\zeta)$ is of no principal value. This is because, according to [177], theoretical simplifications appear due not to the $E^m(\eta,\zeta)$ function itself and relationships (3.51) but to a rational division of angular variables in the azimuthal harmonics $\varrho^m(\eta,\zeta)$ and $\sigma^m(\eta,\zeta)$ realized by introducing new functions Ω_i^m, Θ_i^m (3.37) and polynomials S_i^m and Q_i^m (3.34). In addition, the $E^m(\eta,\zeta)$ function has no physical meaning and, theoretically at least, photometric quantities based on this function are more important. Simplifications when considering linear combinations of classical functions involving their sum and their difference, will be substantiated below.

The Ω_i^m and Θ_i^m functions determined from (3.37) have clear physical (probabilistic) meaning. To define this meaning for of isotropic scattering, probabilities $P(\tau,\eta,\tau_0)$ and $P(\tau_0-\tau,\eta,\tau_0)$ must be determined after Sobolev [184]. These are probabilities that a quantum of light absorbed at an optical depth τ will leave this medium either through the upper ($\tau = 0$) or lower ($\tau = \tau_0$) boundary of the atmosphere at an angle $\theta = \arccos \eta$ to the normal within elementary solid angle $d\omega$ upon multiple scattering. For anisotropic scattering, bearing in mind the well-known probabilistic meaning of traditional functions by Ambartsumyan [184] and relationships (3.37), the probabilistic meaning of the new functions Ω_i^m and Θ_i^m is easily found.

Then, instead of the X^m and Y^m functions by Chandrasekhar ($m = 0, 1, \ldots, N$), we chall consider their sum and difference, respectively [177]:

$$W^m = X^m + Y^m \;, \qquad U^m = X^m - Y^m \;. \tag{3.52}$$

Then, instead of (3.38) and (3.39), we obtain

$$\Omega_i^m(\eta) = P_m^m(\eta)[W^m(\eta)Q_i^m(\eta^{2k}) + U^m(\eta)Q_i^m(\eta^{2k+1})] \;, \tag{3.53}$$

$$\Theta_i^m(\eta) = P_m^m(\eta)[W^m(\eta)S_i^m(\eta^{2k+1}) + U^m(\eta)S_i^m(\eta^{2k})] \;. \tag{3.54}$$

Polynomials $Q_i^m(\eta^{2k})$, $S_i^m(\eta^{2k})$ and $Q_i^m(\eta^{2k+1})$, $S_i^m(\eta^{2jk+1})$ in (3.53 – 3.54) contain, respectively, either even or odd powers only of the angular variable η.

For W^m and U^m functions, one can easily write independent linear singular integral equations using the respective equations for X^m and Y^m functions by Chandrasekhar [187]. The techniques to solve these equations have been described in [140, 184, 187]. However, practically, these functions can be found using integral relationships which follow from (3.50 – 3.51):

$$W^m(\zeta) = 1 + \exp(-\tau_0/\zeta) + \int_0^{\tau_0} [\phi^m(\tau, \tau_0) + \phi^m(\tau_0 - \tau, \tau_0)] e^{-\tau/\zeta} d\tau \;, \tag{3.55}$$

$$U^m(\zeta) = 1 - \exp(-\tau_0/\zeta) + \int_0^{\tau_0} [\phi^m(\tau, \tau_0) - \phi^m(\tau_0 - \tau, \tau_0)] e^{-\tau/\zeta} d\tau \;. \tag{3.56}$$

Equations (3.55 – 3.56) show that the sum and the difference of Sobolev's functions $\phi^m(\tau, \tau_0)$ and $\phi^m(\tau_0 - \tau, \tau_0)$ are critical in assessing the W^m and U^m functions.

To find Sobolev's functions $\phi^m(\tau, \tau_0)$, integral Eq. (3.28) was solved by the numerically iterative method for a 3-term phase function (3.19). Here, the nuclear function $K^m(\tau)$ ($m = 0, 1, 2$) is

$$\begin{aligned} K^0(\tau) &= \frac{\Lambda}{2}\left(1 + \frac{x_2}{4}\right) E_1(\tau) + \frac{\Lambda}{2}\left[(1-\Lambda)x_1 - \frac{x_2}{4}(1-\Lambda)(3-\Lambda x_1) - \frac{3}{4}x_2\right] \\ &\quad \times E_{i3}(\tau) + \frac{3}{8}\Lambda x_2(1-\Lambda)(3-\Lambda x_1)E_{i5}(\tau) \;, \\ K^1(\tau) &= \frac{\Lambda}{4}x_1 E_{i1}(\tau) - \frac{\Lambda}{4}[x_1 - x_2(3-\Lambda x_1)]E_{i3}(\tau) - \frac{\Lambda}{4}(3-\Lambda x_1)x_2 E_{i5}(\tau) \;, \\ K^2(\tau) &= \frac{3}{16}\Lambda x_2[E_{i1}(\tau) - 2E_{i3}(\tau) + E_{i5}(\tau)] \;, \end{aligned} \tag{3.57}$$

where $E_{in}(\tau)$ is the integral exponential function.

Table 3.6. Estimates of ϕ^m functions by Sobolev (m = 0, 1, 2) for an approximating 3-term phase function ($\tau_0 = 0.3$; $\Lambda = 1$; $x_1 = 1.269$; $x_2 = 0.909$)

τ	ϕ^0	ϕ^1	ϕ^2	τ	ϕ^0	ϕ^1	ϕ^2
0	5.1998	2.5677	1.2382	0.16	1.2988	0.5207	0.1642
0.01	2.8807	1.3430	0.5851	0.17	1.2603	0.5027	0.1566
0.02	2.4835	1.1419	0.4718	0.18	1.2233	0.4856	0.1496
0.03	2.2550	1.0190	0.4069	0.19	1.1875	0.4698	0.1431
0.04	2.0942	0.9324	0.3619	0.20	1.1528	0.4541	0.1369
0.05	1.9699	0.8656	0.3276	0.21	1.1190	0.4394	0.1312
0.06	1.8682	0.8112	0.3002	0.22	1.0859	0.4253	0.1258
0.07	1.7820	0.7654	0.2774	0.23	1.0534	0.4117	0.1208
0.08	1.7069	0.7259	0.2580	0.24	1.0213	0.3986	0.1159
0.09	1.6401	0.5909	0.2413	0.25	0.9894	0.3858	0.1113
0.10	1.5798	0.6597	0.2265	0.26	0.9574	0.3734	0.1069
0.11	1.5246	0.6315	0.2134	0.27	0.9252	0.3612	0.1028
0.12	1.4736	0.6058	0.2016	0.28	0.8923	0.3492	0.09878
0.15	1.3391	0.5398	0.1724	0.30	0.8176	0.3240	0.09101

A specific feature of the core $K^m(\tau)$ at $\tau = 0$ was considered in the ε-vicinity of the $\tau = 0$ point by subsequent estimation of the radius of circumference τ_ε. The circumference radius τ_ε was chosen, because the known asymptotic relationship for $E_{i1}(\tau)$ at $\tau \to 0$ is fulfilled with a given accuracy (10^{-5}):

$$E_{i1}^{as}(\tau) = -\ln\tau - c \tag{3.58}$$

where the Euler constant is $c = 0.57721$. With five significant figures coinciding in the left- and right-hand parts of the asymptotic Eq. (3.56), we have $\tau_\varepsilon = 10^{-5}$.

The $K^m(\tau)$ nuclear function [the $K^m(0)$ value being substituted by $K^m(\tau_\varepsilon)$] was chosen as zero approximation for $\phi^m_{(0)}$ in an iterative solution for (3.29). The integrals were calculated after Simpson's method, the $[0, \tau_0]$ integral being divided into 600 parts. The process of successive approximation stopped upon reaching the value 10^{-5} for the remaining term. In the calculations of $W^m(\eta, \tau_0)$ and $U^m(\eta, \tau_0)$ functions using (3.55 – 3.56), the features of the subintegral functions at zero ($\tau = 0$) were considered as just described. Table 3.6 gives values of Sobolev's functions $\phi^m(\tau, \tau_0)$, and Table 3.7 estimates $W^m(\eta, \tau_0)$ and $U^m(\eta, \tau_0)$ functions for purely scattering media ($\Lambda = 1$, $\tau_0 = 0.3$) and for an approximating 3-term phase function $\bar{x}(\gamma)$ with the coefficients $x_1 = 1.269$ and $x_2 = 0.909$ [177].

Table 3.8 gives estimates for $\Omega_i^m(\eta, \tau_0)$ and $\Theta_i^m(\eta, \tau_0)$ functions for the same optical characteristics; spherical polydisperse particles with refraction index $m = 1.5$; average radius $\bar{r} = 0.5$; and Yunge-distribution parameters $\beta = 2$ and $\alpha = 0$.

Use of new functions W^m and U^m, Ω_i^m and Θ_i^m makes it possible to simplify the basic integral Eq. (3.35 – 3.36) for Q_i^m and S_i^m polynomials and expressions for ϱ^m and σ^m azimuthal harmonics. Now instead of (3.35) and (3.36), we have, considering (3.52 – 3.54):

Table 3.7. Estimates of U^m and W^m functions ($m = 0, 1, 2$) for an approximating three-term phase function ($\tau_0 = 0.3$; $\Lambda = 1$; $x_1 = 1.269$; $x_2 = 0.909$)

η	W^0	U^0	W^1	U^1	W^2	U^2
0	1.0000	1.0000	1.0000	1.0000	1.0000	1.0000
0.1	1.3441	1.0273	1.1756	0.9886	1.0953	0.9683
0.2	1.6921	0.8451	1.4218	0.8108	1.2936	0.7928
0.3	1.9371	0.6887	1.6086	0.6602	1.4529	0.6453
0.4	2.1046	0.5750	1.7396	0.5512	1.5667	0.5387
0.5	2.2239	0.4919	1.8340	0.4714	1.6494	0.4606
0.6	2.3126	0.4290	1.9047	0.4111	1.7117	0.4017
0.7	2.3810	0.3800	1.9596	0.3642	1.7601	0.3559
0.8	2.4352	0.3410	2.0031	0.3267	1.7985	0.3193
0.9	2.4791	0.3091	2.0386	0.2962	1.8300	0.2894
1.0	2.5154	0.2826	2.0679	0.2709	1.8561	0.2647

$$
\begin{aligned}
Q_i^m(\eta) = {} & \frac{\Lambda}{2}\eta \int_0^1 P_m^m(\eta')A_m^m(\eta',\eta') \\
& \times \frac{W^m(\eta')[Q_i^m(\eta'^{2k}) - Q_i^m(\eta^{2k})] + U^m(\eta')[Q_i^m(\eta'^{2k+1}) - Q_i^m(\eta^{2k+1})]}{\eta' - \eta} d\eta' \\
& - \frac{\Lambda}{2}\eta \int_0^1 P_m^m(\eta')[W^m(\eta')Q_i^m(\eta'^{2k}) + U^m(\eta')Q_i^m(\eta'^{2k+1})] \\
& \times \frac{A_m^m(\eta',\eta') - A_m^m(\eta,\eta')}{\eta' - \eta} d\eta' + R_{im}^m(\eta) \ ,
\end{aligned} \tag{3.59}
$$

$$
\begin{aligned}
S_i^m(\eta) = {} & \frac{\Lambda}{2}\eta \int_0^1 P_m^m(\eta')A_m^m(\eta',\eta') \\
& \times \frac{W^m(\eta')[S_i^m(\eta'^{2k+1}) - S_i^m(\eta^{2k+1})] + U^m(\eta')[S_i^m(\eta'^{2k}) - S_i^m(\eta^{2k})]}{\eta' - \eta} d\eta' \\
& - \frac{\Lambda}{2}\eta \int_0^1 P_m^m(\eta')[W^m(\eta')S_i^m(\eta'^{2k+1}) + U^m(\eta')S_i^m(\eta'^{2k})] \\
& \times \frac{A_m^m(\eta',\eta') - A_m^m(\eta,\eta')}{\eta' - \eta} d\eta' + R_{im}^m(\eta) \quad (m = 0, 1, 2, \ldots, N) \ .
\end{aligned} \tag{3.60}
$$

In this analysis, a phase function $x(\gamma)$ was considered to be arbitrary. Therefore, a solution for integral Eqs. (3.59) and (3.60) becomes:

$$
Q_i^m(\eta) = R_{im}^m(0) + \sum_{k=1}^{N} c_{ik}^m \eta^m \ ; \qquad S_i^m(\eta) = R_{im}^m(0) + \sum_{k=1}^{N} d_{ik}^m \eta^k \ . \tag{3.61}
$$

Substituting (3.61) into (3.59) and (3.60) and then equalizing the coefficients by similar powers η in their right- and left-hand parts, we obtain an $(N+1-m)$ system of linear algebraic equations to find coefficients c_{ik}^m and

Table 3.8. Estimates of Ω_i^m and Θ_i^m (m = 0, 1, 2) functions for an approximating three-term phase function

η	Ω_0^0	Ω_1^0	Ω_2^0	Θ_0^0	Θ_1^0	Θ_2^0
0	1.0000	0	−0.5000	1.0000	0	−0.5000
0.1	1.3113	0	−0.1840	1.0131	0.1004	−0.4938
0.3	1.8712	0	0.01284	0.2706	0.4343	−0.2571
0.5	2.1454	0	0.05638	0.3441	0.8310	−0.07916
0.7	2.2961	0	0.07535	0.2035	1.2455	0.05994
1.0	2.4252	0	0.08495	0.0162	1.8798	0.2359
	Ω_1^1	Ω_2^1	Θ_1^1	Θ_2^1	Ω_2^2	Θ_2^2
0	1.0000	0	1.0000	0	3.0000	3.0000
0.1	1.1495	0.1972	0.9750	0.3066	3.2529	2.8758
0.3	1.4958	0.3789	0.5957	1.2066	3.9066	1.7616
0.5	1.5464	0.4093	0.3495	2.0815	3.7111	1.0365
0.7	1.3620	0.3651	0.1877	1.5675	2.6928	0.5445
1.0	0	0	0	0	0	0

d_{ik}^m (m = 0, ..., N). It must be emphasized that, regarding the initial system of linear algebraic equations suggested by Sobolev to determine polynomials q_i^m and s_i^m [186], the rank of the system of linear algebraic equations to determine new polynomials Q_i^m and S_i^m is halved [177], i.e., it is equal to N.

Now we express the ϱ^m and σ^m azimuthal harmonics through W^m and U^m functions. Substituting (3.53) and (3.54), respectively, into (3.40) and (3.41), we obtain, upon simple transformations:

$$\begin{aligned}\varrho^m(\eta,\zeta) = \frac{\Lambda P_m^m(\eta) P_m^m(\zeta)}{8(\eta+\zeta)} \Bigg\{ & W^m(\eta) W^m(\zeta) \sum_{i=m}^{N} c_i^m(-1)^{i+m} \\ & \times [S_i^m(\zeta^{2k+1}) Q_i^m(\eta^{2k}) + S_i^m(\eta^{2k+1}) Q_i^m(\zeta^{2k})] \\ & + U^m(\eta) U^m(\zeta) \sum_{i=m}^{N} c_i^m(-1)^{i+m} [S_i^m(\zeta^{2k}) Q_i^m(\eta^{2k+1}) \\ & + S_i^m(\eta^{2k}) Q_i^m(\zeta^{2k+1})] + [W^m(\eta) U^m(\zeta) + W^m(\zeta) U^m(\eta)] \\ & \times \sum_{i=m}^{N} [Q_i^m(\eta^{2k}) S_i^m(\zeta^{2k}) + Q_i^m(\zeta^{2k+1}) S_i^m(\eta^{2k+1})]\, c_i^m(-1)^{i+m} \Bigg\} ,\end{aligned} \tag{3.62}$$

$$\begin{aligned}\sigma^m(\eta,\zeta) = \frac{\Lambda P_m^m(\eta) P_m^m(\zeta)}{8(\eta-\zeta)} \Bigg\{ & W^m(\eta) W^m(\zeta) \sum_{i=m}^{N} c_i^m(-1)^{i+m} \\ & \times [Q_i^m(\eta^{2k}) S_i^m(\zeta^{2k+1}) - S_i^m(\eta^{2k+1}) Q_i^m(\zeta^{2k})] \\ & + U^m(\eta) U^m(\zeta) \sum_{i=m}^{N} c_i^m(-1)^{i+m} [Q_i^m(\eta^{2k+1}) S_i^m(\zeta^{2k})\end{aligned}$$

$$- Q_i^m(\zeta^{2k+1}) S_i^m(\eta^{2k})] + [W^m(\zeta) U^m(\eta) - U^m(\zeta) W^m(\eta)]$$

$$\left. \times \sum_{i=m}^{N} c_i^m (-1)^{i+m} [Q_i^m(\zeta^{2k+1}) S_i^m(\eta^{2k+1}) - S_i^m(\zeta^{2k}) Q_i^m(\eta^{2k})] \right\} . \tag{3.63}$$

Note, that to simplify recording of (3.62 – 3.63), their summation and subtraction should be substituted by one function $E^m(\eta, \zeta)$, with subsequent use of (3.51). Thus, use of (3.62 – 3.63) and (3.50) enables one to obtain a more compact formula for the $E^m(\eta, \zeta)$ function than for the $R^m(\eta, \zeta)$ and $r^m(\eta, \zeta)$ functions:

$$E^m(\eta, \zeta) = \frac{\Lambda P_m^m(\eta) P_m^m(\zeta)}{4} \left\{ W^m(\eta) W^m(\zeta) \right.$$

$$\times \sum_{i=m}^{N} c_i^m (-1)^{i+m} S_i^m(\zeta^{2k+1}) Q_i^m(\eta^{2k}) + U^m(\eta) U^m(\zeta)$$

$$\times \sum_{i=m}^{N} c_i^m (-1)^{i+m} S_i^m(\zeta^{2k}) Q_i^m(\eta^{2k+1}) + W^m(\eta) U^m(\zeta)$$

$$\times \sum_{i=m}^{N} c_i^m (-1)^{i+m} S_i^m(\zeta^{2k}) Q_i^m(\eta^{2k}) + W^m(\zeta) U^m(\eta)$$

$$\left. \times \sum_{i=m}^{N} c_i^m (-1)^{i+m} S_i^m(\eta^{2k+1}) Q_i^m(\zeta^{2k+1}) \right\} . \tag{3.64}$$

However, when calculating this algorithm, use of (3.64) has not advantage over (3.62 – 3.63). This is because principal theoretical simplifications introduced by W^m, U^m functions and by S_i^m and Q_i^m polynomials have already been included in (3.62 – 3.63).

Thus, to find the azimuthal harmonics of ϱ^m and σ^m brightness coefficients, according to the suggested algorithm, it is necessary, first, to calculate the $W^m(\eta, \tau_0)$ and $U^m(\eta, \tau_0)$ functions and then to determine the $Q_i(\eta, \tau_0)$ and $S^m(\eta, \tau_0)$ polynomials.

3.6 Linear Integral Equations for Basic Functions

Based on results obtained by Sobolev [187] and Mallikin [339 – 341], we obtain linear integral equations for functions $\Omega_i^m(\eta, \tau_0)$ and $\Theta_i^m(\eta, \tau_0)$, $R^m(\eta, \zeta)$ and $r^m(\eta, \zeta)$, $W^m(\eta, \tau_0)$ and $U^m(\eta, \tau_0)$ introduced in [177]. According to [183, 187], $\varrho^m(\eta, \zeta, \tau_0)$ and $\sigma^m(\eta, \zeta, \tau_0)$ brightness coefficients satisfy the following system of linear integral equations:

$$\varrho^m(\eta,\zeta,\tau_0)T^m(\eta) = \frac{\Lambda}{2}\int_0^1 \frac{\varrho^m(\eta',\zeta,\tau_0)A^m(\eta,\eta')\,d\eta'}{\nu'-\nu}$$

$$-\frac{\Lambda}{2}e^{-\tau_0/\eta}\int_0^1 \frac{\sigma^m(\eta',\zeta,\tau_0)A^m(\eta,-\eta')\eta'd\eta'}{\eta'+\eta}$$

$$+A^m(\eta,-\zeta)\varrho_1^m(\eta,\zeta) \ , \tag{3.65}$$

$$\sigma^m(\eta,\zeta,\tau_0)T^m(\eta) = \frac{\Lambda}{2}\int_0^1 \frac{\sigma^m(\eta',\zeta,\tau_0)A^m(\eta,\eta')\eta'd\eta'}{\eta'-\eta}$$

$$-\frac{\Lambda}{2}e^{-\tau_0/\eta}\int_0^1 \frac{\varrho^m(\eta',\zeta,\tau_0)A^m(\eta,-\eta')\eta'd\eta'}{\eta'+\eta}$$

$$+A^m(\eta,\zeta)\sigma_1^m(\eta,\zeta) \ , \tag{3.66}$$

where $T^m(\eta)$ function is

$$T^m(\eta) = 1-\eta\int_{-1}^{1}\frac{\psi^m(\eta')}{\eta-\eta'}d\eta' = 1+\frac{\Lambda}{2}\eta\int_{-1}^{1}\frac{A^m(\eta',\eta')}{\eta'-\eta}d\eta' \ . \tag{3.67}$$

Functions $\varrho_1^m(\eta,\zeta)$ and $\sigma_1^m(\eta,\zeta)$, that determine the contribution of single scattering to the azimuthal harmonics of brightness coefficients $\varrho^m(\eta,\zeta,\tau_0)$ and $\sigma^m(\eta,\zeta,\tau_0)$ are

$$\varrho_1^m(\eta,\zeta) = \frac{\Lambda}{4(\eta+\zeta)}\left[1-e^{-\tau_0(1/\eta+1/\zeta)}\right] \ , \quad \sigma_1^m(\eta,\zeta)\frac{\Lambda}{4}\frac{e^{-\tau_0/\eta}-e^{-\tau_0/\zeta}}{\eta-\zeta} \ . \tag{3.68}$$

Summing and extracting (3.65) and (3.66) in accordance with (3.42) and (3.43), we obtain for $R^m(\eta,\zeta)$ and $r^m(\eta,\zeta)$ functions independent linear integral equations with almost similar cores but different free terms:

$$R^m(\eta,\zeta)T^m(\eta) = \frac{\Lambda}{2}\int_0^1 \frac{R^m(\eta',\zeta)A^m(\eta,\eta')\eta'd\eta'}{\eta'-\eta}$$

$$-\frac{\Lambda}{2}e^{-\tau_0/\eta}\int_0^1 \frac{R^m(\eta',\zeta)A^m(\eta,-\eta')\eta'd\eta'}{\eta+\eta'} + R_1^m(\eta,\zeta) \ , \tag{3.69}$$

$$r^m(\eta,\zeta)T^m(\eta) = \frac{\Lambda}{2}\int_0^1 \frac{r^m(\eta',\zeta)A^m(\eta,-\eta')\eta'd\eta'}{\eta'-\eta}$$

$$+\frac{\Lambda}{2}e^{-\tau_0/\eta}\int_0^1 \frac{r^m(\eta',\zeta)A^m(\eta,-\eta')\eta'd\eta'}{\eta+\eta'} + r_1^m(\eta,\zeta) \ , \tag{3.70}$$

where $R_1^m(\eta,\zeta)$ and $r_1^m(\eta,\zeta)$ functions are

$$\begin{aligned} R_1^m(\eta,\zeta) &= A^m(\eta,-\zeta)\varrho_1^m(\eta,\zeta) + A^m(\eta,\zeta)\sigma_1^m(\eta,\zeta) \\ r_1^m(\eta,\zeta) &= A^m(\eta,-\zeta)\varrho_1^m(\eta,\zeta) - A^m(\eta,\zeta)\sigma_1^m(\eta,\zeta) \ . \end{aligned} \tag{3.71}$$

Equations (3.69) and (3.70), like Eqs. (3.65) and (3.66), may not have single solutions [339–341]. To obtain single solutions, with physical meaning, they must be enhanced with some additional relationships. The type of relationships is determined by the existence of roots ($0 \leqslant k_m < 1$) of the characteristic Eq.

$$T^m\left(\frac{1}{k_m}\right) = 0 . \tag{3.72}$$

With $m = 0$, Eq. (3.72) always has a core k_0 [126, 318]. With $m \geqslant 1$, the presence or absence of the cores is determined in each individual case by the type of $A^m(\eta, \eta')$ function. This problem is considered separately, for 3- and 4-term approximating phase functions.

Taking $\eta = 1/k_m$ in (3.69) and (3.70) and mindful of (3.72), we obtain following additional conditions:

$$R_1^m\left(\frac{1}{k_m}, \zeta\right) = \frac{\Lambda}{2} k_m \int_0^1 A^m\left(\frac{1}{k_m}, \eta'\right) \frac{R^m(\eta', \zeta)\eta' d\eta'}{1 - k_m \eta'} + \frac{\Lambda}{2} k_m e^{-k_m \tau_0} \int_0^1 A^m\left(\frac{1}{k_m}, -\eta'\right) \frac{R^m(\eta', \zeta)\eta' d\eta'}{1 + k_m \eta'} , \tag{3.73}$$

$$r_1^m\left(\frac{1}{k_m}, \zeta\right) = \frac{\Lambda}{2} k_m \int_0^1 A^m\left(\frac{1}{k_m}, \eta'\right) \frac{r^m(\eta', \zeta)\eta' d\eta'}{1 - k_m \eta'} - \frac{\Lambda}{2} k_m e^{-k_m \tau_0} \int_0^1 A^m\left(\frac{1}{k_m}, -\eta'\right) \frac{r^m(\eta', \zeta)\eta' d\eta'}{1 + k_m \eta'} . \tag{3.74}$$

Let us assume that the considered atmosphere of a finite optical thickness is located in deep layers of a semi-infinite medium with the same physical properties. Then, by analogy with results in [187] instead of (3.73–3.74), we find the following relationships:

$$\begin{aligned} &2\int_0^1 R^m(\eta, \eta') i^m(\eta')\eta' d\eta' + 2e^{-k_m \tau_0} \int_0^1 R^m(\eta, \eta') i^m(-\eta')\eta' d\eta' \\ &\quad = [1 - e^{-\tau_0(k_m + 1/\eta)}] i^m(-\eta) + [e^{-k_m \tau_0} - e^{-\tau_0/\eta}] i^m(\eta) , \\ &2\int_0^1 r^m(\eta, \eta') i^m(\eta')\eta' d\eta' - 2e^{-k_m \tau_0} \int_0^1 r^m(\eta, \eta') i^m(-\eta')\eta' d\eta' \\ &\quad = [1 - e^{-\tau_0(k_m + 1/\eta)}] i^m(-\eta) - [e^{-k_m \tau_0} - e^{-\tau_0/\eta}] i^m(\eta) , \end{aligned} \tag{3.75}$$

where $i^m(\eta)$ function is

$$i^m(\eta) = \frac{A^m(1/k_m, \eta)}{1 - k_m \eta} . \tag{3.76}$$

Note, that (3.69–3.70) can be written as one linear integral equation in the form of a matrix:

$$\hat{R}^m(\eta,\zeta)T^m(\eta) = \frac{\Lambda}{2}\int_0^1 \hat{R}^m(\eta',\zeta)\,\frac{A^m(\eta,\eta')\eta'd\eta'}{\eta'-\eta}$$
$$-\hat{\beta}\,\frac{\Lambda}{2}e^{-\tau_0/\eta}\int_0^1 \hat{R}^m(\eta',\zeta)\,\frac{A^m(\eta,-\eta')\eta'd\eta'}{\eta+\eta'} + \hat{R}_1^m(\eta,\zeta) \tag{3.77}$$

where matrix functions $\hat{R}^m$, $\hat{R}_1^m$ and $\hat{\beta}$ are

$$\hat{R}^m(\eta,\zeta) = \begin{pmatrix} R^m(\eta,\zeta) \\ r^m(\eta,\zeta) \end{pmatrix}, \qquad \hat{\beta} = \begin{pmatrix} 1 & 0 \\ 0 & -1 \end{pmatrix},$$
$$\hat{R}_1^m(\eta,\zeta) = \begin{pmatrix} A^m(\eta,-\zeta)\varrho_1^m(\eta,\zeta) + A^m(\eta,\zeta)\sigma_1^m(\eta,\zeta) \\ A^m(\eta,-\zeta)\varrho_1^m(\eta,\zeta) - A^m(\eta,\zeta)\sigma_1^m(\eta,\zeta) \end{pmatrix}. \tag{3.78}$$

Using (3.80), the $E^m(\eta,\zeta)$ function can be transformed into the following matrix:

$$\hat{E}^m(\eta,\zeta) = \begin{pmatrix} \eta & 0 \\ 0 & \zeta \end{pmatrix}\begin{pmatrix} R^m(\eta,\zeta) \\ r^m(\eta,\zeta) \end{pmatrix} = \begin{pmatrix} \eta & 0 \\ 0 & \zeta \end{pmatrix}\hat{R}^m(\eta,\zeta)\ . \tag{3.79}$$

Thus, the determination of the $E^m(\eta,\zeta)$ function is totally equivalent to the determination of the matrix function $\hat{R}^m(\eta,\zeta)$.

Using the determination of functions Ω_i^m and Θ_i^m, W^m and U^m, and systems of linear integral equations, corresponding to the Ambartsumyan φ_i^m and ψ_i^m functions and the Chandrasekhar X^m and Y^m functions [339–341], we obtain, by transformations similar to (3.77):

$$\hat{\Omega}_i^m(\eta)T^m(\eta) = \frac{\Lambda}{2}\eta\int_0^1 \hat{\Omega}_i^m(\eta')\,\frac{A^m(\eta,\eta')d\eta'}{\eta'-\eta}$$
$$-\hat{\beta}\,\frac{\Lambda}{2}\eta e^{-\tau_0/\eta}\int_0^1 \hat{\Omega}_i^m(\Omega_i^m(\eta')\,\frac{A^m(\eta,-\eta')d\eta'}{\eta+\eta'} + \hat{f}_i^m(\eta)\ , \tag{3.80}$$

where matrix functions $\hat{\Omega}_i^m$ and $\hat{f}_i^m$ are

$$\hat{\Omega}_i^m(\eta) = \begin{pmatrix} \Omega_i^m(\eta) \\ \Theta_i^m(\eta) \end{pmatrix} \hat{f}_i^m(\eta) = R_i^m(\eta)\begin{pmatrix} 1+(-1)^{i+m}e^{-\tau_0/\eta} \\ 1-(-1)^{i+m}e^{-\tau_0/\eta} \end{pmatrix}. \tag{3.81}$$

Linear integral equations as matrices for W^m and U^m functions, are written in the following way:

$$\hat{X}^m(\eta)T^m(\eta) = \eta\int_0^1 \hat{X}^m(\eta')\frac{\psi^m(\eta')d\eta'}{\eta'-\eta}$$
$$-\hat{\beta}\eta e^{-\tau_0/\eta}\int_0^1 \hat{X}^m(\eta')\frac{\psi^m(\eta')d\eta'}{\eta+\eta'} + \hat{\varepsilon}\ , \tag{3.82}$$

where $\hat{X}^m$ and $\hat{\varepsilon}$ matrix functions are

$$\hat{X}^m(\eta) = \begin{pmatrix} W^m \\ U^m \end{pmatrix}, \qquad \hat{\varepsilon} = \begin{pmatrix} 1+e^{-\tau_0/\eta} \\ 1-e^{-\tau_0/\eta} \end{pmatrix}. \tag{3.83}$$

Similarly to (3.73 – 3.74), additional conditions needed for single solutions of (3.80) and (3.82), are:

$$\begin{aligned}&\frac{\Lambda}{2}\int_0^1 \Omega_i^m(\eta)i^m(\eta)d\eta+\frac{\Lambda}{2}e^{-k_m\tau_0}\int_0^1 \Omega_i^m(\eta)i^m(-\eta)d\eta\\&\quad=R_i^m\left(\frac{1}{k_m}\right)[1+(-1)^{i+m}e^{-k_m\tau_0}]\ ,\\&\frac{\Lambda}{2}\int_0^1 \Theta_i^m(\eta)i^m(\eta)d\eta-\frac{\Lambda}{2}e^{-k_m\tau_0}\int_0^1 \Theta_i^m(\eta)i^m(-\eta)d\eta\\&\quad=R_i^m\left(\frac{1}{k_m}\right)[1-(-1)^{i+m}e^{-k_m\tau_0}]\ .\end{aligned} \tag{3.84}$$

$$\begin{aligned}&\int_0^1 W^m(\eta)\frac{\psi^m(\eta)}{1-k_m\eta}d\eta+e^{-k_m\tau_0}\int_0^1 W^m(\eta)\frac{\psi^m(\eta)}{1+k_m\eta}d\eta=1+e^{-k_m\tau_0}\\&\int_0^1 U^m(\eta)\frac{\psi^m(\eta)}{1-k_m\eta}d\eta-e^{-k_m\tau_0}\int_0^1 U^m(\eta)\frac{\psi^m(\eta)}{1+k_m\eta}d\eta=1-e^{-k_m\tau_0}\ .\end{aligned} \tag{3.85}$$

The linear integral equations obtained for the basic functions Θ_i^m and Ω_i^m, R^m and r^m, W^m and U^m have in practice the same structure which is determined by their kernels. In fact, they do not contain singularities and can be solved with the help to technique described, for instance, in [34, 140]. In particular, to solve them efficiently, integral terms can be substituted by the Gauss sum and iterative schemes can be used, where free terms of the obtained integral equations are used as zero approximation. For instance, in the numerical solution of (3.72) this iterative scheme is

$$\hat{R}_{(0)}^m(\eta,\zeta)=\hat{R}_i^m(\eta,\zeta)\ , \tag{3.86}$$

$$\begin{aligned}&\left[T^m(\eta)-\frac{\Lambda}{2}\eta\int_0^1\frac{A^m(\eta,\eta')}{\eta'-\eta}d\eta'\right]\hat{R}_{(n+1)}^m(\eta,\zeta)\\&\quad=\hat{R}_i^m(\eta,\zeta)+\frac{\Lambda}{2}\int_0^1 A^m(\eta,\eta')\hat{R}_{(n)}^m(\eta',\zeta)d\eta'\\&\quad\quad+\frac{\Lambda}{2}\eta\int_0^1 A^m(\eta,\eta')\frac{\hat{R}_{(n)}^m(\eta',\zeta)-\hat{R}_{(n)}^m(\eta,\zeta)}{\eta'-\eta}d\eta'\\&\quad\quad-\beta\frac{\Lambda}{2}e^{-\tau_0/\eta}\int_0^1 A^m(\eta,-\eta')\hat{R}_{(n)}^m(\eta',\zeta)\frac{\eta' d\eta'}{\eta+\eta'}\ .\end{aligned} \tag{3.87}$$

In addition, $A^m(\eta,\zeta)$ functions can be determined not from an expansion of a given phase function $x(\gamma)$ in Legendre polynomials but by solutions of the equation [187]:

$$A^m(\eta,\zeta)=p^m(\eta,\zeta)-\frac{\Lambda}{2}\eta\int_{-1}^1 A^m(\eta,\eta')\frac{p^m(\eta,\zeta)-p^m(\eta',\zeta)}{\eta-\eta'}d\eta' \tag{3.88}$$

where p^m functions are

$$p^m(\eta,\zeta) = \frac{1}{2\pi}\int_0^{2\pi} x(\gamma)\cos m\varphi\, d\varphi \ . \qquad (3.89)$$

Note, that linear integral equations of the kind shown above can also be obtained from $E^m(\eta,\zeta)$ functions. Substituting the upper line of (3.51) into (3.65–3.66), summing them and making some transformations, we obtain:

$$\begin{aligned} E^m(\eta,\zeta)T^m(\eta) &= [A^m(\eta,-\zeta)\varrho_1(\eta,\zeta)(\eta+\zeta) + A^m(\eta,\zeta)\sigma_1(\eta,\zeta)(\eta-\zeta)] \\ &+\frac{\Lambda}{2}\int_0^1 A^m(\eta,\eta')\left[\zeta\frac{E^m(\eta',\zeta)\zeta - E^m(\zeta,\eta')\eta'}{\zeta^2-\eta'^2} + \frac{E^m(\eta',\zeta)\eta}{\eta'-\eta}\right]d\eta' \\ &-\frac{\Lambda}{2}e^{-\tau_0/\eta}\int_0^1 A^m(\eta,-\eta')\left[\zeta\frac{E^m(\zeta,\eta')\eta' - E^m(\eta',\zeta)\zeta}{\zeta^2-\eta'^2} + \frac{E^m(\eta',\zeta)\eta}{\eta'+\eta}\right]d\eta' \ . \end{aligned} \qquad (3.90)$$

Regulating the right-hand part of the unique equation (3.90), we obtain:

$$\begin{aligned} E^m(\eta,\zeta)&\left[T^m(\eta) - \frac{\Lambda}{2}\eta\int_0^1 \frac{A^m(\eta,\eta')}{\eta'-\eta}d\eta'\right] \\ &= [A^m(\eta,-\zeta)\varrho_1(\eta,\zeta)(\eta+\zeta) + A^m(\eta,\zeta)\sigma_1(\eta,\zeta)(\eta-\zeta)] \\ &+\frac{\Lambda}{2}\int_0^1 A^m(\eta,\eta')\left[\zeta\frac{E^m(\eta',\zeta)\zeta - E^m(\zeta,\eta')\eta'}{\zeta^2-\eta'^2}\right. \\ &\left.+\eta\frac{E^m(\eta',\zeta) - E^m(\eta,\zeta)}{\eta'-\eta}\right]d\eta' - \frac{\Lambda}{2}e^{-\tau_0/\eta}\int_0^1 A^m(\eta,-\eta') \\ &\times\left[\zeta\frac{E^m(\zeta,\eta')\eta' - E^m(\eta',\zeta)\zeta}{\zeta^2-\eta'^2} + \frac{E^m(\eta',\zeta)\eta}{\eta'+\eta}\right]d\eta' \ . \end{aligned} \qquad (3.91)$$

We obtain similar equations if the bottom line of (3.51) is substituted into linear integral Eqs. (3.69–3.70). To ensure unique solutions for (3.90–3.91), they should be enhanced with additional relationships, as shown in [341]. For this purpose, in (3.90–3.91) $\eta = 1/k_m$ should be substituted and $T^m(1/k_m)$ taken to be zero. The iterative scheme of numerical solution of (3.90–3.91), is similar to that just shown. For instance, for (3.91) it is

$$\begin{aligned} E^m_{(n+1)}(\eta,\zeta)&\left[T^m(\eta) - \frac{\Lambda}{2}\eta\int_0^1 \frac{A^m(\eta,\eta')}{\eta'-\eta}d\eta'\right] \\ &= [A^m(\eta,-\zeta)\varrho_1(\eta,\zeta)(\eta+\zeta) + A^m(\eta,\zeta)\sigma_1(\eta,\zeta)(\eta-\zeta)] \\ &+\frac{\Lambda}{2}\int_0^1 A^m(\eta,\eta')\left[\zeta\frac{E^m_{(n)}(\eta',\zeta)\zeta - E^m_{(n)}(\zeta,\eta')\eta'}{\zeta^2-\eta'^2}\right. \end{aligned}$$

$$+\eta\,\frac{E^m_{(n)}(\eta',\zeta)-E^m_{(n)}(\eta,\zeta)}{\eta'-\eta}\Bigg]\,d\eta'-\frac{\Lambda}{2}e^{-\tau_0/\eta}\int_0^1 A^m(\eta,\eta')$$

$$\times\left[\zeta\,\frac{E^m_{(n)}(\zeta,\eta')\eta'-E^m_{(n)}(\eta',\zeta)\zeta}{\zeta^2-\eta'^2}+\frac{E^m_{(n)}(\eta',\zeta)\eta}{\eta'+\eta}\right]d\eta' \tag{3.92}$$

where zero approximation $E^m_{(0)}(\eta,\zeta)$ is chosen to be

$$E^m_{(0)}(\eta,\zeta)=A^m(\eta,\zeta)\varrho^m_1(\eta,\zeta)(\eta+\zeta)+A^m(\eta,\zeta)\sigma_1(\eta,\zeta)(\eta-\zeta)\ . \tag{3.93}$$

Calculations of the iterative scheme (3.92 – 3.93) differs principally from (3.86 – 3.87). Calculation of iterations (3.92 – 3.93) can lead to larger computer time expenditures but to a more acceptable presentation of numerical data compared to symmetrical functions $R^m_{(n)}(\eta,\zeta)$ and $r^m_{(n)}(\eta,\zeta)$. This problem must be considered separately, based on definite calculations of these functions.

3.7 Application of the Specular Reflection Principle in the Theory of Radiation Transfer

The analysis just described does not answer the major question: why does an introduction of algebraic combinations of summation and extraction of classical functions lead to a simpler and rational form of presentation of this theory, simultaneously simplifying its calculation algorithms? The answer to this question is of great interest for the whole radiation transfer theory. Thus, we shall consider a flat homogeneous layer of a finite optical thickness τ_0, with an arbitrary location of isotropic external and internal sources, the power of which depends only on the optical depth (Fig. 3.2).

Let us select the line of geometrical symmetry of this layer at a level of $1/2\,\tau_0$. We shall also consider an arbitrary level of optical thickness τ and a level of optical thickness $\tau_0-\tau$ symmetrical to it (in the middle of the layer).

Various problems of classical radiation transfer theory are solved using a technique proposed by Sobolev [184] and based on the concept of photons' exit from a medium, after multiple scattering. Based on probability theory, Sobolev developed a general statistical interpretation concerning the phenomenological theory of radiation transfer. In particular, he solved the problem of radiation diffusion in a flat homogeneous layer for isotropic scattering with any sources where power depends only on depth [187]. Minin [129, 130] applied probability techniques to consider the problem of radiation diffusion in a flat homogeneous layer with non-isotropic scattering. According to [130], we introduce probability $p(\tau,\eta',\ \eta,\varphi'-\varphi,\tau_0)d\omega$ that a photon, moving in a direction with an azimuth φ' at an angle $\theta_0=\arccos\eta'$ to the external normal at the upper boundary ($\tau=0$) and absorbed at an optical depth τ, will leave the medium through the boundary $\tau=0$ at an angle $\theta=\arccos\eta$

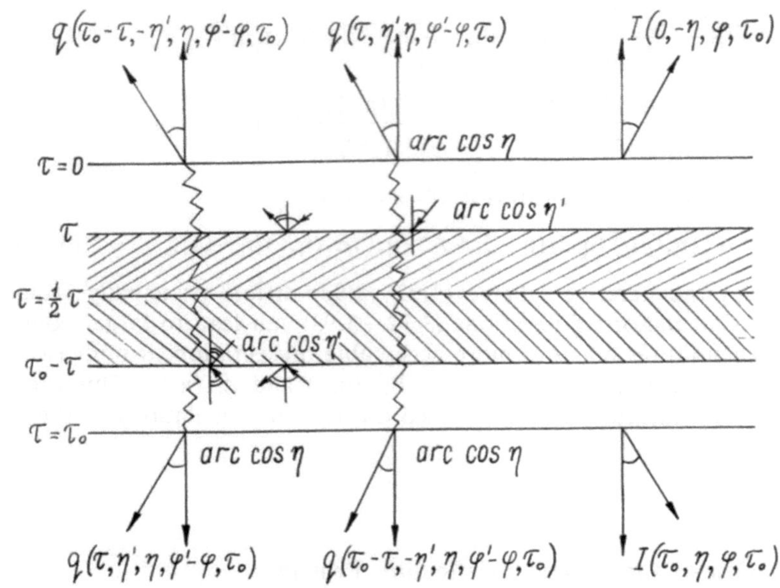

Fig. 3.2. Probability equivalence of symmetrical levels τ and $\tau_0-\tau$ in mirrored directions

to the normal and with an azimuth φ within the solid angle $d\omega$. Consider also probability $q(\tau,\eta',\eta,\varphi'-\varphi,\tau_0)d\omega$ that a photon emitted at an optical thickness τ in the direction determined by an angle arc cos η' with respect to the external normal, with an azimuth φ', will leave the medium, upon multiple scattering, through the upper boundary $\tau=0$ at an angle arc cos η to the normal and with an azimuth φ within the solid angle $d\omega$. Considering the meaning in probability terms of phase function $x(\gamma)$ and Λ in [130], relationships were obtained between $p(\tau,\eta',\eta,\varphi'-\varphi,\tau_0)$ and $q(\tau,\eta',\eta,\varphi'-\varphi,\tau_0)$:

$$p(\tau,\eta',\eta,\varphi'-\varphi,\tau_0)=\frac{\Lambda}{4\pi}\int_0^{2\pi}d\varphi''\int_{-1}^{+1}x(\gamma)q(\tau,\eta'',\eta,\varphi''-\varphi,\tau_0)d\eta'' , \tag{3.94}$$

$$q(\tau,\eta',\eta,\varphi'-\varphi,\tau_0)=\delta(\eta'-\eta)\delta(\varphi'-\varphi)e^{-\tau/\eta} + \int_0^{\tau}e^{-(\tau-\tau')/\eta'}p(\tau',\eta',\eta,\varphi'-\varphi,\tau_0)\frac{d\tau'}{\eta'} \quad (\eta'>0) , \tag{3.95}$$

$$q(\tau,\eta',\eta,\varphi'-\varphi,\tau_0)=-\int_{\tau}^{\tau_0}e^{-(\tau-\tau')/\eta'}p(\tau',\eta',\eta,\varphi'-\varphi,\tau_0)\frac{d\tau'}{\eta'} \quad (\eta'<0) . \tag{3.96}$$

Let us now consider for each of symmetrical levels τ and $\tau_0-\tau$ the q_{up} and q_{down} probabilities that photons emitted at these levels in reflected directions will leave the medium, respectively, through the upper ($\tau=0$) and lower ($\tau=\tau_0$) boundaries, also in reflected directions, relative to level $1/2\,\tau_0$ (Fig. 3.2).

Bearing in mind the specular symmetry of the problem at level $1/2\,\tau_0$, we obtain the following relationships of the probability equivalency at symmetrical levels τ and $\tau_0-\tau$:

$$q_{\text{up}}(\tau,\eta',\eta,\varphi'-\varphi,\tau_0)=q_{\text{down}}(\tau_0-\tau,\,-\eta',\eta,\varphi'-\varphi,\tau_0)\ , \tag{3.97}$$

$$q_{\text{down}}(\tau,\eta',\eta,\varphi'-\varphi,\tau_0)=q_{\text{up}}(\tau_0-\tau,\,-\eta',\eta,\varphi'-\varphi,\tau_0)\ . \tag{3.98}$$

It follows from (3.97 – 3.98) that photons emitted in specular directions at symmetric levels τ and $\tau_0-\tau$, have similar total probability of exit in specular directions (Fig. 3.2):

$$\begin{aligned}&q_{\text{up}}(\tau,\eta',\eta,\varphi'-\varphi,\tau_0)+q_{\text{down}}(\tau,\eta',\eta,\varphi'-\varphi,\tau_0)\\&\quad=q_{\text{down}}(\tau_0-\tau,\,-\eta',\eta,\varphi'-\varphi,\tau_0)+q_{\text{up}}(\tau_0-\tau,\,-\eta',\eta,\varphi'-\varphi,\tau_0)=q_1^+\ .\end{aligned} \tag{3.99}$$

Equation (3.99) determines the first probability invariant of the problem. The second invariant, which follows from (3.97 – 3.98),

$$\begin{aligned}&q_{\text{up}}(\tau,\eta',\eta,\varphi'-\varphi,\tau_0)+q_{\text{up}}(\tau_0-\tau,\,-\eta',\eta,\varphi'-\varphi,\tau_0)\\&\quad=q_{\text{down}}(\tau,\eta',\eta,\varphi'-\varphi,\tau_0)+q_{\text{down}}(\tau_0-\tau,\,-\eta',\eta,\varphi'-\varphi,\tau_0)\end{aligned} \tag{3.100}$$

means that total probabilities of exit in specular directions through the lower ($\tau=0$) or upper ($\tau=\tau_0$) boundaries of the medium for the photons emitted in specular directions at symmetric levels τ and $\tau_0-\tau$ are equal.

From (3.100) there follow the third and the fourth probability invariants q_1^- and q_2^-:

$$\begin{aligned}&q_{\text{up}}(\tau,\eta',\eta,\varphi'-\varphi,\tau_0)-q_{\text{down}}(\tau,\eta',\eta,\varphi'-\varphi,\tau_0)\\&\quad=q_{\text{down}}(\tau_0-\tau,\,-\eta',\eta,\varphi'-\varphi,\tau_0)-q_{\text{up}}(\tau_0-\tau,\,-\eta',\eta,\varphi'-\varphi,\tau_0)=q_1^-\\&q_{\text{up}}(\tau,\eta',\eta,\varphi'-\varphi,\tau_0)-q_{\text{up}}(\tau_0-\tau,\,-\eta',\eta,\varphi'-\varphi,\tau_0)\\&\quad=q_{\text{down}}(\tau_0-\tau,\,-\eta',\eta,\varphi'-\varphi,\tau_0)-q_{\text{down}}(\tau,\eta',\eta,\varphi'-\varphi,\tau_0)=q_2^-\ .\end{aligned} \tag{3.101}$$

Equation (3.101) mean that the differences of the probabilities of exit for the photons emitted in specular directions at arbitrary symmetrical levels τ and $\tau_0-\tau$ are equal in module for both the lower and upper boundaries in specular directions of photon exit. In (3.97 – 3.101) an angle $\arccos\eta$ at the upper ($\tau=0$) and lower ($\tau=\tau_0$) boundaries of the medium is measured from the external normal to the layer.

A commutative invariance of (3.97 – 3.98) in the permutation results from the spatial optical symmetry of a homogeneous flat layer (or a layer with the symmetrical position of inhomogeneities) in the permutation of the specular levels τ and $\tau_0-\tau$. Thus, probability invariants $q_1^+, q_2^+, q_1^-, q_2^-$ are caused by the use of this method to determine the optical fields in the media with a plane (line) of optical and spatial specular symmetries.

Apparently, if a 3-D medium is characterized by specular symmetry of its physical properties, then the invariance of total probability q_1^+ and probability difference q_1^- does not depend on the geometry of the medium, its inhomogeneity, location of energy sources and their power. For the $q_1^+(\tau,\eta',\eta,\varphi'-\varphi,\tau_0)$ function, the following principle of probabilistic equivalency can be formulated: total probability of photon's exit in specular directions will not change at symmetrical levels τ and $\tau_0-\tau$, if, simultaneously, a layer with a small optical thickness $\Delta\tau$ is either added to or extracted from these levels from the external or internal side. These considerations hold also for a probability function $p(\tau,\eta',\eta,\varphi'-\varphi,\tau_0)$ and for the respective probability invariants p_1^+, p_2^+, p_1^-, and p_2^-.

Major probability invariants q_1^+ and q_1^- simplify their determination, compared to traditional probabilities, of reaching the upper or lower boundary of the medium by a photon. The respective photometric invariants equal to the sum and difference of diffusively transmitted and reflected radiation in specular directions at symmetrical levels τ and $\tau_0-\tau$ are determined through probability invariants q_1^+ and q_1^-.

In the initial problem of radiation transfer (Fig. 3.2), the photometric invariants are not shown explicitly. In fact, at levels τ and $\tau_0-\tau$ diffused radiation has different angular structure. As a result, the diffusively reflected radiation $I(0,-\eta,\varphi,\tau_0)$ goes out at the upper boundary of the medium $(\tau=0)$, and the diffusively transmitted radiation $I(\tau_0,\eta,\varphi,\tau_0)$ leaves the medium at the lower boundary $(\tau=\tau_0)$:

$$I(0,-\eta,\varphi,\tau_0)=\int_0^{\tau_0}\int_{-1}^{1}\int_0^{2\pi} g(\tau,\eta',\varphi')q(\tau,\eta',\eta,\varphi'-\varphi,\tau_0)\,d\eta'\,d\varphi\,\frac{d\tau}{\eta} \tag{3.102}$$

$$I(\tau_0,\eta,\varphi,\tau_0)=\int_0^{\tau_0}\int_{-1}^{1}\int_0^{2\pi} g(\tau,\eta',\varphi')q(\tau_0-\tau,-\eta',\eta,\varphi'-\varphi,\tau_0)\,d\eta'\,d\varphi'\,\frac{d\tau}{\eta} \tag{3.103}$$

where: $g(\tau,\eta,\varphi)\,d\tau\,d\omega$ is an amount of radiative energy emitted per unit of time within a solid angle $d\omega$ by radiation sources at an optical depth τ; at an angle $\arccos\eta$ to the external normal to the layer; and at an azimuth φ [130]. In (3.102) and (3.103), the angle $\arccos\eta$ is read from the external normal to the layer at $\tau=0$ and $\tau=\tau_0$. In specular directions concerning the symmetry axis, the values of intensities $I(0,-\eta,\varphi,\tau_0)$ and $I(\tau_0,\eta,\varphi,\tau_0)$ are not equal. Summing and extracting (3.102) and (3.103), we obtain (Fig. 3.3):

$$\begin{aligned} I^+ &= I(0,-\eta,\varphi,\tau_0)+I(\tau_0,\eta,\varphi,\tau_0)\\ &=\int_0^{\tau_0}\int_{-1}^{1}\int_0^{2\pi} g(\tau,\eta',\varphi')q_1^+(\tau,\eta',\eta,\varphi'-\varphi,\tau_0)\,d\eta'\,d\varphi'\,\frac{d\tau}{\eta}\,, \end{aligned} \tag{3.104}$$

$$\begin{aligned} I^- &= I(0,-\eta,\varphi,\tau_0)-I(\tau_0,\eta,\varphi,\tau_0)\\ &=\int_0^{\tau_0}\int_{-1}^{1}\int_0^{2\pi} g(\tau,\eta',\varphi')q_1^-(\tau,\eta',\eta,\varphi'-\varphi,\tau_0)\,d\eta'\,d\varphi'\,\frac{d\tau}{\eta}\,. \end{aligned} \tag{3.105}$$

Formulas (3.104 – 3.105) show how probability invariants q_1^+ and q_1^- are related to photometric invariants I^+ and I^-. It will be shown below that

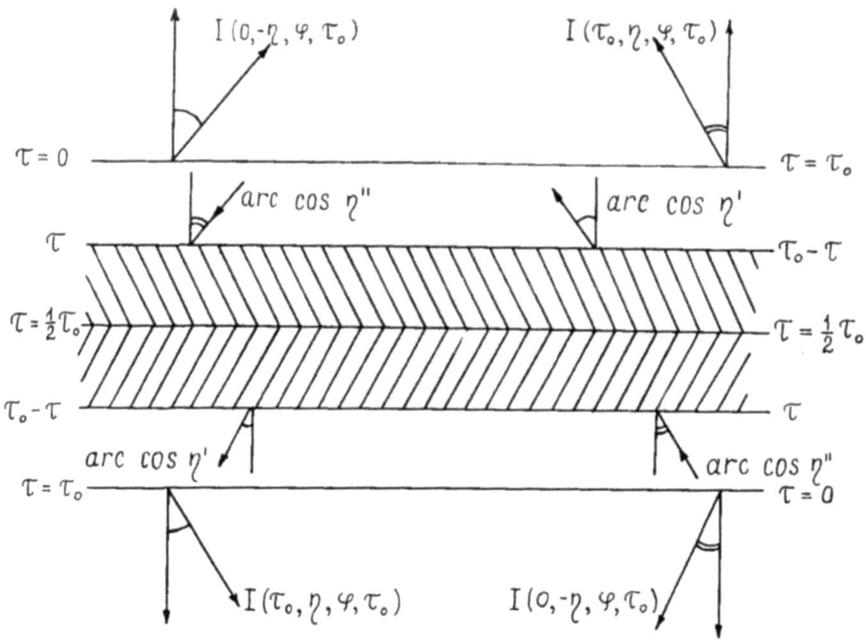

Fig. 3.3. Photometric symmetrization of the problem of radiation transfer in a plane-parallel atmosphere of a finite optical thickness τ_0

photometric invariants I^+ and I^-, in contrast to intensities $I(0,-\eta,\varphi,\tau_0)$ and $I(\tau_0,\eta,\varphi,\tau_0)$, are determined from independent linear integral equations.

Considering the initial problem of radiation transfer for a planetary atmosphere where its upper boundary is illuminated by parallel sun rays at an angle arc $\cos\zeta$; we obtain the sum of the intensities of the upward and downward radiation fluxes as and the module of their difference at levels τ and $\tau_0-\tau$, for specular directions, then the invariants at a mutual inversion of symmetrical levels in a flat layer are:

$$\begin{cases} I^+(\tau,\eta,\zeta,\varphi,\tau_0) = I(\tau,\eta,\zeta,\varphi,\tau_0) + I(\tau_0-\tau,\,-\eta,\zeta,\varphi,\tau_0) \\ I^-(\tau,\eta,\zeta,\varphi,\tau_0) = I(\tau,\eta,\zeta,\varphi,\tau_0) - I(\tau_0-\tau,\,-\eta,\zeta,\varphi,\tau_0) \end{cases} \tag{3.106}$$

$$\begin{cases} I^+(\tau,\eta,\zeta,\varphi,\tau_0) = I^+(\tau_0-\tau,\,-\eta,\zeta,\varphi,\tau_0) \\ I^-(\tau,\eta,\zeta,\varphi,\tau_0) = -I^-(\tau_0-\tau,\,-\eta,\zeta,\varphi,\tau_0) \end{cases} \tag{3.107}$$

Relationships similar to (3.106 – 3.107) hold also for the source function $B(\tau,\eta,\zeta,\varphi,\tau_0)$ of the initial problem:

$$\begin{cases} B^+(\tau,\eta,\zeta,\varphi,\tau_0) = B(\tau,\eta,\zeta,\varphi,\tau_0) + B(\tau_0-\tau,\,-\eta,\zeta,\varphi,\tau_0) \\ B^-(\tau,\eta,\zeta,\varphi,\tau_0) = B(\tau,\eta,\zeta,\varphi,\tau_0) - B(\tau_0-\tau,\,-\eta,\zeta,\varphi,\tau_0) \end{cases} \tag{3.108}$$

$$\begin{cases} B^+(\tau,\eta,\zeta,\varphi,\tau_0) = B^+(\tau_0-\tau,\,-\eta,\zeta,\varphi,\tau_0) \\ B^-(\tau,\eta,\zeta,\varphi,\tau_0) = -B^-(\tau_0-\tau,\,-\eta,\zeta,\varphi,\tau_0)\;. \end{cases} \tag{3.109}$$

Thus, photometric invariants I^+ and $|I^-|$ are the sought-for intensities in the photometrically symmetrical problems of radiation transfer.

Using linear integral equations for the azimuthal harmonics of the source function $B(\tau,\eta,\zeta,\varphi,\tau_0)$ obtained by Mallikin [340], we obtain the following independent equation for the invariants B_m^+ and B_m^- (3.108):

$$\begin{aligned} B_m^+(\tau,\eta,\zeta,\tau_0)T^m(\zeta) &= \frac{\Lambda S}{4}[A^m(\zeta,\eta)e^{-\tau/\zeta}+A^m(\zeta,-\eta)e^{-(\tau_0-\tau)/\zeta}] \\ &+\frac{\Lambda}{2}\zeta\int_0^1\frac{A^m(\zeta,\eta')}{\eta'-\eta}B_m^+(\tau,\eta,\eta',\tau_0)d\eta' \\ &-\frac{\Lambda}{2}\zeta e^{-\tau_0/\zeta}\int_0^1\frac{A^m(\zeta,-\eta')}{\eta'+\zeta}B_m^+(\tau,\eta,\eta',\tau_0)d\eta' \end{aligned} \quad (3.110)$$

$$\begin{aligned} B_m^-(\tau,\eta,\zeta,\tau_0)T^m(\zeta) &= \frac{\Lambda S}{4}[A^m(\zeta,\eta)e^{-\tau/\zeta}-A^m(\zeta,-\eta)e^{-(\tau_0-\tau)/\zeta}] \\ &+\frac{\Lambda}{2}\zeta\int_0^1\frac{A^m(\zeta,\eta')}{\eta'-\zeta}B_m^-(\tau,\eta,\eta',\tau_0)d\eta' \\ &+\frac{\Lambda}{2}\zeta e^{-\tau_0/\zeta}\int_0^1\frac{A^m(\zeta,-\eta')}{\eta'+\zeta}B_m^-(\tau,\eta,\eta',\tau_0)d\eta' \end{aligned} \quad (3.111)$$

It is sufficient to determine the $B_m^\pm$ functions, like other photometric invariants, only for the interval $[0,1/2\tau_0]$ at $\eta\in[-1,1]$ or for the interval $[0,\tau_0]$ but at $\eta\in[0,1]$. From (3.110–3.111), simply transformed, we obtain independent linear equations for the azimuthal harmonics of photometric invariants I_m^+ and I_m^- (3.106):

$$\begin{aligned} I_m^+(\tau,\eta,\zeta,\tau_0)T^m(\zeta) &= f_m^+(\tau,\eta,\zeta,\tau_0)+\frac{\Lambda}{2}\zeta\int_0^1\frac{A^m(\zeta,\eta')}{\eta'-\zeta}I_m^+(\tau,\eta,\eta',\tau_0)d\eta' \\ &-\frac{\Lambda}{2}\zeta e^{-\tau_0/\zeta}\int_0^1\frac{A^m(\zeta,-\eta')}{\eta'+\zeta}I_m^+(\tau,\eta,\eta',\tau_0)d\eta' , \\ I_m^-(\tau,\eta,\zeta,\tau_0)T^m(\zeta) &= f_m^-(\tau,\eta,\zeta,\tau_0)+\frac{\Lambda}{2}\zeta\int_0^1\frac{A^m(\zeta,\eta')}{\eta'-\zeta}I_m^-(\tau,\eta,\eta',\tau_0)d\eta' \\ &+\frac{\Lambda}{2}\zeta e^{-\tau_0/\zeta}\int_0^1\frac{A^m(\zeta,-\eta')}{\eta'+\zeta}I_m^-(\tau,\eta,\eta',\tau_0)d\eta' \end{aligned} \quad (3.112)$$

where functions $f_m^\pm(\tau,\eta,\zeta,\tau_0)$ are:

$$f_m^\pm(\tau,\eta,\zeta,\tau_0) = f^m(\tau,\eta,\zeta,\tau_0)\pm f^m(\tau_0-\tau,-\eta,\zeta,\tau_0)$$

$$f^m(\tau,\eta,\zeta,\tau_0) = \begin{cases} \dfrac{\Lambda S}{4}A^m(\zeta,\eta)\zeta\dfrac{e^{-\tau/\zeta}-e^{-\tau/\eta}}{\zeta-\eta} , & \eta>0 \\ \dfrac{\Lambda S}{4}A^m(\zeta,\eta)\dfrac{\zeta}{\zeta-\eta}e^{-\tau/\zeta}[1-e^{-(1/\zeta-1/\eta)(\tau_0-\tau)}] , & \eta<0 . \end{cases} \quad (3.113)$$

Using a correlation of the photon exit $p(\tau,\eta',\eta,\varphi'-\varphi,\tau_0)$ with the source function $B(\tau,\eta',\eta,\varphi'-\varphi,\tau_0)$ applied in considering the radiance of a flat layer with an optical thickness τ_0 illuminated at the upper boundary by the flux of parallel rays [130], we obtain:

$$\begin{aligned} B(\tau,\eta',\eta,\varphi'-\varphi,\tau_0) &= \pi S p(\tau,\eta',\eta,\varphi'-\varphi,\tau_0) \\ &= \pi S p^0(\tau,\eta',\eta,\tau_0) + 2\pi S \sum_{m=1}^{N} p^m(\tau,\eta',\eta,\tau_0)\cos m(\varphi'-\varphi) \ . \end{aligned} \tag{3.114}$$

From (3.114) based on [340], we easily obtain the following dependent linear integral equations for the azimuthal harmonics of the probabilities of exit $p^m(\tau,\eta',\eta,\tau_0)$ and $p^m(\tau_0-\tau,\eta',\eta,\tau_0)$:

$$\begin{aligned} p^m(\tau,\eta',\eta)T^m(\eta) &= \frac{\Lambda}{4\pi}A^m(\eta,\eta')e^{-\tau/\eta} \\ &\quad - \frac{\Lambda}{2}\eta\int_0^1 \frac{A^m(\eta,\eta'')}{\eta-\eta''}p^m(\tau,\eta',\eta'')d\eta'' - \frac{\Lambda}{2}\eta e^{-\tau_0/\eta} \\ &\quad \times \int_0^1 \frac{A^m(\eta,-\eta'')}{\eta+\eta''}p^m(\tau_0-\tau,-\eta',\eta'')d\eta'' \ , \end{aligned} \tag{3.115}$$

$$\begin{aligned} p^m(\tau_0-\tau,-\eta',\eta)T^m(\eta) &= \frac{\Lambda}{4\pi}A^m(\eta,-\eta')e^{-(\tau_0-\tau)/\eta} \\ &\quad - \frac{\Lambda}{2}\eta\int_0^1 \frac{A^m(\eta,\eta'')}{\eta-\eta''}p^m(\tau_0-\tau,\eta',\eta'')d\eta'' \\ &\quad - \frac{\Lambda}{2}\eta e^{-\tau_0/\eta}\int_0^1 \frac{A^m(\eta,-\eta'')}{\eta+\eta''}p^m(\tau,\eta',\eta'')d\eta'' \ . \end{aligned} \tag{3.116}$$

Summing and extracting (3.115) and (3.116), we obtain independent linear integral equations for probability invariants p_m^+ and p_m^- determined by relationships:

$$\begin{aligned} p_m^+(\tau,\eta,\eta',\tau_0) &= p^m(\tau,\eta',\eta,\tau_0) + p^m(\tau_0-\tau,-\eta',\eta,\tau_0) \\ p_m^-(\tau,\eta,\eta',\tau_0) &= p^m(\tau,\eta',\eta,\tau_0) - p^m(\tau_0-\tau,-\eta',\eta,\tau_0) \\ p_m^+(\tau,\eta,\eta',\tau_0) &= p_m^+(\tau_0-\tau,\eta,-\eta',\tau_0) \\ p_m^-(\tau,\eta,\eta',\tau_0) &= -p_m^-(\tau_0-\tau,\eta,-\eta',\tau_0) \ . \end{aligned} \tag{3.117}$$

Introducing the matrix

$$\hat{p}^m(\tau,\eta,\eta',\tau_0) = \begin{pmatrix} p_m^+ \\ p_m^- \end{pmatrix} \tag{3.118}$$

we obtain from (3.115 – 3.116) one matrix linear integral equation for $\hat{p}(\tau,\eta,\eta',\tau_0)$:

$$
\begin{aligned}
T^m(\eta)\hat{p}^m(\tau,\eta,\eta',\tau_0) = \hat{L}^m(\tau,\eta,\eta',\tau_0) & \\
&-\frac{\Lambda}{2}\eta\int_0^1 \frac{A^m(\eta,\eta'')}{\eta-\eta''}\hat{p}^m(\tau,\eta'',\eta',\tau_0)\,d\eta'' \\
&-\hat{\beta}^m\frac{\Lambda}{2}\eta e^{-\tau_0/\eta}\int_0^1 \frac{A^m(\eta,-\eta'')}{\eta+\eta''}\hat{p}^m(\tau,\eta'',\eta',\tau_0)\,d\eta''
\end{aligned} \tag{3.119}
$$

where matrix $\hat{\beta}^m$ is determined, using (3.78), and matrix $\hat{L}^m$ is:

$$
\hat{L}^m(\tau,\eta,\eta',\tau_0) = \frac{\Lambda}{4\pi}\begin{pmatrix} A^m(\eta,\eta')e^{-\tau/\eta} + A^m(\eta,-\eta')e^{-(\tau_0-\tau)/\eta} \\ A^m(\eta,\eta')e^{-\tau/\eta} - A^m(\eta,-\eta')e^{-(\tau_0-\tau)/\eta}\end{pmatrix} . \tag{3.120}
$$

Equations (3.110 – 3.111) and (3.119) are reduced to the Fredholm type [34], and in solving these one can efficiently substitute the integral terms with the Gauss sum or use the iterative schemes considered earlier. Additional relationships to ensure unique solutions are obtained similarly to (3.73 – 3.74). Using a relation of I and q [130]

$$
q(\tau,\eta',\eta,\varphi'-\varphi,\tau_0) = \frac{1}{\pi S} I(\tau,\eta',\eta,\varphi'-\varphi,\tau_0) \tag{3.121}
$$

independent linear integral equations for probability invariants q_+^m and q_-^m can easily be obtained from (3.112 – 3.113). Like other probability invariants, it is enough to determine them either for the interval $[0, \frac{1}{2}\tau_0]$ at $\eta\in[-1,1]$ or for $[0,\tau_0]$ at $\eta\in[0,1]$.

Note, that an asymmetrical function $E^m(\eta,\zeta)$ determined from (3.50), in accordance with its meaning, should lead to one photometric parameter – intensity I_Σ^m characterized by its photometric invariant $I_m^\pm$ and by one asymmetrical probability function P_Σ^m. It follows from (3.50) that functions I_Σ^m, $I_m^\pm$, E^m, and P_Σ^m are interrelated in the following way:

$$
I_\Sigma^m(\tau,\eta,\zeta,\tau_0) = \eta I_m^+(\tau,\eta,\zeta,\tau_0) + \zeta I_m^-(\tau,\eta,\zeta,\tau_0) \tag{3.122}
$$

$$
I_\Sigma^m(0,\eta,\zeta,\tau_0) = S\zeta E^m(\eta,\zeta,\tau_0) = \pi S\int_0^{\tau_0} e^{-\tau/\zeta} P_\Sigma^m(\tau,\eta,\zeta,\tau_0)\frac{d\tau}{\eta} \tag{3.123}
$$

where the probability function P_Σ^m is

$$
\begin{aligned}
P_\Sigma^m(\tau,\eta,\zeta,\tau_0) &= \eta\,[p^m(\tau,-\zeta,\eta,\tau_0) + p^m(\tau_0-\tau,\zeta,\eta,\tau_0)] \\
&\quad + \zeta\,[p^m(\tau,-\zeta,\eta,\tau_0) - p^m(\tau_0-\tau,\zeta,\eta,\tau_0)] \\
&= \eta p_m^+(\tau,\eta,\zeta,\tau_0) + \zeta p_m^-(\tau,\eta,\zeta,\tau_0) .
\end{aligned} \tag{3.124}
$$

Functions I_Σ^m and P_Σ^m make it possible to combine most rationally the photometric and probability invariants. Summing (3.124) into Fourier series, we introduce a single probability function P_Σ which generalizes the concept of probability of a photon's exit for a homogeneous plane-parallel planetary atmosphere

$$\begin{aligned}P_\Sigma(\tau,\eta,\zeta,\varphi,\tau_0) &= \eta\,[p(\tau,-\zeta,\eta,\varphi,\tau_0)+p(\tau_0-\tau,\zeta,\eta,\varphi,\tau_0)]\\ &\quad +\zeta\,[p(\tau,-\zeta,\eta,\varphi,\tau_0)-p(\tau_0-\tau,\zeta,\eta,\varphi,\tau_0)]\\ &= \eta\,p^+(\tau,\eta,\zeta,\varphi,\tau_0)+\zeta\,p^-(\tau,\eta,\zeta,\varphi,\tau_0)\ . \end{aligned}\tag{3.125}$$

The first item in (3.125) determines total probability of a photon's exit. The second item characterizes an asymmetry of the probability of a photon's exit through the lower and upper boundaries. A single probability function (3.125) can be transformed into a matrix

$$\hat{P}_\Sigma(\tau,\eta,\zeta,\varphi,\tau_0) = \begin{pmatrix}\eta, & 0\\ 0, & \zeta\end{pmatrix}\ , \tag{3.126}$$

where probability invariants p^+ and p^- are:

$$\begin{aligned}p^+(\tau,\eta,\zeta,\varphi,\tau_0) &= p(\tau,-\zeta,\eta,\varphi,\tau_0)+p(\tau_0-\tau,\zeta,\eta,\varphi,\tau_0)\\ p^-(\tau,\eta,\zeta,\varphi,\tau_0) &= p(\tau,-\zeta,\eta,\varphi,\tau_0)-p(\tau_0-\tau,\zeta,\eta,\varphi,\tau_0)\ .\end{aligned}\tag{3.127}$$

Thus, a single probability function $P_\Sigma(\tau,\eta,\zeta,\varphi,\tau_0)$ is totally determined by the values of its probability invariants $p^+(\tau,\eta,\zeta,\varphi,\tau_0)$ and $p^-(\tau,\eta,\zeta,\varphi,\tau_0)$. On the other hand, with the properties of probability invariants at an inversion of angular variables η and ζ known, a single probability function $P_\Sigma(\tau,\eta,\zeta,\varphi,\tau_0)$ totally determines the $p^\pm(\tau,\eta,\zeta,\varphi,\tau_0)$ values.

Now we obtain equations for $I^m_\Sigma(\tau,\eta,\zeta,\tau_0)$ function. We use Eqs. (3.112) and relationships between photometric invariants $I^\pm_m(\tau,\eta,\zeta,\tau_0)$, with an inversion of angular variables η and ζ:

$$I^\pm_m(\tau,\eta,\zeta,\tau_0) = L^\pm_m(\tau,\eta,\zeta,\tau_0)\,I^\pm_m(\tau,\zeta,\eta,\tau_0) \tag{3.128}$$

where $L^\pm_m$ are some functionals whose presentation depends on the symmetry of solutions for the boundary problems of the radiation transfer theory. Substituting (3.128) into (3.122) and solving the system of two equations obtained, we have

$$\begin{aligned}I^+_m(\tau,\eta,\zeta,\tau_0) &= \frac{\eta\,L^-_m\,I^m_\Sigma(\tau,\eta,\zeta,\tau_0)-\zeta\,I^m_\Sigma(\tau,\zeta,\eta,\tau_0)}{\eta^2L^-_m(\tau,\zeta,\eta,\tau_0)-\zeta^2L^+_m(\tau,\zeta,\eta,\tau_0)}\\ I^-_m(\tau,\eta,\zeta,\tau_0) &= \frac{\zeta\,L^+_m\,I^m(\tau,\eta,\zeta,\tau_0)-\eta\,I^m_\Sigma(\tau,\zeta,\eta,\tau_0)}{\zeta^2L^+_m(\tau,\zeta,\eta,\tau_0)-\eta^2L^-_m(\tau,\zeta,\eta,\tau_0)}\ .\end{aligned}\tag{3.129}$$

Substituting (3.129) into (3.112), summing them and making some transformations, we obtain a linear integral equation for intensity I^m_Σ. An accurate analytical presentation of $L^\pm_m$ functionals in (3.128) is difficult to obtain for lack of a-priori information on the symmetry of photometric invariants $I^\pm_m$ at an arbitrary optical depth τ, except for $\tau=0$ and $\tau=\tau_0$, since at these levels, due to the known angular symmetry of brightness coefficients ϱ and σ, the $L^\pm_m$ functionals are:

$$L^\pm_m(0,\eta,\zeta,\tau_0) = |L^\pm_m(\tau_0,\eta,\zeta,\tau_0)| = \eta/\zeta\ . \tag{3.130}$$

Here, equations for I_Σ^m intensities follow from (3.123) and from (3.120)–(3.121). For instance, using (3.121), we find an equation for $I_\Sigma^m(0,\eta,\zeta,\tau_0)$:

$$
\begin{aligned}
& I_\Sigma^m(0,\eta,\zeta,\tau_0)\left[T^m(\eta)-\frac{\Lambda}{2}\eta\int_0^1\frac{A^m(\eta,\eta')}{\eta'-\eta}d\eta'\right] \\
&= S\zeta[A^m(\eta,-\zeta)\varrho_1(\eta,\zeta)(\eta+\zeta)+A^m(\eta,\zeta)\sigma_1(\eta,\zeta)(\eta-\zeta)] \\
&+\frac{\Lambda}{2}\int_0^1 A^m(\eta,\eta')\left[\zeta^2\frac{I_\Sigma^m(0,\eta',\zeta,\tau_0)-I_\Sigma^m(0,\zeta,\eta',\tau_0)}{\zeta^2-\eta'^2}\right. \\
&\left.+\frac{I_\Sigma^m(0,\eta',\zeta,\tau_0)-I_\Sigma^m(0,\eta,\zeta,\tau_0)}{\eta'-\eta}\right]d\eta'-\frac{\Lambda}{2}e^{-\tau_0/\eta}\int_0^1 A^m(\eta,-\eta') \\
&\times\left[\zeta^2\frac{I_\Sigma^m(0,\zeta,\eta',\tau_0)-I_\Sigma^m(0,\eta',\zeta,\tau_0)}{\zeta^2-\eta'^2}+I_\Sigma^m(0,\eta',\zeta,\tau_0)\frac{\eta}{\eta'+\eta}\right]d\eta' \qquad (3.131)
\end{aligned}
$$

Another equation for $I_\Sigma^m(0,\eta,\zeta,\tau_0)$ can be obtained using some results from [130]. According to [130], we have:

$$
\left\{
\begin{aligned}
&(\eta+\zeta)\varrho^m(\eta,\zeta)=\pi[p^m(0,-\zeta,\eta,\tau_0)-e^{-\tau_0/\zeta}p^m(\tau_0,-\zeta,\eta,\tau_0)] \\
&\quad+2\pi\zeta\int_0^1[p^m(\eta',\zeta)p^m(0,\eta',\eta,\tau_0)-\sigma^m(\eta',\zeta)p^m(\tau_0,-\eta',\eta,\tau_0)]d\eta' \\
&(\eta-\zeta)\sigma^m(\eta,\zeta)=\pi[p^m(\tau_0,\zeta,\eta,\tau_0)-e^{-\tau_0/\zeta}p^m(0,\zeta,\eta,\tau_0)] \\
&\quad+2\pi\zeta\int_0^1[\varrho^m(\eta',\zeta)p^m(\tau_0,-\eta',\eta,\tau_0)-\sigma^m(\eta',\zeta)p^m(0,\eta',\eta,\tau_0)]d\eta'
\end{aligned}
\right. \qquad (3.132)
$$

Summing Eqs. (3.132), we find, in accordance with (3.50) and (3.123):

$$
\begin{aligned}
I_\Sigma^m(0,\eta,\zeta,\tau_0)&=\pi S\zeta[p_m^+(0,\eta,\zeta,\tau_0)-e^{-\tau_0/\zeta}p_m^+(\tau_0,\eta,\zeta,\tau_0)] \\
&+2\pi\zeta^2\int_0^1\frac{I_\Sigma^m(0,\eta',\zeta,\tau_0)-I_\Sigma^m(0,\zeta,\eta',\tau_0)}{\zeta^2-\eta'^2}p_m^+(0,\eta,\eta',\tau_0)d\eta'\ . \qquad (3.133)
\end{aligned}
$$

For the internal layers of a scattering medium, an accurate determination of functions in (3.128) can be replaced with their approximate presentation $L_m^\pm \simeq \tilde{L}_m^\pm$. Use of $\tilde{L}_m^\pm(\tau,\eta,\zeta,\tau_0)$ functions is an introduction into (3.128) of expressions for $L_m^\pm(\tau,\eta,\zeta,\tau_0)$ which are obtained from an approximate solution of the problem of radiation transfer in a flat homogeneous layer with a finite optical thickness τ_0. Explicit analytical expressions for the introduced functions $\tilde{L}_m^\pm(\tau,\eta,\zeta,\tau_0)$ are simply obtained, if they are presented as a relationship:

$$
\tilde{L}_m^\pm(\tau,\eta,\zeta,\tau_0)=\frac{\tilde{I}_m^\pm(\tau,\eta,\zeta,\tau_0)}{\tilde{I}_m^\pm(\tau,\zeta,\eta,\tau_0)} \qquad (3.134)
$$

where photometric invariants $\tilde{I}_m^{\pm}$ can be found using different approximation techniques (Sobolev's method [184], single scattering approximation, etc.). With an explicit presentation of the introduced functions $\tilde{L}_m^{\pm}$ determined, the solution of the respective equation for I_Σ^m and the use of (3.129) makes it possible to find approximate values of photometric invariants $I_m^{\pm}$, i.e., to determine the intensities of the upward and downward radiation fluxes in specular directions at symmetrical levels τ and $\tau_0-\tau$. Finally, an introduction, according to (3.114) and (3.121), of relationships

$$Q_\Sigma = \eta q_1^+ + \zeta q_1^- = \frac{1}{\pi S}[\eta I^+ + \zeta I^-] = \frac{1}{\pi S} I_\Sigma \tag{3.135}$$

$$P_\Sigma = \eta p^+ + \zeta p^- = \frac{1}{\pi S}[\eta B^+ + \zeta B^-] = \frac{1}{\pi S} B_\Sigma \tag{3.136}$$

closes a system of single probability and photometric parameters of the symmetrized theory of radiation transfer in a flat homogeneous layer of a finite optical thickness.

The presentation of photometric parameters as a sum and the difference of solutions for the non-symmeterized radiation transfer problem has also been used in [138, 204, 284–287]. However, these publications lack an idea of probabilistic equivalency of the specular levels in a flat layer of a finite optical thickness and the resulting probability invariants in non-symmetrical problems of classical radiation transfer theory. This probabilistic equivalency makes it possible to present the radiation transfer theory in a simpler way and to solve its basic problems more rationally.

Thus, the presence of the probabilistic equivalency of specular levels τ and $\tau_0-\tau$ in the initial photometrically non-symmetrized problem of the radiation transfer theory (Fig. 3.2) is determined by the spatial symmetry of the medium and by the definition of the probability of a photon's exit. The probability symmetry of specular levels τ and $\tau_0-\tau$ results in probability invariants p^+, q^+ and p^-, q^-, which exist independently of the photometric aspect of the problem. The specular photometric equivalence of symmetric levels is emphasized by the presence of the probability invariants and, by itself, is of no principal importance. Hence, a symmetrization of basic classical functions first made in [177] and exemplified by solving the problem of light single scattering in a homogeneous flat layer of a finite optical thickness τ_0, reflects the essence of the probability treatment of the phenomenological theory of radiation transfer itself.

3.8 The Meaning in Probability Terms of Basic Ω_i^m and Θ_i^m Functions

Basic Ω_i^m and Θ_i^m functions introduced, according to (3.37), have a precise physical meaning. To clarify this, we present a probability $p(\tau, \eta', \eta, \varphi'-\varphi, \tau_0)$

as (3.114), that is similar to an expansion of the given phase function $x(\gamma)$ in Fourier series by φ azimuth. Using expressions of intensities of the diffusively reflected $I(0,-\eta,\zeta,\varphi,\tau_0)$ and diffusively transmitted $I(\tau_0,\eta,\zeta,\varphi,\tau_0)$ light in terms of probabilities $p(\tau,-\zeta,\eta,\varphi,\tau_0)$ and $p(\tau_0-\tau,\zeta,\eta,\varphi,\tau_0)$, we obtain [184]:

$$I^m(0,-\eta,\zeta,\tau_0) = S\zeta\varrho^m(\eta,\zeta,\tau_0) = \pi S \int_0^{\tau_0} e^{-\tau/\zeta} p^m(\tau,-\zeta,\eta,\tau_0)\,\frac{d\tau}{\eta} \tag{3.137}$$

$$I^m(\tau_0,\eta,\zeta,\tau_0) = S\zeta\sigma^m(\eta,\zeta,\tau_0) = \pi S \int_0^{\tau_0} e^{-\tau/\zeta} p^m(\tau_0-\tau,\zeta,\eta,\tau_0)\,\frac{d\tau}{\eta}\ . \tag{3.138}$$

Summing and extracting (3.137) and (3.138), in accordance with definitions (3.42 – 3.43) and (3.106), we find:

$$I_m^+(0,\eta,\zeta,\tau_0) = S\zeta R^m(\eta,\zeta) = \pi S \int_0^{\tau_0} e^{-\tau/\zeta} p_m^+(\tau,\eta,\zeta,\tau_0)\,\frac{d\tau}{\eta} \tag{3.139}$$

$$I_m^-(0,\eta,\zeta,\tau_0) = S\zeta r^m(\eta,\zeta) = \pi S \int_0^{\tau_0} e^{-\tau/\zeta} p_m^-(\tau,\eta,\zeta,\tau_0)\,\frac{d\tau}{\eta} \tag{3.140}$$

where functions $p_m^+(\tau,\eta,\zeta,\tau_0)$ and $p_m^-(\tau,\eta,\zeta,\tau_0)$ are determined from (3.117). The p_m^- function characterizes an asymmetry of the medium at levels τ and $\tau_0-\tau$ for the probability of a photon's exit through the upper and lower boundaries. Knowledge of p_m^+ and p_m^- functions enables the determination of $p_m(\tau,-\zeta,\eta,\tau_0)$ and $p_m(\tau_0-\tau,\zeta,\eta,\tau_0)$ functions:

$$p^m(\tau,-\zeta,\eta,\tau_0) = \tfrac{1}{2}[p_m^+(\tau,\eta,\zeta,\tau_0) + p_m^-(\tau,\eta,\zeta,\tau_0)]\ , \tag{3.141}$$

$$p^m(\tau_0-\tau,\zeta,\eta,\tau_0) = \tfrac{1}{2}[p_m^+(\tau,\eta,\zeta,\tau_0) - p_m^-(\tau,\eta,\zeta,\tau_0)]\ . \tag{3.142}$$

Now we use relationships for $p^m(0,\zeta,\eta,\tau_0)$ and $p^m(\tau_0,\zeta,\eta,\tau_0)$ found by Minin [130]:

$$p^m(0,\zeta,\eta,\tau_0) = \frac{\Lambda}{4\pi}\sum_{i=m}^{N} c_i^m P_i^m(\zeta)\varphi_i^m(\eta)\ , \tag{3.143}$$

$$p^m(\tau_0,\zeta,\eta,\tau_0) = \frac{\Lambda}{4\pi}\sum_{i=m}^{N} c_i^m P_i^m(\zeta)\psi_i^m(\eta)\ . \tag{3.144}$$

Relationships (3.143 – 3.144) make it possible to define the physics of functions $\Omega_i^m(\eta)$ and $\Theta_i^m(\eta)$. Defining these functions (3.37) and relationships (3.45), we find from (3.143 – 3.144):

$$\begin{aligned} p_m^+(0,\eta,\zeta,\tau_0) &= p^m(0,\zeta,\eta,\tau_0) + p^m(\tau_0,-\zeta,\eta,\tau_0) \\ &= \frac{\Lambda}{4\pi}\sum_{i=m}^{N} c_i^m P_i^m(\zeta)\,\Omega_i^m(\eta)\ , \end{aligned} \tag{3.145}$$

$$\begin{aligned} p_m^-(0,\zeta,\eta,\tau_0) &= p^m(0,\zeta,\eta,\tau_0) - p^m(\tau_0,-\zeta,\eta,\tau_0) \\ &= \frac{\Lambda}{4\pi}\sum_{i=m}^{N} c_i^m P_i^m(\zeta)\Theta_i^m(\eta)\ . \end{aligned} \tag{3.146}$$

Thus, the $\Omega_i^m(\eta)$ functions characterize complete probability of a photon's exit at its absorption in boundary layers $\tau = 0$ and $\tau = \tau_0$, and $\Theta_i^m(\eta)$ functions characterize an asymmetry of this probability due to different probabilities of photon's reaching the upper and lower boundaries of the medium. In other words, functions $\Omega_i^m(\eta)$ and $\Theta_i^m(\eta)$ are determined, respectively, by probability invariants $p_m^+(\tau, \eta, \zeta, \tau_0)$ and $p_m^-(\tau, \eta, \zeta, \tau_0)$ at $\tau = 0$.

We use the following equations for Ω_i^m and Θ_i^m functions aftr Ambartsumyan [130] to obtain new relationships characterizing the properties of φ_i^m and ψ_i^m functions:

$$\varphi_i^m(\eta) = P_i^m(\eta) + \sum_{k=m}^{N} (-1)^{i+k} P_\kappa^m(\eta) \int_0^{\tau_0} e^{-\tau/\eta} \Phi_{ki}^m(\tau, \tau_0)\, d\tau \tag{3.147}$$

$$\psi_i^m(\eta) = P_i^m(\eta) e^{-\tau_0/\eta} + \sum_{k=m}^{N} P_k^m(\eta) \int_0^{\tau_0} e^{\tau/\eta} \Phi_{ki}^m(\tau_0 - \tau, \tau_0)\, d\tau \tag{3.148}$$

where functions Φ_{ki}^m are determined from the formula

$$\Phi_{ki}^m(\tau, \tau_0) = \pi \int_0^1 P_k^m(\eta) F_i^m(\tau, \eta, \tau_0) \frac{d\eta}{\eta} \, . \tag{3.149}$$

Functions $F_i^m(\tau, \eta, \tau_0)$ are introduced in the following way [130]:

$$p^m(\tau, \eta, \zeta, \tau_0) = \sum_{i=m}^{N} P_i^m(\zeta) F_i^m(\tau, \eta, \tau_0) \, . \tag{3.150}$$

In accordance with the determination of functions Ω_i^m and Θ_i^m from (3.147) and (3.148), we find relationships:

$$\Omega_i^m(\eta, \tau_0) = P_i^m(\eta)[1 + (-1)^{i+m} e^{-\tau_0/\eta}] + \sum_{k=m}^{N} (-1)^{i+k} P_k^m(\eta) \int_0^{\tau_0} e^{-\tau/\eta} [\Phi_{ki}^m(\tau, \tau_0) + \Phi_{ki}^m(\tau_0 - \tau, \tau)]\, d\tau \, , \tag{3.151}$$

$$\Theta_i^m(\eta, \tau_0) = P_i^m(\eta)[1 - (-1)^{i+m} e^{-\tau_0/\eta}] + \sum_{k=m}^{N} (-1)^{i+k} P_k^m(\eta) \int_0^{\tau_0} e^{-\tau/\eta} [\Phi_{ki}^m(\tau, \tau_0) - \Phi_{ki}^m(\tau_0 - \tau, \tau)]\, d\tau \, . \tag{3.152}$$

It follows from (3.151 – 3.152) that in the determination of the internal fields of radiation the symmetrized theory of radiation transfer must comprise the difference and the sum of functions $\Phi_{ki}^m(\tau, \tau_0)$ and $\Phi_{ki}^m(\tau_0 - \tau, \tau_0)$, which are determined by the respective linear combination of integrals of probability functions $p^m(\tau, \zeta, \eta, \tau_0)$ and $p^m(\tau_0 - \tau, \zeta, \eta, \tau_0)$, i.e., by probability invariants p_m^+ and p_m^-. Basic functions must be symmetrized to use the invariance principle after Ambartsumyan and when the transfer of polarized radiation is considered (in a flat layer of finite optical thickness). The efficiency of symmetrization will be exemplified next be classical problem of diffused reflection and transmission of light by a plane-parallel planetary atmosphere ($\tau_0 < \infty$).

3.9 Accurate Expressions for the Azimuthal Harmonics of Brightness Coefficients ϱ^m and σ^m for an Approximating Three-Term Phase Function (m = 0, 1, 2)

The transformation of the basic function of the multiple anisotropic light scattering theory described makes it possible to solve more efficiently a classical problem of the determination of the azimuthal harmonics of brightness coefficients ϱ^m and σ^m for flat layer of a finite optical thickness ($\tau_0 < \infty$) with an arbitrary, height-independent phase function. First, we obtain accurate expressions for the $S_i^m(\eta)$ and $Q_i^m(\eta)$ polynomials using an approximating three-term phase function (3.19) [177].

Zero Harmonics. At m = 0, according to recursive relationships (3.33), we have

$$R_{00}^0(\eta) = 1 \ , \qquad R_{10}^0(\eta) = (1-\Lambda)\eta \ , \tag{3.153}$$

$$R_{00}^0(\eta) = \tfrac{1}{2}[(3-\Lambda x_1)(1-\Lambda)\eta^2 - 1] \tag{3.154}$$

considering relationships

$$c_0^0 = 1 \ , \qquad c_1^0 = x_1 \ , \qquad c_2^0 = x_2 \tag{3.155}$$

$$P_0^0(\eta) = 1 \ , \qquad P_1^0(\eta) = \eta \ , \qquad P_2^0(\eta) = \tfrac{1}{2}[3\eta^2 - 1) \tag{3.156}$$

from (3.32) and (3.153 – 3.154) it follows

$$A_0^0(\eta,\eta') = 1 + x_1(1-\Lambda)\eta\eta' + \frac{x_2}{4}[(3-\Lambda x_1)(1-\Lambda)\eta^2 - 1](3\eta'^2 - 1) \ . \tag{3.157}$$

Let the required polynomials $Q_i^0(\eta)$ and $S_i^0(\eta)$ (i = 0, 1, 2) be written as

$$Q_i^0(\eta) = R_{i0}^0(0) + c_{i1}\eta + c_{i2}\eta^2 \ , \tag{3.158}$$

$$S_i^0(\eta) = R_{i0}^0(0) + d_{i1}\eta + d_{i2}\eta^2 \ . \tag{3.159}$$

The coefficients of expansion of $R_{i0}^0(\eta)$ polynomials in η powers we write in terms of m_i and l_i:

$$R_{i0}^0(\eta) = R_{i0}^0(\eta) + m_i\eta + l_i\eta^2 \ , \tag{3.160}$$

where

$$R_{00}^0(0) = 1 \ , \quad R_{10}^0(0) = 0 \ , \quad R_{20}^0(0) = -1/2 \ , \tag{3.161}$$

$$m_0 = 0 \ , \quad m_1 = 1 - \Lambda \ , \quad m_2 = 0 \ , \tag{3.162}$$

$$l_0 = 0 \ , \quad l_1 = 0 \ , \quad l_2 = \tfrac{1}{2}(3-\Lambda x_1)(1-\Lambda) \ . \tag{3.163}$$

Substituting (3.158) and (3.159) into (3.59) and (3.60), we obtain the following systems of algebraic equations to find c_{ik} and d_{ik} (i = 0, 1, 2):

$$\begin{cases} M_{01}c_{i1} + M_{02}c_{i2} = M_{00}R_{i0}^0(0) + m_i \ , \\ N_{01}c_{i1} + N_{02}c_{i2} = N_{00}R_{i0}^0(0) + l_i \ , \end{cases} \tag{3.164}$$

$$\begin{cases} M_{11}\,d_{i1} + M_{12}\,d_{i2} = M_{10}\,R_{i0}^{0}(0) + m_i \\ N_{11}\,d_{i1} + N_{12}\,d_{i2} = N_{10}\,R_{i0}^{0}(0) + l_i \ . \end{cases} \tag{3.165}$$

The coefficients of these systems are expressed in [88] and [177] through moments α_k^0 and β_k^0 of X^0 and Y^0 functions:

$$\alpha_k^0 = \int_0^1 X^0(\eta)\eta^k d\eta \ , \quad \beta_k^0 = \int_0^1 Y^0(\eta)\eta^k d\eta \ . \tag{3.166}$$

Solutions for equation systems (3.164) and (3.165) are written as:

$$\begin{cases} \Delta_0 c_{i1} = N_{02}\,[M_{00}\,R_i(0) + m_i] - M_{02}\,[N_{00}\,R_{i0}^{0}(0) + l_i] \\ \Delta_0 c_{i2} = M_{01}\,[N_{00}\,k_i(0) + l_i] - N_{01}\,[M_{00}\,R_{i0}^{0}(0) + m_i] \ , \end{cases} \tag{3.167}$$

$$\begin{cases} \Delta_1\,d_{i1} = N_{12}\,[M_{10}\,R_i(0) + m_i] - M_{12}\,[N_{10}\,R_{i0}^{0}(0) + l_i] \\ \Delta_1\,d_{i2} = M_{11}\,[N_{10}\,R_i(0) + l_i] - N_{11}\,[M_{10}\,R_{i0}^{0}(0) + m_i] \ , \end{cases} \tag{3.168}$$

where the determinants of systems Δ_i are

$$\Delta_1 = M_{i1}\,N_{i2} - M_{i2}\,N_{i1} \quad (i = 0,\ 1) \tag{3.169}$$

Substituting $i = 0, 1, 2$ in (3.167) and (3.168), we find successively all coefficients c_{ik} and d_{ik}. Substituting them into (3.158) and (3.159), we obtain the required expressions for $Q_i^0(\eta)$ and $S_i^0(\eta)$ polynomials when real absorption ($\Lambda < 1$) occurs:

$$Q_0^0(\eta) = 1 + \frac{1}{\Delta_0}\,[(M_{00}\,N_{02} - M_{02}\,N_{00})\eta + (M_{01}\,N_{00} - M_{00}\,N_{01})\eta^2] \ , \tag{3.170}$$

$$Q_1^0(\eta) = \frac{1-\Lambda}{\Delta_0}\,(N_{02}\,\eta - N_{01}\,\eta^2) \ , \tag{3.171}$$

$$Q_2^0(\eta) = -\frac{1}{2}\,Q_0(\eta) - \frac{(3 - \Lambda x_1)(1-\Lambda)}{2\Delta_0}\,(M_{02}\,\eta - M_{01}\,\eta^2) \ , \tag{3.172}$$

$$S_0^0(\eta) = 1 + \frac{1}{\Delta_1}\,[(M_{10}\,N_{12} - M_{12}\,N_{10})\eta + (M_{11}\,N_{10} - M_{10}\,N_{11})\eta^2] \ , \tag{3.173}$$

$$S_1^0(\eta) = \frac{1-\Lambda}{\Delta_1}\,(N_{12}\,\eta - N_{11}\,\eta^2) \ , \tag{3.174}$$

$$S_2^0(\eta) = -\frac{1}{2}\,S_0^0(\eta) - \frac{(3 - \Lambda x_1)(1-\Lambda)}{2\Delta_1}\,(M_{12}\,\eta - M_{11}\,\eta^2) \ . \tag{3.175}$$

The Case of Pure Scattering ($\Lambda = 1$). In the physics of planetary atmospheres, it is often assumed that the coefficient of real absorption is much below the coefficient of scattering, i.e., the probability of quantum survival at elementary scattering is close to unity ($\Lambda \sim 1$). Therefore, for practical applications, it is important to consider an extreme case of pure scattering ($\Lambda = 1$).

We use integral properties of X^0 and Y^0 Chandrasekhar functions [187] valid at $\Lambda = 1$ [340]:

$$\frac{1}{2}\int_0^1 A^0(\eta,\eta)[X^0(\eta)+Y^0(\eta)] = 1 \ , \tag{3.176}$$

$$\int_0^1 A^0(\eta,\eta)[X^0(\eta)-Y^0(\eta)]\,\eta\,d\eta = \tau_0\int_0^1 A^0(\eta,\eta)\,Y^0(\eta)\,d\eta \ . \tag{3.177}$$

Then for $Q_i^0(\eta)$ and $S_i^0(\eta)$ polynomials at $\Lambda\to 1$ we obtain the following expressions:

$$Q_0^0(\eta) = 1+\frac{2-\alpha_0^0-\beta_0^0}{\alpha_1^0-\beta_1^0}\eta \ , \tag{3.178}$$

$$Q_1^0(\eta) = 0 \ , \tag{3.179}$$

$$Q_2^0(\eta) = -\frac{1}{2}+\frac{\alpha_0^0+\beta_0^0}{2(\alpha_1^0-\beta_1^0)}\eta \ , \tag{3.180}$$

$$S_0^0(\eta) = 1+\frac{(3-x_1)\tau_0(2-\alpha_0+\beta_0)-6(\alpha_1-\beta_1)}{(3-x_1)\tau_0(\alpha_1+\beta_1)+6(\alpha_2+\beta_2)} \ , \tag{3.181}$$

$$S_1^0(\eta) = \frac{4\eta}{(3-x_1)\tau_0(\alpha_1^0+\beta_1^0)+6(\alpha_2^0+\beta_2^0)} \ , \tag{3.182}$$

$$S_2^0(\eta) = -\frac{1}{2}+\frac{(3-x_1)\tau_0(\alpha_0^0-\beta_0^0)+6(\alpha_1^0-\beta_1^0)}{2[(3-x_1)\tau_0(\alpha_1^0+\beta_1^0)+6(\alpha_2^0+\beta_2^0)]}\eta \ . \tag{3.183}$$

In an asymptotic case $\tau_0\to\infty$, formulas (3.178 – 3.183) are transformed into relationships obtained earlier by Sobolev [185]. Note, that expressions (3.178 – 3.183) can be obtained in another way. For this purpose, some integral relationships between $\varphi_i^0(\eta)$ and $\psi_i^0(\eta)$ functions should be used [187, 340].

These formulas and relationships can now be used to find accurate expressions for the azimuthal harmonics of the brightness coefficients $\varrho^m(\eta,\zeta)$ and $\sigma^m(\eta,\zeta)$. When $m=0$, ϱ^0 and σ^0 functions become

$$\begin{aligned}\varrho^0(\eta,\zeta) = \frac{1}{8}\Big\{&[X^0(\eta)X^0(\zeta)+Y^0(\eta)Y^0(\zeta)](C+D)-[X^0(\eta)Y^0(\zeta)\\ &+X^0(\zeta)Y^0(\eta)](C-D)+\frac{2}{\eta+\zeta}[X^0(\eta)X^0(\zeta)\\ &-Y^0(\eta)Y^0(\zeta)]\left(1+\frac{x_2}{4}+\frac{3}{4}x_2\eta\zeta\right)\Big\} \ ,\end{aligned} \tag{3.184}$$

$$\begin{aligned}\sigma^0(\eta,\zeta) = \frac{1}{8}\Big\{&[X^0(\eta)X^0(\zeta)+Y^0(\eta)Y^0(\zeta)](C-D)-[X^0(\eta)Y^0(\zeta)\\ &+X^0(\zeta)Y^0(\eta)](C+D)+\frac{2}{\eta-\zeta}[X^0(\zeta)Y^0(\eta)\\ &-X^0(\eta)Y^0(\zeta)]\left(1+\frac{x_2}{4}-\frac{3}{4}x_2\eta\zeta\right)\Big\} \ .\end{aligned} \tag{3.185}$$

Coefficients C and D in (3.184) and (3.185) are

$$C = \frac{2 - \left(1 + \frac{x_2}{4}\right)(\alpha_0^0 + \beta_0^0)}{\alpha_1 - \beta_1}, \tag{3.186}$$

$$D = \frac{2\tau_0(3 - x_1) - \left(1 + \frac{x_2}{4}\right)(3 - x_1)\tau_0(\alpha_0^0 - \beta_0^0) + 6(\alpha_1^0 - \beta_1^0)}{(3 - x_1)\tau_0(\alpha_1^0 + \beta_1^0) + 6(\alpha_2^0 + \beta_2^0)}. \tag{3.187}$$

Substituting (3.52) into (3.184) and (3.185), we obtain finite formulas for the azimuthal harmonics of the coefficients of reflection ϱ^0 and transmission σ^0 when $m = 0$:

$$\varrho^0(\eta, \zeta) = \frac{1}{8}\left[U^0(\eta)U^0(\zeta)C + W^0(\eta)W^0(\zeta)D + \frac{W^0(\eta)U^0(\zeta) + W^0(\zeta)U^0(\eta)}{\eta + \zeta}\left(1 + \frac{x_2}{4} + \frac{3}{4}x_2\eta\zeta\right)\right], \tag{3.188}$$

$$\sigma^0(\eta, \zeta) = \frac{1}{8}\left[U^0(\eta)U^0(\zeta)C - W^0(\eta)W^0(\zeta)D + \frac{W^0(\eta)U^0(\zeta) - W^0(\zeta)U^0(\eta)}{\eta - \zeta}\left(1 + \frac{x_2}{4} - \frac{3}{4}x_2\eta\zeta\right)\right]. \tag{3.189}$$

Table 3.9 gives accurate estimates of zero azimuthal harmonics ϱ^0 and σ^0, calculated in [177] using formulas (3.181 – 3.182) for $\tau_0 = 0.351$; $\Lambda = 0.872$ and an approximating 3-term phase function ($x_1 = 1.500$, $x_2 = 1.500$). Taking into account the symmetry of brightness coefficients and their azimuthal harmonics by angular variables η and ζ, estimates of ϱ^0 are given in the upper diagonal of the table, and estimates of σ^0 in the lower diagonal.

With expressions (3.188), (3.189) and (3.186) and (3.187), accurate asymptotic formulas for brightness coefficients $\varrho^0(\eta, \zeta)$ and $\sigma^0(\eta, \zeta)$ at $\tau_0 \gg 1$ can

Table 3.9. Accurate values of zero azimuthal harmonics of $\varrho^0(\eta, \zeta)$ and $\sigma^0(\eta, \zeta)$ brightness coefficients for an approximating 3-term phase function (the Elterman-1968 model, $\lambda = 600$ nm, $\tau_0 = 0.351$, $\lambda = 0.872$, $x_1 = 1.500$, $x_2 = 1.500$)

	0	0.1	0.3	0.5	0.7	
	–	3.2975	1.0976	0.5762	0.3130	0
		1.8358	0.8791	0.4995	0.2788	0.1
0	–		0.4815	0.2776	0.1527	0.3
0.1	0.3542	–		0.1595	0.0894	0.5
0.3	0.5101	0.6673	–		0.0565	0.7
0.5	0.3896	0.4673	0.3587	–		
0.7	0.2648	0.3108	0.2556	0.2164	–	

easily be obtained. Here, $W^0(\eta)\rightarrow H^0(\eta)$ and $U^0(\eta)\rightarrow H^0(\eta)$, where $H^0(\eta)$ is the Chandrasekhar function at $m = 0$ [205]. The required asymptotic formulas are written as:

$$\varrho^0_{as}(\eta,\zeta) = \frac{H^0(\eta)H^0(\zeta)}{4(\eta+\zeta)}\left(1+\frac{x_2}{4}+\frac{3}{4}x_2\eta\zeta\right) + \frac{H^0(\eta)H^0(\zeta)}{2h^0_1}\left[1-\frac{1}{2}\left(1+\frac{x_2}{4}\right)h^0_0 - \frac{2}{(3-x_1)\tau_0 h^0_1 + 6h^0_2}\right], \quad (3.190)$$

$$\sigma^0_{as}(\eta,\zeta) = \frac{H^0(\eta)H^0(\zeta)}{h^0_1[(3-x_1)\tau_0 h^0_1 + 6h^0_2]} \quad (3.191)$$

where h^0_k are the moments of the Chandrasekhar function H^0:

$$h^0_k = \int_0^1 H^0(\eta)\eta^k d\eta \ . \quad (3.192)$$

Substituting expressions for $Q^0_i(\eta)$ and $S^0_i(\eta)$ polynomials at $\Lambda = 1$ in (3.53) and (3.54), we find $\Omega^0_i(\eta)$ and $\Theta^0_i(\eta)$ functions $(i = 0,1,2)$:

$$\Omega^0_0(\eta) = W^0(\eta) + U^0(\eta)\frac{2-\alpha^0_0+\beta^0_0}{\alpha^0_1-\beta^0_1}\eta \ , \quad (3.193)$$

$$\Theta^0_0(\eta) = U^0(\eta) + W^0(\eta)\frac{(3-x_1)(2-\alpha^0_0+\beta^0_0)\tau_0 - 6(\alpha^0_1-\beta^0_1)}{(3-x_1)(\alpha^0_1+\beta^0_1)\tau_0 + 6(\alpha^0_2+\beta^0_2)}\eta \ , \quad (3.194)$$

$$\Omega^0_1(\eta) = 0 \ , \quad \Theta^0_1(\eta) = W^0(\eta)\frac{4\eta}{(3-x_1)(\alpha^0_1+\beta^0_1)\tau_0 + 6(\alpha^0_2+\beta^0_2)} \ , \quad (3.195)$$

$$\Omega^0_2(\eta) = -\frac{1}{2}W^0(\eta) + U^0(\eta)\frac{\alpha^0_0+\beta^0_0}{2(\alpha^0_1-\beta^0_1)}\eta \ , \quad (3.196)$$

$$\Theta^0_2(\eta) = -\frac{1}{2}U^0(\eta) + W^0(\eta)\frac{(3-x_1)\tau_0(\alpha^0_0-\beta^0_0) + 6(\alpha^0_1-\beta^0_1)}{2[(3-x_1)\tau_0(\alpha^0_1+\beta^0_1) + 6(\alpha^0_2+\beta^0_2)]}\eta \ . \quad (3.197)$$

Expressions for Ω^0_i and Θ^0_i functions $(i = 0,1,2)$ with real absorption in the medium $(\Lambda < 1)$ are given in [88].

The First Harmonic (m = 1). When $m = 1$, we find

$$A^1_1(\eta,\eta') = \tfrac{1}{2}(1-\eta'^2)^{1/2}[x_1 + x_2\eta\eta'(3-\Lambda x_1)] \ . \quad (3.198)$$

The R^1_{i1} polynomials $(i = 1,2)$ are:

$$R^1_{i1}(\eta) = R^1_{i1}(0) + m^1_i\eta \quad (i = 1,2) \quad (3.199)$$

where

$$R^1_{11}(\eta) = 1 \ , \quad R^1_{21}(\eta) = (3-\Lambda x_1)\eta \ , \quad m^1_1 = 0 \ , \quad m^1_2 = (3-\Lambda x_1) \ . \quad (3.200)$$

Solutions for integral equations (3.35) and (3.36) at $N = 2$ and $m = 1$, are as follows [177]:

$$Q_i^1(\eta) = R_{i1}^1(0) + c_{i1}^1(\eta) \ , \quad S_i^1(\eta) = R_{i1}^1(0) + d_{i1}^1 \quad (i = 1,2) \ . \tag{3.201}$$

Substitute (3.198 – 3.199) into (3.35) and (3.36) at m = 1. Equalizing the expressions by equal powers η in the left- and right-hand parts of these equations, we find coefficients c_{i1}^1 and d_{i1}^1 (i = 1,2). Substituting them into (3.199), we find accurate expressions for polynomials $Q_i^1(\eta)$ and $S_i^1(\eta)$ (i = 1,2) for an approximating 3-term phase function (3.19) and $\Lambda \leqslant 1$:

$$\begin{aligned} Q_1^1(\eta) &= 1 + \frac{\Lambda x_2(3-\Lambda x_1)[(\alpha_3^1+\beta_3^1)-(\alpha_1^1+\beta_1^1)]}{4-\Lambda x_1[(\alpha_0^1-\beta_0^1)-(\alpha_2^1-\beta_2^1)]}\eta \ , \\ Q_2^1(\eta) &= \frac{4(3-\Lambda x_1)}{4-\Lambda x_1[(\alpha_0^1-\beta_0^1)-(\alpha_2^1-\beta_2^1)]}\eta \ , \end{aligned} \tag{3.202}$$

$$\begin{aligned} S_1^1(\eta) &= 1 + \frac{\Lambda x_2(3-\Lambda x_2)[(\alpha_3^1-\beta_3^1)-(\alpha_1^1-\beta_1^1)]}{4-\Lambda x_1[(\alpha_0^1+\beta_0^1)-(\alpha_2^1-\beta_2^1)]}\eta \ , \\ S_2^1(\eta) &= \frac{4(3-\Lambda x_1)}{4-\Lambda x_1[(\alpha_0^1+\beta_0^1)-(\alpha_2^1+\beta_2^1)]}\eta \ . \end{aligned} \tag{3.203}$$

Using (3.53), (3.54) and (3.202 – 3.203), we obtain accurate expressions for $\Omega_i^1(\eta)$ and Θ_i^1 functions (i = 1,2):

$$\begin{aligned} \Omega_1^1(\eta) &= \left[W^1(\eta) + U^1(\eta)\frac{\Lambda x_2(3-\Lambda x_1)(\alpha_3^1+\beta_3^1-\alpha_1^1-\beta_1^1)\eta}{4-\Lambda x_1(\alpha_0^1-\beta_0^1-\alpha_2^1+\beta_2^1)}\right]\sqrt{1-\eta^2} \ , \\ \Omega_2^1(\eta) &= U^1(\eta)\frac{4(3-\Lambda x_1)\eta\sqrt{1-\eta^2}}{4-\Lambda x_1(\alpha_0^1-\beta_0^1-\alpha_2^1+\beta_2^1)} \ , \\ \Theta_1^1(\eta) &= \left[U^1(\eta) + W^1(\eta)\frac{\Lambda x_2(3-\Lambda x_1)(\alpha_3^1-\beta_3^1-\alpha_1^1+\beta_1^1)\eta}{4-\Lambda x_1(\alpha_0^1+\beta_0^1-\alpha_2^1-\beta_2^1)}\right]\sqrt{1-\eta^2} \ , \\ \Theta_2^1(\eta) &= W^1(\eta)\frac{4(3-\Lambda x_1)\eta\sqrt{1-\eta^2}}{4-\Lambda x_1(\alpha_0^1-\alpha_2^1+\beta_0^1-\beta_2^1)} \ . \end{aligned} \tag{3.204}$$

By substituting (3.204) into (3.40 – 3.41) or (3.202) and (3.203) into (3.62 – 3.63), we find at m = 1 accurate expressions for the first azimuthal harmonics ϱ^1 and σ^1 considering especially a 3-term phase function (3.19):

$$\begin{aligned} \varrho^1(\eta,\zeta) &= \frac{\Lambda(1-\eta^2)^{1/2}(1-\zeta^2)^{1/2}}{8(\eta+\zeta)} \\ &\times \left\{ W^1(\eta)W^1(\zeta)\frac{x_1}{2}(\eta+\zeta)\frac{\Lambda x_2(3-\Lambda x_1)(\alpha_3^1-\alpha_1^1-\beta_3^1+\beta_1^1)}{4-\Lambda x_1[(\alpha_0^1-\alpha_2^1)+(\beta_0^1-\beta_2^1)]} \right. \\ &+ U^1(\eta)U^1(\zeta)\frac{x_1}{2}(\eta+\zeta)\frac{\Lambda x_2(3-\Lambda x_1)(\alpha_3^1-\alpha_1^1+\beta_3^1-\beta_1^1)}{4-\Lambda x_1[(\alpha_0^1-\alpha_2^2)-(\beta_0^1-\beta_2^1)]} \end{aligned}$$

$$+[W^1(\eta)U^1(\zeta)+W^1(\zeta)U^1(\eta)]\left[\frac{x_1}{2}\right.$$

$$\left.+\frac{\frac{x_1}{2}\Lambda^2x_2^2\eta\zeta(3-\Lambda x_1)^2[(\alpha_3^1-\alpha_1^1)^2-(\beta_3^1-\beta_1^1)^2]-\frac{8}{3}x_2\eta\zeta(3-\Lambda x_1)^2}{[4-\Lambda x_1(\alpha_0^1-\alpha_2^1)]^2-\Lambda^2x_1^2(\beta_0^1-\beta_2^1)^2}\right]\Bigg\}, \tag{3.205}$$

$$\sigma^1(\eta,\zeta)=\frac{\Lambda(1-\eta^2)^{1/2}(1-\zeta^2)^{1/2}}{8(\eta-\zeta)}$$

$$\times\Bigg\{W^1(\eta)W^1(\zeta)\frac{x_1}{2}(\zeta-\eta)\frac{\Lambda x_2(3-\Lambda x_1)(\alpha_3^1-\alpha_1^1-\beta_3^1+\beta_1^1)}{4-\Lambda x_1[(\alpha_0^1-\alpha_2^1)+(\beta_0^1-\beta_2^1)]}$$

$$+U^1(\eta)U^1(\zeta)\frac{x_1}{2}(\eta-\zeta)\frac{\Lambda x_2(3-\Lambda x_1)(\alpha_3^1-\alpha_1^1+\beta_3^1-\beta_1^1)}{4-\Lambda x_1[(\alpha_0^1-\alpha_2^1)-(\beta_0^1-\beta_2^1)]}$$

$$+[W^1(\zeta)U^1(\eta)-U^1(\zeta)W^1(\eta)]\left[-\frac{x_1}{2}\right.$$

$$\left.+\frac{\frac{x_1}{2}\Lambda^2x_2^2(3-\Lambda x_1)^2\eta\zeta[(\alpha_3^1-\alpha_1^1)^2-(\beta_3^1-\beta_1^1)^2]-\frac{8}{3}x_2\eta\zeta(3-\Lambda x_1)^2}{[4-\Lambda x_1(\alpha_0^1-\alpha_2^1)]^2-\Lambda^2x_1^2(\beta_0^1-\beta_2^1)^2}\right]\Bigg\}. \tag{3.206}$$

In Table 3.10, accurate estimates are given of the first azimuthal harmonics ϱ^1 and σ^1 calculated in [177] using (3.205 – 3.206) for $\tau_0 = 0.351$, $\Lambda = 0.872$ and an approximating 3-term phase function ($x_1 = 1.500$, $x_2 = 1.500$). The ϱ^1 estimates are given in the upper diagonal of Table 3.10, the σ^1 estimates in the lower diagonal.

The Second Harmonic (m = 2). With m = 2 we have:

$$A_2^2(\eta,\eta')=\frac{x_2}{8}(1-\eta'^2)\ ,\quad R_{22}^2(\eta)=1\ . \tag{3.207}$$

Table 3.10. Exact expressions for the first azimuthal harmonics $\varrho^1(\eta,\zeta)$ and $\sigma^1(\eta,\zeta)$ of brightness coefficients for an approximating three-term phase function (the Elterman-1968 model, $\lambda = 600$ nm, $\tau_0 = 0.351$, $\Lambda = 0.872$, $x_1 = 1.500$, $x_2 = 1.500$)

	0	0.1	0.3	0.5	0.7	
	–	1.6925	0.5527	0.2987	0.1732	0
		0.8762	0.4010	0.2265	0.1298	0.1
0	–		0.1925	0.0963	0.0446	0.3
0.1	1.3446	–		0.0349	0.0051	0.5
0.3	0.2382	0.3256	–		−0.0119	0.7
0.5	0.1928	0.2434	0.2067	–		
0.7	0.1364	0.1696	0.1488	0.1207	–	

Table 3.11. Exact estimates of the second azimuthal harmonics $\varrho^2(\eta,\zeta)$ and $\sigma^2(\eta,\zeta)$ of brightness coefficients for an approximating 3-term phase function (the Elterman-1968 model, $\lambda = 600$ nm, $\tau_0 = 0.351$, $\Lambda = 0.872$, $x_1 = 1.500$, $x_2 = 1.500$)

	0	0.1	0.3	0.5	0.7	
	–	1.2384	0.3888	0.1938	0.0945	0
		0.6386	0.2946	0.1614	0.0821	0.1
0	–		0.1617	0.0927	0.0482	0.3
0.1	0.0778	–		0.0538	0.0281	0.5
0.3	0.1489	0.1934	–		0.0147	0.7
0.5	0.1088	0.1251	0.0843	–		
0.7	0.0626	0.0684	0.0450	0.0269	–	

The $Q_2^2(\eta)$ and $S_2^2(\eta)$ polynomials are equal:

$$Q_2^2(\eta) = S_2^2(\eta) = R_{22}^2(0) = 1 \ . \tag{3.208}$$

Using (3.208), we obtain exact expressions for $\Omega_2^2(\eta)$ and $\Theta_2^2(\eta)$ functions [177]:

$$\Omega_2^2(\eta) = 3(1-\eta^2)W^2(\eta) \ , \quad \Theta_2^2(\eta) = 3(1-\eta^2)U^2(\eta) \ . \tag{3.209}$$

By substituting (3.209) into (3.40) and (3.41), we find at $m = 2$ exact expressions for the second azimuthal harmonics ϱ^2 and σ^2 in the case of a 3-term phase function (3.19):

$$\varrho^2(\eta,\zeta) = \frac{3\Lambda x_2(1-\eta^2)(1-\zeta^2)}{64(\eta+\zeta)}[W^2(\eta)U^2(\zeta)+W^2(\zeta)U^2(\eta)] \tag{3.210}$$

$$\sigma^2(\eta,\zeta) = \frac{3\Lambda x_2(1-\eta^2)(1-\zeta^2)}{64(\eta-\zeta)}[W^2(\eta)U^2(\zeta)-W^2(\zeta)U^2(\eta)] \ . \tag{3.211}$$

In this case the function $E^2(\eta,\zeta)$ is:

$$E^2(\eta,\zeta) = \frac{3\Lambda x_2(1-\eta^2)(1-\zeta^2)}{32}W^2(\eta)U^2(\zeta) \ . \tag{3.212}$$

Table 3.11 gives exact values of the second azimuthal harmonics for ϱ^2 and σ^2 calculated in [177] with formulas (3.210–3.211) for $\tau_0 = 0.351$, $\Lambda = 0.872$ and an approximating 3-term phase function ($x_1 = 1.500$, $x_2 = 1.500$). Due to a symmetry of these harmonics, the ϱ^2 values are given in the upper diagonal part of the table, and the σ^2 estimates in the lower part.

3.10 Atmospheric Albedo, Spherical Albedo and Irradiance of the Planetary Surface

Now we consider the estimation of the planetary atmosphere albedo $A_1(\zeta,\tau_0)$, spherical albedo $A_*(\tau_0)$ and irradiance of the planetary surface $E(\zeta,\tau_0)$, all of which are important from a practical point of view. For an arbitrary, height-independent phase function $x(\gamma)$, these estimates are [187]:

$$A_1(\zeta, \tau_0) = -\frac{1}{\zeta}\varphi_1^0(\zeta, \tau_0) \ , \tag{3.213}$$

$$A_*(\tau_0) = 1 - 2\int_0^1 \varphi_1^0(\zeta, \tau_0)\, d\zeta \ , \tag{3.214}$$

$$E(\zeta, \tau_0) = \pi S \psi_1^0(\zeta, \tau_0) \ . \tag{3.215}$$

Summing and extracting $\Omega_i^0(\eta)$ and $\Theta_i^0(\eta)$ functions determined by relationships (3.38 – 3.39) and then using definitions (3.52) for $W^0(\eta)$ and $U^0(\eta)$ functions and some integral relationships [187], we obtain, for pure scattering (3.19) exact expressions for φ_1^0 and ψ_1^0 functions by Ambartsumyan:

$$\varphi_1^0(\eta) = \Psi_1^0(\eta) = \frac{2\eta W^0(\eta)}{(3 - x_1)\tau_0(\alpha_1^0 + \beta_1^0) + 6(\alpha_2^0 + \beta_2^0)} \ . \tag{3.216}$$

Thus, instead of (3.213 – 3.215), using (3.216), we have exact reference formulas to determine $A_1(\zeta)$, A_*, and E values for a 3-term phase function:

$$A_1(\zeta, \tau_0) = 1 - \frac{2W^0(\eta)}{(3 - x_1)(\alpha_1^0 + \beta_1^0)\tau_0 + 6(\alpha_2^0 + \beta_2^0)} \ , \tag{3.217}$$

$$A_*(\tau_0) = 1 - \frac{4}{(3 - x_1)\tau_0 + 6\dfrac{\alpha_2^0 - \beta_2^0}{\alpha_1^0 - \beta_1^0}} \ , \tag{3.218}$$

$$E(\zeta) = \frac{2\pi S W^0(\eta)}{(3 - x_1)\tau_0(\alpha_1^0 + \beta_1^0) + 6(\alpha_2^0 + \beta_2^0)} \ . \tag{3.219}$$

Note, that a transition from a 2-term phase function ($x_2 = 0$) to a three-term (3.19) does not change formulas (3.217 – 3.219), i.e., they are independent of the x_2 coefficient. Hence, for practical and important Rayleigh phase function $x(\gamma) = 3/4\ (1 + \cos^2\gamma)$ when $x_0 = 1$, $x_1 = 0$, and $x_2 = 1/2$, formulas (3.217 – 3.219) allow an exact estimation of the albedo of a molecular atmosphere, its spherical albedo, and irradiance of the planetary surface [177]:

$$A_1(\zeta, \tau_0) = 1 - \frac{2W^0(\eta)}{3\tau_0(\alpha_1^0 + \beta_1^0) + 6(\alpha_2^0 + \beta_2^0)} \ , \tag{3.220}$$

$$A_*(\tau_0) = 1 - \frac{4}{3\tau_0 + 6\dfrac{\alpha_2^0 + \beta_2^0}{\alpha_1^0 + \beta_1^0}} \ , \tag{3.221}$$

$$E(\zeta, \tau_0) = \frac{2\pi S W^0(\eta)}{3\tau_0(\alpha_1^0 + \beta_1^0) + 6(\alpha_2^0 + \beta_2^0)} \ . \tag{3.222}$$

The results obtained enable exact asymptotic formulas for $A_1(\zeta, \tau_0)$, $A_*(\tau_0)$ and $E(\zeta, \tau_0)$ to be written for a planetary atmosphere with a large optical thickness $\tau_0 \gg 1$ and an arbitrary three-term phase function $\bar{x}(\gamma)$. Substituting in (3.213 – 3.215) the function $W^0(\eta)$ at $\tau_0 \to \infty$ with the $H^0(\eta)$ function after Chandrasekhar, we obtain:

$$A_{1as}(\zeta, \tau_0) = 1 - \frac{2H^0(\eta)}{(3-x_1)h_1^0\tau_0 + 6h_2^0}, \tag{3.223}$$

$$A_*^{as}(\tau_0) = 1 - \frac{4}{(3-x_1)\tau_0 + 6h_2^0/h_1^0}, \tag{3.224}$$

$$E_{as}(\zeta, \tau_0) = \frac{2\pi S H^0(\zeta)}{(3-x_1)h_1^0\tau_0 + 6h_2^0}. \tag{3.225}$$

Exact asymptotic formulas (3.223 – 3.225) have been obtained in [348]. For a simpler two-term phase function, formula (3.223) has been defined first for pure scattering ($\Lambda = 1$) [64]. For the same phase function but with pure scattering ($\Lambda < 1$), the diffused reflection and transmission of light by an optically thick layer has been studied in [226]. With an assumption that the given phase function $x(\gamma)$ is expanded in Legendre polynomials $P_n(\cos\gamma)$, the angular distribution of radiation leaving an optically thick medium, has been outlined by Ambartsumyan [7].

Among classical problems of radiation transfer theory, an important problem is the determination of the spectral, spatial and angular distribution of radiation intensity reflected by a medium of an infinitely large optical thickness $\tau_0 = \infty$ (drainage basins, cloudy atmospheres of the Earth, Venus and Jupiter).

As a rule, the phase functions of such media have a marked maximum towards small angles $\gamma \sim 0°$ [43, 216, 276]. Since an expansion of strongly stretched phase functions $x(\gamma)$ in Legendre polynomials $P_n(\cos\gamma)$ requires consideration of hundreds of terms, strict techniques producing exact solutions in the transfer theory are inefficient. For an approximate account of the angular anisotropy of real media, a technique has been proposed based on similarity ratios (3.18), careful consideration single scattering with an arbitrary phase function $x(\gamma)$ and an approximate description of the diffused radiation field for an approximating 4-term phase function (3.20).

Express the coefficient of reflection $\varrho(\eta, \zeta, \varphi)$ through a medium of an infinitely large optical thickness with a four-term phase function (3.20) as part of a Fourier series in azimuth φ:

$$\varrho(\eta, \zeta, \varphi) = \varrho^0(\eta, \zeta) + 2\sum_{m=1}^{3} \varrho^m(\eta, \zeta)\cos m\varphi. \tag{3.226}$$

The azimuthal harmonics $\varrho^m(\eta, \zeta)$ in (3.226) are:

$$\varrho^m(\eta, \varphi) = \frac{\Lambda}{4}\sum_{i=m}^{3}(-1)^{i+m} c_i^m \frac{\varphi_i^m(\eta)\varphi_i^m(\zeta)}{\eta+\zeta}. \tag{3.227}$$

As Sobolev has shown [187], the Ambartsumyan $\varphi_i^m(\eta)$ functions are related to the Chandrasekhar $H^m(\eta)$ functions in case of an arbitrary phase function $x(\gamma)$ in the following way:

$$\varphi_i^m(\eta) = P_m^m(\eta) q_i^m(\eta) H^m(\eta) \quad (m = 1,2,3,\ldots,N, \; i = m, m+1, \ldots, N) \; , \tag{3.228}$$

where q_i^m polynomials have a power $(N-m)$

$$q_i^m(\eta) = R_{im}^m(0) + \sum_{k=1}^{N} a_{ik}^m \eta^k \quad (m = 1,2,3,\ldots,N, \; i = m, m+1, \ldots, N) \; . \tag{3.229}$$

The $R_{im}^m(\eta)$ polynomials are determined from recurrent relationships (3.33). The Chandrasekhar $H^m(\eta)$ functions satisfy a non-linear integral equation [205]:

$$H^m(\eta) = 1 + \eta H^m(\eta) \int_0^1 \frac{\psi^m(\eta') H^m(\eta')}{\eta + \eta'} d\eta' \quad (m = 1,2,3,\ldots,N) \; . \tag{3.230}$$

Linear integral equations for these functions have been obtained in [187]. The core functions $\psi^m(\eta)$ in (3.230) are determined according to (3.31 – 3.32).

Thus, to obtain exact expressions for the coefficient of reflection $\varrho(\eta,\zeta,\varphi)$ and its azimuthal harmonics $\varrho^m(\eta,\zeta)$ for an approximating 4-term phase function (3.20), first, it is necessary to find expressions for $q_i^m(\eta)$ polynomials at $m = 0,1,2,3$ from an integral equation given in [187]. Then, linear equation (3.230) must be solved numerically and moments h_k^m of the Chandrasekhar H^m function must be found:

$$h_k^m = \int_0^1 H^m(\eta) \eta^k d\eta \; . \tag{3.231}$$

Then, using (3.226 – 3.228), we determine the total coefficient of reflection $\varrho(\eta,\zeta,\varphi)$ and its azimuthal terms $\varrho^m(\eta,\zeta)$. In [370], exact expressions have been obtained for the azimuthal harmonics at $m = 0,1,2,3$ and an approximating phase function (3.20).

To determine the contribution of single scattering into the azimuthal harmonics $\varrho^m(\eta,\zeta)$ of the total reflection coefficient $\varrho(\eta,\zeta,\varphi)$, the respective terms $\varrho_{(1)}^m(\eta,\zeta)$ were selected in (3.227) and calculated from an exact formula:

$$\varrho_{(1)}^m(\eta,\zeta) = \frac{\Lambda S p^m(\eta,\zeta)}{4(\eta+\zeta)} \; , \tag{3.232}$$

where $p^m(\eta,\zeta)$ are coefficients of expansion of the given phase function $x(\gamma)$ in Fourier series by φ azimuth:

$$x(\gamma) = p^0(\eta,\zeta) + 2 \sum_{m=1}^{N} p^m(\eta,\zeta) \cos m\varphi \; . \tag{3.233}$$

3.11 Approximate Expressions for the Coefficients of Diffused Reflection and Transmission with an Arbitrary Phase Function

The strict expressions obtained for the azimuthal harmonics of brightness coefficients ϱ^m and σ^m make it possible, by summing the respective Fourier series (3.22 – 3.23), to obtain exact formulas for total coefficients of diffused reflection ϱ and transmission σ in using similarity ratios (3.18) and an approximating three-term phase function (3.19). It follows from the physics of similarity ratios that these formulas approximately consider the effect on the radiation field of the angular anisotropy of an initial phase function $x(\gamma)$, with a maximum towards $\gamma \sim 0°$.

Table 3.12 gives exact estimates of total brightness coefficients $\varrho(\eta, \zeta, \varphi)$ and $\sigma(\eta, \zeta, \varphi)$ calculated for $\tau_0 = 0.3$, $\Lambda = 1$ and an approximating three-term phase function ($x_1 = 1.269$, $x_2 = 0.909$) with azimuth $\varphi = 0°$. Based on the symmetry of coefficients of reflection ϱ and transmission σ, from variables η and ζ, values of $\varrho(\eta, \zeta, \varphi)$ are given in the upper diagonal part of the table, and values of $\sigma(\eta, \zeta, \varphi)$ in the lower part.

To raise the efficiency of the proposed technique for the determination of brightness coefficients ϱ and σ, their finite calculation formulas we should consider the contributions of single scattering with a given atmospheric phase function $x(\gamma)$ [177]:

$$\varrho_{\text{calc}}(\eta, \zeta, \varphi) = \varrho(\tau_0^*, \eta, \zeta) - \frac{\Lambda_* S x(\gamma)}{4(\eta + \zeta)} [1 - e^{-\tau_0^*(1/\eta + 1/\zeta)}] + \frac{\Lambda S x(\gamma)}{4(\eta + \zeta)} [1 - e^{-\tau_0(1/\eta + 1/\zeta)}] , \tag{3.234}$$

$$\sigma_{\text{calc}}(\eta, \zeta, \varphi) = \sigma(\tau_0^*, \eta, \zeta) - \frac{\Lambda_* S \bar{x}(\gamma)}{4(\zeta - \eta)} [e^{-\tau_0^*/\zeta} - e^{-\tau_0^*/\eta}] + \frac{\Lambda S x(\gamma)}{4(\zeta - \eta)} [e^{-\tau_0/\zeta} - e^{-\tau_0/\eta}] . \tag{3.235}$$

When atmospheric phase functions $x(\gamma)$ are approximated well with 3-term interpolation expressions (3.19) without using similarity ratios (3.18) and

Table 3.12. Exact estimates of brightness coefficients $\varrho(\eta, \zeta, \varphi)$ and $\sigma(\eta, \zeta, \varphi)$ for an approximating three-term phase function

	0	0.1	0.3	0.5	0.7	
		10.6949	3.2910	1.6734	0.7100	0
0	–	3.5946	2.4520	1.0891	0.6896	0.1
0.1	0.8792		1.2617	0.6629	0.3239	0.3
0.3	1.2339	1.5933		0.3328	0.1370	0.5
0.5	0.9692	1.1202	0.8878	–	0.1370	0.7
0.7	0.7134	0.7058	0.3958			

without selecting a major maximum of phase function at $\gamma = 0°$, the proposed approximate technique gives almost exact results. However, for real aerosol phase functions with a marked peak towards small angles of scattering, such approximation is of little value [43]. Therefore, according to [177], it is necessary to eliminate a maximum from the given strongly stretched phase function $x(\gamma)$ at some small angle $\gamma = \gamma_{cut}$. Also in solving an initial equation of radiation transfer from (3.14), to use a similarity ratio (3.18), by approximating the remaining phase function $\bar{x}(\gamma)$ with the help of two interpolation coefficients $|x_1| < 3$ and $|x_2| < 5$. From the physical point of view, this approach is justified because diffused radiation is characterized by a decrease (especially with growing τ) of the angular anisotropy compared with singly scattered radiation. However, to improve the reliability and potential of practical application of the proposed technique, strict quantitative estimates of this effect are needed including real optical models of an aerosol atmosphere. To achieve this calculations were made of the coefficients of reflection ϱ and transmission σ for the real terrestrial atmosphere in the wavelength interval $\lambda = 400$ to 850 nm, using an advanced method of spherical harmonics [77]. An aerosol model by Elterman (1968) closed with respect to atmospheric phase function, was used in calculations (see Table 1.3). Expanding an initial phase function $x(\gamma)$ in Legendre polynomials $P_n(\cos\gamma)$, 3 and 50 terms were left. The calculation of brightness coefficients ϱ and σ was made in both cases, for: a homogeneous atmosphere with a height-constant phase function $x(\gamma)$ and quantum survival probability Λ; and an inhomogeneous atmosphere whose phase function and quantum survival probability depend on height. At $N = 3$ and $N = 50$, the first-order scattering was considered using the respective formulas of the theory of radiation transfer [184]. This made it possible to accurately assess the upper limit of uncertainty of the modified method by Sobolev. This states that: the first-order scattering is considered for an arbitrary phase function $x(\gamma)$; the contribution of multiple scattering is described approximately based on similarity ratios; and on the use of exact solutions of the theory of radiation transfer for an approximating 3-term phase function (3.19). The results of comparison are given in Table 3.13.

For each of the optical models shown in Table 3.13, the upper line of estimates corresponds to reflection coefficient ϱ, the lower line to the coefficient of transmission σ. Bearing in mind optimal conditions of spectrophotometry of natural formations from space [68], we calculated the coefficient of reflection $\varrho(\eta, \zeta, \varphi)$ for $\zeta = 1$ (the sun in zenith) and $\eta = 1$ (nadir viewing), and the coefficient of transmission $\sigma(\eta, \zeta, \varphi)$ for $\zeta = 1$ and $\eta = 0.9$. An analysis of the respective numerical data of Table 3.13 suggests two conclusions:

1. Consideration of the vertical dependence of optical parameters $x(\gamma)$ and Λ is unimportant (within 2 – 3%) for the determination of brightness coefficients of the terrestrial atmosphere in the visible spectrum.
2. The spectrum-averaged uncertainty of the proposed calculation technique based on a modified technique by Sobolev does not exceed 5%.

Table 3.13. Comparison of results of model calculations of brightness coefficients ϱ and σ of an aerosol terrestrial atmosphere using an advanced method of spherical harmonics and a modified method by Sobolev (the Elterman-1968 model; $\lambda = 400$ to 850 ml)

Wavelength λ, nm	400	450	550	650	750	850
Homogeneous model, real phase function (N = 50)	0.08423 0.37940	0.08099 0.35243	0.05691 0.29825	0.04729 0.23924	0.03972 0.19206	0.03383 0.16915
Inhomogeneous model, real phase function (N = 50)	0.08794 0.39543	0.08253 0.35856	0.05874 0.30734	0.04920 0.24848	0.04159 0.20082	0.03440 0.17171
Inhomogeneous model, 3-term phase function (N = 3)	0.07466 0.33570	0.07237 0.31443	0.04742 0.24812	0.04243 0.21432	0.03656 0.17653	0.03246 0.16203
Inhomogeneous model, real phase function, single scattering (N = 50)	0.04344 0.19533	0.04017 0.17451	0.03903 0.20420	0.03691 0.18642	0.03498 0.16888	0.03290 0.16424
Inhomogeneous model, 3-term phase function single scattering (N = 3)	0.03489 0.15303	0.03374 0.14277	0.03047 0.15563	0.03209 0.15835	0.03124 0.14717	0.03179 0.15507

These conclusions are important for numerical solutions of radiation transfer theory problems. In particular, they are of special importance for the spherical harmonics technique, since they make it possible to omit the vertical dependences of atmospheric optical parameters. Thus, no more than 3 terms remain in an expansion of an initial phase function $x(\gamma)$ in Legendre polynomials $P_n(\cos\gamma)$. This is still more important in solving inverse problems of remote sensing from optical survey of the Earth from space. It enables the use of 3-order matrices instead of the matrices of the 50 – 100th order.

The expressions obtained for ϱ^m in a 4-term phase function make it possible, by summing Fourier series (3.226), to find an exact formula for the total coefficient of diffuse reflection $\varrho(\eta,\zeta,\varphi)$ using similarity ratios (3.18) and an approximating 4-term phase function (3.20). The upper diagonal part of Table 3.14 gives exact estimates of the $\varrho(\eta,\zeta,\varphi)$ function, calculated in [176] using formulas (3.226 – 3.227) for pure scattering $\Lambda = 1$ for a 4-term phase function ($x_1 = 1.615$, $x_2 = 1.266$, $x_3 = 0.432$) and $\varphi = 0°$.

The problem of the accuracy of different approximate expressions for the coefficient of reflection $\varrho(\eta,\zeta,\varphi)$ by a medium with an infinitely large optical thickness $\tau_0 \gg 1$ when using the multiple light scattering theory in practice gains importance. For instance, an approximate formula for $\varrho(\eta,\zeta,\varphi)$ defined after the classical method by Sobolev [184] is now widely used:

$$\varrho_{appr}(\eta,\zeta,\varphi) = \frac{1}{2} + \frac{1}{4}x(\gamma)\,\frac{1}{\eta+\zeta} + \frac{\eta\zeta(3+x_1)}{4(\eta+\zeta)} \,. \qquad (3.236)$$

Table 3.14. Estimates of reflection coefficients $\varrho(\eta, \zeta, \varphi)$ and $\varrho_{\text{app}\eta}(\eta, \zeta, \varphi)$ for pure scattering $\Lambda = 1, \tau_0 = \infty$ and an approximating four-term phase function

	0.1	0.3	0.5	0.7	1.0	
	6.66	3.47	2.19	1.43	0.65	0.1
		2.43	1.76	1.30	0.82	0.3
0.1	5.75		1.43	1.18	0.94	0.5
0.3	2.90	1.94		1.10	1.03	0.7
0.5	1.84	1.41	1.15		1.12	1.0
0.7	1.24	1.06	1.00	0.97		
1.0	0.67	0.81	0.92	1.01	1.10	

The exact expression for $\varrho(\eta, \zeta, \varphi)$ just obtained permits an accurate estimation of the reliability of Sobolev's method in its classical version [187]. In the lower diagonal part of Table 3.14, estimates were obtained using (3.236). A comparison with exact estimates of $\varrho_{\text{appr}}(\eta, \zeta, \varphi)$ in the upper part shows that the simplicity of (3.236) is not supplemented with a high accuracy of approximate estimates of $\varrho_{\text{appr}}(\eta, \zeta, \varphi)$ for all angles η and ζ.

To advance the effectiveness of the proposed modification of Sobolev's technique in a finite formula for reflection coefficient $\varrho(\eta, \zeta, \varphi)$ it is necessary to examine thoroughly single scattering at a given phase function $x(\gamma)$:

$$\varrho_{\text{calc}}(\eta, \zeta, \varphi) = \varrho(\eta, \zeta, \varphi) - \frac{\Lambda_* S x(\gamma)}{4(\eta + \zeta)} + \frac{\Lambda S x(\gamma)}{4(\eta + \zeta)} . \tag{3.237}$$

With an initial phase function $x(\gamma)$ well approximated by 4-term interpolation expressions (3.18) and selecting maximum of phase function at $\gamma = 0°$, the proposed technique gives almost accurate results [88, 89]. When this approximation does not function well it is necessary to have a maximum at a small angle $\gamma = \gamma_{\text{cut}}$ off the given phase function $x(\gamma)$ and to use a similarity ratio in solving the initial transfer equation (3.14) by approximating the remaining phase function $\bar{x}(\gamma)$ with three interpolation coefficients: $|x_1| < 3$, $|x_2| < 5$, $|x_3| < 7$. Note, that even with one approximating coefficient x_1, the use of similarity ratios in the simplest version of a two-term phase function $x(\gamma) = 1 + x_1 \cos \gamma$ ensures a highly accurate determination of the radiation intensity reflected by a flat layer with an infinitely large optical thickness $\tau_0 \gg 1$. This has been exemplified [20] by calculations of the phase curve of the brightness of the Venusian cloudy atmosphere in the visible spectrum.

These exact and approximate expressions for basic functions of multiple light scattering theory in a flat layer with a finite and infinitely large optical thickness can be used successfully. For example in calculations of the brightness coefficients for two-layer media considering optical refraction phenomena or mirror-effects at the atmosphere-sea interface. They are useful in theoretical determination of the Earth phase brightness observed from space at large distances ($\sim 10^4 \sim 10^5$ km).

The expressions obtained for coefficients of diffused reflection and transmission of a medium with a finite or infinitely large optical thickness enable calculation of the radiation field of the atmosphere-sea system. Here, it is necessary to use the results from [40, 86, 87]. In these works the coefficient of brightness of a vertically inhomogeneous two-layer medium has been expressed through the respective brightness coefficients of one-layer media, and the effects of reflection and refraction of light at the atmosphere-sea interface have been included.

3.12 Determination of the Parameters of the Radiation Field with the Use of an Advanced Method of Spherical Harmonics

The spherical harmonics technique is a useful way of solving the radiation transfer equation when the practical problems are so complicated that an exact analytical solution is impossible [382].

However, current computer capacity makes it impossible to realize a classical presentation of the spherical harmonics technique without double checking calculation accuracy. This is because at the irradiated upper boundary of the atmosphere, the incident radiation has a high angular singularity. Since this singularity is preserved within the whole atmospheric layer H, one cannot expect a rapid convergence of Fourier series for total radiation intensity. In addition, for spectral intervals with strongly scattered radiation, calculations using this method require an additional accuracy check because of the need to solve systems of high-order equations. Very large computer memories are needed as well as high speed operation. Further improvements were required in the standard calculation procedure used in the spherical harmonics technique.

Let us consider these improvements for zero azimuthal harmonic of solar radiation intensity (m = 0). The radiation transfer calculation scheme for both reflection non-orthotropicity and non-zero azimuthal harmonics will be considered generally in the next chapter.

Photons of total radiation are divided into two classes: those involved in at least one internal collision (scattered radiation photons) and those entering from outside (direct solar radiation photons):

$$I(z,\zeta,\eta) = I^{(1)}(z,\zeta,\eta) + \frac{S}{2} e^{-1/\zeta \int_z^H \alpha(z')dz'} \delta(\zeta-\eta) \ . \tag{3.238}$$

Photons of these two classes are presented in formula (3.238), respectively, by the first and second items. Substituting (3.238) into the equation of initial boundary problem (3.14) – (3.16), we obtain:

$$\eta \frac{\partial I^{(1)}(z,\zeta,\eta)}{\partial z} + \alpha(z) I^{(1)}(z,\zeta,\eta) = \sum_{p'=1}^{P'} \sigma_S^{p'}(z) \sum_{k=0}^{2K+1} B_k^{p'}(z) \frac{2k+1}{2}$$

$$\times P_k(\eta) \left[\int_{-1}^{1} I^{(1)}(z,\zeta,\eta') P_k(\eta') d\eta' + \frac{S}{2} e^{-1/\zeta \int_z^H \alpha(z')dz'} P_k(\zeta) \right] , \tag{3.239}$$

$$I^{(1)}(0,\eta^+) = 2A \int_0^1 I^{(1)}(0,\eta^-)\eta^- d\eta^- + \frac{D_0}{\pi} A , \tag{3.240}$$

$$I^{(1)}(H,\eta^-) = 0 , \tag{3.241}$$

where:

$$D_0 = \pi S \zeta e^{-[\int_0^H \alpha(z')dz']/\zeta} = \pi S \zeta e^{-[\sum_{i=1}^{I'+1} \alpha^i (z_i - z_{i-1})]/\zeta} . \tag{3.242}$$

Since $I^{(1)}(z,\eta)$ does not contain a high singularity, in contrast to $I(z,\eta)$, the solution of boundary problem (3.239 – 3.242) can be approximated by part of the low-order Fourier series:

$$I^{(1)}(z,\eta) \approx (4\pi)^{-1} \sum_{n=0}^{2N+1} (2n+1) \xi_n(z) P_n(\eta) . \tag{3.243}$$

Upon substituting (3.243) into (3.239), using the orthogonal properties of Legendre polynomials and recurrent formulas for these polynomials, the integrated differential equation (3.239) is reduced to a system of linear differential equations:

$$(n+1) \frac{d\xi_{n+1}(z)}{dz} + n \frac{d\xi_{n-1}(z)}{dz} + \alpha(z)[1 - C_n(z)](2n+1)\xi_n(z)$$

$$= \pi S \alpha(z) C_n(z)(2n+1)(-1)^n P_n(\zeta) e^{-1/\zeta \int_z^H \alpha(z')dz'} ,$$

$$(n = 0, 1, \ldots, 2N+1) \tag{3.244}$$

$$\frac{d\xi_{2N+2}}{dz} = 0 , \tag{3.245}$$

where

$$C_n(z) = \frac{1}{\alpha(z)} \sum_{p'=1}^{P'} \sigma_S^{p'}(z) B_k^{p'}(z) .$$

Boundary conditions (3.240) and (3.241) presenting, in fact, conditions of transformation of some functions $f_1(\eta)$ and $f_2(\eta)$ into zero at $\eta = \eta^+$ and $\eta = \eta^-$, a final P_{2N+1} approximation can be satisfied only $N+1$ times, for instance, by their transformation into the form after Marshak [50]:

$$\int_0^1 f_1(\eta^+) P_{2n-1}(\eta^+) d\eta^+ = 0 ,$$

$$\int_{-1}^0 f_2(\eta^-) P_{2n-1}(\eta^-) d\eta^- = 0 , \qquad (n = 1, 2, \ldots, N+1) .$$

Upon integration, we obtain:

$$\vec{\xi}_{odd}(0) = D_0 \frac{4A}{1+A} \vec{R} + \left(-\hat{P}_5 + \frac{4A}{1+A} \hat{P}_6 \right) \vec{\xi}_{even}(0) \ , \tag{3.246}$$

$$\vec{\xi}_{even}(H) = \hat{P}_5 \vec{\xi}_{even}(H) \ , \tag{3.247}$$

where $\hat{P}_5$ and $\hat{P}_6$ are square matrices of the order $N+1$, and $\vec{\xi}_{odd}(0)$, $\vec{\xi}_{even}(0)$, $\vec{\xi}_{odd}(H)$, $\vec{\xi}_{even}(H)$ are second-order vectors;

$$(\hat{P}_5)_{n,k} = (4n-3)\chi_{2n-1,2k-1} \ ,$$

$$(\hat{P}_5)_{n,k} = (4n-3)\chi_{0,2k-1}\chi_{2n-2,1} \ ,$$

$$\chi_{n,1} = \int_0^1 P_n(\eta) P_1(\eta)\, d\eta = \begin{cases} \dfrac{1}{2n+1} \ , & n = 1 \\ 0, n = 1 \ , & (n-1)-\text{even} \\ \dfrac{(-1)^{1/2(n+1-1)} n!\, 1!}{2^{n+1-1}(1-n)(n+1+1)\left[\left(\frac{n}{2}\right)!\left(\frac{1-1}{2}\right)!\right]^2} \ , & \text{n-even, l-odd} \end{cases}$$

$$[\vec{\xi}_{even}(0)]_n = \vec{\xi}_{2n-2}(0) \ , \qquad [\vec{\xi}_{odd}(0)]_n = \vec{\xi}_{2n-1}(0) \ ,$$

$$[\vec{\xi}_{odd}(H)]_n = \vec{\xi}_{2n-2}(H) \ , \qquad [\vec{\xi}_{odd}(H)]_n = \vec{\xi}_{2n-1}(H) \ .$$

The components of $\vec{R}$ vector are determined from equations:

$$(\vec{R})_n = \chi_{0,2n-1} \ , \qquad (n = 1,2,\ldots,N+1) \ . \tag{3.248}$$

If an atmospheric layer H is divided into so many zones $I'+1$ that, within the accuracy limits, the integral of an arbitrary $\xi_n(z)$ function over any zone i can be calculated from the formula of trapezoids, then integrating (3.244) over the zone with lower z_{i-1} and upper z_i boundaries leads to a system of linear algebraic equations:

$$\xi_{n+1}(z_{i-1}) - (1 - C_n^i)\Delta_{i-1}^i \alpha^i \frac{2n+1}{2(n+1)} \xi_n(z_{i-1}) + \frac{n}{n+1} \xi_{n-1}(z_{i-1})$$
$$= \xi_{n+1}(z_i) + (1 - C_n^i)\Delta_{i-1}^i \alpha^i \frac{2n+1}{2(n+1)} \xi_n(z_i) + \frac{n}{n+1} \xi_{n-1}(z_i) - Q_n^{(1)} \tag{3.249}$$

where

$$Q_n^{(i)} = S\zeta C_n^i (-1)^n P_n(\zeta) \frac{2n+1}{n+1} (g_{i+1} - g_i)$$

$$\Delta_{i-1}^i = (z_i - z_{i-1}) \ ,$$

$$g_i = e^{-\sum_{j=i}^{I'+1} \Delta_{j-1}^j \alpha^j 1/\zeta} \ , \qquad \text{if} \quad i = 1,2,\ldots,I'+1 \ ; \qquad g_{I'+2} = 1 \ .$$

With optically thick atmospheric layers, individual zones of strong scattering or absorption of radiation are subdivided into optically thinner sub-zones.

Thus, if vector $\vec{\xi}(z_i)$ is known, three-diagonal system of equations (3.249) is easily solved with respect to $\vec{\xi}(z_{i-1})$. The vector-matrix form of this system is:

$$\hat{T}_i^{(3)}\vec{\xi}(z_i) = \hat{T}_i\vec{\xi}(z_{i-1}) + \vec{Q}^i \ , \qquad i = 1,2,\ldots,z'+1 \ ,$$
$$\vec{\xi}(z_i) = \hat{P}^i\vec{\xi}(z_{i-1}) + [\hat{T}_i^{(3)}]^{-1}\vec{Q}^i \tag{3.250}$$

where $\hat{T}_i^{(3)}$, $\hat{T}_i^{(4)}$, $\hat{P}^i = [\hat{T}_i^{(3)}]^{-1}\hat{T}_i^{(4)}$ are square matrices of the order $2N+2$, and $Q_{n-1}^{(i)}$ is the vector of the same order with components $[\vec{Q}^{(i)}]_n = Q_{n-1}^{(i)}$.

Excluding in (3.250) the vectors of moments of angular distribution continuous at the internal boundaries of the zones, we obtain the necessary equations:

$$\vec{\xi}(H) = \hat{P}\vec{\xi}(0) + \vec{Q} \ , \tag{3.251}$$

$$\text{where } \vec{\xi}_{\text{even}}(H) = \hat{P}_1\vec{\xi}_{\text{even}}(0) + \hat{P}_2\vec{\xi}_{\text{odd}}(0) + \vec{Q}_{\text{even}} \ ,$$
$$\vec{\xi}_{\text{odd}}(H) = \hat{P}_3\vec{\xi}_{\text{even}}(0) + \hat{P}_4\vec{\xi}_{\text{even}}(0) + \vec{Q}_{\text{odd}} \ , \tag{3.252}$$

where

$$\hat{P} = \hat{P}^{I'+1}\hat{P}^{I'}\hat{P}^{I'-1}\ldots\hat{P}^2\hat{P}^1 \ , \tag{3.253}$$

$$\vec{Q} = \sum_{i=1}^{I'-1}(\hat{P}^{I'+1}\hat{P}^{I'}\ldots\hat{P}^1)[\hat{T}_i^{(3)}]^{-1}\vec{Q}^{(i)} + \hat{P}^{I'+1}[\hat{T}_{I'}^{(3)}]^{-1}Q^{(I')}$$
$$+ [\hat{T}_{I'+1}^{(3)}]\vec{Q}^{I'+1} \ . \tag{3.254}$$

Matrices $\hat{P}_1, \hat{P}_2, \hat{P}_3, \hat{P}_4$, vectors $\vec{Q}_{\text{even}}, \vec{Q}_{\text{odd}}, \vec{\xi}_{\text{even}}(0), \vec{\xi}_{\text{odd}}(0), \vec{\xi}_{\text{even}}(H), \vec{\xi}_{\text{odd}}(H)$, are constructed, respectively, from matrices $\hat{P}$ and vectors $\vec{Q}, \vec{\xi}(0), \vec{\xi}(H)$ following the rules described above. This system of Eqs. (3.249) is three-diagonal, which permits the substantial reduction of computer time expenditure on the realization of the spherical harmonics technique. It follows from the vector-matrix form (3.250) that all terms of $\hat{P}$ matrix can be found from (3.253), in contrast to the classical method of spherical harmonics, not by converting high-order matrices but by writing a system of vectors $\vec{\xi}(z_{i-1})$ as $\{1,0,0,\ldots\}$, $\{0,1,0,\ldots 0\}$, $\{0,0,0,\ldots,1\}$.

This enables the reduction of calculations for a fast-operating procedure to solve 3-diagonal systems for matrices $\hat{P}$ and vectors $\vec{\xi}$ calculation at the boundaries of atmospheric layers with different optical properties. This procedure has provided a two-order decrease of computer time expenditure compared to the classical method of spherical harmonics.

These are basic improvements to the method of spherical harmonics. Equations (3.246), (3.247), (3.251) and (3.252) form a closed system of four unknown quantities. Solving it, we find:

$$\vec{\xi}(0) = -\frac{4A}{1+A}D_0\hat{P}_7^{-1}(\hat{P}_4 - \hat{P}_5\hat{P}_2)\vec{R} - \hat{P}_7^{-1}(\vec{Q}_{\text{odd}} - \hat{P}_5\vec{Q}_{\text{even}}) \ , \tag{3.255}$$

where matrix $\hat{P}_7$ is determined from the formula

$$\hat{P}_7 = \hat{P}_3 - \hat{P}_5\hat{P}_1 - (\hat{P}_4 - \hat{P}_5\hat{P}_2)\left(\hat{P}_5 - \frac{4A}{1-A}\hat{P}_6\right) . \tag{3.256}$$

Substituting (3.255) into (3.246), we determine $\vec{\xi}_{odd}(0)$. Knowing $\vec{\xi}_{even}(0)$ and $\vec{\xi}_{odd}(0)$, we calculate $\vec{\xi}_{even}(H)$ and $\vec{\xi}_{odd}(H)$ using (3.251) and (3.252).

With the use of vectors determined from equations:

$$\begin{aligned}(\vec{\Pi}_{even})_n &= (4n-3)P_{2N-2}(\eta) , \\ (\vec{\Pi}_{odd})_n &= (4n-1)P_{2N-1}(\eta)\end{aligned} \tag{3.257}$$

from the angular distribution of intensity at an arbitrary z_i level of the atmosphere, can be represented as a sum of two scalar products:

$$I(z_i, \eta) = (4\pi)^{-1}[(\vec{\xi}_{even}(z_i), \vec{\Pi}_{even}) + (\vec{\xi}_{odd}(z_i), \vec{\Pi}_{odd}) . \tag{3.258}$$

Note, that at $\eta = \eta^+$ and vectors $\vec{\xi}_{odd}(H)$ and $\vec{\xi}_{even}(H)$, calculated from (3.251) and (3.252), the latter equation can be considered as one of the possible forms of the functional dependence between spectral surface albedo and the outgoing shortwave radiation measured at the atmospheric top. This circumstance can be used to transform the outgoing radiation intensity recorded from satellites into surface albedo. For transition from satellite-recorded quantities to albedos of homogeneous natural objects, analytical methods can also be used to solve transfer equations [178, 184, 187]. These permit a determination of the reflected radiation field for objects with albedo $A > 0$ based on solution of the problem of radiation transfer with albedo $A = 0$.

The technique described above has been realized on BESM-6 and ES-1022 computers. Figure 3.4 compares the calculated intensities of reflected radiation (the right-hand part) and transmitted radiation (the left-hand part) with classical results of Chandrasekhar [205] for diffused reflection and transmission of solar radiation by a non-absorbing Rayleigh atmosphere. A comparison of approximation orders shows that an order P_{2N+1} describes these laws with a high accuracy. Such comparisons were made for all optical thicknesses of the atmosphere given in [205]. Due to radiation polarization, the curves of Fig. 3.4 taken from [205] change by not more than 3%. The compared results show possibilities in applying the technique suggested for calculations of radiation transfer to optically thick atmospheric layers. With observation angles less than 60° vertically, an error in the determination of radiation intensity, including leaving the atmosphere at the top and incident onto the surface, does not exceed 1% at a selected order of approximation $2N+1 = 71$.

Figure 3.5 shows results of comparing the outgoing radiation intensity calculated after the method of successive approximations [156] for viewing azimuths $\varphi = 0, \pi$ and after an advanced method of spherical harmonics (oscillations have been smoothed down after the running mean method in the angular distribution of intensity).

Results of comparison show that the proposed method for calculations reproduces the laws obtained in the publications previously mentioned. Cal-

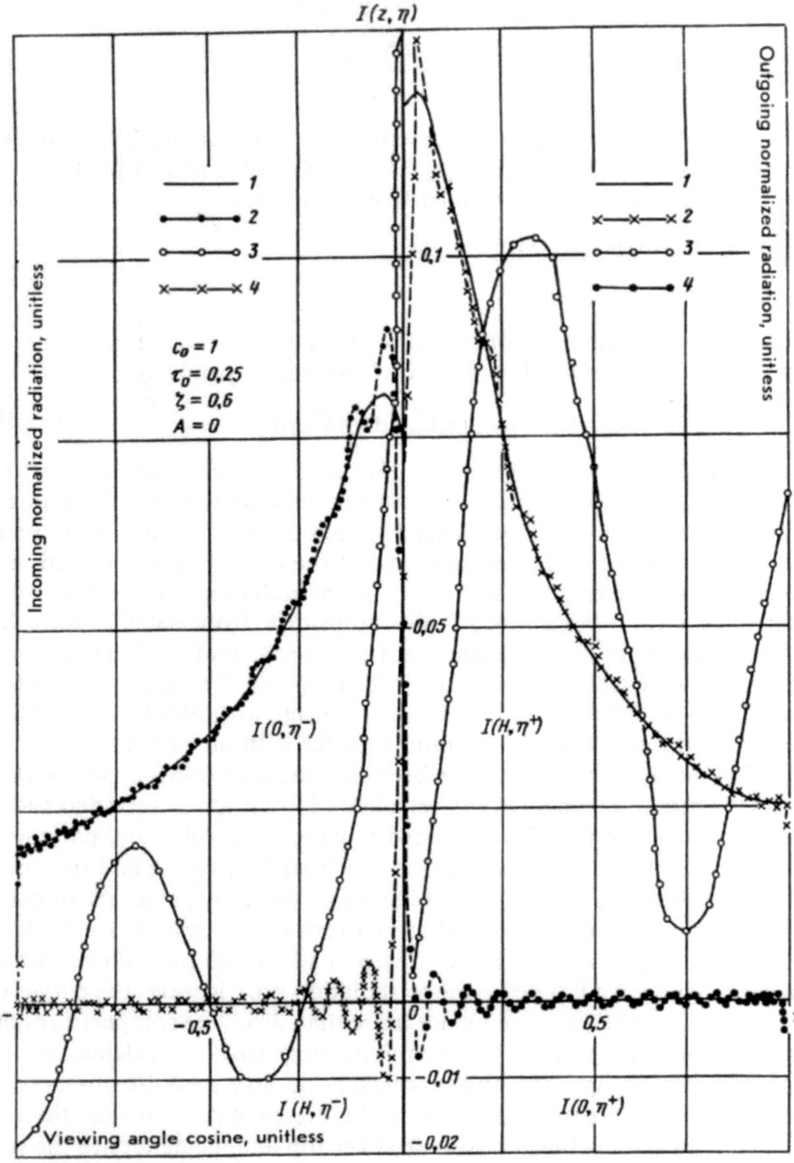

Fig. 3.4. Reflected (*right-hand side*) and transmitted (*left-hand side*) radiances of the Rayleigh non-absorbing atmosphere: *1* the Chandrasekhar method; *2* advanced spherical harmonics technique ($2N+1 = 71$); *3* advanced spherical harmonics technique ($2N+1 = 5$); *4* diffusively reflected and transmitted radiation

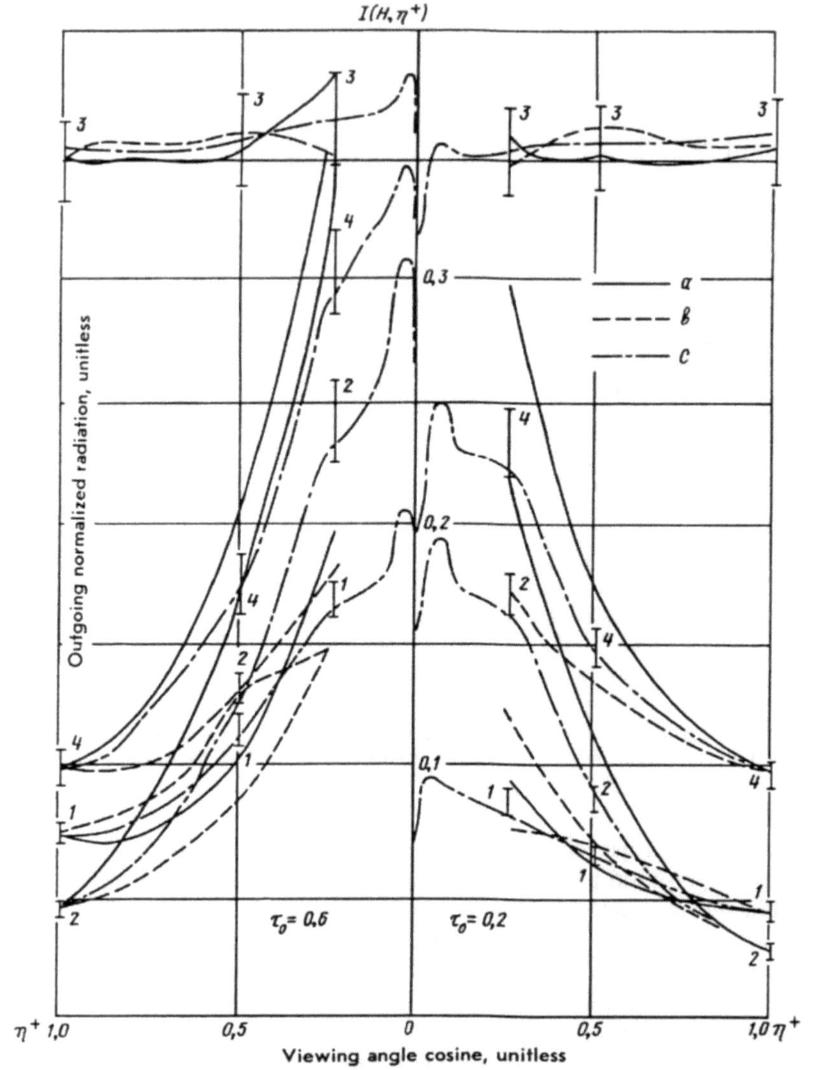

Fig. 3.5. Angular distributions of reflected radiance for a 2-layer non-absorbing aerosol atmosphere: *a* successive approximations technique [156], $\varphi = 0°$; *b* successive approximations technique [156], $\varphi = \pi$; *c* advanced spherical harmonics technique ($2N+1 = 71$), $\varphi = 0°$: *1* $A = 0$; $\theta_0 = 30°$; *2* $A = 0$; $\theta_0 = 75°$; *3* $A = 0.4$; $\theta_0 = 30°$; *4* $A = 0.4$; $\theta_0 = 75°$; $C_0 = 1$, C_n ($n = 1, \ldots, 10$) are taken from [156]; $C_n = 0$ ($n > 10$). Limits are shown to the accuracy of the techniques of successive approximations for intensities at $\varphi = 0°$ and $\varphi = \pi$

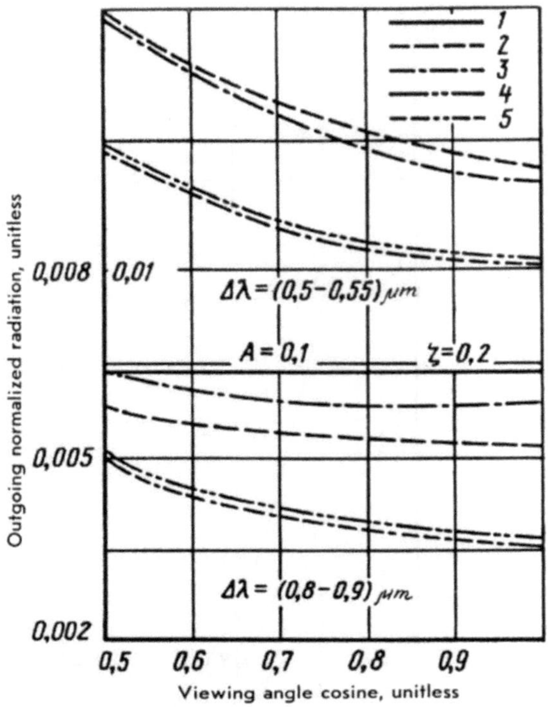

Fig. 3.6. Angular distributions of reflected radiances in two spectral intervals: *1* no atmosphere; *2* real multilayer atmosphere; *3* real 1-layer atmosphere; *4, 5* Rayleigh atmosphere for curves 2 and 3

culations with this technique for the azimuth-averaged intensity, phase functions VI (the right-hand part) of Fig. 3.5 and VIII (the left-hand part) from [156] are within the accuracy limits of the method of successive approximations, assuming monotonic change of calculated quantities in the transition from $\varphi = 0$ to $\varphi = \pi$. At angles θ less than 60°, a 1% accuracy of the method is preserved for a real stratified atmosphere, with an order of approximation not exceeding 80. Note that the azimuthal dependence is rather substantial (Fig. 3.5) especially for horizontal viewing directions. Figures 3.6 shows the extent to which the vertical inhomogeneity of the atmosphere can be taken into account.

Figure 3.7 gives the angular distributions of the outgoing radiation intensity calculated for a vertically inhomogeneous atmosphere model using Monte-Carlo simulations [3] and the proposed advanced method of spherical harmonics. In the first case, the atmosphere is spherical, in the second case, it is plane-parallel. Calculations were made for $\lambda = 0.85\ \mu m$ and two spectral albedos $A = 0$ and $A = 0.48$ (approximately, grass cover [3]).

Figure 3.7 confirms that the error of the proposed calculation technique does not exceed 5% at angles θ less than 60°. A similar conclusion holds for all wavelengths of the near IR spectrum given in [3].

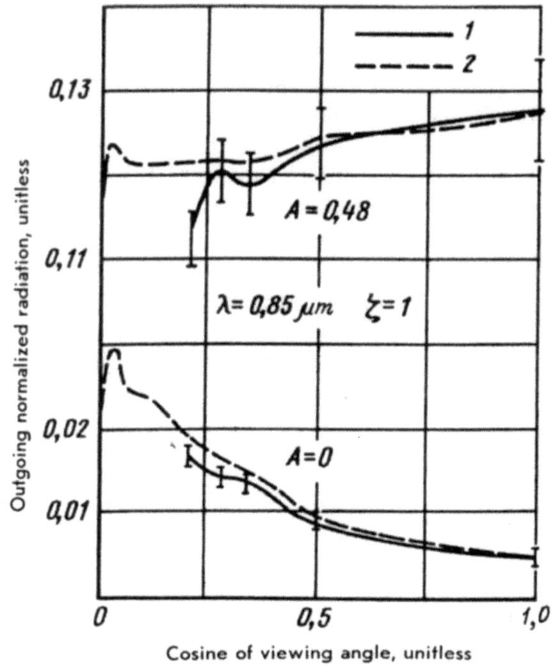

Fig. 3.7. Angular distributions of reflected intensities for the multilayer aerosol atmosphere model: *1* Monte-Carlo simulation, spherical atmosphere model [3]; *2* advanced spherical harmonics technique, plane-parallel atmosphere model

Table 3.15. A comparison of the angular distributions of reflected radiation intensities

η^+	0.4	0.6	0.8	1
$\Lambda = 1$	0.05422	0.03123	0.02165	0.01769
	0.05414	0.03117	0.02156	0.01761
	0.05370	0.03126	0.02172	0.01782
	0.05240	0.03070	0.02125	0.01680
$\Lambda = 0.9$	0.03685	0.02215	0.01617	0.01375
	0.03683	0.02210	0.01610	0.01374
	0.03687	0.02218	0.01620	0.01390
	0.03780	0.02240	0.01600	0.01345

Note: For each of the Λ values in the first line, calculations were made using the spherical harmonics method. Other lines are taken from [322]

Table 3.15 shows results of calculations of the angular distributions of reflected radiation intensities method and other methods for $\tau_0 = 1$, $\theta_0 = 0°$, and $A = 0$.

A comparison of the advanced method of spherical harmonics with a modification of Sobolev's method based on obtaining exact solutions for the initial boundary problem (3.14–3.16) for approximating model phase functions, on using similarity ratios (3.18) and on considering single scattering for the real atmospheric phase function has already been made.

Note also, that with the use of basic numerical methods of the theory of radiation transfer (discrete ordinates, spherical harmonics, the Hunt-Grant matrix method, the Case method), it is easier to find photometric invariants (the sum and difference of intensities at arbitrary symmetrical levels in specular directions) than their standard constituents [380].

Chapter 4 A One-Dimensional Scalar Transfer Function of the Atmosphere

4.1 Theoretical Assessment of the Spectral Transfer Function for the Brightnesses of a Horizontally Homogeneous Surface

To reduce the spectrophotometric information obtained from the Earth's survey from space, a theoretical determination of the transfer function for the brightness of the atmosphere-object-background system is important [31, 94, 273, 288]. Using the spatial inhomogeneity of spectrophotometered surfaces (backgrounds), parts of which are "cut out" by the space-born spectrophotometer's slit during the survey from a space-borne platform, the system "object-background" can be divided conditionally into the following types (Fig. 4.1): (1) a horizontally homogeneous, infinitely extended surface (background); (2) a small (point) object against a horizontally-homogeneous, infinitely-extended background; (3) a set of randomly distributed small (point) objects against an extended background; (4) an interface between two horizontally homogeneous surfaces.

First, we shall consider a classical case of a horizontally-homogeneous, infinitely-extended surface (background), since it is a basic case. In this case we have $I_B = I_{ob} = I(\tau_0, \eta, \zeta, \varphi)$ and $I_h^B = I_h^{ob} = I_h(\tau_0, \tau, \eta, \zeta, \varphi)$. Thus, instead of (Eq. 3.1) we obtain one equation for reflected radiation intensity measured from space:

$$I(0, \eta, \zeta, \varphi, \tau_0) = I_h(\eta, \zeta, \varphi, \tau_0) + e^{-\tau_0/\eta} I_0(\zeta, \tau_0) \ . \tag{4.1}$$

It follows from (3.2) that $\Pi_B = \Pi_{ob} = \Pi(\eta, \zeta, \varphi, \tau_0)$. According to (3.1 – 3.2), the scalar transfer function $\Pi(\eta, \zeta, \varphi, \tau_0)$ in this case is [94]:

$$\Pi = \frac{I_0(\zeta, \tau_0)}{I_0(\zeta, \tau_0) T(\tau_0, \eta) + I_h(\eta, \zeta, \varphi, \tau_0)} \ . \tag{4.2}$$

Since the surface is horizontally homogeneous, the problem of the brightness contrast of an extended surface or above-water object with the background does not arise ($K = K_0 = 0$). Now we express the scalar transfer function Π [4.2] and its constituents I_0 and I_h through basic functions used in the theory of multiple scattering. By definition of the coefficients of diffuse reflection $\bar{\varrho}(\eta, \zeta, \varphi, \tau_0)$ and transmission $\bar{\sigma}(\eta, \zeta, \varphi, \tau_0)$ with the albedo of the basal $A \neq 0$, intensities $I(0, \eta, \zeta, \varphi)$ and $I(\tau_0, \eta, \zeta, \varphi)$ in (4.1) and (4.2) are:

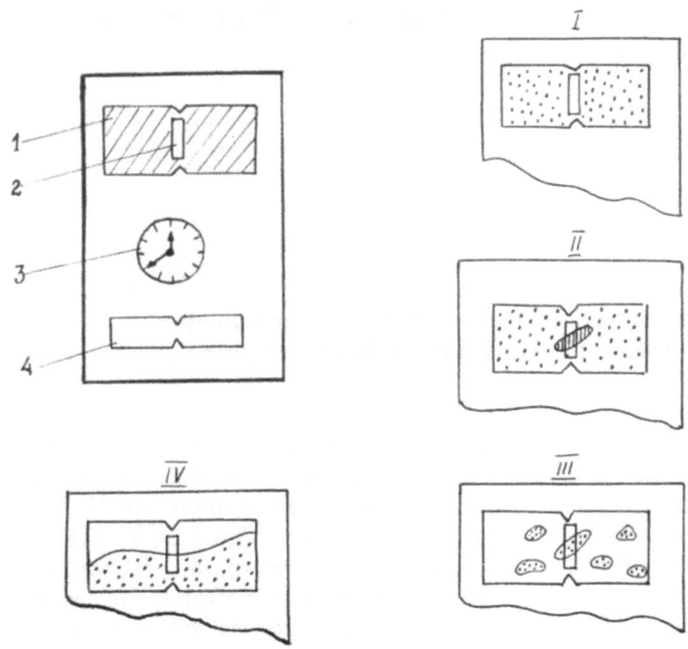

Fig. 4.1. Classification of the object-background system by the spatial inhomogeneity of the spectrophotometered underlying surface. A still made by RSS-2 spectrograph (*1* photoreferencing; *2* a surface site projected to the spectrograph's slit; *3* the dial image; *4* a reference spectrogram notation; *I* A horizontally homogeneous underlying surface; *II* a small (point) object; *III* a set of randomly distributed small (point) objects; *IV* the interface between two horizontally homogeneous surfaces

$$I(0,-\eta,\zeta,\varphi)=S\zeta\bar{\varrho}(\eta,\zeta,\varphi,\tau_0) \tag{4.3}$$

$$I(\tau_0,\eta,\zeta,\varphi)=S\zeta\bar{\sigma}(\eta,\zeta,\varphi,\tau_0)\ . \tag{4.4}$$

The brightness coefficients $\bar{\varrho}$ and $\bar{\sigma}$ in the case of an arbitrary phase function $x(\gamma)$ and orthotropic reflecting surface with albedo A at the atmospheric base level are easily expressed through respective quantities ϱ and σ at $A=0$ [183, 404]:

$$\bar{\varrho}(\eta,\zeta,\varphi,\tau_0)=\varrho(\eta,\zeta,\varphi,\tau_0)+\frac{A\mu(\eta,\tau_0)\mu(\zeta,\tau_0)}{1-AC(\tau_0)} \tag{4.5}$$

$$\bar{\sigma}(\eta,\zeta,\varphi,\tau_0)=\sigma(\eta,\zeta,\varphi,\tau_0)+\frac{A_0(\eta,\tau_0)\mu(\zeta,\tau_0)}{1-AC(\tau_0)}\ . \tag{4.6}$$

Functions $\mu(\zeta,\tau_0), \nu(\zeta,\tau_0)$ and $C(\tau_0)$ are determined in the following way:

$$\mu(\zeta,\tau_0)=e^{-\tau_0/\zeta}+\frac{1}{\pi}\int_0^{2\pi}d\varphi\int_0^1\sigma(\eta,\zeta,\varphi,\tau_0)\eta\,d\eta\ , \tag{4.7}$$

$$\nu(\zeta,\tau_0) = \frac{1}{\pi}\int_0^{2\pi} d\varphi \int_0^1 \varrho(\eta,\zeta,\varphi,\tau_0)\eta\, d\eta \ , \tag{4.8}$$

$$C(\tau_0) = \frac{2}{\pi}\int_0^{2\pi} d\varphi \int_0^1 \eta\, d\eta \int_0^1 \varrho(\eta,\zeta,\varphi,\tau_0)\zeta\, d\zeta \tag{4.9}$$

and have a clear physical meaning [187].

For the homogeneous surface brightness $I_0(\zeta,\tau_0)$ from (4.2) the following expression holds:

$$I_0 = AS\zeta e^{-\tau_0/\zeta} + \frac{A}{\pi}\int_0^{2\pi} d\varphi \int_0^1 I(\eta,\zeta,\varphi,\tau_0)\eta\, d\eta \ . \tag{4.10}$$

Substituting into (4.10) relationships (4.4) and (4.6) and taking into account a correlation of functions $C(\tau_0)$ and $\nu(\zeta,\tau_0)$:

$$C(\tau_0) = 2\int_0^1 \nu(\zeta,\tau_0)\zeta\, d\zeta \ , \tag{4.11}$$

we obtain, upon respective transformations:

$$I_0 = \frac{AS\zeta\mu(\zeta,\tau_0)}{1 - AC(\tau_0)} \ . \tag{4.12}$$

Considering (4.12) as a sum of the terms of decreasing geometrical progression with the denominator $q = AC(\tau_0) < 1$ and the first term $a_1 = AS\zeta\mu(\varphi,\tau_0)$, we find [97]:

$$I_0(\zeta,\tau_0) = AS\zeta\mu(\zeta,\tau_0) + A^2C(\tau_0)S\zeta\mu(\zeta,\tau_0) + A^3C^2(\tau_0) \times S\zeta\mu(\zeta,\tau_0) + \dots \ . \tag{4.13}$$

The first, second, ..., n-th terms in the converging series (4.13) represent the brightness of a homogeneous surface, respectively, at single, two-fold, ..., n-fold reflection of incident solar radiation. A representation of the brightness of the reflecting base as a sum of series (4.13) is of great importance in obtaining subsequently approximate formulas to assess the brightnesses of a horizontally inhomogeneous orthotropic surface (a small object against an extended background, an interface between two homogeneous media) as well as in case of non-orthotropically reflecting background polarizing the incident radiation.

Using relationships (4.1), (4.4 – 4.6) and (4.11), we find (for the brightness of atmospheric haze I_h measured from space) the following expression:

$$I_h = S\zeta\left[\varrho(\eta,\varphi,\zeta,\tau_0) + \frac{A\mu(\zeta,\tau_0)T_\sigma(\eta,\tau_0)}{1 - AC(\tau_0)}\right] \ , \tag{4.14}$$

where diffuse transparency $T_\sigma(\eta,\tau_0)$ is

$$T_\sigma = \frac{1}{\pi}\int_0^{2\pi} d\varphi \int_0^1 \sigma(\eta,\eta',\varphi,\tau_0)\eta'\, d\eta' = 2\int_0^1 \sigma^0(\eta,\eta')\eta'\, d\eta' \ . \tag{4.15}$$

Coefficients of reflection ϱ and transmission σ in (4.5) through (4.9) are determined according to (3.21).

Substituting (4.12) and (4.14) into (4.1), we obtain, upon transformations, the basic relationship which will be used in analysis of related satellite experimental data:

$$I = S\zeta\varrho(\eta,\zeta,\varphi,\tau_0) + I_0(\zeta,\tau_0)\,[T(\eta,\tau_0) + T_\sigma(\eta,\tau_0)] \ . \tag{4.16}$$

The sum of direct $T(\eta,\tau_0)$ and diffuse $T_\sigma(\eta,\tau_0)$ atmospheric transparencies in the direction $\theta = \arccos\eta$ is equal to function $\mu(\eta,\tau_0)$:

$$T(\eta,\tau_0) + T_0(\eta,\tau_0) = \mu(\tau_0,\eta) \ . \tag{4.17}$$

As will be shown the identity of relationships (4.1) and (4.16) is important for the determination of a transfer function and its constituents I_h and μ from combined satellite and aircraft experiments.

Now, substituting (4.12) and (4.14) into (4.2), we find a finite expression for a transfer function Π for a horizontally homogeneous and orthotropic surface (background):

$$\Pi = \frac{A\mu(\zeta,\tau_0)}{[1 - AC(\tau_0)]\,\varrho(\eta,\zeta,\varphi,\tau_0) + A\mu(\eta,\tau_0)\,\mu(\zeta,\tau_0)} \ . \tag{4.18}$$

In viewing the planetary surface in directions close to nadir ($\eta \sim 1$), for small zenith solar angles ($\zeta \sim 1$) and large albedo ($A = 0.5$) the reflection from surface in the visible ($\lambda = 400-800$ nm) contributes much more to the outgoing emission than multiple scattering in the atmosphere, i.e.,

$$\varrho(\eta,\zeta,\varphi,\tau_0) \ll \frac{A\mu(\eta,\tau_0)\,\mu(\zeta,\tau_0)}{1 - AC(\tau_0)} \ . \tag{4.19}$$

Inequality (4.19) makes it possible to obtain from (4.18) in the visible and near IR spectrum a simple approximate formula for Π having an apparent physical meaning:

$$\Pi \simeq \frac{1}{\mu(\eta,\tau_0)} = \frac{1}{T(\eta,\tau_0) + T_\sigma(\eta,\tau_0)} \ . \tag{4.20}$$

In the case of a weakly reflecting surface (sea, plough-land) an opposite inequality to (4.19) is valid.

From (4.18) we obtain the following approximate expression for a transfer function Π:

$$\Pi \simeq \frac{A\mu(\eta,\tau_0)}{[1 - AC(\tau_0)]\,\varrho(\eta,\zeta,\varphi)} \ . \tag{4.21}$$

Formula (4.19) permits a simple generalization for non-orthotropic reflection of solar radiation from the surface (background). In making this generalization as for the polarization transfer function, the contribution of photons singly reflected by the surface and multiply re-reflected by it is characterized by, respectively, coefficient of reflection $r(\eta,\zeta,\varphi)$ and albedo $A(\zeta)$. Then, instead of (4.19), we obtain the following expression for a transfer function [174]:

$$\Pi = \frac{r(\eta,\zeta,\varphi)\,\mu(\zeta,\tau_0)}{[1 - AC(\tau_0)]\,\varrho(\eta,\zeta,\varphi,\tau_0) + r(\eta,\zeta,\varphi)\,\mu(\zeta,\tau_0)\,\mu(\eta,\tau_0)} \; . \tag{4.22}$$

The coefficient r and albedo A in (4.22) are related in the following way:

$$A(\zeta) = \int_0^{2\pi}\int_0^1 r(\eta,\zeta,\varphi)\,\eta\, d\eta\, d\varphi \; . \tag{4.23}$$

Formula (4.16) and (4.22) were obtained on the assumption that the atmospheric phase function $x(\gamma)$ and the probability of quantum survival Λ do not change with height [42]. These formulas can be generalized for an arbitrary height-dependence of the atmospheric optical parameters based on results obtained in [188 and 210], the writing of (4.16) and (4.22) remaining unchanged. However, the physical meaning of the constituents of the transfer function $\mu(\eta,\tau_0)$ and $C(\tau_0)$ changes because of the necessity to introduce two additional coefficients of brightness $\bar{\varrho}$ and $\bar{\sigma}$ at solar irradiance of the medium, not from above but from beneath [188]. This generalization for the transfer function Π will not be considered. This is because, according to Table 3.13, the effect of the vertical optical inhomogeneity of an aerosol atmosphere on the field of reflected radiation, with respect to such parameters as phase function $x(\gamma)$ and probability of quantum survival Λ can be neglected in the visible ($\lambda = 400-800$ nm) outside the atmospheric gas absorption bands, zenith location of the sun ($\zeta = 1$) and nadir viewing ($\eta = 1$). This conclusion for the simplest case of isotropic scattering $x(\gamma) = 1$ has also been obtained in [33]. In addition, it must be emphasized that an arbitrary dependence of functions $x(\gamma)$ and Λ on height z can always be accurately considered in (4.18) and (4.22) in the single scattering approximation, leaving the height-constant parameters $x(\gamma)$ and Λ to describe the diffuse part of total radiation field.

It follows from (4.18) and (4.22) that, in order to calculate the scalar transfer function $\Pi(\eta,\zeta,\varphi,\tau_0)$, it is necessary to tabulate functions $\mu(\eta,\tau_0)$, $\nu(\eta,\tau_0)$ and $C(\tau_0)$ for the optically closed aerosol models of the atmosphere. As shown in [97], the constituents of the transfer function μ,ν and $C(\tau_0)$, with an arbitrary transfer function $x(\gamma)$ and probability of quantum survival Λ are easily expressed through φ_i^m and Ψ_i^m functions by Ambartsumian and, hence, through basic functions θ_i^m and Ω_i^m already introduced, according to (3.38 – 3.39):

$$\mu(\eta,\tau_0) = \Psi_1^0(\eta,\tau_0)\frac{1}{\eta} = \frac{1}{2\eta}[\theta_1^0(\eta,\tau_0) - \Omega_1^0(\eta,\tau_0)] \tag{4.24}$$

$$\nu(\eta,\tau_0) = 1 - \frac{\varphi_1^0(\eta,\tau_0)}{\eta} = 1 - \frac{1}{2\eta}[\Omega_1^0(\eta,\tau_0) - \theta_1^0(\eta,\tau_0)]\, d\eta \tag{4.25}$$

$$C(\tau_0) = 1 - \int_0^1 [\Omega_1^0(\eta,\tau_0) - \theta_1^0(\eta,\tau_0)]\, d\eta \; . \tag{4.26}$$

Thus, the problem of determination of 1-dimensional transfer functions (4.18) and (4.22) for given optical parameters of the environment (background) is reduced to tabulating the basic functions $\theta_i^m(\eta,\tau_0)$ and $\Omega_i^m(\eta,\tau_0)$ at $i = 0,1,2,\ldots,N$; $m = i, i+1,\ldots,N$.

4.2 On the Solution of Inverse Problems of Space-Based Atmospheric Optics for Model Phase Functions

As has been shown in [97], the problem of analytical determination of scalar transfer functions (4.21) and (4.22) for given optical parameters of the environment (background) is reduced to tabulating the basic functions $\theta_i^m(\eta,\tau_0)$, $\Omega_i^m(\eta,\tau_0)$ and $C(\tau_0)$, $D(\tau_0)$. The optical parameters of the atmosphere and surface are not known beforehand when surveying the Earth from space. Thus it is worthwhile, firstly, to consider the problem of their estimation from satellite-derived measurements of the spatial, angular and spectral distributions of the outgoing radiation intensity at the atmospheric top [94, 154, 178, 181, 263, 305–309, 376].

A correct assessment of the atmospheric transfer functions Π and P is based on preliminary solutions of inverse problems of space atmospheric optics. These problems are connected with assessing such parameters of the medium as atmospheric phase function $x(\gamma)$, probability of quantum survival Λ, atmospheric optical thickness τ_0, and surface albedo A [114, 263]. The space-derived coefficient of diffuse reflection $\bar{\varrho}(\eta,\zeta,\varphi,\tau_0)$ is assumed to be modelled as a result of solving the direct problem of the radiation transfer theory using similarity ratios for the model phase function $x(\gamma)$ containing an arbitrary but finite number of terms in an expansion in Legendre polynomials $P_n(\cos\gamma)$. Bearing in mind that the proposed analytical method of determining the brightness coefficients for the environment is based on exact solutions of the boundary problem of the transfer theory for an approximating 3-term phase function, we shall consider this case first.

Let us set the problem of how, using the angular dependences of the field of reflected radiation on θ, θ_0 and φ and making relevant measurements of atmospheric brightness in the visible ($\lambda = 400-800$ nm) from satellite, to estimate such optical parameters of the atmosphere as τ_0, $x(\gamma)$, Λ and A? Using the azimuthal dependence of reflected radiation intensity, we present the function in the considered model (N = 3) as part of a Fourier series by φ angle:

$$\bar{\varrho}(\eta,\zeta,\varphi,\tau_0) = \bar{\varrho}^0(\eta,\zeta,\tau_0) + 2\bar{\varrho}^1(\eta,\zeta,\tau_0)\cos\varphi + 2\bar{\varrho}^2(\eta,\zeta,\tau_0)\cos 2\varphi \;. \quad (4.27)$$

Azimuthal harmonics $\bar{\varrho}^0$, $\bar{\varrho}^{-1}$ and $\bar{\varrho}^2$ using (4.5) are expressed through φ_i^m and Ψ_i^m – functions by Ambartsumian [187]:

$$\bar{\varrho}^0(\eta,\zeta,\tau_0) = \frac{\Lambda}{4}\left[\frac{\varphi_0^0(\eta)\varphi_0^0(\zeta) - \Psi_0^0(\eta)\Psi_0^0(\zeta)}{\eta+\zeta}\right.$$

$$- x_1 \frac{(\varphi_1^0(\eta)\varphi_1^0(\zeta) - \Psi_1^0(\eta)\Psi_1^0(\zeta)}{\eta+\zeta}$$

$$\left. + x_2 \frac{\varphi_2^0(\eta)\varphi_2^0(\zeta) - \Psi_2^0(\eta)\Psi_2^0(\zeta)}{\eta+\zeta}\right]$$

$$+ \frac{A(\zeta)\mu(\eta,\tau_0)\mu(\zeta,\tau_0)}{1 - AC(\tau_0)} , \tag{4.28}$$

$$\bar{\varrho}^1(\eta,\zeta,\tau_0) = \frac{\Lambda}{8} x_1 \frac{\varphi_1^1(\eta)\varphi_1^1(\zeta) - \Psi_1^1(\eta)\Psi_1^1(\zeta)}{\eta+\zeta}$$

$$- \frac{\Lambda}{24} x_2 \frac{\varphi_2^1(\eta)\varphi_2^1(\zeta) - \Psi_2^1(\eta)\Psi_2^1(\zeta)}{\eta+\zeta} , \tag{4.29}$$

$$\bar{\varrho}^2(\eta,\zeta,\tau_0) = \frac{\Lambda}{96} x_2 \frac{\varphi_2^2(\eta)\varphi_2^2(\zeta) - \Psi_2^2(\eta)\Psi_2^2(\zeta)}{\eta+\zeta} . \tag{4.30}$$

At $A = 0$, we have $\bar{\varrho}^0 = \varrho^0$, $\bar{\varrho}^1 = \varrho^1$, $\bar{\varrho}^2 = \varrho^2$. Suppose, there are results of measurements of the azimuthal dependence of the total brightness coefficient $\bar{\varrho}$ made from space at different viewing angles and solar irradiance θ. Selecting three arbitrary values of azimuth, for instance, $\varphi = 0, \pi/2, \pi$, instead of (4.27) we obtain a system of three-order linear algebraic equations with respect to unknown azimuthal harmonics $\bar{\varrho}^m$ $(m = 0,1,2)$, with a determinant of the system Δ other than zero $(\Delta = -16)$. Solving this system, we obtain [376]:

$$\bar{\varrho}^0(\eta,\zeta) = \frac{1}{4}\,[\bar{\varrho}(\eta,\zeta,0,\tau_0) + 2\bar{\varrho}(\eta,\zeta,\pi/2,\tau_0) + \varrho(\eta,\zeta,\pi,\tau_0)] , \tag{4.31}$$

$$\bar{\varrho}^1(\eta,\zeta) = \frac{1}{4}\,[\bar{\varrho}(\eta,\zeta,0,\tau_0) - \bar{\varrho}(\eta,\zeta,\pi,\tau_0)] , \tag{4.32}$$

$$\bar{\varrho}^2(\eta,\zeta) = \frac{1}{8}\,[\bar{\varrho}(\eta,\zeta,0,\tau_0) - 2\bar{\varrho}(\eta,\zeta,\pi/2,\tau_0) + \bar{\varrho}(\eta,\zeta,\pi,\tau_0)] . \tag{4.33}$$

From a practical point of view, the solution of inverse problems of space-derived atmospheric optics is of interest in three cases: $A \simeq 0$; $A(\theta_0)$, orthotropic horizontally inhomogeneous surface; $A(\theta_0, x)$, orthotropic horizontally inhomogeneous surface along the x coordinate. First, we shall consider $A = 0$ which is approximately realized over sea surfaces. To simplify expressions (4.28 – 4.29), we use the properties of φ_i^m and Ψ_i^m functions by Ambartsumian, valid for arbitrary atmospheric phase functions containing a finite number of terms in their expansion in Legendre polynomials $P_n(\cos\gamma)$ [178]:

$$\varphi_0^0(0) = \varphi_1^1(0) = 1 \ , \qquad \varphi_2^2(0) = 3 \ , \qquad \varphi_2^0(0) = -\frac{1}{2}$$

$$\varphi_1^0(0) = \varphi_2^1(0) = \Psi_0^0(0) = \Psi_2^0(0) = \Psi_1^1(0) = \Psi_2^1(0) = \Psi_2^2(0) = 0 \ . \tag{4.34}$$

Now, taking $\zeta = 0$ or $\eta = 0$ in (4.28 – 4.30), following (4.34), we have

$$\begin{cases} \bar{\varrho}^0(\eta,0) = \varrho^0(\eta,0) = \dfrac{\Lambda}{4}\dfrac{\varphi_0^0(\eta,\tau_0)}{\eta} - \dfrac{\Lambda x_2}{8}\dfrac{\varphi_2^0(\eta,\tau_0)}{\eta} \ , \\ \bar{\varrho}^1(\eta,0) = \varrho^1(\eta,0) = \dfrac{\Lambda x_1}{8}\dfrac{\varphi_1^1(\eta,\tau_0)}{\eta} \ , \\ \bar{\varrho}^2(\eta,0) = \varrho^2(\eta,0) = \dfrac{\Lambda x_2 \varphi_2^2(\eta)}{32\eta} \ . \end{cases} \tag{4.35}$$

Note, that the φ_0^0, φ_1^1, and φ_2^2 functions in (4.35) are related to azimuthal harmonics ϱ^0, ϱ^1 and ϱ^2 which, according to (4.31 – 4.33), are expressed through the total reflection coefficient $\bar{\varrho}(\eta,\zeta,\varphi,\tau_0)$ measured from space:

$$\varphi_0^0(\eta) = 1 + 2\eta\int_0^1 \varrho^0(\eta,\eta')\,d\eta' \ ,$$

$$\varphi_2^0(\eta) = P_2(\eta) + 2\eta\int_0^1 P_2(\eta')\varrho^0(\eta,\eta')\,d\eta' \ , \tag{4.36}$$

$$\varphi_1^1(\eta) = P_1^1(\eta) + 2\eta\int_0^1 P_1^1(-\eta')\varrho^1(\eta,\eta')\,d\eta' \ , \tag{4.37}$$

$$\varphi_2^2(\eta) = P_2^2(\eta) + 2\eta\int_0^1 P_2^2(-\eta')\varrho^2(\eta,\eta')\,d\eta' \ . \tag{4.38}$$

Now, solving systems (4.35) for the remaining unknown parameters Λ, x_1 and x_2, we find:

$$\Lambda = \frac{4\varrho^0(\eta,0)\eta}{1 + 2\eta\int_0^1 \varrho^0(\eta,\eta')\,d\eta} + \frac{16\varrho^2(\eta,0)\eta\left[P_2(\eta) + \int_0^1 P_2(\eta')\varrho^0(\eta,\eta')\,d\eta'\right]}{\left[1 + 2\eta\int_0^1 \varrho^0(\eta,\eta')\,d\eta'\right]\left[P_2^2(\eta) + 2\eta\int_0^1 P_2^2(-\eta')\varrho^2(\eta,\eta')\,d\eta'\right]} \tag{4.39}$$

$$x_1 = \frac{8\varrho^1(\eta,0)\eta}{\Lambda\left[P_2^2(\eta) + 2\eta\int_0^1 P_2^2(-\eta')\varrho^2(\eta,\eta')\,d\eta'\right]} \tag{4.40}$$

$$x_2 = \frac{32\varrho^2(\eta,0)\eta}{\Lambda\left[P_2^2(\eta) + 2\eta\int_0^1 P_2^2(-\eta')\varrho^2(\eta,\eta')d\eta'\right]} \ . \tag{4.41}$$

In the case of isotropic scattering $x(\gamma) = 1$ the probability of quantum survival is:

$$\Lambda = \frac{4\varrho^0(\eta,0)\eta}{1 + 2\eta\int_0^1 \varrho^0(\eta,\eta')d\eta'} \ . \tag{4.42}$$

Now we find the atmospheric optical thickness τ_0, which is equivalent to determining direct atmospheric transparency in a given viewing direction. In solving this problem we use an equation for the highest azimuthal harmonic ($m = 2$) of total reflection coefficient ϱ. Taking the equality $\eta = \zeta$ in (4.30) to be satisfied, and expressing the function $\Psi_2^2(\eta,\tau_0)$ through the second azimuthal harmonic $\varrho^2(\eta,\eta,\tau_0)$ measured according to (4.33) and (4.38), we have

$$\Psi_2^2(\eta,\tau_0) = \sqrt{[\varphi_2^2(\eta,\tau_0)]^2 - \frac{192\varrho^2(\eta,\eta,\tau_0)\eta}{\Lambda x_2}} \ . \tag{4.43}$$

We express total coefficient of diffuse transmission at $A = 0$ in a way similar to (4.27):

$$\begin{aligned}\tilde{\sigma}(\eta,\zeta,\varphi,\tau_0) = \sigma(\eta,\zeta,\varphi,\tau_0) &= \sigma^0(\eta,\zeta,\tau_0) + 2\sigma^1(\eta,\zeta,\tau_0)\\ &\times\cos\varphi + 2\sigma^2(\eta,\zeta,\tau_0)\cos 2\varphi \ .\end{aligned} \tag{4.44}$$

Then for a 3-term phase function ($N = 2$) the highest azimuthal harmonic ($m = 2$) of total coefficient of transmission $\sigma^2(\eta,\zeta,\tau_0)$ is

$$\sigma^2(\eta,\zeta,\tau_0) = \frac{\Lambda x_2}{96}\,\frac{\varphi_2^2(\zeta)\Psi_2^2(\eta) - \varphi_2^2(\eta)\Psi_2^2(\zeta)}{\eta-\zeta} \ . \tag{4.45}$$

Using (4.34), instead of (4.45), at $\eta = 0$ or $\zeta = 0$ we obtain an expression

$$\sigma^2(\eta,0,\tau_0) = \frac{1}{32}\Lambda x_2\frac{\Psi_2^2(\eta,\tau_0)}{\eta} \tag{4.46}$$

where $\Psi_2^2(\eta,\tau_0)$ function is determined through space-derived quantity $\varrho^2(\eta,\eta)$ according to (4.43), (4.33) and (4.38). Now we use an apparent dependence of $\Psi_2^2(\eta,\tau_0)$ function by Ambartsumian on the atmospheric optical thickness τ_0 and on the second azimuthal harmonic $\sigma^2(\eta,\zeta,\tau_0)$:

$$\Psi_2^2(\eta,\tau_0) = P_2^2(\eta)e^{-\tau_0/\eta} + 2\eta\int_0^1 P_2^2(\eta')\sigma^2(\eta,\eta')d\eta' \ . \tag{4.47}$$

Using (4.43) and (4.47), we find an exact expression for atmospheric optical thickness τ_0:

$$\tau_0 = \eta \ln \left| \frac{P_2^2(\eta)}{\sqrt{[\varphi_2^2(\eta,\tau_0)]^2 - \frac{192\varrho^2(\eta,\eta,\tau_0)\eta}{\Lambda x_2} - 2\eta \int\limits_0^1 P_2^2(\eta')\sigma^2(\eta,\eta')\,d\eta'}} \right| . \tag{4.48}$$

Following (4.33), (4.38), (4.43) and (4.45), the right-hand part of (4.48) involves only the optical quantities measured from space.

In the visible spectrum ($\lambda = 400-800$ nm), the absorption by atmospheric gases, except for the ozone absorption band centered at $\lambda = 600$ nm, is small ($\Lambda \simeq 1$). In this case an estimation of τ_0 value can be made with the use of some integral relationships between φ_i^m and Ψ_i^m functions by Ambartsumian, coefficient of phase function x_1, and optical thickness τ_0 [187]:

$$\begin{aligned}
&\int\limits_0^1 \varphi_{2i}^0(\eta)\left[\left(1-\frac{x_1}{3}\right)\tau_0+\eta\right]d\eta - \int\limits_0^1 \Psi_{2i}^0(\eta)\eta\,d\eta = 2\left(1-\frac{x_1}{3}\right)\tau_0\delta_{2i,0} ,\\
&\int\limits_0^1 \varphi_{2i+1}^0(\eta)\left[\left(1-\frac{x_1}{3}\right)\tau_0+\eta\right]d\eta + \int\limits_0^1 \Psi_{2i+1}^0(\eta)\eta\,d\eta = \frac{2}{3}\delta_{2i+1,1} ,\\
&\int\limits_0^1 [\varphi_i^0(\eta)+(-1)^i\Psi_i^0(\eta)]\,d\eta = 2\delta_{i0} .
\end{aligned} \tag{4.49}$$

For pure isotropic scattering ($\Lambda = 1$, $x_1 = 0$) from (4.48) we obtain, upon transformations, the following expression for optical thickness τ_0 [178]:

$$\tau_0 = \frac{\int\limits_0^1 [\varphi(\eta)-\Psi(\eta)]\eta\,d\eta}{1-\frac{1}{2}\int\limits_0^1 [\varphi(\eta)-\Psi(\eta)]\,d\eta} = \frac{\int\limits_0^1 [\varphi(\eta)-\Psi(\eta)]\eta\,d\eta}{\int\limits_0^1 \Psi(\eta)\,d\eta} , \tag{4.50}$$

where

$$\begin{aligned}
&\varphi(\eta,\tau_0) = 4\eta\varrho(\eta,0,\tau_0)\\
&\Psi(\eta,\tau_0) = \sqrt{8\eta\{2\eta[\varrho(\eta,0,\tau_0)^2] - \varrho(\eta,\eta,\tau_0)\}} .
\end{aligned}$$

Substituting (4.50) into (4.49), we find a finite expression for at $x(\gamma) = 1$ and $\Lambda = 1$:

$$\tau_0 = \frac{\int\limits_0^1 \{4\eta\varrho(\eta,0,\tau_0) - \sqrt{8\eta\{2\eta[\varrho(\eta,0,\tau_0)]^2 - \varrho(\eta,\eta,\tau_0)\}}\}\eta\,d\eta}{1-\frac{1}{2}\int\limits_0^1 \{4\eta\varrho(\eta,0,\tau_0) - \sqrt{8\eta\{2\eta[\varrho(\eta,0,\tau_0)]^2 - \varrho\eta,\eta,\tau_0)\}}\}\,d\eta} . \tag{4.52}$$

Now we estimate the albedo $A(\zeta)$ of an orthotropic horizontally homogeneous surface. A special feature of the inverse problem of space atmospheric optics in the presence of a reflecting base at the atmospheric lower boundary level is that the quantity of $A(\theta_0)$ albedo is contained only in the zero harmonic $\bar{\varrho}_0$ of the total reflection coefficient (4.27). This makes it possible, in pure scattering ($\Lambda = 1$), to use the remaining harmonics $\bar{\varrho} = \varrho^1$ and $\bar{\varrho}^2 = \varrho^2$ to determine coefficients x_1, x_2 and optical thickness τ_0, according to the algorithm described ($A = 0$). Upon assessing the optical parameters of the atmosphere x_1, x_2 and τ_0, it becomes possible to calculate all basic functions in (4.28) for zero harmonic $\bar{\varrho}^0$ following a scheme for solving the direct problem of the transfer theory in the case of a 3-term phase function. The $A(\theta_0)$ albedo (our second problem) is then found from (4.28):

$$A(\zeta) = \frac{\bar{\varrho}^0(\eta,\zeta) - \varrho^0(\eta,\zeta)}{[\bar{\varrho}^0(\eta,\zeta) - \varrho^0(\eta,\zeta)]\,C(\tau_0) + \mu(\eta)\,\mu(\zeta)}\,, \tag{4.53}$$

where the reflection coefficient $\bar{\varrho}^0(\eta,\zeta)$ is estimated experimentally from formulas (4.31), and $\bar{\varrho}^0(\eta,\zeta)$, $\mu(\eta)$ and $C(\tau_0)$ functions are calculated from the formulas of the multiple scattering theory for a 3-term phase function (3.49) and for atmospheric optical parameters x_1, x_2, τ_0 found from solving the inverse problem.

The algorithm for estimating the optical parameters of the atmosphere and reflective characteristics of an orthotropic horizontally homogeneous surface (our third problem) is generalized for an arbitrary height-independent phase function $x(\gamma)$ containing a finite number of terms in an expansion in Legendre polynomials $P_n(\cos\gamma)$.

4.3 Tabulating Basic Constituents of the Scalar Transfer Function of an Aerosol Atmosphere

Suppose, inverse problems of space-based spectrophotometry of natural formations, connected with an assessment of the optical parameters of the atmosphere and surfaces (backgrounds) have been solved, using the algorithm just described. Then the atmospheric radiative correction of space-derived spectral brightnesses of horizontally homogeneous surfaces (backgrounds) can efficiently be realized in the visible spectrum ($\Lambda = 400-800$ nm) with the help of 1-D transfer functions $\Pi(\zeta,\eta,\varphi,\tau_0)$ for given optical parameters of the medium. Note, that the closed volume of real optical characteristics of the atmosphere-surface system which could be determined by converting space-derived spectrophotometric data, has not yet been obtained. Therefore, to tabulate transfer functions, it is worthwhile using a simulation modeling technique based on calculations of basic constituents $\varrho(\eta,\zeta,\varphi,\tau_0)$, $\mu(\eta,\tau_0)$ and $C(\tau_0)$ for the most representative optical models of the terrestrial atmosphere.

Optical closing of the Elterman-1968 aerosol model performed in Section 1.3 for atmospheric phase function $x(\gamma)$ makes it possible to use this model to tabulate the scalar transfer function Π and its constituents ϱ, μ, C in the visible spectrum ($\lambda = 400-800$ nm).

In Fig. 4.2 graphs in polar coordinates are given of the monochromatic atmospheric phase function $x(\gamma)$, of normalized residual phase function

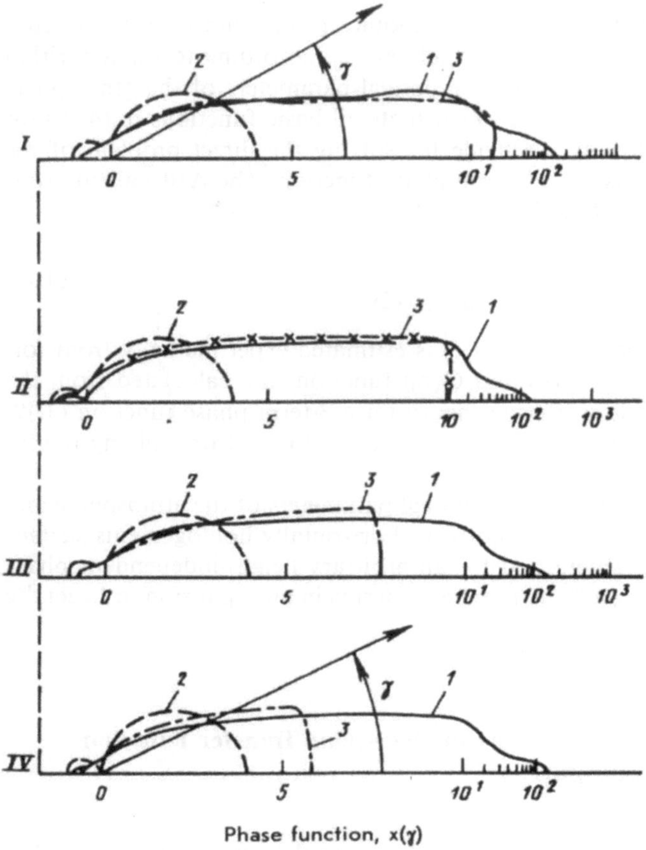

Fig. 4.2. Monochromatic atmospheric phase functions x(γ) corresponding to the 1968 Elterman aerosol model for $\lambda = 700$ nm; *1* atmospheric phase function (Table 1.3); *I* $\gamma_{cut} = 5°$; *II* $\gamma_{cut} = 10°$; *III* $\gamma_{cut} = 15°$; *IV* $\gamma_{cut} = 20°$; *2* approximating three-term phase function x(γ); *3* normalized residual phase function x(γ)

$\tilde{x}(\gamma)/1-0.5\,L$, and a three-term phase function $\tilde{x}(\gamma) = 1 + x_1 \cos\gamma + x_2 P_2(\cos\gamma)$ approximating it at different angles of cutting for the Elterman-1968 aerosol model and $\lambda = 700$ nm. Interpolation coefficients $|x_1| < 3$ and $|x_2| < 5$ were chosen proceeding from the condition of best approximation of a normalized residual phase function $\tilde{x}(\gamma)/1-0.5\,L$ of the functions of the (3.19) type. The choice of x_1 and x_2 is limited to where they are the first two coefficients of expansion of a real atmospheric phase function x(γ) in Legendre polynomials $P_n(\cos\gamma)$:

$$x_n = \frac{2n+1}{2} \int_0^{\pi} x(\gamma) P_n(\cos\gamma) \sin\gamma \, d\gamma \ , \quad n = 1,2 \ . \tag{4.54}$$

Table 4.1. Monochromatic estimates of $\phi^m(\tau, \tau_0)$ functions by Sobolev for the Elterman-1968 aerosol model and the respective aerosol phase function (Table 1.3) at $\lambda = 550$ nm ($\tau_0 = 0.369$, $\Lambda = 0.902$, $x_1 = 1.460$, $x_2 = 1.540$, $\gamma_{cut} = 5°$)

0.01	2.8742	1.4872	0.9120	0.17	1.2334	0.5843	0.2553
0.02	2.4693	1.2645	0.7397	0.18	1.1952	0.5648	0.2437
0.03	2.2370	1.1354	0.6412	0.19	1.1605	0.5472	0.2324
0.04	2.0734	1.0442	0.5725	0.20	1.1266	0.5304	0.2237
0.05	1.9474	0.9737	0.5203	0.21	1.0940	0.5143	0.2147
0.06	1.8395	0.9134	0.4763	0.22	1.0619	0.4988	0.2061
0.07	1.7528	0.8651	0.4416	0.23	1.0308	0.4839	0.1980
0.08	1.6776	0.8234	0.4121	0.24	0.9999	0.4694	0.1903
0.09	1.6106	0.7863	0.3864	0.25	0.9697	0.4554	0.1829
0.10	1.5504	0.7532	0.3638	0.26	0.9396	0.4417	0.1759
0.11	1.4952	0.7231	0.3436	0.27	0.9097	0.4283	0.1692
0.12	1.4446	0.6957	0.3254	0.28	0.8795	0.4151	0.1628
0.13	1.3971	0.6702	0.3089	0.29	0.8473	0.4014	0.1562
0.14	1.3530	0.6467	0.2938	0.30	0.8157	0.3883	0.1501
0.15	1.3110	0.6246	0.2800	0.31	0.7811	0.3746	0.1440
0.16	1.2715	0.6039	0.2672				

Table 1.3 has listed monochromatic values of atmospheric phase functions $x(\gamma)$ calculated with formulas of Mie theory for the Elterman-1968 aerosol model.

Table 4.1 presents values of optical parameters L, τ_0^*, Λ_* from similarity ratio (3.18) for the Elterman-1968 aerosol model, spectral interval $\lambda = 400-700$ nm and angles of cutting $\gamma_{cut} = 5°$ and 20°. For each wavelength λ the upper line of estimates refers to the angle $\gamma_{cut} = 5°$, and the lower line to $\gamma_{cut} = 20°$. An exception is $\lambda = 600$ nm for which maximum absorption by atmospheric ozone is observed in the Chappui band. For this wavelength the upper line gives the L, τ_0^* and Λ_* parameters for angles $\gamma_{cut} = 5°$ and 10°, and the lower line for angles $\gamma_{cut} = 15°$ and 20°.

At the bottom of Table 4.1 for an approximating three-term phase function $\bar{x}(\gamma) = 1 + x_1 \cos\gamma + x_2 P_2(\cos\gamma)$ estimates are given of interpolation coefficients x_1 and x_2 corresponding to a threshold angle of cutting a maximum of an initial atmospheric phase function at $\gamma_{cut} = 10°$.

Using similarity ratio (3.18), and exact and approximate solutions for the basic boundary problem of the radiation transfer theory for an approximating 3-term and arbitrary phase function (see Tables 4.1 and 1.3), basis functions can be calculated to tabulate the scalar transfer function $\Pi(\zeta, \eta, \varphi, \tau_0, A)$ and its constituents $\varrho(\zeta, \eta, \varphi, \tau_0)$, $\sigma(\zeta, \eta, \varphi, \tau_0)$, $\nu(\eta, \tau_0)$, $C(\tau_0)$, $\mu(\eta, \tau_0)$. In this case the single scattering is taken for a given atmospheric phase function exactly and the field of diffuse radiation is calculated for an approximating 3-term phase function exactly. Table 4.2 lists monochromatic estimates of $\phi^m(\tau, \tau_0)$ function by Sobolev ($m = 0, 1, 2$) for the Elterman-1968 closed aerosol model.

With the use of similarity ratios (3.18) and with an exact account of single scattering for an aerosol phase function (Table 1.2, Elterman-1968 model), Table 4.3 gives estimates for monochromatic coefficients of diffuse reflection

Table 4.2. Optical parameters of the Elterman-1968 closed aerosol model in the visible spectrum ($\lambda = 400-700$ nm) with the use of similarity ratios

λ, nm	400	450	500	550	600	700
L	0.113 0.198	0.032 0.241	0.048 0.278	0.060 0.307	0.068 0.175 0.244 0.329	0.076 0.356
Λ^*	0.960 0.956	0.974 0.971	0.956 0.951	0.902 0.890	0.859 0.852 0.847 0.841	0.975 0.976
τ_0^*	0.676 0.615	0.501 0.449	0.411 0.365	0.369 0.326	0.341 0.324 0.314 0.301	0.248 0.213
x_1	1.300	1.420	1.420	1.460	1.504	
x_2	1.660	1.580	1.580	1.540	1.496	1.492

Table 4.3. Monochromatic estimates of the coefficients of diffuse reflection $\varrho(\eta, \zeta, \varphi, \tau_0)$ and $\sigma(\eta, \zeta, \varphi, \tau_0)$ for the Elterman-1968 closed aerosol model ($\lambda = 550$ nm, $\Lambda = 0.369$, $x_1 = 1.460$, $x_2 = 1.540$)

	0.1	0.2	0.3	0.4	0.5	0.6	0.7	0.8	0.9	1.0	
	9.9882	4.4922	2.5274	1.5794	1.0485	0.7196	0.4967	0.3419	0.2254	0.1195	0.1
0.1		2.5282	1.5746	1.0438	0.7243	0.5120	0.3642	0.2569	0.1752	0.1005	0.2
0.2	5.7876		1.0320	0.7113	0.5046	0.3633	0.2638	0.1896	0.1327	0.0823	0.3
0.3	2.8146	5.7356		0.5008	0.3612	0.2843	0.1951	0.1429	0.1025	0.0691	0.4
0.4	1.7612	2.4172	4.3955		0.2641	0.1968	0.1471	0.1099	0.0813	0.0599	0.5
0.5	1.1949	1.4172	1.7833	3.2423		0.1484	0.1130	0.0862	0.0662	0.0536	0.6
0.6	0.8444	0.9248	1.0207	1.2819	2.3746		0.0878	0.0691	0.0554	0.0493	0.7
0.7	0.6033	0.6336	0.6595	0.7246	0.9006	1.7088		0.0565	0.0479	0.0468	0.8
0.8	0.4276	0.4316	0.4319	0.4518	0.6076	0.6320	1.1494		0.0433	0.0465	0.9
0.9	0.2907	0.2832	0.2750	0.2770	0.2943	0.3338	0.4139	0.6787		0.0578	1.0
1.0	0.1459	0.1355	0.1226	0.1169	0.1163	0.1220	0.1348	0.1621	0.2265		

$\varrho(\eta, \zeta, \varphi, \tau_0)$ and transmission $\sigma(\eta, \zeta, \varphi, \tau_0)$. Due to the symmetry of brightness coefficients ϱ and σ with angular variables η and ζ in the upper diagonal part of Table 4.3 estimates of $\varrho(\eta, \zeta, \varphi, \tau_0)$ are given, and in the lower part those of $\sigma(\eta, \zeta, \varphi, \tau_0)$.

Finally, Table 4.4 lists estimates of major basis functions $\mu(\eta, \tau_0)$, $\nu(\eta, \tau_0)$, and $C(\tau_0)$, these make it possible to carry out a numerical modeling of the scalar transfer function of the atmosphere $\Pi(\tau_0, \eta, \zeta, \varphi, A)$ when surveying the Earth from space, for given reflective characteristics of a horizontally homogeneous surface, with the use of similarity ratio (3.18) and the Elterman-1968 closed aerosol model. As will be shown, the data of Tables 4.3 and

Table 4.4. Estimates of basic constituents η, λ and C of atmospheric transfer functions for the Elterman-1968 aerosol model and the respective atmospheric phase function $x(\gamma)$ (Table 1.2, $\gamma_{cut} = 10°$)

	λ, nm						
	450	500	550	600	650	700	800
0.1	0.4548 0.6599	0.5362 0.5246	0.5102 0.4596	0.4835 0.4118	0.5713 0.4367	0.6366 0.4583	0.6663 0.4526
0.2	0.5365 0.5784	0.6491 0.4175	0.6345 0.3542	0.6167 0.3104	0.7124 0.3181	0.7769 0.3267	0.8065 0.3176
0.3	0.6094 0.5060	0.7438 0.3302	0.7364 0.2733	0.7295 0.2360	0.8129 0.2368	0.8687 0.2416	0.8922 0.2355
0.4	0.6671 0.4487	0.8152 0.2651	0.8118 0.2151	0.8011 0.1836	0.8818 0.1819	0.9292 0.1857	0.9470 0.1831
0.5	0.7122 0.4040	0.8693 0.2163	0.8680 0.1724	0.8584 0.1458	0.9310 0.1431	0.9714 0.1468	0.9847 0.1472
0.6	0.7478 0.2687	0.9111 0.1787	0.9111 0.1400	0.9021 0.1175	0.9677 0.1144	1.0024 0.1183	1.0119 0.1211
0.7	0.7765 0.3403	0.9442 0.1491	0.9451 0.1148	0.9364 0.0956	0.9960 0.0924	1.0261 0.0966	0.0326 0.1014
0.8	0.8000 0.3171	0.9710 0.1252	0.9725 0.0947	0.9639 0.0782	1.0185 0.0751	1.0448 0.0795	1.0488 0.0861
0.9	0.8195 0.2978	0.9931 0.1056	0.9951 0.0783	0.9866 0.0642	1.0368 0.0611	1.0598 0.0658	1.0618 0.0737
1	0.8360 0.2815	1.0117 0.0993	1.0140 0.0647	1.0055 0.0526	0.0520 0.0496	1.0722 0.0545	1.0725 0.0636
$C(\tau_0)$	0.3684	0.1813	0.1438	0.1218	0.1202	0.1251	0.1284

4.4 enable one to estimate both scalar and polarization transfer functions for the brightnesses and brightness contrasts of small-sized (point) objects, horizontally inhomogeneous surfaces and backgrounds, spectrophotometered from space in the visible spectrum ($\lambda = 400 - 800$ nm).

4.4 The Effect of the Non-Orthotropic Surface on the Radiation Field

Use of experimental reflective characteristics of natural formations for the interpretation of space-derived information is limited. This is because the non-orthotropic characteristics depend on illumination conditions, clouds and other factors, which require, along with the coefficients of spectral brightness of natural objects, knowledge of these, difficult-to-control factors, at the moment of survey. An advanced method for spherical harmonics described in

Section 3.12 is interesting because the form of presentation of calculation results is preserved for non-zero harmonics of the expansion of intensity in Fourier series too. Only the number of 3-diagonal systems of algebraic equations (3.249) for the moments of the angular distributions of intensity increases. In this situation some universal form of presentation can be used for the reflective characteristics of various objects [149]. We introduce a definition of a phase function $x_R(\eta^+, \varphi, \eta^-, \varphi')$ as a probability of the transition of reflected radiation from a direction (η^-, φ') to (η^+, φ).

We make use of the fact that the basic types of radiation reflection by various objects can be modeled using the following type of reflection phase function [149]:

$$x_R(\eta^+, \varphi, \eta^-, \varphi') = \frac{d}{1+d}\left\{1 + \frac{1-g^2}{d[1+g^2-2g(\vec{\Omega}\vec{\Omega}')]^{3/2}}\right\} . \tag{4.55}$$

The g and d parameters in (4.55) can be chosen for different types of natural formations in the process of numerical experiments after a closed scheme, using space-derived multi-zonal information and ground measurements of the phase functions of reflected radiation. Note, that expression (4.55) describes all basic types of reflection characteristic of natural objects.

With respect to a reflection phase function we suppose it to satisfy the reciprocity theorem

$$x_n(\eta^+, \varphi, \eta^-, \varphi') = x_n(\eta^-, \varphi', \eta^+, \varphi)$$

and to be normalized according to the condition

$$\frac{1}{4\pi}\int_0^{2\pi} d\varphi \int_{-1}^{+1} x_R(\vec{\Omega}\vec{\Omega}')\, d\eta = 1 .$$

We use the following expansion [196]:

$$\frac{1-g^2}{(1+g^2-2g\vec{\Omega}\vec{\Omega}')^{3/2}} = \sum_{n=0}^{\infty} (2n+1) g^n P_n(\vec{\Omega}\vec{\Omega}') , \quad |g| < 1 , \quad |\vec{\Omega}, \vec{\Omega}'| \leqslant 1 . \tag{4.56}$$

In contrast to Lambertian reflection, the non-orthotropicity affects non-zero azimuthal harmonics of reflected radiation. We make an assumption that only an expansion of a phase function with the same m-index as that of the outgoing radiation intensity contributes to each given harmonic of this intensity. Mathematically it means that k = m always in the expressions used.

An account of the anisotropy of reflected radiation on the surface is reduced, on the one hand, to an approximation of the reflective characteristics of natural objects by analytical dependence (4.55), and, on the other hand, to a presentation of reflection phase function as a dual series:

$$x_R(\eta^+, \varphi, \eta^-, \varphi') = h \frac{d}{1+d} \left\{ 1 + \frac{1}{d} \sum_{n=0}^{\infty} \left[P_n(\eta^+) P_n(\eta^-) + 2 \sum_{l=1}^{n} \frac{(n-l)!}{(n+l)!} P_n^l(\eta^+) P_n^l(\eta^-) \cos l(\varphi - \varphi') \right] \times (2n+1) g^n \right\}, \quad (4.57)$$

where h is the normalizing coefficient. A presentation of reflection phase functions in the form of (4.57) is suitable because there is no need to use a pair of numbers (g, d) involved in the expression (4.55) in the calculation.

In accordance with the given form of presentation of reflection phase functions, all types of reflecting surfaces are located between two geometric figures shown in Fig. 4.3. The problem of the catalogued a priori information needed to recognize the spectral images of natural objects is reduced to the finding of the shape of surface typical of a given type of natural formation. Using the tabulated reflective characteristics with a set of input data, in the form of angles of incidence and reflection of radiation, we can estimate a pair of numbers (g, d) for the phase function of a given object, for instance, using a least squares technique. An advantage of the presentation of the reflection

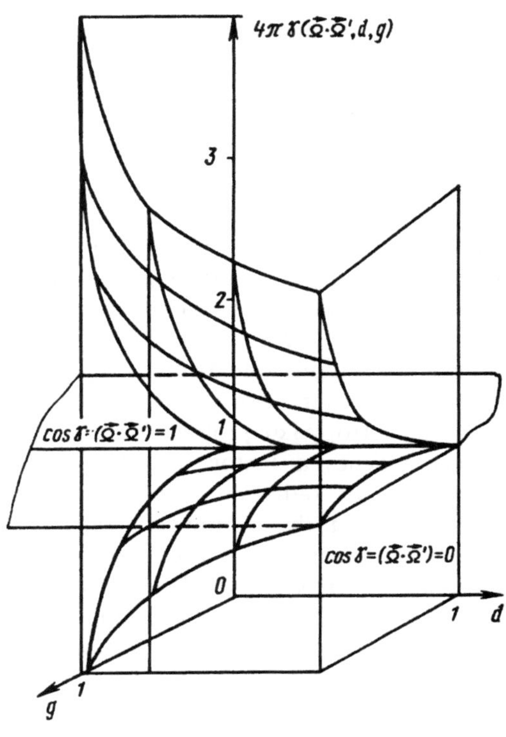

Fig. 4.3. The reflection indicatrix as a function of a pair of numbers (g, d) for two extreme values of the scattering angle $\gamma = (\vec{\Omega}, \vec{\Omega}') = 1$ and 0

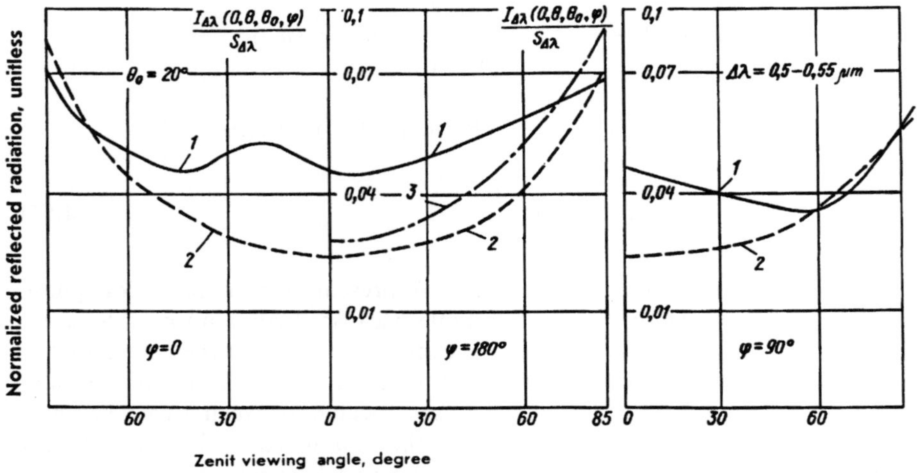

Fig. 4.4. The effect of the azimuthal dependence on reflected radiance: *1* reflection following the Fresnel law; *2* Lambertian surface, A = 0; *3* Lambertian surface, A = 0, $\varphi = 0°$

phase function is that it has a probabilistic character, and, therefore, only its angular change and not absolute estimates are needed from the experiment.

The effect of the azimuthal dependence on the outgoing radiation intensity is absolutely essential and is illustrated in Fig. 4.4. Near nadir, the calculated intensities of radiation reflected by the water surface following the laws of Fresnel and Lambert (outgoing off the atmospheric top) can differ by a factor of 2, which corresponds to the data in [4].

The angular change of integral phase functions for reflected radiation is, in general, similar for spectral intervals [6]. Therefore, Figs. 4.5 and 4.6 show the angular dependences of total coefficients of reflection for different natural formations [254] for viewing azimuths $\varphi = 0°$ and $\varphi = 180°$ and spectral intervals 0.5 – 0.55 μm (Fig. 4.5) and 0.9 – 1.0 μm (Fig. 4.6). Figure 4.5 shows the angular distributions of the outgoing radiation intensity also including non-orthotropic reflection, Lambertian surface and with averaging of reflected radiation over azimuth (using zero harmonic of expansion only). Substantial differences between intensities are observed for objects with strong anisotropy [207] (object 7 in Fig. 4.5). For instance, near nadir the difference reaches a factor of 1.5. However, the effect of the azimuthal dependence of radiation repeatedly scattered by the atmosphere is more substantial than that of the reflection anisotropy.

It follows from Fig. 4.6 that not only the surface but also the atmosphere substantially transforms radiation over the spectrum: in the spectral interval 0.9 – 1.0 μm with strong absorption of radiation by atmospheric components, special features of radiation reflection by the surface are less noticeable; and in the interval 0.5 – 0.55 μm, with weak absorption, they are seen more markedly.

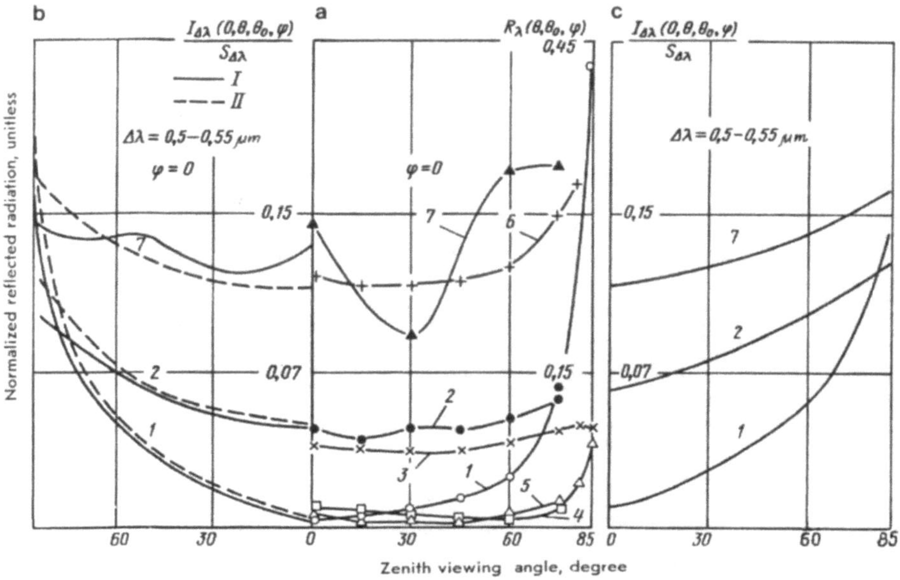

Fig. 4.5 a – c. Angular variations of the reflection coefficients for different natural objects (**a**): *1* calm water surface; *2* dry grass cover; *3* marshed grass cover; *4* mixed forest; *5* pine forest; *6* trampled ground; *7* sand surface. The angular distributions of reflected radiation at the atmospheric top level for objects *1, 2, 7*; *I* ignoring reflection anisotropy; *II* with reflection anisotropy (**b**); values similar to **b** for Lambertian surfaces and $\varphi = 0°$ (**c**)

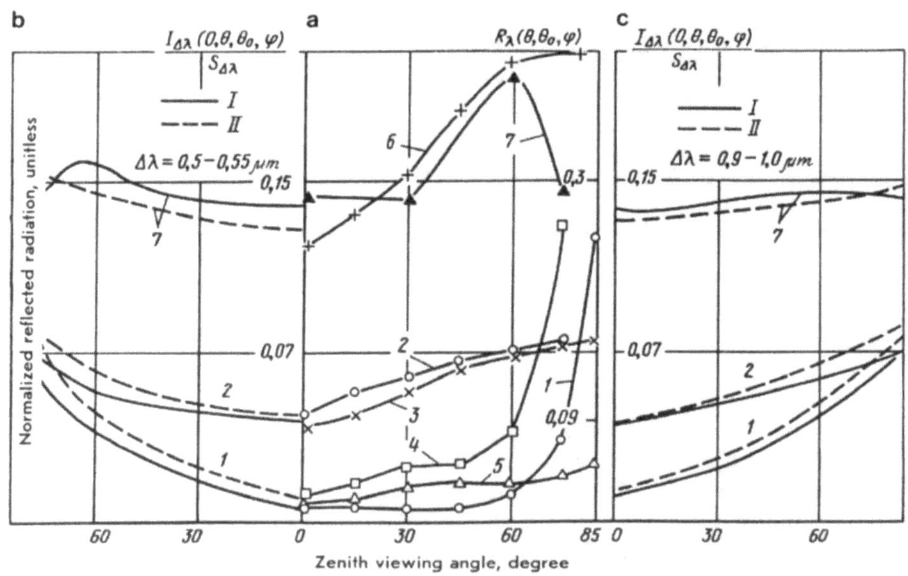

Fig. 4.6 a – c. Angular variations of the reflection coefficients for natural objects shown in Fig. 4.5 a; **b** and **c** are similar to Fig. 4.5 c for $\varphi = 180°$. *b* $\Delta\lambda = (0.5-55)$ µm; *c* $\Delta\lambda = (0.9-1.0)$ µm

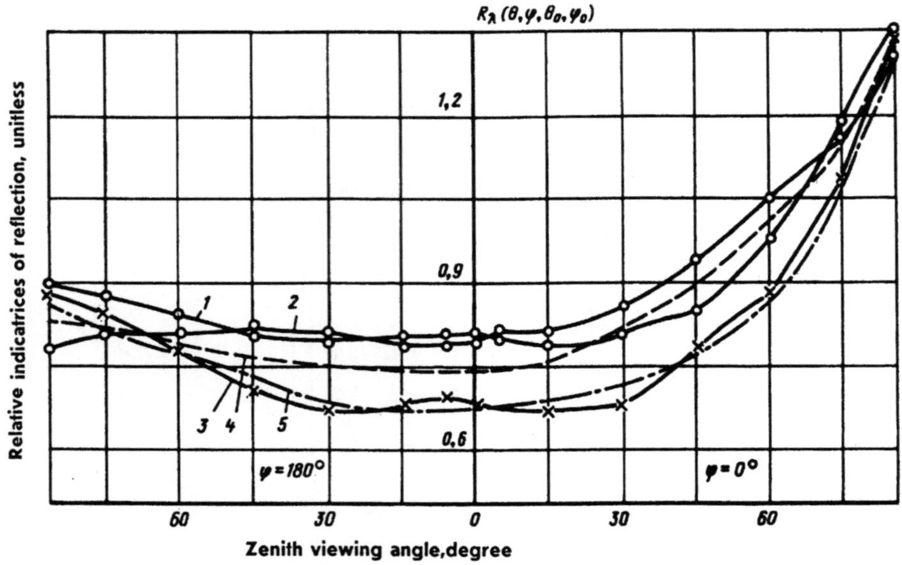

Fig. 4.7. The vertical change of snow cover reflection indicatrices: *1* experimental data; z = 200 m; A = 0.82; $\theta_0 = 62.9°$; $\lambda = 0.509\ \mu m$; *2* z = 3 km; A = 0.84; $\theta_0 = 66.4°$; $\lambda = 0.509\ \mu m$; *3* z = 8 km; A = 0.74; $\theta_0 = 68.4°$; $\lambda = 0.509\ \mu m$; *4* calculated data [76] for $\Delta\lambda = (0.5-0.55)\ \mu m$; *5* calculated data [207] for $\Delta\lambda = (0.5-0.55)\ \mu m$

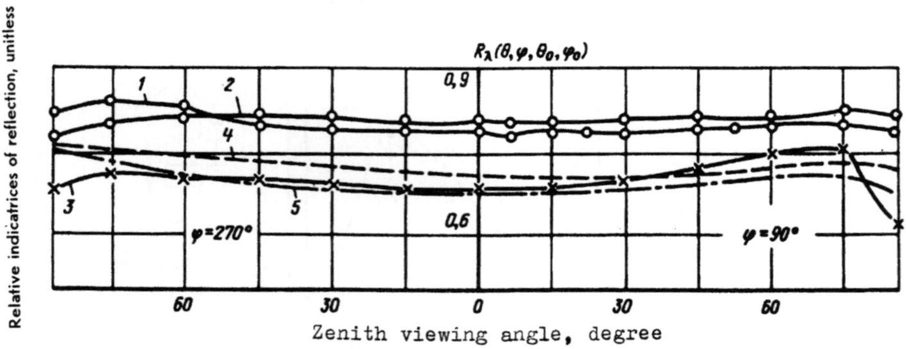

Fig. 4.8. The vertical change of the snow cover reflection phase functions (see Fig. 4.7) for $\varphi = 90°$ and $\varphi = 270°$

Figures 4.7–4.8 illustrate results of comparison of the angular change of calculated and measured coefficients of reflection for the snow surface – atmosphere system [6]. Measurements were made with the spectral albedometer and scanning radiometer on board the MGO flying laboratory [109].

The results of calculations and measurements are reasonably consistent, which confirms the possibility of using the calculation scheme that considers the non-orthotropic reflection in the interpretation of experimental data. The coincidence of calculated and measured estimates points to a smooth change of radiation intensity in the spectral intervals under consideration.

Thus, the non-orthotropicity of reflection is an important factor of the formation of a real reflection field and is efficiently realized by the calculation scheme of atmospheric correction of aerospace measurements using the forms and types of reflection represented in (4.55–4.57).

4.5 Results of Numerical Modeling of the Background Spectral Transfer Functions

Tabulating the principal components of the scalar background transfer function $\Pi(\eta, \zeta, \varphi, \tau_0)$ makes it possible to calculate its monochromatic values in the visible spectrum for different types of natural and artificial extended backgrounds formed by horizontally homogeneous surfaces and by their covers [31, 68, 94, 97]. The dependence of the background spectral transfer function Π_λ and its constituents ϱ, σ, ν on the optical model of the atmosphere, reflective characteristics of the surface, sun illumination conditions, viewing direction, and wavelength is totally determined by the respective solutions of the problem of multiple scattering of light in a flat layer of a finite optical thickness with a horizontally homogeneous base [321]. Figure 4.9 shows the graphs of spectral transfer functions for some types of natural horizontally-homogeneous extended backgrounds. The calculated estimates of Π_λ were obtained with formula (4.18) for the Elterman-1968 closed aerosol model when spectrophotometering the Earth from space at nadir ($\eta = 1$) and zenith solar angles ($\zeta = 1$).

Table 4.5 lists monochromatic estimates of transfer functions calculated in the same way for a definite set of artificial covers in the wavelength interval $\lambda = 500-800$ nm.

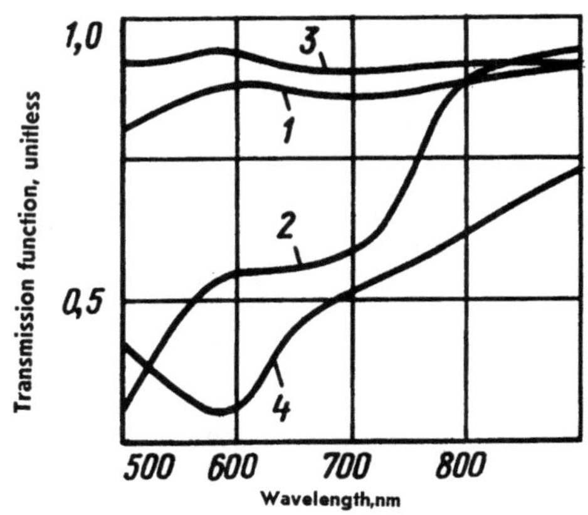

Fig. 4.9. Theoretical values of the spectral transmission function of the atmosphere for the 1968 Elterman closed aerosol model for $\zeta = \eta = 1$: *1* white (river) sand; *2* thick green grass cover; *3* dry fresh snow; *4* black soil

Table 4.5. The spectral transfer function of the atmosphere $\Pi(\eta, \zeta, \varphi, \tau_0)$ for the brightness of horizontally-homogeneous extended backgrounds (objects) when spectrophotometering the Earth from space in nadir ($\eta = 1$) (the Elterman-1968 closed aerosol model)

ζ	Water surface (calm)						
	λ, nm						
	400	500	550	600	650	700	800
0.8	0.1963	0.2569	0.3149	0.3691	0.4137	0.4522	0.5209
0.9	0.2985	0.2412	0.2961	0.3464	0.3831	0.4299	–
A_λ (0.8)	0.0250	0.0248	0.0247	0.0246	0.0243	0.0242	0.0240
A_λ (0.9)	0.0246	0.0216	0.0216	0.0215	0.0211	0.0210	–

ζ	Concrete (light-colored slabs)					
	λ, nm					
	500	550	600	650	700	800
	A_λ					
	0.130	0.220	0.290	0.340	0.380	0.400
0.1	0.2201	0.4042	0.5600	0.5897	0.5945	0.6274
0.2	0.3004	0.5383	0.7254	0.7323	0.7212	0.7519
0.3	0.3779	0.6435	0.8335	0.8151	0.7913	0.8159
0.4	0.4354	0.7076	0.8880	0.8550	0.8265	0.8470
0.5	0.4712	0.7390	0.9064	0.8699	0.8421	0.8470
0.6	0.4899	0.7469	0.9088	0.8717	0.8466	0.8648
0.7	0.4952	0.7453	0.9051	0.8649	0.8455	0.8629
0.8	0.4916	0.7333	0.8807	0.8473	0.8381	0.8574
0.9	0.4820	0.7166	0.8596	0.8388	0.8291	0.8496
1.0	0.4679	0.6969	0.8371	0.8228	0.8183	0.8350

ζ	Unmetalled highway (dry)					
	λ, nm					
	500	550	600	650	700	800
	A_λ					
	0.080	0.100	0.120	0.130	0.150	0.220
0.1	0.1455	0.2268	0.3250	0.3355	0.3493	0.4653
0.2	0.2049	0.3298	0.4824	0.4773	0.4778	0.6030
0.3	0.2659	0.4255	0.6127	0.5802	0.5640	0.6827
0.4	0.1337	0.4908	0.6893	0.6370	0.6124	0.7240
0.5	0.3448	0.5255	0.7201	0.6597	0.6352	0.7428
0.6	0.3611	0.5368	0.7209	0.6624	0.6419	0.7484
0.7	0.3659	0.5327	0.7151	0.6521	0.6355	0.7458
0.8	0.3626	0.5190	0.6785	0.6357	0.6292	0.7382
0.9	0.3541	0.6006	0.6483	0.6134	0.6161	0.7225
1.0	0.3417	0.4795	0.6174	0.5909	0.6009	0.7077

Table 4.5 (continued)

ζ	Unmetalled highway (wet)					
	λ, nm					
	500	550	600	650	700	800
	A_λ					
	0.070	0.090	0.090	0.100	0.120	0.190
0.1	0.1293	0.2081	0.2624	0.2774	0.2985	0.4266
0.2	0.1833	0.3057	0.4052	0.4083	0.4193	0.5637
0.3	0.2396	0.3980	0.5325	0.5090	0.5042	0.6457
0.4	0.2842	0.4640	0.6115	0.5668	0.5532	0.6889
0.5	0.9135	0.4963	0.6442	0.5910	0.5767	0.7087
0.6	0.3290	0.5075	0.6451	0.5932	0.5836	0.7146
0.7	0.3335	0.5034	0.6389	0.5825	0.5666	0.7119
0.8	0.3304	0.4899	0.6002	0.5552	0.5705	0.7039
0.9	0.3223	0.4717	0.5688	0.5426	0.5570	0.6925
1.0	0.3106	0.4508	0.5373	0.5197	0.5414	0.6718

ζ	Concrete highway					
	λ, nm					
	500	550	600	650	700	800
	A_λ					
	0.124	0.136	0.153	0.151	0.166	0.223
0.1	0.3008	0.3563	0.4124	0.4385	0.4756	0.5694
0.2	0.3666	0.4403	0.5123	0.5426	0.5829	0.6766
0.3	0.4367	0.5211	0.5998	0.6251	0.6619	0.7460
0.4	0.4970	0.5852	0.6648	0.6830	0.7160	0.7908
0.5	0.5461	0.6338	0.7112	0.7245	0.7543	0.8211
0.6	0.5844	0.6699	0.7446	0.7530	0.7808	0.8422
0.7	0.6136	0.6960	0.7680	0.7735	0.8000	0.8576
0.8	0.6349	0.7145	0.7836	0.7894	0.8126	0.8675
0.9	0.6475	0.7225	0.7899	0.7910	0.8174	0.8715
1.0	0.6201	0.6882	0.7519	0.7532	0.7825	0.8431

ζ	Asphalted highway				
	λ, nm				
	550	600	650	700	800
	A_λ				
	0.130	0.160	0.180	0.190	0.200
0.1	0.2785	0.3959	0.4162	0.4078	0.4401
0.2	0.3945	0.5628	0.5660	0.5413	0.5775
0.3	0.4968	0.6907	0.6166	0.6266	0.6588

Table 4.5 (continued)

ζ	Asphalted highway				
	λ, nm				
	550	600	650	700	800
	A_λ				
	0.130	0.160	0.180	0.190	0.200
0.4	0.5639	0.7620	0.7195	0.6730	0.7013
0.5	0.5987	0.7872	0.7402	0.6946	0.7208
0.6	0.6098	0.7907	0.7426	0.7008	0.7267
0.7	0.6058	0.7854	0.7333	0.6954	0.7240
0.8	0.5923	0.7522	0.7091	0.6889	0.7161
0.9	0.5738	0.7243	0.6978	0.6766	0.7050
1.0	0.5524	0.6052	0.6766	0.6621	0.6846

The results of numerical modeling of Π_λ show that, to ensure a minimum effect of the radiative factors of the atmosphere on the spectrophotometric information obtained from space, maximum values of solar zenith angle θ_0 must not exceed 30°. With the growing solar zenith angle and decreasing brightness of an extended surface (above-water) background, the transforming and reducing effect of atmospheric thickness increases substantially. With the increasing wavelength the reducing effect of the atmosphere outside the absorption bands weakens, determined by the reduced scattering properties of its molecular and aerosol components. Therefore, the optimal spectral interval $\Delta\lambda$ for space-based spectrophotometry of natural formations (backgrounds) is in the "red" region of the visible interval $\lambda = 700 - 850$ nm.

The spectral, angular and spatial behavior of monochromatic transfer functions $\Pi_\lambda(\eta, \zeta, \varphi, \tau_0)$ is determined by the relationship between the brightness of atmospheric haze $I_h(\eta, \zeta, \varphi, \tau_0)$ and the brightness of background (object) at the surface level $I_0(\eta, \zeta, \varphi, \tau_0)$. The brightness of atmospheric haze I_h and the coefficient of diffuse reflection ϱ decrease, but the brightness of the surface I_0 increases with the growing wavelength I_0. When the solar zenith distance θ_0 grows, the brightnesses I_{haze} and I_0 decrease, and the coefficient of diffuse reflection ϱ grows. Such influences on the principal components of transfer function Π_λ lead to the fact that the scattering and absorbing layer of the atmosphere intensifies the brightness of weakly reflecting natural formations, objects and backgrounds spectrophotometered from space. The brightness of strongly reflecting backgrounds (objects) when surveying the Earth from space, in nadir at high sun elevations decreases, as a rule, compared with surface (above-water) conditions of their observation.

In a molecular, purely scattering (without ozone) atmosphere with albedo $A_\lambda = 0.1 - 0.8$ the brightness of atmospheric haze $I_{h\lambda}$, is less and the surface brightness $I_{0,\lambda}$ is greater compared to a general aerosol model of the atmosphere [31, 68]. This difference manifests itself particularly with increased wavelength $I_{h,\lambda}$, which is determined by increasing optical activity of the aerosol in the near IR spectral region. The presence of ozone reduces brightness of $I_{haze,\lambda}$ and $I_{0,\lambda}$ compared to that of a purely scattering molecular atmosphere [31, 68, 97]. In this case at $A_\lambda = 0.78$ (snow) with the interval of zenith solar distances $\theta_0 = 0 - 30°$ the transfer function $\Pi_\lambda(1, \zeta, \varphi, \tau_0)$ is greater than unity. With increased wavelength it decreases towards unity. At $\theta_0 > 30°$ this function grows with wavelength, also approaching unity. For smaller albedos ($A = 0.24$, sand) the transfer function is $\Pi_\lambda(1, \zeta, \varphi, \tau_0) < 1$ in the whole spectral interval considered.

Here a general regularity consists in that, the Π_λ values below unity grow and decrease, respectively, with increased wavelength and solar zenith distance θ_0. With increasing surface albedo A_λ the values of transfer function Π_λ (when it is below unity) increase too. The presence of atmospheric ozone in this case leads to an increase of Π_λ function, decreasing, the transforming effect of the atmosphere on the spectral brightness of horizontally homogeneous backgrounds. With increasing wavelength the function grows monotonically. For large albedos ($A_\lambda = 0.78$) the Π_λ function first grows with increasing wavelength, reaching a maximum near $\lambda = 600$ nm, and then decreases. In this case there are two spectral regions (450 – 500 and 700 – 800 nm) in which $\Pi_\lambda \approx 1$, i.e., the conditions of the experiment on the spectrophotometry of natural formations (backgrounds) from space are optimal. With decreasing surface albedo the values of transfer functions $\Pi_\lambda(1, \zeta, \varphi, \tau_0)$ decrease.

A comparison of transfer functions Π_λ calculated for different optical models shows that an addition of aerosol to a molecular atmosphere decreases the values of Π_λ function when it is below unity, and raises them when $\Pi_\lambda > 1$ for all albedos A_λ in the whole spectral interval under consideration.

4.6 Diagrams of an Optimization of the Transforming Effect of the Atmosphere when Surveying the Earth from Space

According to the graphical and tabulated data, the values of the spectral transfer function Π_λ for the brightness of natural formations (backgrounds) can be above or below unity, depending on surface albedo A_λ, solar zenith angle θ_0 and wavelength λ. Bearing in mind the continuity of the Π_λ transfer function with these parameters varying, the question arises, for which values of albedo A_λ, solar zenith angle θ_0 and wavelength λ this function will be either equal to unity or close to it? To answer this question, many calculations are required and numerous graphs and tables are needed. This problem is solved more clearly and efficiently using a diagram graphic technique similar to that suggested in [236]. With one of the base calculated parameters fixed and laid on horizontal axis, we write in a vertical column the calculated estimates

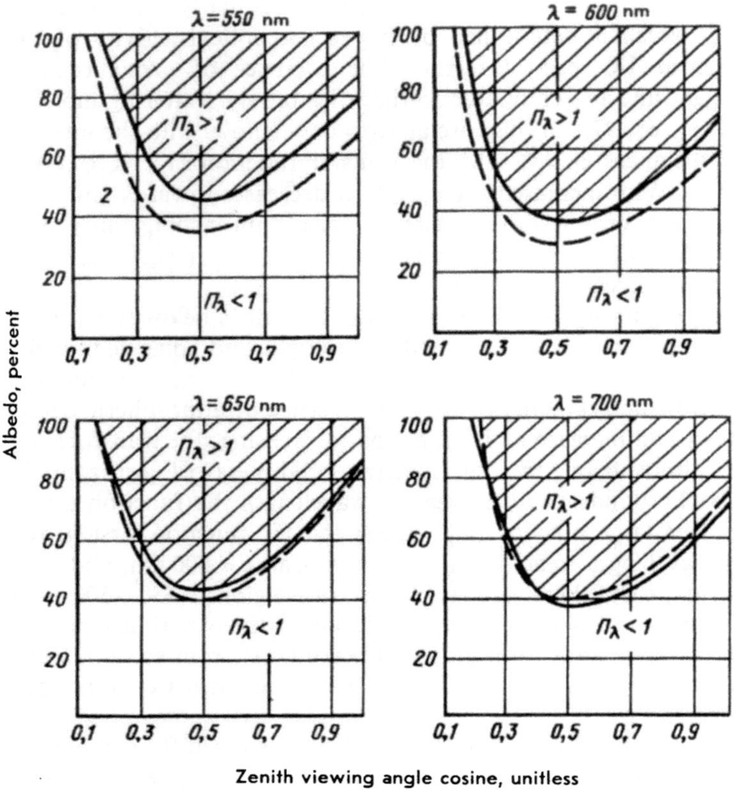

Fig. 4.10. Diagrams of the optimization of the monochromatic transfer function of the atmosphere Π_λ for the 1968 Elterman closed aerosol model and for the three-term phase function (see Table 4.1): *1* without similarity conversion; *2* with similarity conversion

of the Π_λ transfer function, as a function of the parameter on the ordinate axis. Linking the estimates $\Pi_\lambda = 1$ (if any) by a continuous curve, we obtain an idea of optimal conditions of the survey of the environment for the fixed and varied parameters of the problem. Figures 4.10 and 4.11 show the respective results obtained with formula (4.18) and the data of Tables 4.3 and 4.4.

4.7 The Geographical and Seasonal Distribution of the Transfer Function Monochromatic Values

For the practical realization of remote sensing of the Earth from space quantitative estimates of the geographical and seasonal distribution of spectral transfer function in the visible $\lambda = 400-800$ nm are of great interest.

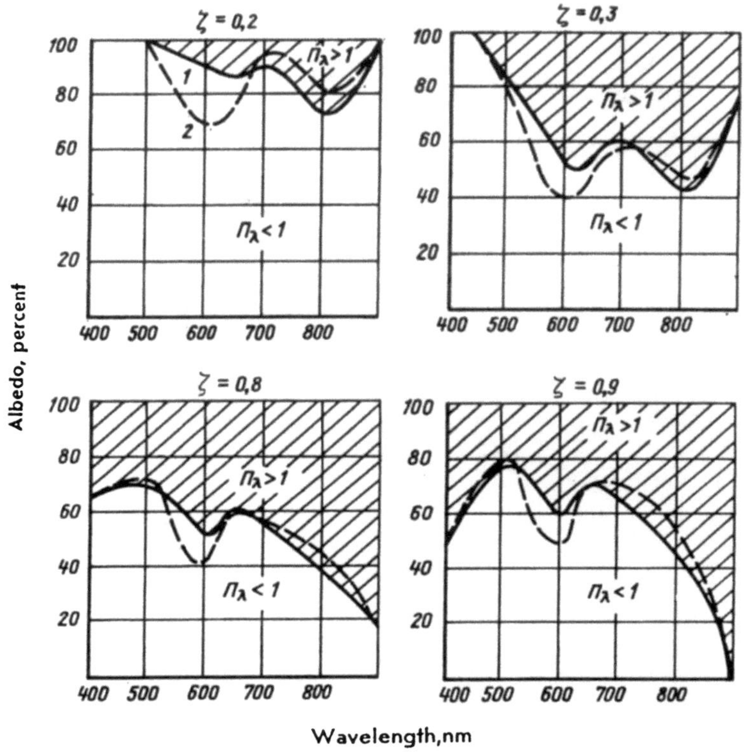

Fig. 4.11. Optimization diagrams for the spectral transfer function of the atmosphere Π_λ for the 1968 Elterman closed aerosol model and the three-term phase function (see Table 4.1): *1* without similarity conversion; *2* with similarity conversion

Monochromatic values of the background transfer functions Π_λ corresponding to summer and winter conditions were calculated from total albedos for large areas [145], from tabulated estimates of basic functions (Tables 4.3–4.4), and from formula (4.18).

To obtain initial values of albedo A needed for calculation, the globe was divided into $10° \times 10°$ squares, and the albedo values were taken latitude by latitude. The latitudinal distribution of albedo A obtained was used to calculate the monochromatic transfer functions Π_λ for a horizontally homogeneous Lambertian surface, the Elterman closed aerosol model and nadir viewing of the Earth. The solar zenith angle θ_0 was taken as minimum for each latitude, i.e., the height of the sun over the horizon was calculated for conditions of local (regional) zenith. The respective results for summer and winter conditions are shown in Figs. 4.12 and 4.13 by isolines at $\lambda = 500$ nm. On the left-hand side of each figure are shown the values of latitude, and on the right-hand side the respective minimum zenith solar distance θ_0.

The features of the geographical and seasonal distribution of Π_λ over the globe correspond to the geographical and seasonal distribution of albedos for

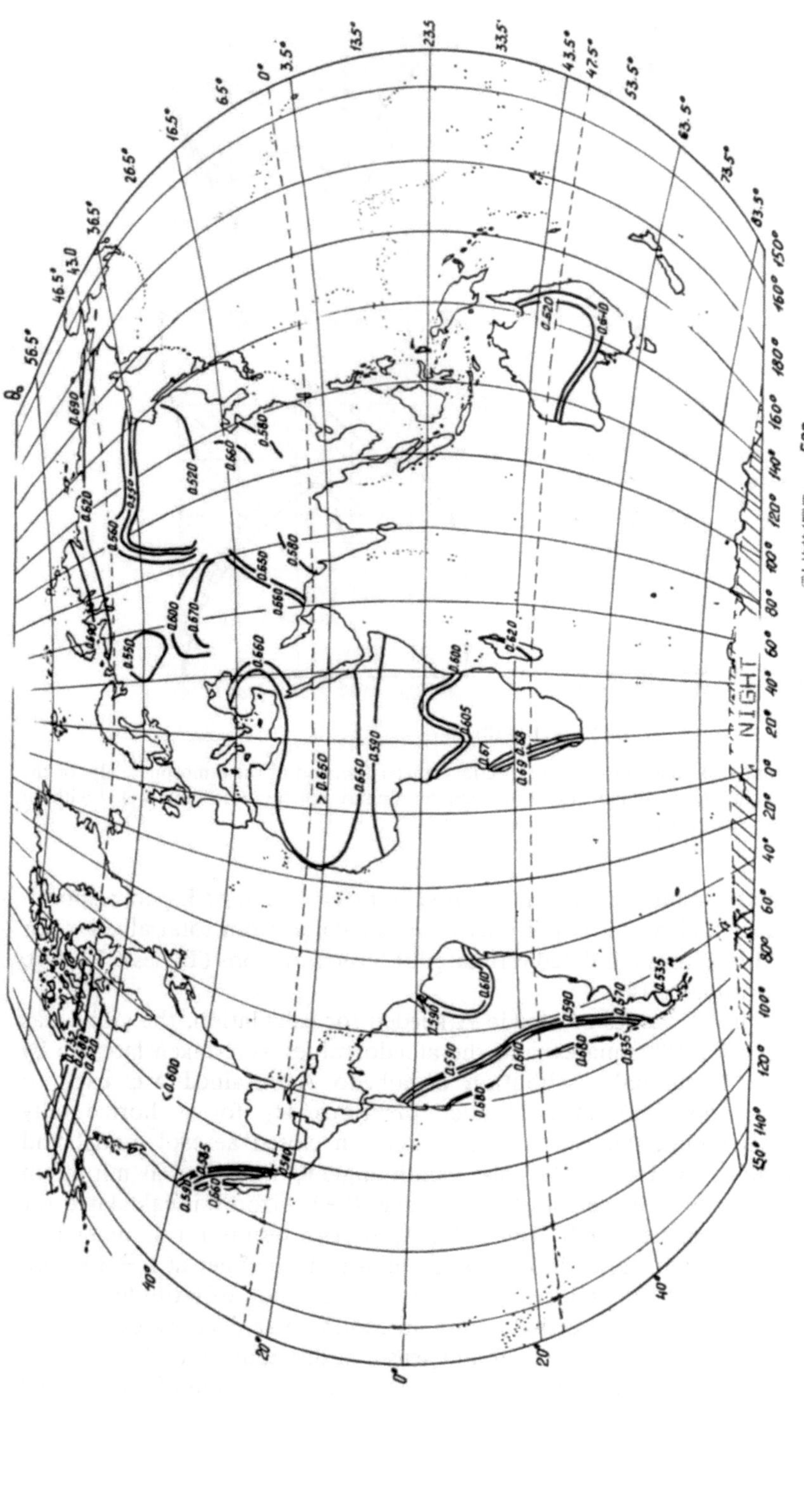

Fig. 4.12. The geographical and seasonal distributions of the monochromatic transfer function of the terrestrial atmosphere Π_λ (the 1968 Elterman aerosol model; nadir viewing $\eta = 1$, local solar zenith, $\lambda = 500$ nm, summer)

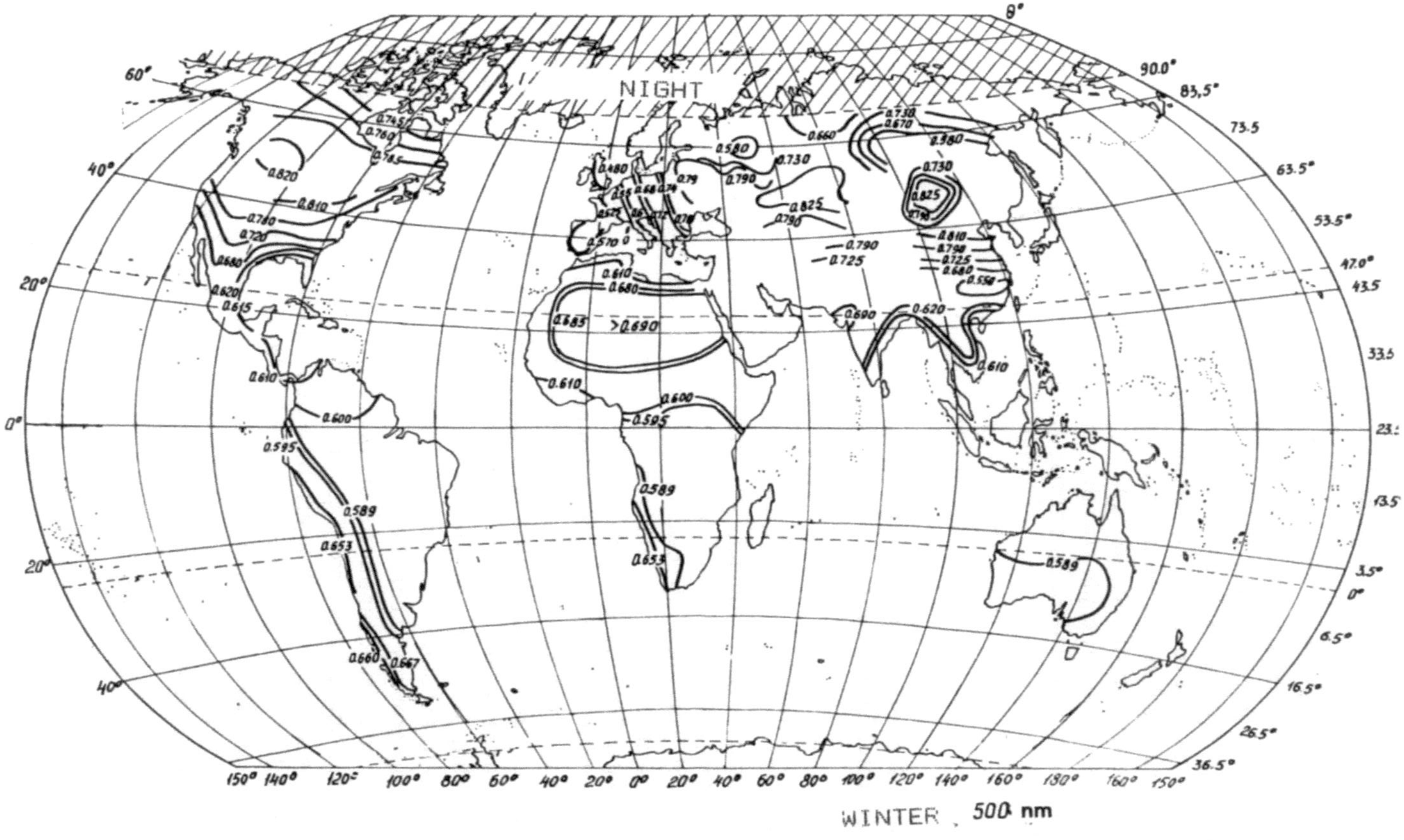

Fig. 4.13. The geographical and seasonal distributions of the monochromatic transfer function of the terrestrial atmosphere (the 1968 Elterman aerosol model; nadir viewing, $\eta = 1$, local solar zenith, $\lambda = 500$ nm, winter)

continents, oceans, and seas. They also depend on latitudinal variations of maximum height of the sun over the horizon. In the summer maximum Π_λ values are observed in the northern tropics. With increasing latitude they decrease monotonically both northward and southward from the northern tropics. In winter maximum Π_λ values correspond to the zone of tundra where snow remains clean and ensures a high surface (background) albedo.

Note, that "average" isolines shown in Figs. 4.12–4.13 were obtained for a stationary atmospheric model not taking seasonal and regional variations of atmospheric aerosol into account. Therefore, real data on the geographical and seasonal distribution of the monochromatic values of the Π_λ transfer function may differ in some details from the results of theoretical modeling because of spatial and temporal variations of the optical characteristics of atmospheric aerosol.

4.8 The Effect of Atmospheric Haze on the Colour of Natural Formations

Observations from space have shown that various natural formations, objects and backgrounds (in particular, sea surface in different geographical and climatic zones) may be identified according to their colorimetric characteristics [68, 112, 148]. This opens up certain possibilities for solving various problems of remote sensing based on analysis of experimental and theoretical colorimetric data. Since there is always an absorbing and scattering atmospheric layer between the on-board optical instruments and the planetary surface, it is necessary to assess the transforming effect of atmospheric haze on the color of the surface (above-water) objects and backgrounds either observed or spectrophotometered from space. A generalized transfer function of the atmosphere enabling assessment of the respective colorimetric characteristics at the surface level from color tones and color intensity of surface (above-water) objects and backgrounds is considered next.

To obtain complete quantitative and qualitative characteristics of the color of surface and above-water objects (backgrounds), three numerical parameters must be known: brightness of object (background) I, color tone characterized by wavelength λ, and color intensity P. The brightness of object (background) I is either assessed experimentally or calculated theoretically. To find the color tone and color intensity, relative color coefficients X, Y, Z are usually used. Now we introduce the colorimetric transfer functions of the atmosphere $\Pi_{x,0b}$, $\Pi_{y,0b}$, $\Pi_{z,0b}$ and $\Pi_{x,B}$, $\Pi_{y,B}$, $\Pi_{z,B}$ for the color coefficients of surface (above-water) object or background X, Y, Z according to [100]:

$$
\begin{aligned}
X_{0b}^0 &= \Pi_{x,0b} X_{0b} , & Y_{0b}^0 &= \Pi_{y,0b} Y_{0b} , & Z_{0b}^0 &= \Pi_{z,0b} Z_{0b} , \\
X_B^0 &= \Pi_{x,B} X_B , & Y_B^0 &= \Pi_{y,B} Y_B , & Z_B^0 &= \Pi_{z,B} Z_B ,
\end{aligned}
\tag{4.58}
$$

where X^0, Y^0, Z^0 and X, Y, Z are the color coefficients of object (background), respectively, at the level of surface and atmospheric top. The use of (4.58) and of scalar transfer function Π for the spectral brightness of object (background) assessed in [100] makes it possible to introduce a generalized matrix transfer function of the atmosphere according to relationship:

$$\begin{pmatrix} I^0_{0b} \\ X^0_{0b} \\ Y^0_{0b} \\ Z^0_{0b} \end{pmatrix} = \begin{pmatrix} \Pi_{0b} & 0 & 0 & 0 \\ 0 & \Pi_{x,0b} & 0 & 0 \\ 0 & 0 & \Pi_{y,0b} & 0 \\ 0 & 0 & 0 & \Pi_{z,0b} \end{pmatrix} \begin{pmatrix} I_{0b} \\ X_{0b} \\ Y_{0b} \\ Z_{0b} \end{pmatrix} \tag{4.59}$$

For numerical illustration of these expressions calculations were made in [100] of the colorimetric transfer function Π_x and Π_y for the case of an orthotropic horizontally homogeneous surface with nadir viewing ($\theta = 0°$), solar zenith angle $\theta_0 = 0 - 80°$ and $\bar{A} = 0.78, 0.24, 0.084$. The chosen values of the albedo characterize average spectral albedos of, respectively, snow, sand and green plants in the visible. An optical model of an aerosol atmosphere and numerical scheme for calculations of auxiliary functions in the assessment of Π_x and Π_y have been taken from [97]. A characteristic feature of the numerical data obtained is that an aerosol atmosphere at $\theta = 0°$ and $\theta_0 = 30°$ transforms the colorimetric characteristics of natural formations little when surveying the Earth from space. In a wide range of variations of the solar zenith distance θ_0, in the case of a strongly reflecting surface, a basic contribution to this transformation is made, not by atmospheric aerosol, but by ozone.

4.9 An Assessment of the Polarization Transfer Function of the Atmosphere

An important additional way of enlarging the information content of the Earth's surface survey from space, is to consider the polarization of reflected and scattered radiation [237] recorded by the on-board instruments at the atmospheric top. Confining ourselves to the case of linear polarization of radiation [43], we introduce the polarization transfer function of the atmosphere $P(\gamma, \tau_0)$ and relate it to the scalar transfer function Π [96].

The components of total spectral brightness of the medium I_λ, measured from space, in planes parallel and perpendicular to the principal plane of scattering, are denoted respectively, through $I_{||,\lambda}(\eta, \zeta, \varphi, \tau_0)$ and $I_{\perp,\lambda}(\eta, \zeta, \varphi, \tau_0)$.

The plane of single light scattering we take as the principal plane of scattering. First, we suppose that the medium is limited from beneath by an infinitely-extended horizontally-homogeneous surface with a coefficient of reflection $r(\eta,\zeta,\varphi,\tau_0)$:

$$r(\eta,\zeta,\varphi,\tau_0)=r_{||}(\eta,\zeta,\varphi,\tau_0)+r_{\perp}(\eta,\zeta,\varphi,\tau_0)\ , \tag{4.60}$$

where $r_{||}$ and $r_{\perp}$ are the components measured, respectively, in the planes parallel and perpendicular to the principal plane of scattering. We assume radiation polarization to take place only at a single scattering in the atmosphere and at a single reflection of photons from the surface (object or background). A secondary polarization of reflected radiation on its way from the surface to atmospheric top is neglected. The planes of scattering for radiation reflected from the surface and scattered in the atmosphere are assumed to coincide. Then, at the atmospheric top level ($\tau=0$) the extent of linear polarization of radiation reflected by the planetary surface-atmosphere system at $\varphi=0$ and $\varphi=\pi$ is equal to [96]:

$$\begin{aligned}P_\lambda(\gamma,\tau_0)&=\frac{I_{||,\lambda}(\eta,\zeta,\varphi,\tau_0)-I_{\perp,\lambda}(\eta,\zeta,\varphi,\tau_0)}{I_\lambda(\eta,\zeta,\varphi,\tau_0)}\ ,\\ I_{\lambda,||}(\eta,\zeta,\varphi,\tau_0)&+I_{\lambda,\perp}(\eta,\zeta,\varphi,\tau_0)=I_\lambda(\eta,\zeta,\varphi,\tau_0)\ .\end{aligned} \tag{4.61}$$

The spectral polarization transfer function of the atmosphere $P(\gamma,\tau_0)$ is estimated in the following way [96]:

$$P_{0,\lambda}(\gamma,\tau_0)=P_{\lambda,\mathrm{mol}}(\gamma,\tau_0)P_{\lambda,\mathrm{aer}}(\gamma,\tau_0)\ , \tag{4.62}$$

where $P_{0,\lambda}(\gamma,\tau_0)$ is the extent of linear polarization of radiation by the medium at the surface level,

$$\begin{aligned}P_{0,\lambda}(\gamma,\tau_0)&=\frac{I^0_{||,\lambda}(\eta,\zeta,\varphi,\tau_0)-I_{\perp,\lambda}(\eta,\zeta,\varphi,\tau_0)}{I^0_{||,\lambda}(\eta,\zeta,\varphi,\tau_0)+I^0_{\perp,\lambda}(\eta,\zeta,\varphi,\tau_0)}\ ,\\ I_{0,||}(\eta,\zeta,\varphi,\tau_0)&+I_{0,\perp}(\eta,\zeta,\varphi,\tau_0)=I_{0,\lambda}\ .\end{aligned} \tag{4.63}$$

In (4.63) $I^0_{||,\lambda}$ and $I^0_{\perp,\lambda}$ are the components of total brightness of the surface $I_{0,\lambda}$, respectively, in planes parallel and perpendicular to the principal plane of scattering. Since, according to our assumption, the intensity $I_{0,\lambda}$ gets polarized only at single scattering of photons from the surface (object, background), we have, instead of (4.63):

$$P_{0,\lambda}(\gamma)=\frac{r_{||,\lambda}(\eta,\zeta,\varphi)-r_{\perp,\lambda}(\eta,\zeta,\varphi)}{r_\lambda(\eta,\zeta,\varphi)}\ . \tag{4.64}$$

Now we obtain expressions for the polarization transfer function P_λ for the cases that are interesting from the practical point of view.

For an infinitely extended horizontally homogeneous surface with a coefficient of reflection r_λ, with the assumptions mentioned, the $I_{||,\lambda}$ and $I_{\perp,\lambda}$ components can easily be written as [96]:

$$I_{||,\lambda}(\eta,\zeta,\varphi,\tau_0) = S\zeta[\varrho_{||,\lambda}(\eta,\zeta,\varphi,\tau_0) + I_{||,0,\lambda}(\eta,\zeta,\varphi,\tau_0)\mu_\lambda(\eta,\tau_0)] \ ,$$
$$I_{\perp,\lambda}(\eta,\zeta,\varphi,\tau_0) = S\zeta[\varrho_{\perp,\lambda}(\eta,\zeta,\varphi,\tau_0) + I_{\perp,0,\lambda}(\eta,\zeta,\varphi,\tau_0) \times \mu_\lambda(\eta,\tau_0)] \ , \qquad (4.65)$$

where the $\mu_\lambda(\eta,\tau_0)$ function is determined from relationship (4.24), and $\varrho_{\lambda,||}$, $\varrho_{\lambda,\perp}$ – are the components of total coefficient of reflection $\varrho_\lambda = \varrho_{\lambda,||} + \varrho_{\lambda,\perp}$ in the absence of reflection from the surface $A = 0$, respectively, in planes parallel and perpendicular to the principal plane of scattering. The $\varrho_{||,\lambda}$ and $\varrho_{\perp,\lambda}$ brightnesses are found from the formulas:

$$\varrho_{\lambda,||} = \frac{\Lambda}{4(\eta+\zeta)}[1 - e^{-\tau_0(1/\eta+1/\zeta)}]x_{||}(\gamma) + \Delta\varrho_{\lambda,||}(\eta,\zeta,\varphi,\tau_0) \ ,$$
$$\varrho_{\lambda,\perp} = \frac{\Lambda}{4(\eta+\zeta)}[1 - e^{-\tau_0(1/\eta+1/\zeta)}]x_{\perp}(\gamma) + \Delta\varrho_{\lambda,\perp}(\eta,\zeta,\varphi,\tau_0) \ , \qquad (4.66)$$

where $\Delta\varrho_{\lambda,||} = \Delta\varrho_{\lambda,\perp} = \Delta\varrho_\lambda$ is a non-polarized addition determined by multiple scattering of light in the atmosphere. The polarization components $x_{||}(\gamma)$ and $x_\perp(\gamma)$ of total phase function $x(\gamma) = x_{||}(\gamma) + x_\perp(\gamma)$ are calculated using Mie theory following the technique used in Section 1.2 to derive the phase matrix $P(\gamma,\tau_0)$:

$$x_{||}(\gamma) = \frac{\hat{P}_2(\gamma,\tau_0)}{2} \ , \qquad x_\perp(\gamma) = \frac{\hat{P}_1(\gamma,\tau_0)}{2} \ . \qquad (4.67)$$

Using (4.63) and (4.65), we find

$$P(\gamma,\tau_0) = P_{\text{atm.}}(\gamma,\tau_0)\frac{S\zeta\varrho(\eta,\zeta,\varphi,\tau_0)}{I(\eta,\zeta,\varphi,\tau_0)} + P_0(\gamma)\frac{I_0(\gamma,\tau_0)\mu(\eta,\tau_0)}{I(\eta,\zeta,\varphi,\tau_0)} \ , \qquad (4.68)$$

where the extent of linear polarization of atmospheric brightness is:

$$P_{\text{atm.},\lambda} = \frac{\varrho_{\lambda,||} - \varrho_{\lambda,\perp}}{\varrho_\lambda} \ . \qquad (4.69)$$

It follows from (4.67) that in the absence of the polarization of light the scalar functions I_0, I_h, I and Π are easily interrelated:

$$\frac{S\zeta\varrho(\eta,\zeta,\varphi,\tau_0)}{I(\eta,\zeta,\varphi,\tau_0)} = 1 - \mu(\eta,\tau_0)\Pi(\eta,\zeta,\varphi,\tau_0) \ . \qquad (4.70)$$

Using (4.68 – 4.70), we obtain the principal formula for the extent of radiation polarization $P(\gamma,\tau_0)$ at the upper boundary of a horizontally-homogeneous layer with the reflecting base [96]:

$$P(\gamma,\tau_0) = P_{\text{atm.}}(\gamma,\tau_0)[1 - \mu(\eta,\tau_0)\Pi(\eta,\zeta,\varphi,\tau_0)]$$
$$+ P_0(\gamma,\tau_0)\mu(\eta,\tau_0)\Pi(\eta,\zeta,\varphi,\tau_0) \ . \qquad (4.71)$$

The scalar transfer function Π in (4.71) is determined from (4.18).

With the planetary atmosphere assumed to consist of molecules and aerosols randomly mixed in the environment, the following relationships are valid for total phase function $x(\gamma)$ and its components $x_{\perp}(\gamma)$ and $x_{||}(\gamma)$:

$$x(\gamma) = \frac{3}{4}\delta(1+\cos^2\gamma) + (1-\delta)\frac{P_{1,\,aer.} + P_{2.aer.}}{2} ,$$

$$x_{||}(\gamma) = \frac{P_{2,\,aer.}}{2}(1-\delta) + \frac{3}{4}\delta ; \tag{4.72}$$

$$x_{\perp}(\gamma) = \frac{P_{1,\,aer.}}{2}(1-\delta) + \frac{3}{4}\delta , \tag{4.73}$$

$$\delta = \frac{\tau_{0,\,mol}}{\tau_{0,\,mol} + \tau_{0,\,aer.}} , \qquad x_{aer.} = \frac{P_{1,\,aer.} + P_{2,\,aer.}}{2} . \tag{4.74}$$

Using (4.68), (4.72 – 4.74) for the extent of polarization of the atmospheric haze brightness $P_{atm.}$, we find the following expression [96]:

$$P_{atm.}(\gamma, \tau_0) = \frac{\Lambda\left[\frac{3}{4}\delta\sin^2\gamma + (1-\delta)\frac{P_{2,\,aer.} - P_{1,\,aer.}}{2}\right]}{4(\eta+\zeta)\varrho_{atm.}(\eta, \zeta, \varphi, \tau_0)}[1 - e^{-\tau_0(1/\eta + 1/\zeta)}] , \tag{4.75}$$

$$\varrho_{atm.} = \Lambda\left[\frac{3}{4}\delta(1+\cos^2\gamma) + (1-\delta)\frac{P_{2,\,aer.} + P_{1,\,aer.}}{2}\right]\frac{[1 - e^{-\tau_0(1/\eta + 1/\zeta)}]}{4(\eta+\zeta)} + 2\Delta\varrho(\tau_0, \gamma) . \tag{4.76}$$

Substituting (4.71) and (4.72), we obtain the following relationship for transfer function $P(\gamma, \tau_0)$ [96]:

$$P(\gamma, \tau_0) = \frac{P_0(\gamma, \tau_0)}{P_{atm.}(\gamma, \tau_0)[1 - \mu(\eta, \tau_0)\Pi(\eta, \zeta, \varphi, \tau_0)] + P_0(\gamma, \tau_0)\mu(\eta, \tau_0)\Pi} . \tag{4.77}$$

With P and P_0 functions known, it is easy to find the function $P_{atm.}$

$$P_{atm.} = \frac{P_0(\gamma, \tau_0)[1 - \mu(\eta, \tau_0)P(\gamma, \tau_0)\Pi(\eta, \zeta, \varphi, \tau_0)]}{[1 - \mu(\eta, \tau_0)\Pi(\eta, \zeta, \varphi, \tau_0)]P(\gamma, \tau_0)} . \tag{4.78}$$

Figures 4.14 and 4.15 show results of calculations of the extent of polarization of radiation scattered by an aerosol atmosphere and reflected from a horizontally-homogeneous surface. They consider two types of the surface (sand and turf) of an aerosol horizontally-homogeneous atmosphere model (Elterman 1968) and two wavelengths $\lambda = 4920$ and 6430 Å, solar zenith distance $\theta_0 = \arccos\zeta = 53°$, the angle $\theta = \arccos\eta$ varied within $0 - 90°$. The optical characteristics of these surfaces P_0, albedo A and their dependence on angles θ_0, θ, and φ for $\lambda = 4920$ and 6430 Å have been taken from [238] (see Table 4.6). The aerosol phase matrix of scattering $P(\gamma, \tau_0)$ was calculated

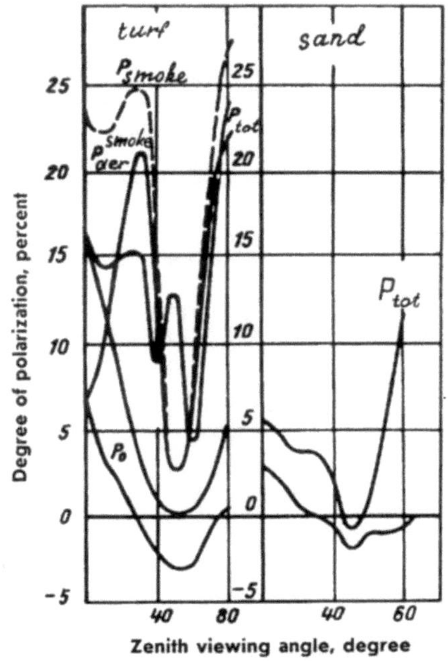

Fig. 4.14. Polarization of radiation scattered by the aerosol atmosphere and reflected from the horizontally homogeneous surface during the spectral-polarized survey of the Earth from space; the 1968 Elterman model; $\lambda = 6430$ Å; $\theta_0 = 0°$; $\theta = 180°$; *sand-turf*

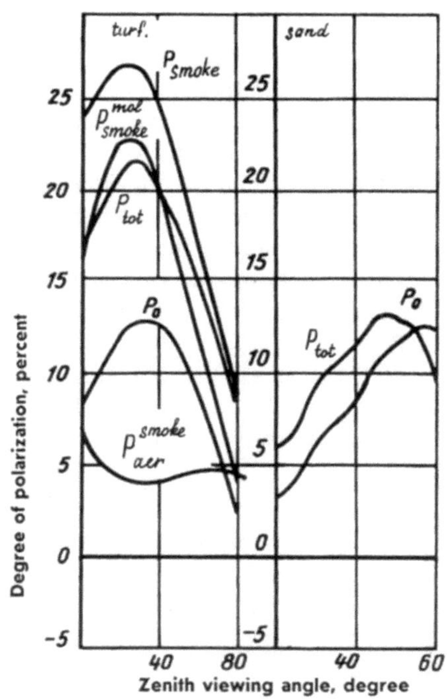

Fig. 4.15. Polarization of radiation scattered by the aerosol atmosphere and reflected from the horizontally homogeneous surface during the spectral-polarized survey of the Earth from space; the 1968 Elterman model; $\lambda = 6430$ Å; $\theta_0 = 0°$; $\theta = 0°$; *turf-sand*

Table 4.6. Monochromatic estimates of the polarization extent P of the brightness of a horizontally homogeneous surface as a function of solar zenith distance $\theta = 53°$ and viewing azimuth φ

φ	Grass		Sand	
	$\lambda = 6430$ Å P_0 (%)	$\lambda = 4920$ Å P_0 (%)	$\lambda = 6430$ Å P_0 (%)	$\lambda = 4920$ Å P_0 (%)
80°	+2.5	+5	+12	+18
70°	+6	+10	+12	+18
60°	+9	+16	+11	17
50°	+11	+22	+10	16
40°	+12.5	+28	+8	14
30°	+12.5	+30	+7	11
20°	+12	+27	+6	9
10°	+10	+23	+4	7
0°	+8	+16	+3	4
10°	+4	+8	+2	3
20°	+2	+3	+0.5	1.5
30°	0	0	0	0
40°	−2	−2	−0.5	−2
50°	−3	−3	−2	−3
60°	−3	−3	−1	−2.5
70°	−1	−1	−1	−2
80°	+0.5	+0.5	−0.5	−1
Surface albedo	A = 4%	A = 6%	A = 31%	A = 23%

from Mie theory for reflection coefficient $m = 1.5$ and generalized gamma distribution [43].

A surface formed of two horizontally-homogeneous planes with coefficients of reflection $r_i (i = 1, 2)$ is not standard and requires a special theoretical study. However, on the supposition that, by its optical properties, the atmosphere is horizontally homogeneous, two extreme cases can be selected which can be considered approximately.

If the polarization measurements of the surface brightness are made from satellite far from the interface between two media (at a distance exceeding the height of a homogeneous atmospheric layer), this case is reduced to that already considered with the help of modified formulas (4.65), (4.66) and (4.77).

If these measurements are made near the interface between two homogeneous media, the extent of polarization $P_i(\gamma, \tau_0)$ can approximately be determined from the formula [96]:

$$P_i(\gamma, \tau_0) = P_{atm.}(\gamma, \tau_0)[1 - \mu(\eta, \tau_0)\bar{\Pi}_i] + P_0 \mu(\eta, \tau_0)\bar{\Pi}_i \ , \tag{4.79}$$

where $\bar{\Pi}_i$ function is found from (4.62) and the extent of linear polarization P_0 is determined either experimentally or theoretically with an average coefficient of reflection $\bar{r} = 1/2(r_1 + r_2)$. In this case for a transfer function P_i and extent of atmospheric polarization brightness $P_{atm.}$ the following expressions are valid:

$$P_i = \frac{P_0}{P_{atm.}[1-\mu\bar{\Pi}_i] + \bar{P}_0\bar{\Pi}_i\mu} , \tag{4.80}$$

$$P_{atm.} = \frac{\bar{P}_0[1-\mu P_i\bar{\Pi}_i]}{1-\mu\bar{\Pi}_i} . \tag{4.81}$$

For a small object against an extended background we consider that, taking polarization into account, the brightness of atmospheric haze over a small object is totally determined by the brightness of atmospheric haze over an extended background $I_{\|\,haze,0b} = I_{\|\,h,B}$, $I_{\perp\,h,0b} = I_{\perp\,h,B}$. Then, for the spectral-polarization brightness of a small object observed from space, the following expressions are valid [96]:

$$\begin{aligned} I_{0b,\|} &= S\zeta\varrho_{\|}(\eta,\zeta,\varphi,\tau_0) + I^0_{B,\|}T_\sigma + I^0_{0b,\|}T , \\ I_{0b,\perp} &= S\zeta\varrho_{\perp}(\eta,\zeta,\varphi,\tau_0) + I^0_{B,\perp}T_\sigma + I^0_{0b,\perp}T . \end{aligned} \tag{4.82}$$

If follows from (4.82) that the extent of linear polarization of the small object brightness P_{0b}, when surveying the Earth from space, is:

$$P_{0b}I_{0b} = P_{atm.}S\zeta\varrho(\eta,\zeta,\varphi,\tau_0) + P_{B,0}I_{0,B}T_\sigma + P_{0,0b}I_{0,0b}T . \tag{4.83}$$

For some applications it is worthwhile introducing the transfer function, not for the polarization extent, but for the polarization constituents of the brightness of a small object ($I_{0b,\|}$ and $I_{0b,\perp}$), or the spectral-polarization contrasts of its brightness with its extended background:

$$\begin{aligned} k_{0b,\|} &= \frac{|I_{0b,\|} - I_{B,\|}|}{I_{0b,\|} + I_{b,\|}} \geqslant 0 \\ k_{0b,\perp} &= \frac{|I_{0b,\perp} - I_{B,\perp}|}{I_{0b,\perp} + I_{B,\perp}} \geqslant 0 . \end{aligned} \tag{4.84}$$

In this case, the generalized matrix expressions for the respective transfer functions will be:

$$\begin{pmatrix} I^0_{0b,0} \\ I^0_{0b,\|} \\ I^0_{0b,\perp} \end{pmatrix} = \begin{pmatrix} \Pi_{0b}, & 0, & 0 \\ 0, & \Pi_{0b,\|}, & 0 \\ 0, & 0, & \Pi_{0b,\perp} \end{pmatrix} \begin{pmatrix} I_{0b} \\ I_{0b,\|} \\ I_{0b,\perp} \end{pmatrix} \tag{4.85}$$

$$\begin{pmatrix} k^0_{0,0b} \\ k^0_{0b,\|} \\ k^0_{0b,\perp} \end{pmatrix} = \begin{pmatrix} P_{0b}, & 0 & 0 \\ 0, & P_{0b,\|}, & 0 \\ 0, & 0, & P_{0b,\perp} \end{pmatrix} \begin{pmatrix} k_{0b} \\ k_{0b,\|} \\ k_{0b,\perp} \end{pmatrix} \tag{4.86}$$

Expressions for P_{0b} and Π_{0b} have been obtained above. Expressions for $\Pi_{0b,\|}$, $\Pi_{0b,\perp}$ and $P_{0b,\|}$, $P_{0b,\perp}$ are easily obtained using (4.65) and (4.82).

The space-based remote spectral-polarization survey of natural formations, surface (above-water) objects and backgrounds has not yet been done.

Nevertheless, it must be emphasized that similarly to λ, the parameters of the polarization transfer function γ, τ_0, $P_i (i = 1, 2)$ can be estimated from combined satellite-aircraft experiments. These use simultaneous measurements of the extents of polarization P_i and $P_{0,i} (i = 1, 2)$, either near to or far from the interface between two homogeneous media, with substantially different, surface polarization characteristics. First we consider the spectral-polarization survey far from the interface between two homogeneous media.

Suppose, atmospheric transparency $\exp(-\tau_0/\eta)$, $\mu(\eta, \tau_0)$ function and $\Pi_i (i = 1, 2)$ scalar transfer function have been estimated experimentally using a scheme described in [97]. In addition, the polarization extent P_i has been measured from satellite, and $P_{0,i}$ from a low-flying aircraft. Then, instead of (4.80 – 4.81), we have:

$$P_i(\gamma, \tau_0, \lambda) = \frac{P_{0,i}^{exp}(A, \gamma, \tau_0, \lambda)}{P_i^{exp}(A, \gamma, \tau_0, \lambda)} \quad (i = 1, 2) \tag{4.87}$$

$$P_{atm.} = \frac{P_{0,i}^{exp}(A, \gamma, \tau_0, \lambda)[1 - \mu(\eta, \tau_0) P_{0,i}^{exp} \Pi_i / P_i^{exp}]}{1 - \mu(\eta, \tau_0) \Pi_i(\eta, \zeta, \varphi, \tau_0)} \quad (i = 1, 2) \; . \tag{4.88}$$

If the experiment is carried out near the interface, the parameters of the polarization transfer function should be estimated from (4.79 – 4.81), but with special estimation of functions $P_{0,i}$ and Π_i typical of this case [96]. No doubt, the P_i function is best found from an experiment of the first type, when the polarization measurements of brightness are made far from the interface between two homogeneous surfaces, with substantially different reflective properties. Note, that with the use of transfer function P_i not only experimental values of P_i are reduced to the surface level $P_{0,i}$ but also the extent of polarization of atmospheric haze is defined. This makes it possible to solve some inverse problems of atmospheric optics based on comparison of experimental data and the respective calculations using Mie theory. For this purpose the following formula is useful:

$$P_{atm.,\lambda}^{exp}(\gamma, A, \tau_0) = \frac{\Lambda \left[\frac{3}{4} \delta \sin^2 \gamma + (1 - \delta) \frac{P_2^{aer.} - P_1^{aer.}}{2} \right] [1 - e^{-\tau_0(1/\eta + 1/\zeta)}]}{4(\eta + \zeta) I_{atm.,\lambda}^{exp}(\eta, \zeta, \varphi, \tau_0, A)} \; , \tag{4.89}$$

where functions $P_{atm.,\lambda}^{exp}$ and $I_{atm.,\lambda}^{exp}$ are determined experimentally, and $P_{1,\lambda}^{aer.}(\gamma)$, $P_{2,\lambda}^{aer.}(\gamma)$ are calculated after Mie theory for a given model of atmospheric polydisperse aerosol. Comparing the $P_{atm.,\lambda}^{exp}$ values with calculated estimates in the right-hand part of (4.89), we can find such characteristics of polydisperse aerosol as refraction index, phase function and parameters of the aerosol particle-size distribution.

4.10 The Effect of the Atmosphere on the Spectral Albedo of Natural Formations

In this section the effect of the radiative properties of a scattering and absorbing aerosol atmosphere on the albedo of the Earth as a planet is considered. The numerical modeling of the solar radiation transformation by the Earth surface – atmosphere system has been made for the Elterman-1968 closed aerosol model based on the transfer function theory for albedo developed in [32, 95, 98, 99].

Determine the spectral transfer function for albedo for a homogeneous surface. The amount of radiation energy reflected by the planetary atmosphere – surface system is characterized, at fixed zenith solar distance θ_0 and wavelength λ, by a plane monochromatic albedo $A_1(\zeta, \tau_0, \lambda)$ equal to the ratio of fluxes of reflected radiation $H^{\uparrow}(0, \tau_0, \zeta)$ and incident radiation $\pi S_\lambda \cos\theta_0$ at the atmospheric top level ($\tau = 0$):

$$A_1(\zeta, \tau_0, \lambda) = \frac{H_\lambda^{\uparrow}(0, \tau_0, \zeta)}{\pi S_\lambda \zeta} = \frac{1}{\pi S_\lambda \zeta} \int_0^{2\pi} d\varphi \int_0^1 I_\lambda(0, \eta, \varphi, \tau_0)\eta\, d\eta \ . \tag{4.90}$$

Introduce the atmospheric transfer function $\Pi_A(\zeta, \tau_0, \lambda)$, relating a plane albedo A_1 to the albedo of the reflecting base A, by relationship [95]:

$$A_\lambda = \Pi_A(\zeta, \tau_0, \lambda) A_1(\zeta, \tau_0, \lambda) \ . \tag{4.91}$$

In the case of an infinitely extended homogeneous surface the intensity of reflected radiation $I_\lambda(0, \eta, \zeta, \varphi, \tau_0)$ is alternatively written as (4.1) or (4.16). Substituting (4.1) and (4.16) into (4.90), we obtain two equivalent presentations of a plane albedo $A_1(\zeta, \tau_0, \lambda)$ as

$$A_1(\zeta, \tau_0, \lambda) = A_{1h}(\zeta, \tau_0, \lambda) + A_\lambda M_\lambda(\zeta, \tau_0) \tag{4.92}$$

$$A_1(\zeta, \tau_0, \lambda) = \nu(\zeta, \tau_0, \lambda) + A_\lambda [M_\lambda(\zeta, \tau_0) + M_\sigma(\zeta, \tau_0, \lambda)] \ . \tag{4.93}$$

The plane albedo of atmospheric haze $A_{1,h}(\zeta, \tau_0, \lambda)$ from (4.91 – 4.92) is

$$A_{1,h}(\zeta, \tau_0, \lambda) = \nu(\zeta, \tau_0, \lambda) + A_\lambda M_\sigma(\zeta, \tau_0, \lambda) \ . \tag{4.94}$$

Thus, the plane albedo $A_{1,\lambda}(\zeta, \tau_0)$ of the atmosphere-surface system is easily expressed through two principal functions $M_\lambda(\zeta, \tau_0)$ and $M_\sigma(\zeta, \tau_0)$:

$$M_\sigma = \frac{2M(\zeta, \tau_0, \lambda)}{1 - A_\lambda C_\lambda(\tau_0)} \int_0^1 T_\sigma(\eta, \tau_0, \lambda)\eta\, d\eta \tag{4.95}$$

$$M_\lambda = \frac{2M(\zeta, \tau_0, \lambda)}{1 - A_\lambda C_\lambda(\tau_0)} E_{i3}(\tau_0, \lambda) \ . \tag{4.96}$$

In (4.96) the $E_{in}(x)$ is the integral transfer function.

Summing the functions M and M_σ, we introduce a new function $D(\tau_0)$ which is an analogue to function $C(\tau_0)$:

Table 4.7. Monochromatic estimates of function $D(\tau_0)$

λ, nm	450	500	550	600	650	700	800
$D(\tau_0)$	0.7482	0.9090	0.9080	0.8991	0.9614	0.9954	0.9961

$$M + M_\sigma = \frac{M(\zeta, \tau_0) D(\tau_0)}{1 - AC(\tau_0)} \tag{4.97}$$

$$D(\tau_0) = 2 \int_0^1 \mu(\eta, \tau_0) \eta \, d\eta \ . \tag{4.98}$$

Table 4.7 lists the monochromatic estimates of function $D(\tau_0)$ calculated using similarity ratios (3.10), with formula (4.98) for the Elterman-1968 closed aerosol model and the respective aerosol phase function $x_{aer.}(\gamma)$ (Table 1.3), at an angle of cutting the maximum of atmospheric phase function $\gamma_{cut} = 10°$ (Table 4.1).

Substitute (4.92) and (4.93) into (4.91). Then the following alternative expressions are obtained for the transfer function of albedo [95]:

$$\Pi_A = \frac{A}{A_{1,h}(\zeta, \tau_0) + AM(\zeta, \tau_0)} = \frac{A}{\nu(\zeta, \tau_0) + A[M(\zeta, \tau_0) + M_\sigma(\zeta, \tau_0)]} \ . \tag{4.99}$$

Using (4.97), we write for Π_A function:

$$\Pi_A = \frac{A[1 - AC(\tau_0)]}{[1 - AC(\tau_0)]\nu(\zeta, \tau_0) + AD(\tau_0)\mu(\zeta, \tau_0)} \ . \tag{4.100}$$

For the interpretation of satellite measurements of the albedo of the Earth as a planet, a theoretical estimation of its albedo $A_*(\tau_0)$ is of great importance:

$$A_*(\tau_0) = \int_0^1 A_1(\zeta, \tau_0) \zeta \, d\zeta \ . \tag{4.101}$$

Integrating the $A_1(\zeta, \tau_0)$ function in ζ according to (4.101), and using (4.91) through (4.98), we find alternative expressions for $A_*(\tau_0)$

$$A_*(\tau_0) = C(\tau_0) + \frac{AD^2(\tau_0)}{1 - AC(\tau_0)} \tag{4.102}$$

$$A_*(\tau_0) = A_h^*(\tau_0) + \frac{2AD(\tau_0)E_{i3}(\tau_0)}{1 - AC(\tau_0)} \ , \tag{4.103}$$

where a spherical albedo of atmospheric haze $A_h^*(\tau_0)$ is

$$A^*(\tau_0) = C(\tau_0) + \frac{2AD(\tau_0) \int_0^1 T_\sigma(\eta) \eta \, d\eta}{1 - AC(\tau_0)} \ . \tag{4.104}$$

Determine the spectral transfer function Π_A^* for a spherical albedo in a way similar to (4.91):

$$\Pi_A^*(\tau_0) = A/A^*(\tau_0) \ . \tag{4.105}$$

Substituting (4.102) into (4.105), we obtain for $\Pi_A^*(\tau_0)$ the following expression:

$$\Pi_A^*(\tau_0) = \frac{A}{C(\tau_0) + AD^2(\tau_0)/[1 - AC(\tau_0)]} \ . \tag{4.106}$$

The tabulating of principal functions $\mu(\eta, \tau_0, \lambda)$, $\nu(\eta,\ \tau_0,\ \lambda)$, $C_\lambda(\tau_0)$ and $D_\lambda(\tau_0)$ made here for the Elterman-1968 closed aerosol model, makes it possible to find for this model the $A_1(\zeta, \tau_0)$, $A_*(\tau_0)$, Π_A and Π_A^* values as functions of the reflecting basal albedo A, atmospheric optical thickness τ_0, solar zenith distance θ_0 and wavelength λ. Table 4.8 lists results of calculations of the transfer function Π_A for different natural surfaces using spectral albedo A given in [6] for water surface. The principal functions Π_A, A_λ, C and D are calculated for a 3-term approximating phase function (Table 4.1; $\gamma_{cut} = 10°$).

The calculations were made in the wavelength interval 400 – 800 nm, without similarity transformation, with an exact account of single scattering, for a given atmospheric phase function $x(\gamma)$ (Table 1.3). For the first-class surfaces (sand, bare soil), with increasing wavelength λ and solar zenith angle θ_0, the Π_A value grows monotonically. For the second-class surfaces (vegetation cover) the Π_A function varies with growing wavelength λ in a similar way, but at $\Lambda = 650 - 700$ nm a minimum is observed due to the chlorophyll absorption in this spectral interval. The character of variations in Π_A with growing solar zenith angle θ_0 is similar to the first-class surfaces. For the third-class surfaces (water surface, snow and cloud covers) the Π_A values decrease on both sides of their maximum in the interval 500 – 700 nm. In this case the dependence of Π_A on solar zenith angle is similar to that already mentioned.

From the viewpoint of the determination of optimal conditions for space-based measurements of the spectral albedo of natural features, of interest are intervals where $\Pi_A > 1$, $\Pi_A < 1$, and $\Pi_A = 1$, depending on wavelength and solar zenith angle. Such controls are shown on the graphs of the respective diagrams of optimization (Fig. 4.16). The diagrams of optimization of the

Table 4.8. Results of calculations of the transfer function Π_A

ζ	λ, nm				
	500	550	600	650	700
0.8	0.1662	0.2173	0.2634	0.2539	0.2250
0.9	0.1706	0.2284	0.2812	0.2678	0.2311

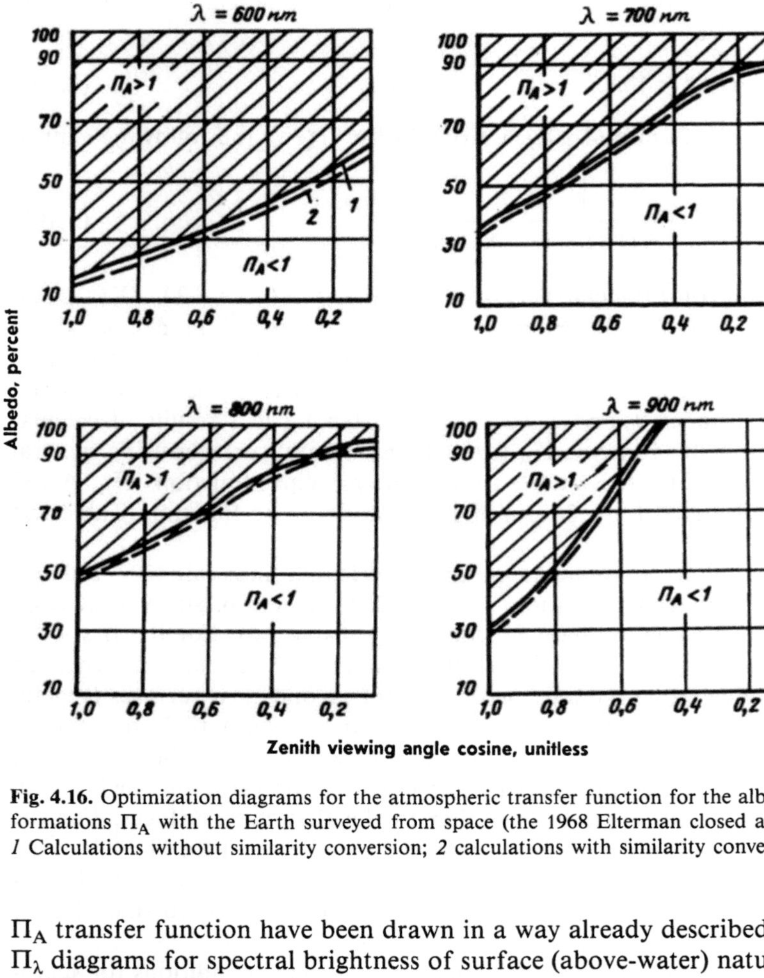

Fig. 4.16. Optimization diagrams for the atmospheric transfer function for the albedo of natural formations Π_A with the Earth surveyed from space (the 1968 Elterman closed aerosol model). *1* Calculations without similarity conversion; *2* calculations with similarity conversion

Π_A transfer function have been drawn in a way already described for similar Π_λ diagrams for spectral brightness of surface (above-water) natural features and backgrounds.

The calculations of Π_A functions were made in the interval $\lambda = 0.4 - 1.06\,\mu m$ for solar zenith angles $\theta = 0 - 90°$ and average total albedos $A = 0.084$ (green plants); 0.24 (sand); 0.78 (clouds). The principal functions μ, ν, C and D were calculated for a 2-term phase function $x(\gamma) = 1 + x_1 \cos\gamma$, with an exact account of real phase function for the single scattering of radiation. An analysis of numerical data obtained in [32, 98, 99] has shown that for strongly reflecting surfaces ($A \geqslant 0.78$) with growing θ_0, a plane albedo A_1, and the albedo of the haze A_h grow through almost the entire interval considered, and the Π_A transfer function decreases. Between $\theta_0 = 10 - 70°$ there is a region where $\Pi_A \approx 1$ for $\lambda = 400 - 500$ nm and $900 - 1060$ nm. With further increasing of λ the wavelength interval in which $\Pi \approx 1$, broadens. In the interval $\theta_0 = 10 - 50°$ a plane albedo A_1 first grows with λ and then decreases (the effect of ozone absorption in the Chappui band), with subsequent growth. At $\theta_0 \geqslant 50°$ a plane albedo A_l first decreases (up to $\lambda = 600$ nm) and then grows. In the θ_0 interval under consideration the Π_A transfer function changes in an opposite way with varying λ. The haze albedo A_h decreases with increased λ due to decreasing intensity of light scattered by the atmosphere and reflected by the surface.

A comparison with models of a purely scattering molecular atmosphere and of a molecular atmosphere including ozone enabled the assessment of the effect of adding the absorbing components to the atmosphere [98]. A comparison with weakly reflecting surfaces ($A = 0.24$ and 0.084) has allowed the study of the effect of surface albedo variations on A_l and Π_A values, with varying wavelength and solar zenith angle θ_0. It should be emphasized that an increase or decrease of the system's albedo value under the influence of a scattering and absorbing aerosol atmosphere depends not only on the chosen spectral interval $\Delta\lambda$ or solar zenith angle θ_0 but also on the value of A albedo of the surface itself. As a rule, weakly reflecting surfaces intensify their brightness when surveyed from space, and strongly reflecting surfaces reduce it.

4.11 The Geographical and Seasonal Change of the Transfer Function for Albedo

For the climatic calculations of the Earth's radiation budget and for the interpretation of satellite measurements of the outgoing radiation fluxes it is important to know the Π_A transfer function over large surface areas in different climatic zones and seasons. Based on satellite data on the geographical and seasonal distribution of total albedo A [145], the Π_A values were calculated with formula (4.100) for different global regions [312]. Based on the results obtained, maps have been drawn of the geographical distribution of Π_A in the wavelength interval 500, 600, and 700 nm for winter and summer conditions, utilizing maximum sun height at a given latitude (Figs. 4.17, 4.18). The geographical distribution of the Π_A transfer function is largely determined by the character of the surface. With growing wavelength λ the Π_A values increase in all latitudes. The transfer function varies most in the winter Northern hemisphere because of albedo variations from 0.8 (in the regions with persistent snow cover) to 0.2 (in the steppe regions). In South America, the geographical distribution of Π_A is of meridional character, which is connected with the location of mountains, tropical forests and grasslands. With increasing wavelength λ the geographical distribution of Π_A becomes more and more latitudinal and the values of Π_A grow. In summer, sharp variations in Π_A are observed only in the regions with the transition from snow cover to forests. In principle, in the summer period the geographical distribution of Π_A is similar to that in winter. In South America and in the mountains of Asia the Π_A is distributed meridionally. Over the rest of the globe the Π_A is distributed mainly latitudinally.

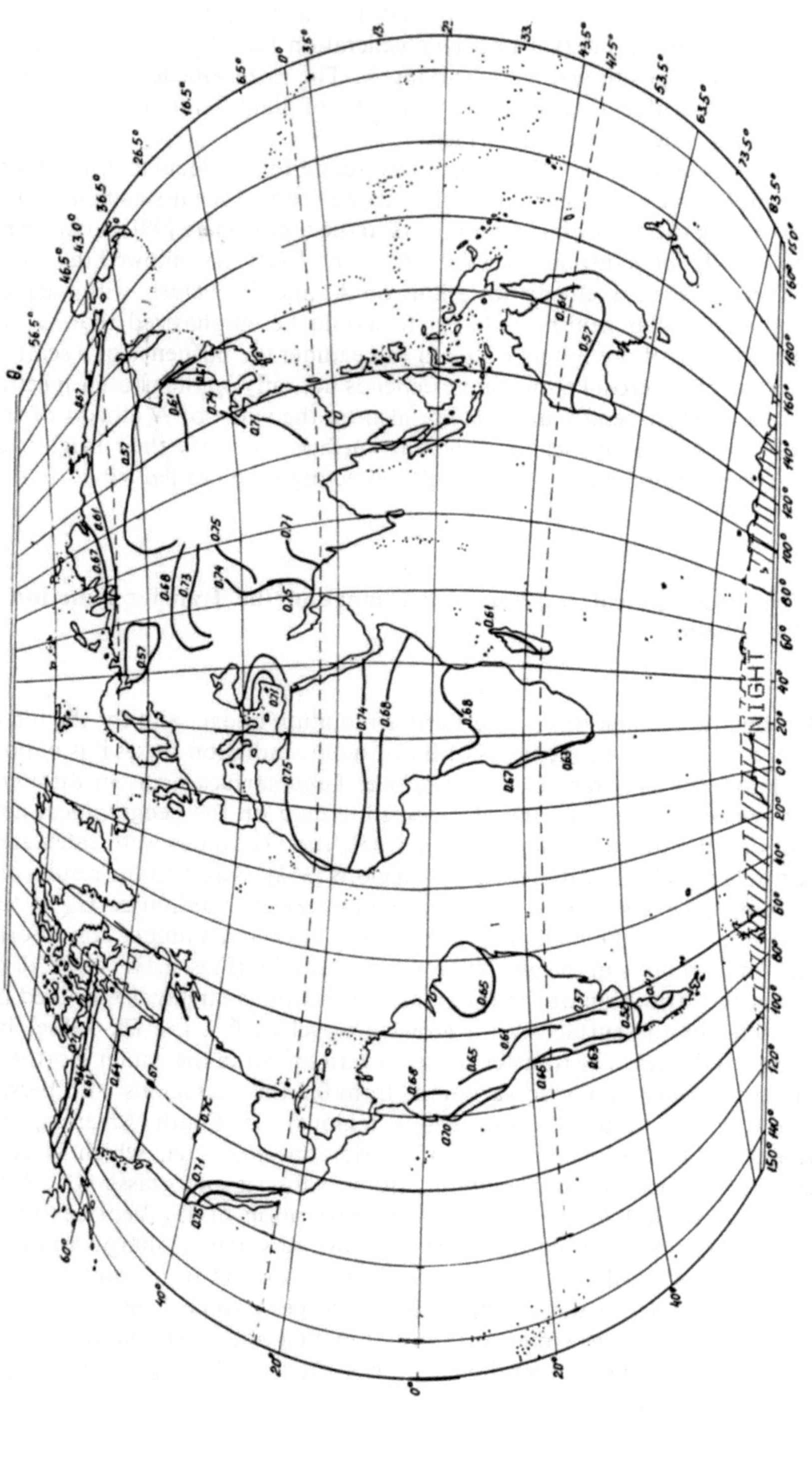

Fig. 4.17. The geographical and seasonal distribution of the atmospheric monochromatic transfer function for the albedo of natural formations Π_A with the Earth surveyed from space; $\lambda = 500$ nm, summer, the 1968 Elterman closed aerosol model

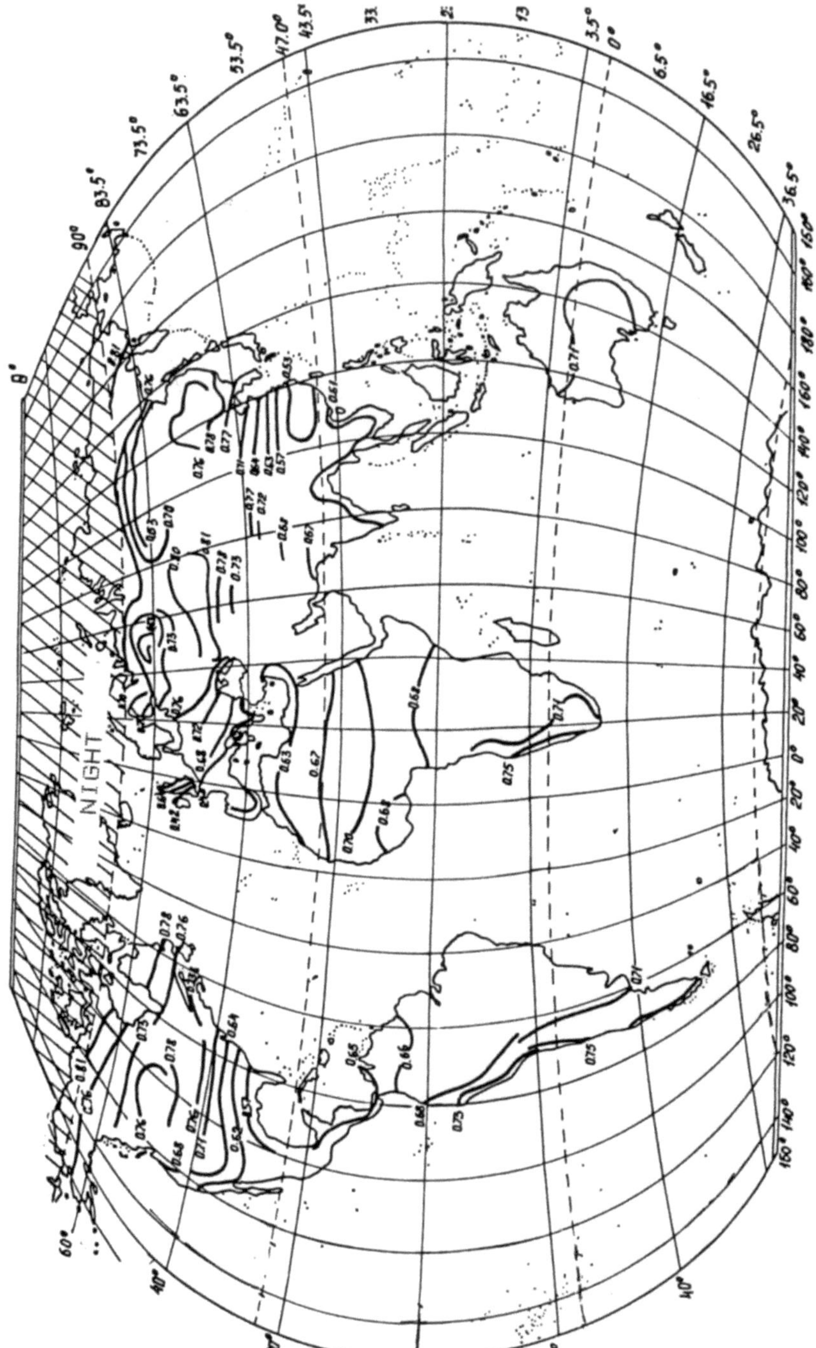

Fig. 4.18. The geographical and seasonal distribution of the atmospheric monochromatic transfer function for the albedo of natural formations Π_A with the Earth surveyed from space; $\lambda = 500$ nm, winter, the 1968 Elterman closed aerosol model

Chapter 5 The Effect of Horizontal Inhomogeneities of the Underlying Surface on the Scalar Transfer Function of the Atmospheric

5.1 Setting the Problem of Theoretical Estimation of a 2-Term Atmospheric Transfer Function

By the definition given in [94], the transfer operator transforms the atmosphere-induced distortions into real brightnesses of objects and of backgrounds. The actual content of a transfer function is much more complicated, since the functions of the transformation of the brightness of an object, viewed at different angles and with different combinations of objects in a given scene, will be different. This is determined by the horizontal inhomogeneity of the atmosphere and Earth's surface as well as by the different extent of non-orthotropicity of natural objects.

The radiative correction of satellite images consists of the transformation of satellite-recorded brightnesses into the optical characteristics of natural objects, constant whatever the viewing conditions, illumination, etc. In the first approximation the space brightnesses reduced to the Earth's surface level can be used to assess its state. In practice, empirical relationships between the coefficients of spectral brightness and parameters of the state of natural objects are most often used. To obtain spectral brightness coefficients (as a ratio of the object's brightness in a given direction to the brightness of an ideal Lambertian reflection) based on aerospace imagery is, apparently, more of a complicated problem than the transition from spectral brightnesses of images to constant characteristics of the physical state of objects (for instance, to complex dielectric penetrability).

With the problem of the satellite information processing thus set, the physics of the transfer operator of the atmosphere is as follows. It is not only an operator of reducing satellite images to the surface level but also an operator for the transformation of initial images into the characteristics of natural objects (albedo, spectral brightness coefficients, distribution of dielectric penetrability, etc.). In this case to find the radiative correction, it is necessary to know the law of radiation reflection and a description of the "environs" in the given surveyed scene. The solution of the respective multidimensional problems in the boundary problems of radiation transfer serves as the mathematical basis for the transformation of aerospace imagery.

In the case of a horizontally-homogeneous atmosphere limited from beneath by a horizontally-inhomogeneous surface, the transfer equation and boundary conditions in a 3-D case are written (ignoring polarization and refraction) as:

$$\eta \frac{\partial I}{\partial z} + \sin\theta\cos\varphi \frac{\partial I}{\partial x} + \sin\theta\sin\varphi \frac{\partial I}{\partial y} = -\alpha(z) I$$

$$+ \frac{\sigma(Z)}{4\pi} \int I x(\gamma, z) d\omega + \frac{\sigma(z)}{4} S x(\gamma_1, z) e^{-1/\zeta \int_0^z \alpha(z') dz'} . \quad (5.1)$$

The boundary conditions for (5.1) are:

$$I(\infty, x, y, \eta, \zeta, \varphi) = 0 , \quad \eta > 0 ; \quad (5.2)$$

$$I(0, x, y, \eta, \zeta, \varphi) = \frac{A(x,y)}{\pi} \int_0^{2\pi} d\varphi \int_0^1 I\eta' d\eta' + A(x,y) S\zeta e^{-1/\zeta \int_0^\infty \alpha(z') dz'} \quad \eta < 0 . \quad (5.3)$$

Equation (5.1) differs from a traditional 1-D transfer equation in a plane medium, by the presence of two first-order partial derivatives of intensity I by horizontal coordinates, x and y. These derivatives determine diffusion of photons in the horizontal plane XOY. Therefore, if the medium is infinitely extended by x and y coordinates, boundary conditions (5.2 – 5.3) must be supplemented by the requirement that the boundary problem (5.1 – 5.3) becomes zero or finite at $x \to \pm\infty$, $y \to \pm\infty$.

One of the ways to solve the boundary problem (5.1 – 5.3) is to carry out a 2-D Fourier transformation of the required intensity by horizontal coordinates x, y. In this case the frequency-contrast characteristic (FCCh) of the atmosphere which connects the Fourier patterns of intensity at the atmospheric top and base, is an analog to the scalar 1-D transfer function introduced according to (3.2).

The boundary problem (5.1 – 5.3) for spectral intervals of a finite width can be solved using either an analytical method or an advanced method of spherical harmonics [77]. A simplified solution for this problem can be obtained based on the use of the mathematical apparatus of the theory of linear systems [199].

Upon a 2-D Fourier transformation of intensity I in horizontal coordinates x and y, we obtain a 1-D boundary problem:

$$\eta \frac{\partial \tilde{\psi}(z, \vec{v}, \vec{S})}{\partial z} + \hat{\sigma}(z, \vec{v}, \vec{S}) \tilde{\psi}(z, \vec{v}, \vec{S}) = \sum_{p'=1}^{P'} \sigma_s^{p'}(z) \int_\Omega \tilde{\psi}(z, \vec{v}, \vec{S}') x_{p'}(z, \vec{S}, \vec{S}') d\vec{S}' ; \quad (5.4)$$

$$\tilde{\psi}(H, \vec{v}, \vec{S}) = 0 , \quad \vec{S} \in \Omega_- ; \quad (5.5)$$

$$\tilde{\psi}(0, \vec{v}, \vec{S}) = \int_{\Omega-} \tilde{R}(\vec{S}, \vec{S}') \tilde{\psi}(0, \vec{v}, S') \eta' d\vec{S}' , \quad \vec{S} \in \Omega_+ . \quad (5.6)$$

Here $\vec{v} = \{v_x, v_y\}$ is a 2-D spatial frequency; $\vec{S} = \{\eta, \sqrt{1-\eta^2}\cos\varphi, \sqrt{1-\eta^2}\sin\varphi\}$ the direction of the viewing line; Ω_+, Ω_- respectively, lower and upper hemisphere; $x_{p'}(z,\vec{S},\vec{S}')$ the phase function of p'-substance; $\sigma(z,\vec{v},\vec{S}) = \alpha(z) - i(\vec{v},\vec{S})$. The remaining notations have been introduced earlier and are given in the notation list.

The techniques for solving the problems (5.4 – 5.6) have been described in [82, 117, 135, 136]. Here we shall not dwell on them, but illustrate results of calculations of the effect of horizontally-inhomogeneous terrestrial surfaces on the formation of the outgoing radiation field on the assumption of no absorption, i.e., $\alpha = \sigma_s = \sigma$. We shall distinguish between the space-frequency characteristic (SFCh) of the entire linearized system of radiation transfer in the atmosphere which is determined as:

$$\begin{aligned} \tilde{Z}(z,\vec{v},\vec{S}) &= |\tilde{\psi}(z,\vec{v},\vec{S})| \; ; \\ \tilde{\theta}(z,\vec{v},\vec{S}) &= \operatorname{arc} \tilde{\psi}(z,\vec{v},\vec{S}) \; , \end{aligned} \tag{5.7}$$

and the SFCh of scattered radiation:

$$\tilde{\psi}(z,\vec{v},\vec{S}) = \begin{cases} \chi(z,\vec{v},\vec{S}) \; , & \vec{S} \in \Omega_- \\ e^{i(\vec{r}_\perp,\vec{v})}[\Lambda^{-1} + \chi(z,\vec{v},\vec{S})] \; , & \vec{S} \in \Omega_+ \; , \end{cases} \tag{5.8}$$

where the coefficient $e^{i(\vec{r}_\perp,\vec{v})}$ with scalar product of vector $\vec{r}_\perp = (H-z)\vec{S}_\perp/\eta$ by $\vec{v}$ determines the direction of observation (viewing); $\Lambda^{-1} = \exp[(-\tau(H)/\eta) - (\tau(z)/\eta)]$ is the factor of radiation reduction;

$$\chi(z,\vec{v},\vec{S}) = \tilde{A}(z,\vec{v},\vec{S})e^{i\tilde{\phi}(z,\vec{v},\vec{S})} \; . \tag{5.9}$$

$\tilde{A}(z,\vec{v},\vec{S})$, by definition of the amplitude-frequency characteristic (AFCh) of scattered radiation; $\tilde{\phi}(z,\vec{v},\vec{S})$ the phase-frequency characteristic (see also [136, 137, 192]).

It is easily seen that:

$$\hat{Z}(z,\vec{v},\vec{S}) = \begin{cases} \tilde{A}(z,\vec{v},\vec{S}) \; , & \vec{S} \in \Omega_- \\ \sqrt{\Lambda^{-2} + \tilde{A}^2(z,\vec{v},\vec{S}) + 2\tilde{A}(z,\vec{v},\vec{S})\Lambda^{-1}\cos\tilde{\phi}(z,\vec{v},\vec{S})} \; , & \vec{S} \in \Omega_+ \; ; \end{cases} \tag{5.10}$$

$$\tilde{\theta}(z,\vec{v},\vec{S}) = \begin{cases} \tilde{\phi}(z,\vec{v},\vec{S}) \; , & \vec{S} \in \Omega_- \\ \operatorname{arctg}\left[\dfrac{\tilde{A}(z,\vec{v},\vec{S})\sin\tilde{\phi}(z,\vec{v},\vec{S})}{\Lambda^{-1} + \tilde{A}(z,\vec{v},\vec{S})\cos\tilde{\phi}(z,\vec{v},\vec{S})}\right] \; , & \vec{S} E \Omega_+ \; . \end{cases}$$

5.2 The Effect of the Atmospheric Optical Parameters and of the Conditions of Survey on a 2-D Transfer Function

Figure 5.1 shows the amplitude-frequency characteristic (AFCh) of surface-reflected and atmosphere-scattered radiation at different levels in the at-

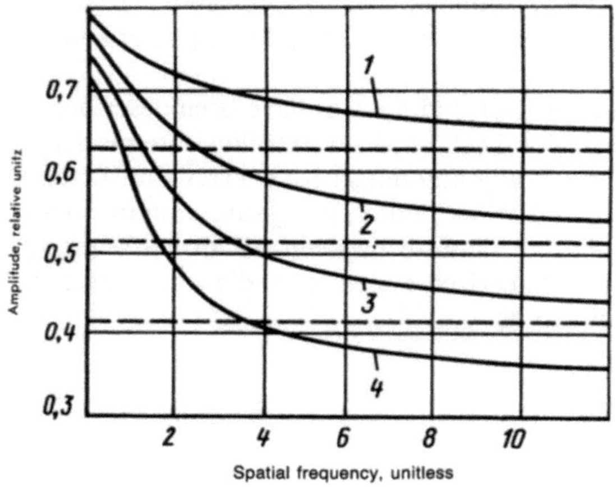

Fig. 5.1. The dependence of the amplitude-frequency characteristics of the atmosphere on the environmental optical thickness: $\tau_0 = 0.8$; $x_1 = 1.57$; $\bar{A} = 0.2$; $\eta = 1$; $\zeta = 0.866$; $\nu_y = 1$; *1* $\tau = 0$; *2* $\tau = 0.2$; *3* $\tau = 0.4$; *4* $\tau = 0.6$

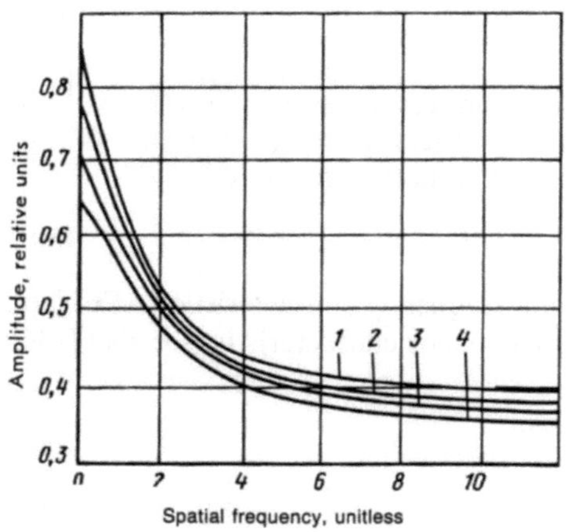

Fig. 5.2. The dependence of the AFCh of the atmosphere on average albedo A: $\tau_0 = 0.8$; $x_1 = 1.57$; $\tau = 0$; $\eta = 1$; $\zeta = 0.866$; $\nu_y = 0$; *1* $\bar{A} = 0.2$; *2* $\bar{A} = 0.4$; *3* $\bar{A} = 0.6$; *4* $\bar{A} = 0.8$

mosphere calculated after Sobolev's technique [145]. The curves show that as the optical thickness of the medium between the reflecting surface and the observer grows, the spatial spectrum of albedo narrows, the asymptotes (dashed lines in Fig. 5.1) become lower following Buger's law and the asymptotic AFCh value grows. It means that the energy of the fine-structure component of the radiation field, with the growing $\tau_0 - \tau$ difference, gradually moves onto large inhomogeneities of the light field. Thus, the fact that images observed through a scattering layer of the atmosphere are blurred is confirmed. Hence, part of the energy of the inhomogeneous light field transforms into the energy of the constant constituent of radiation intensity.

Figure 5.2 illustrates a parametric family of curves $E\tilde{\psi}(\tau, \nu_x, 0, \vec{S})$ at the upper boundary of the layer for different $\bar{A}$. It is seen from the figure that varia-

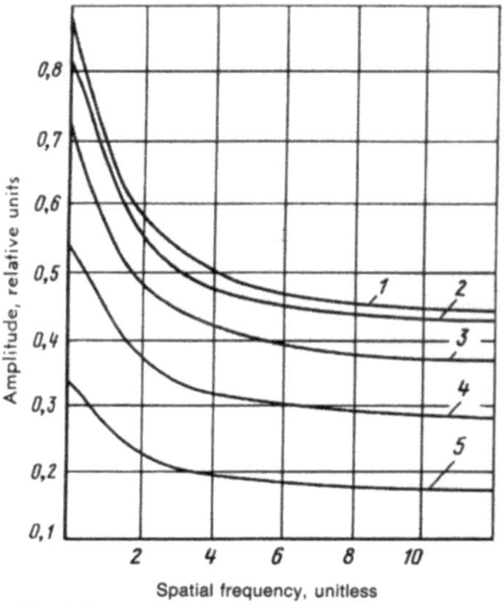

Fig. 5.3

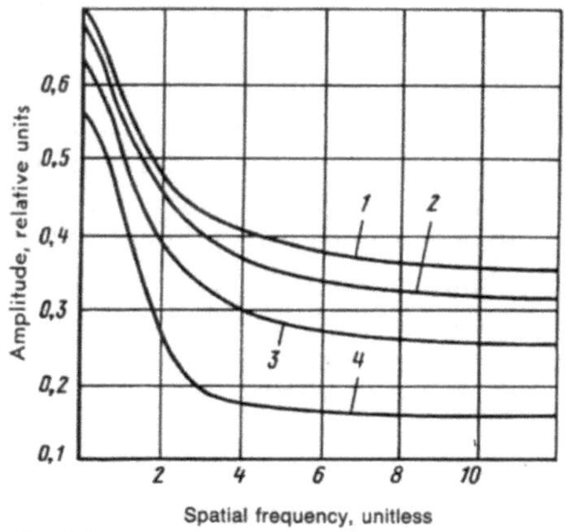

Fig. 5.4

Fig. 5.3. The dependence of the AFCh of the atmosphere on viewing angle θ: $\tau_0 = 0.8$; $\tau = 0$; $x_1 = 1.57$; $\bar{A} = 0.2$; $\eta = 1$; $\nu_y = 0$; *1* $\theta_0 = 0°$; *2* $\theta_0 = 15°$; *3* $\theta_0 = 30°$; *4* $\theta_0 = 45°$; *5* $\theta_0 = 60°$

Fig. 5.4. The dependence of the AFCh of the atmosphere on solar zenith angle θ_0: $\tau_0 = 0.8$; $\tau = 0$; $x_1 = 1.57$; $\bar{A} = 0.2$; $\zeta = 0.866$; $\nu_y = 0$; *1* $\theta = 15°$; *2* $\theta = 30°$; *3* $\theta = 45°$; *4* $\theta = 60°$

tions of average albedo introduce small distortions into the energy spectrum of surface albedo. An exception is frequencies close to zero only. The shifting of the curves is governed by the growth of $\tilde{\psi}(\tau, \vec{0}, \vec{S})$ and E with the increasing parameter $\bar{A}$. The dependence of the AFCh on average albedo shows that with increased surface reflectivity, simultaneously with the growing brightness of the background $I_{\bar{A}}(\tau, \vec{S})$, the transfer of the energy spectrum of albedo becomes slightly better. However, the contrast of the 2-D brightness field becomes worse since the $I_{\bar{A}}(\tau, \vec{S})$ grows faster than the $E\psi(\tau, \nu_x, 0, \vec{S})$. The average irradiance of the terrestrial surface depends on the solar zenith angle θ_0, and, therefore, with the sun approaching zenith the AFCh grows proportionally to E at all the frequencies (Fig. 5.3).

The graphs of the AFCh at the upper boundary of the layer at different slopes of the viewing path are given in Fig. 5.4. At a slant observation the optical thickness between the source and the receiver of radiation increases. Therefore, the AFCh asymptotic values decrease, which is determined by intensified multiple scattering.

A study on the AFCh of the linearized system of the transfer of an optical image in a homogeneous layer depending on parameters $\tau_0 - \tau, x_1, \bar{A}, \eta$ and ζ gives a general idea of the effect of these parameters on the transfer of the energy spectrum of the optical image through the atmosphere. However, in practice the model of a homogeneous layer has a limited sphere of application since the real atmosphere is multi-layered. The AFCh in a non-homogeneous atmosphere, for whose description the simplest model has been chosen with an exponential law of the reduction of the scattering coefficient, will be considered next.

The results of the AFCh calculations with different values of the parameters $\tau_0 - \tau, x_1, \bar{A}, \eta, \zeta$ have shown that the values obtained coincide qualitatively with the values characteristic of a homogeneous layer. A conclusion can be drawn here about the usefulness of a homogeneous layer model to make methodical calculations to reveal principal laws of transformation of the spatial energy spectrum of an optical image in the atmosphere.

However, there is a great difference between absolute AFCh values in homogeneous and non-homogeneous layers at equal values of $\tau_0 - \tau, x_1, \bar{A}, \eta, \zeta$ but different ratios between average coefficient of scattering σ and geometrical thickness H.

Again, examine Fig. 5.1 carefully. We use a dimensional spatial frequency of $\nu_x = \nu'_x \sigma$ km^{-1}. Varying the σ (at $\tau_0 = \text{const}$) we obtain at $\sigma \ll 1$ (a geometrically thick layer with a low specific content of aerosol) strongly compressed curves, and at $\sigma \gg 1$ (a geometrically thin layer with a large specific content of aerosol) they are strongly stretched. Hence, the amplitudinal distortions of an optical image originating from a scattering aerosol layer with $\sigma \ll 1$ lie in the lowest frequency region of the spatial spectrum. For a layer of the same optical thickness with $\sigma \gg 1$ they lie in the low frequency region, with the power slowly varying over the spectrum.

Thus, in calculations of the AFCh as a function of dimensional variables ν_x, ν_y it is necessary to set a geometrical thickness H and a vertical profile of the volume coefficient of scattering, which is equivalent to a setting of the vertical distribution of optical thicknesses.

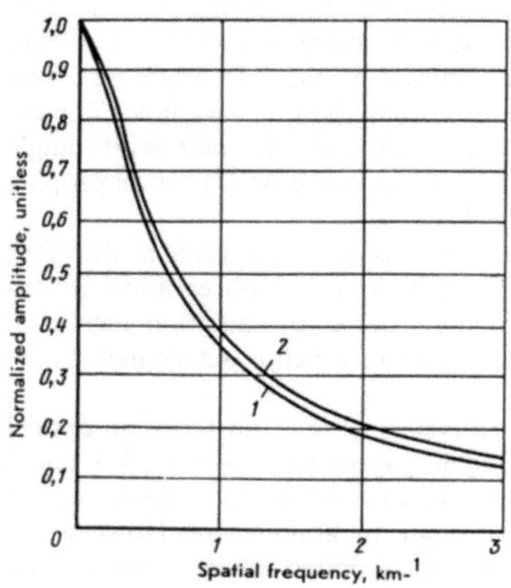

Fig. 5.5. Normalization of the AFCh of radiation in a weakly turbid atmosphere for different average albedos of the surface: $\tau_0 = 0.2$; $\tau = 0$; $x_1 = 0.796$; $\eta = 1$; $\nu_y = 0$; *1* $\bar{A} = 0.6$; *2* $\bar{A} = 0.2$

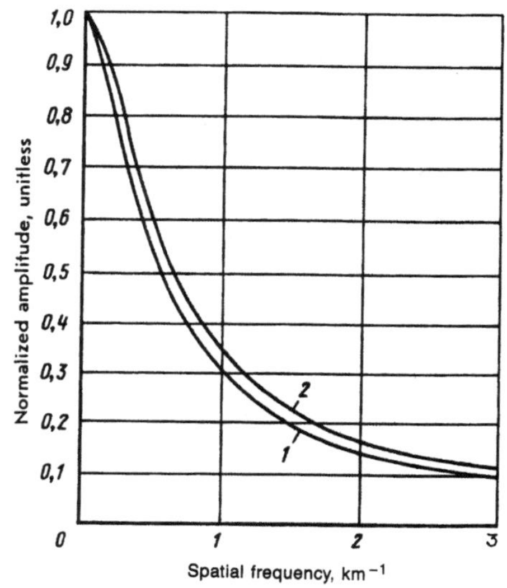

Fig. 5.6. Normalization of the AFCh of radiation in a strongly turbid atmosphere for different average albedos of the surface: $\tau_0 = 0.6$; $\tau = 0$; $x_1 = 1.7$; $\eta = 1$; $\nu_y = 0$; *1* $\bar{A} = 0.6$; *2* $\bar{A} = 0.2$

Figures 5.5, 5.6 illustrate normalized AFCh of the scattered radiation at the atmospheric top for weak and strong turbidity with different values of average surface albedo. As shown in Figs. 5.5, 5.6, the graphs of function $\tilde{A}(z, \vec{\nu}, \vec{S})$ at large values of $\bar{A}$ lie below. This is due to the $\tilde{A}$ increasing with growing $\bar{A}$. In the case of a strongly turbid atmosphere the graphs become steeper toward zero. This is because of the AFCh narrowing and because of $\tilde{A}$ increasing with growing τ_0.

In solving the applied problems requiring a consideration of the spatial-frequency characteristics (SFCh) of the atmosphere, it is better to use an approximation of the SFCh by elementary functions. Such an approximation makes it possible in many cases to write a solution as a simple formula not requiring additional computer time. We confine ourselves to a consideration of the AFCh of the scattered radiation in nadir. Consider exponential, power, and hyperbolic approximations of the AFCh:

$$\tilde{A}(z, \nu, \vec{S}_H) = \tilde{A}_{0H} e^{l|\vec{\nu}|} \; ; \tag{5.11}$$

$$\tilde{A}(z, \vec{\nu}, \vec{S}_H) = \frac{\tilde{A}_{0H}}{\sqrt{1 + l^2 |\vec{\nu}|^2}} \; ; \tag{5.12}$$

$$\tilde{A}(z, \vec{\nu}, \vec{S}_H) = \frac{\tilde{A}_{0H}}{1 + l|\vec{\nu}|} \; . \tag{5.13}$$

Figures 5.7 and 5.8 exemplify approximations (5.11 – 5.13) of functions $\tilde{A}(\vec{\nu})/\tilde{A}_{0H}$ respectively, for homogeneous and inhomogeneous layers. A comparison of the curves has shown that power and hyperbolic approximations describe asymptotic behaviour of the true curve at large $|\vec{\nu}|$ much better than an exponent. Among approximations (5.12) and (5.13) the first must be

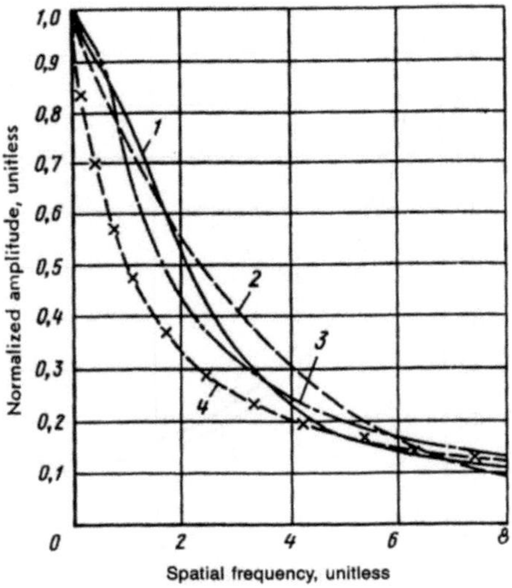

Fig. 5.7. Approximation of the normalized AFCh of the scattered radiation in a homogeneous layer: $\tau_0 = 0.8$; $\tau = 0$; $x_l = 1.894$; $\bar{A} = 0.2$; $\eta = 1$; $v_y = 0$; *1* real curve; *2* approximation by the exponent; $l = 0.3$; *3* approximation by the power function, $l = 1$; *4* approximation by the hyperbola, $l = 1$

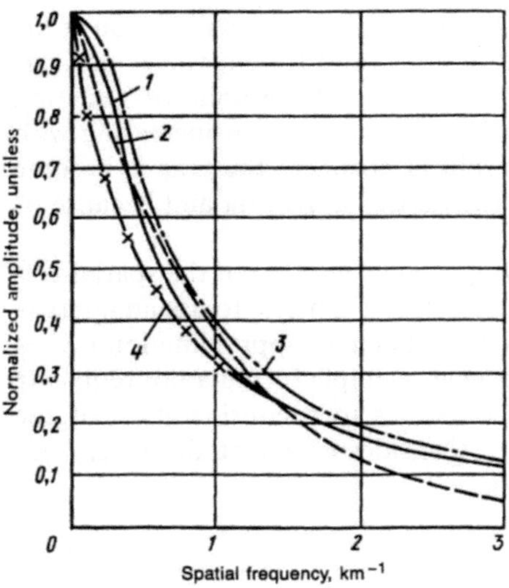

Fig. 5.8. Approximation of the normalized AFCh of the scattered radiation in a non-homogeneous layer: $\tau_0 = 0.8$; $\tau = 0$; $x_l = 1.894$; $\bar{A} = 0.2$; $\eta = 1$; $\sigma_0 = 0.4$; $v_y = 0$; *1* real curve; *2* approximation by the exponent; *3* approximation by the power function, $l = 6.2$; *4* approximation by the hyperbola, $l = 2.25$

favored, since the power function (5.12) describes better an initial part of the true AFCh curve.

Since each state of the atmosphere has a set of initial parameters of its own and, hence, there is a functional dependence $l = l(\bar{A}, \tau, \tau_0, x_l)$, tables can be composed comparing each set $\{\bar{A}, \tau, \tau_0, x_l\}$ with a set $\{l\}$. With these tables the AFCh can be calculated for a given state of the atmosphere using (5.12), without using the transfer equation solution. Thus, the approximation (5.12) and the tables will make any calculation, requiring solutions of a 3-D transfer equation, much easier.

5.3 The Effect of the Horizontally Inhomogeneous Constituent of the Scattered Radiation

The function of point blurring is a component of the pulse transfer function of a linearized system of the transfer of an optical image in the atmosphere. It describes the horizontally-inhomogeneous component of the scattered radiation intensity, created in the atmosphere, by a stationary point source of light located on the surface. The function of blurring along with the SFCh, is an optical constant independent of the spatial inhomogeneities of surface albedo and determined from the formula:

$$Q(z,\vec{r}_\perp,\vec{S}) = \frac{1}{(2\pi)^2} \iint_{-\infty}^{\infty} \tilde{A}(z,\vec{v},\vec{S}) \exp\left[-i(\vec{v},\vec{r}_\perp) + i\tilde{\phi}(z,\vec{v},\vec{S})\right] d\vec{v}$$

$$= \frac{1}{(2\pi)^2} \iint_{-\infty}^{\infty} \tilde{A}(z,\vec{v},\vec{S}) \cos\tilde{\phi}(z,\vec{v},\vec{S}) \cos(\vec{v},\vec{r}_\perp) d\vec{v} \ . \tag{5.14}$$

The solution for the initial boundary problem (5.1 – 5.3) for $I(z,x,y,\eta,\varphi) \equiv I(z,\vec{r}_\perp,\vec{S})$ at $\vec{S} \in \Omega_+$ is expressed through the function of blurring in the following way [134]:

$$I(z,\vec{r}_\perp,\vec{S}) = E\left\{A(\vec{r}_\perp - \vec{\bar{r}}_\perp)\Lambda^{-1} + \iint_{-\infty}^{\infty} Q(z,\vec{r}_\perp - \vec{\bar{r}}_\perp - \xi,\vec{S})A(\xi)d\xi\right\}. \tag{5.15}$$

Expression (5.15) shows that the radiation intensity constituent dependent of the coordinates, is presented with two items. The first item describes the variable constituent of the brightness of the terrestrial surface attenuated by the atmosphere. The second item is the integration of the variable constituent of the brightness with the point-blurring function describing a source of the amplitude-phase distortions of the reflected radiation intensity. For the nadir direction $S = S_H(\theta = 0°)$ the function of blurring is independent of the phase characteristic and hence:

$$Q(z,\vec{r}_\perp,\vec{S}_H) = \frac{1}{(2\pi)^2} \iint_{-\infty}^{\infty} \tilde{A}(z,\vec{v},\vec{S}_H) \cos(\vec{v},\vec{r}_\perp) d\vec{v} \ . \tag{5.16}$$

Analysis of the function of blurring at $\vec{S} = \vec{S}_H$ is, apparently, equivalent to a study of the AFCh.

Figure 5.9 and 5.10 show the graphs of the functions of blurring constituting the AFCh of the scattered radiation. The graphs confirm a weak dependence of the function $Q(z,\vec{r},\vec{S})$ on average surface albedo. The curves corresponding to smaller $\bar{A}$ values lie below. This is due to an increase of $\tilde{A}_H$ when the $\bar{A}$ decreases. The function of blurring at the atmospheric top at $\tau_0 = 0.6$ is much broader than at $\tau_0 = 0.2$. Since there is a single-power source on the surface, the growth of multiple scattering in the medium will cause a blurring of the source image. The function of blurring makes it possible to calculate the areas of the effect of the horizontally-inhomogeneous constituent

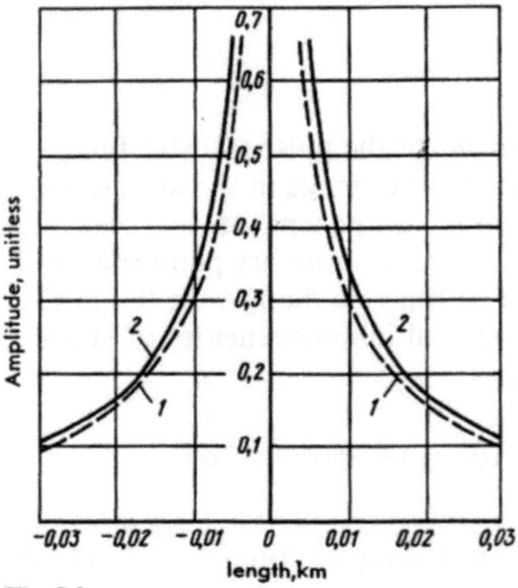

Fig. 5.9

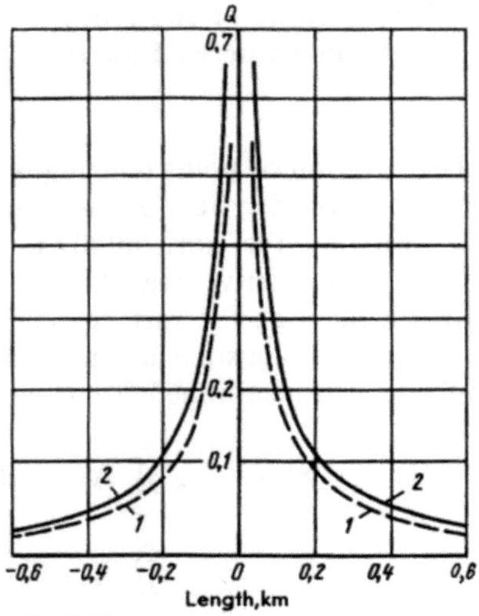

Fig. 5.10

Fig. 5.9. The function of point blurring in a weakly turbid atmosphere with different average albedos of the surface: $\tau_0 = 0.2$; $\tau = 0$; $x_1 = 0.796$; $\eta = 1$; $\zeta = 0.866$; *1* $\bar{A} = 0.2$; *2* $\bar{A} = 0.6$

Fig. 5.10. The function of point blurring in a strongly turbid atmosphere with different average albedos of the surface: $\tau_0 = 0.6$; $\tau = 0$; $x_1 = 1.7$; $\eta = 1$; $\zeta = 0.866$; *1* $\bar{A} = 0.2$; *2* $\bar{A} = 0.6$

Table 5.1. The areas of the effect of the horizontally inhomogeneous constituent of the scattered radiation for the normalization of the point-blurring function

τ_0	A	
	0.2	0.6
0.2	4.8	5.6
0.6	6.4	7.1

of the scattered radiation on the radiative images of natural objects. Results of calculations are given in Table 5.1.

The calculations of the function of blurring Q permit a quantitative estimate of atmospheric distortions of an optical image of natural objects. For instance, the curves in Fig. 5.11 show that the jumps in the brightness of the image at the boundaries of the object, resulting from an account of the AFCh of the atmosphere, remain unchanged.

The absence of the boundary jumps in the brightness of a real image is explained by a finite resolution of the recording optical system. Thus, a combined effect of the atmosphere and the recording instruments leads to a smoothing of the images of natural objects. The profiles of the brightness of objects of arbitrary shape and size can be calculated using formula (5.15).

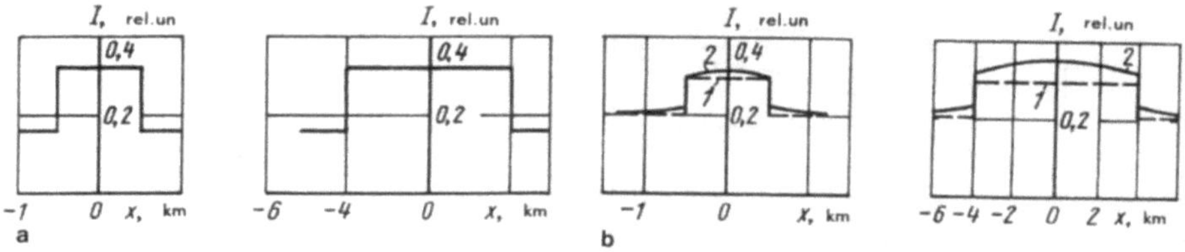

Fig. 5.11 a, b. Profiles of the brightness of the finite-extended objects: $\tau_0 = 0.6$; $\tau = 0$; $x_1 = 1.7$; $\bar{A} = 0.2$; $\eta = 1$; $\zeta = 0.866$; $\Delta A = 0.2$; **a** at the atmospheric base; **b** at the atmospheric top; *1* ignoring the spatial-frequency characteristic; *2* with the spatial-frequency characteristic

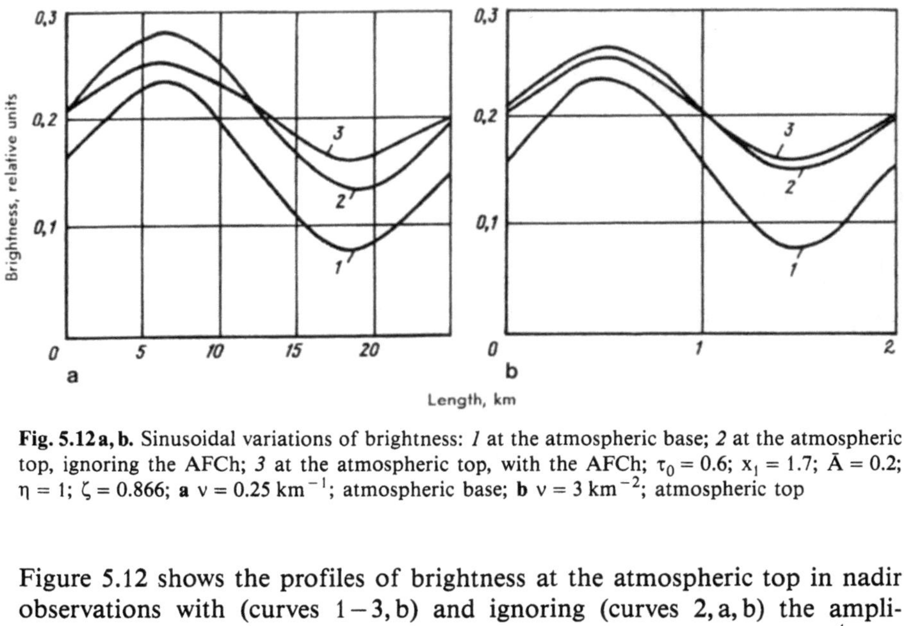

Fig. 5.12 a, b. Sinusoidal variations of brightness: *1* at the atmospheric base; *2* at the atmospheric top, ignoring the AFCh; *3* at the atmospheric top, with the AFCh; $\tau_0 = 0.6$; $x_1 = 1.7$; $\bar{A} = 0.2$; $\eta = 1$; $\zeta = 0.866$; **a** $\nu = 0.25\ \mathrm{km}^{-1}$; atmospheric base; **b** $\nu = 3\ \mathrm{km}^{-2}$; atmospheric top

Figure 5.12 shows the profiles of brightness at the atmospheric top in nadir observations with (curves 1–3, b) and ignoring (curves 2, a, b) the amplitudinal distortions at $A = 0.2 + 0.1 \sin \nu x$ in two cases $\nu_1 = 0.25\ \mathrm{km}^{-1}$ and $\nu_2 = 3\ \mathrm{km}^{-1}$.

Curves 2 and 3 in Fig. 5.13 refer to the profiles of the upward radiation brightness. They have been calculated at $A = 0.2 + 0.1 \sin \nu x$ with and without the SFCh, respectively. They show that the brightnesses of strongly reflecting areas of the surface, calculated theoretically with an account of the SFCh, are underestimated, and those of weakly reflecting surfaces are overestimated.

Consider now a general case of an inhomogeneous surface albedo widely used in studies of the spatial structure of the background of natural features [69]. Let the horizontal variations of albedo be set by a stationary random process with a correlation function [209]

$$\tilde{B}(x) = \sum_{k=1}^{N} c_k e^{l_2 |x|} \cos \nu_k x \ . \qquad (5.17)$$

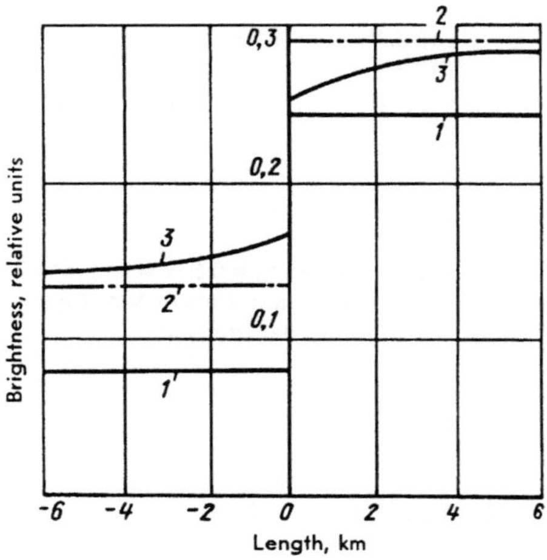

Fig. 5.13. The profile of brightness near the interface between two surface sites with different albedos: $\tau_0 = 0.6$; $x_1 = 1.7$; $\bar{A} = 0.2$; $\eta = 1$; $\zeta = 0.866$; $\tilde{A} = 0.41$; $E = 0.805$; $\tau = 0$. *1* Initial "step"; *2* ignoring the SFCh; *3* with the SFCh

First, consider a partial case (5.17):

$$\tilde{b}(x) = c\,e^{-l|x|}\cos\nu x \ . \tag{5.18}$$

The normalized transfer function of the system of the transfer of an optical image in the atmosphere for stationary input processes (5.18) is written as:

$$|\tilde{\psi}(\nu)|^2 = \Lambda^{-2} + \tilde{A}^2(\nu) + 2\Lambda^{-1}\tilde{A}(\nu)\cos\tilde{\phi}(\nu) \ . \tag{5.19}$$

Omitting elementary samples, we obtain:

$$\tilde{b}(x) = c\,E\,e^{-l|x|}\,|\tilde{\psi}(\nu)|^2\cos\nu x \ . \tag{5.20}$$

In a similar way and taking (5.19) into account, we obtain a correlation function at the output of the scattering layer for a general case (5.17):

$$\tilde{B}(x) = E^2\sum_{k=1}^{N} c_k e^{-l_k|x|}\,[\Lambda^{-2} + 2\Lambda^{-1}\tilde{A}(\nu_k)\cos\phi(\nu_k) + \tilde{A}^2(\nu_k)]\cos\nu_k x \ . \tag{5.21}$$

Comparing (5.21) and (5.17), we draw a conclusion that the correlation function of a stationary process is smoothed by an atmospheric layer. A dispersion of the process decreases after the law:

$$\tilde{B}(0) = E^2\sum_{k=1}^{N} c_k\,[\Lambda^{-2} + 2\Lambda^{-1}\tilde{A}(\nu_k)\cos\tilde{\phi}(\nu_k) + \tilde{A}^2(\nu_k)] \ . \tag{5.22}$$

Here the estimates of dispersion turn out to exceed the respective estimates obtained ignoring the SFCh.

These examples show that an estimation of the effect of horizontal inhomogeneities of the Earth's surface is a necessary stage in developing a mathematical model of the formation of the outgoing radiation field. The results will be used in Chapter 6 in the solution of the inverse problem of the 2-D filtration of atmospheric distortions of the reflected radiation.

5.4 An Assessment of the Effect of Horizontal Diffusion of Photons on the Spectral Transfer Function

Analysis has been made above of the applied and theoretical problems connected with an assessment of 2-D transfer functions of the atmosphere in the visible spectrum (400 – 800 nm). Experimental data on the spectrophotometry of the environment near the land-sea interface obtained from the MS Soyuz-9 make it possible to study the effect of real horizontal diffusion of solar radiation on spectral brightnesses and brightness contrasts of a real horizontally-inhomogeneous surface formed by two homogeneous half-planes with different reflecting properties. Analysis of this problem and drawing of the transfer function for the brightness of a small-sized surface (above-water) object, spectrophotometered from space, shows that a strict theoretical modeling of the effect of atmospheric radiative factors on space-derived spectrophotometric information, must be made considering horizontal diffusion of solar radiation in 2- or 3-D medium [181, 220]. In the real natural medium, a diffusion of photons occurs due to substantial horizontal-inhomogeneity of the atmosphere-object-background system. As follows from the last section, this requires a special study using techniques other than those used traditionally in the 1-D theory of radiation transfer [166].

In publications [174, 178] an account of horizontal diffusion is suggested, using an interative technique, based on the supposition that, in the first approximation, the atmosphere uniformly reduces the brightnesses of all the spatial frequencies of the horizontally inhomogeneous reflecting base. A simple physical analysis of solutions for a 2-D equation of radiation transfer shows that the horizontal diffusion of photons in a horizontally homogeneous atmosphere with the horizontally inhomogeneous Lambertian base, must grow with increasing optical thickness of the atmosphere τ_0, solar zenith angle θ_0, viewing angle θ and azimuth φ. It also grows with increasing variations in local optical inhomogeneities of the surface. Thus, under optimal conditions of space-based survey of natural formations (maximum sun elevation over the horizon for a given latitude, viewing in nadir, $\theta = 0°$, an aerosol cloud-free atmosphere with an optical thickness of $\tau_0 \simeq 0.2 - 0.6$) the contribution of the horizontal diffusion of photons to the formation of space-derived images of surface or above-water objects, will be minimal. The highest- and lowest-frequency details of the surface image, when surveyed from space, must be attenuated uniformly by the atmosphere, independent of spatial frequency. In these extreme asymptotic cases, to draw a transfer function and to realize atmospheric radiative correction of a space-derived image, it is enough to use a simpler, 1-D theory of radiation transfer. A supposition of a uniform (by frequency) distorting effect of the atmosphere is supplemented with an exact account of the real horizontal inhomogeneity of the surface for solar radiation singly reflected from it.

The 1-D asymptotic transfer functions already introduced (Chapter 3) for the brightnesses and brightness contrasts of natural formations, surface (above-water) objects and backgrounds are generalized next, based on an approximate value of horizontal diffusion of photons in the system "aerosol atmosphere – horizontally-inhomogeneous surface". To construct a theoretical model of the radiation field, similarity ratios and exact solutions of the 1-D theory of radiation transfer, are used for an arbitrary three-term phase function (see Chap. 4). We shall search for a solution for the boundary problem (5.1 – 5.3) in the form of two items [174]:

$$I(z,x,y,\eta,\zeta,\varphi)=\hat{I}_0(z,x,y,\zeta,\varphi,\eta)+\tilde{I}(x,y,z,\eta,\zeta,\varphi)\ . \qquad (5.23)$$

The intensity $\hat{I}_0$ is a solution for the boundary problem (5.1 – 5.3) in which the horizontal diffusion of photons is not considered:

$$\eta\frac{\partial\hat{I}_0}{\partial z}=-\alpha(z)\hat{I}_0+\frac{\sigma(z)}{4\pi}\int\hat{I}_0 x(\nu,z)\,d\omega+\frac{\sigma(z)}{4}Sx(\nu_1,z)e^{-1/\zeta\int_0^z\alpha(z')dz'}\ ; \qquad (5.24)$$

$$\hat{I}_0(x,y,\eta,\zeta,\varphi)=0\ ,\quad \eta>0\ ; \qquad (5.25)$$

$$\hat{I}_0(x,y,0,\eta,\zeta,\varphi)=\frac{A(x,y)}{\pi}\int_0^{2\pi}\hat{I}_0\,d\varphi\int_0^1\eta\,d\eta+A(x,y)S\zeta e^{-1/\zeta\int_0^\infty\alpha(z')dz'}\ ,\quad \eta<0\ . \qquad (5.26)$$

Hence, the brightnesses of the areas of the horizontally-inhomogeneous surface with different spatial frequency, in the first approximation are equally attenuated vertically by the atmosphere. The intensity $\tilde{I}$, determined by diffusion of photons in the horizontal plane XOY, is an addition to the intensity $\hat{I}_0$. It is this diffusion occurring in the medium because of the horizontal-inhomogeneity of the surface, that leads to "blurring" of space-derived images of natural formations in the XOY plane. Use of (5.1 – 5.3) reveals that an additional intensity $\tilde{I}$ satisfies a solution for the following boundary problem:

$$\eta\frac{\partial\tilde{I}}{\partial z}+\sin\theta\cos\varphi\frac{\partial\tilde{I}}{\partial x}+\sin\theta\sin\varphi\frac{\partial\tilde{I}}{\partial y}=-\alpha(z)\tilde{I}+\frac{\alpha(z)}{4\pi}\int\tilde{I}x(\nu,z)\,d\omega\ ; \qquad (5.27)$$

$$\tilde{I}(\infty,\eta,\zeta)=0\ ,\quad \eta>0\ ; \qquad (5.28)$$

$$I(0,\eta,\zeta)=A(x,y)S\zeta e^{-1/\zeta\int_0^\infty\alpha(z')dz'}\ ,\quad \eta<0\ . \qquad (5.29)$$

Note, that with the solution for the boundary problem (5.24 – 5.26) obtained, horizontal diffusion in the medium can be taken into account using the converging iterative process following a scheme:

$$\eta\frac{\partial\tilde{I}_n}{\partial z}+\sin\theta\cos\varphi\frac{\partial\tilde{I}_{n-1}}{\tilde{\partial}x}+\sin\theta\sin\varphi\frac{\partial I_{n-1}}{\tilde{\partial}y}=-\alpha(z)\tilde{I}_n$$
$$+\frac{\sigma(z)}{4\pi}\int\tilde{I}_n x(\nu,z)\,d\omega+\frac{\sigma(z)}{4\pi}Sx(\nu_1,z)e^{-1/\zeta\int_0^\infty\alpha(z')dz'}$$
$$(n=0,1,2,\dots)\ . \qquad (5.30)$$

Now we obtain solutions for the boundary problems (5.24 – 5.26) and (5.27 – 5.29) for the atmospheric top level, on the assumption, that photons singly reflected from the surface and then directly and diffusively going to the atmospheric top, are reflected by the real horizontally-inhomogeneous surface with the true albedo $A(x,y)$. With respect to photons repeatedly reflected from the surface, we shall consider them to be re-reflected from some model horizontally-homogeneous surface with an average albedo $\overline{A(x,y)} = A(\theta_0)$ and then to be directly and diffusively transferred to the atmospheric top.

Firstly, we shall consider additional features of the problem, connected with its possible vertical optical inhomogeneity by such parameters as atmospheric phase function $x(\gamma,z)$ and probability of quantum survival $\Lambda(z)$. In contrast to the homogeneous case, for an inhomogeneous atmosphere at $A=0$, it is necessary, apart from coefficients of brightness ϱ and σ, where illumination is from above, to introduce also coefficients of reflection $\tilde{\varrho}$ and transmission $\tilde{\sigma}$ of the atmosphere where illumination is from beneath. Bearing in mind results from [188, 210, 212] and the assumptions mentioned above, we obtain an expression similar to (4.13) for the surface brightness at the atmospheric base level:

$$\begin{aligned} I_0(\infty,\eta,\zeta,\varphi) = {} & A(x,y)S\zeta\mu(\zeta,\tau_0) + A(x,y)\bar{A}S\zeta\mu(\zeta,\tau_0)\tilde{C}(\tau_0) \\ & + \ldots + A(x,y)\bar{A}^2 S\zeta\mu(\zeta,\tau_0)\tilde{C}^2(\tau_0) \\ & + A(x,y)A^3 S\zeta\mu(\zeta,\tau_0)\tilde{C}^3(\tau_0) + \ldots \; . \end{aligned} \tag{5.31}$$

The first, second, and subsequent terms of the series (5.31) determine the radiation intensities for the respective number of reflections of light quanta from the surface. Average albedo $\bar{A}$ in (5.31) is determined from the formula:

$$\bar{A}(\theta_0) = \frac{1}{S}\iint\limits_D A(x,y,\theta_0)\,ds \tag{5.32}$$

where S is the maximum area of the space-derived frame determined by the angular size of the field of view of the on-board optical instruments and by the height of the space carrier's orbit. By summing (5.31) as an infinitely decreasing geometrical progression with the denominator $q = \bar{A}\tilde{C}(\tau_0) < 1$ and first term $A(x,y)S\zeta\mu(\zeta,\tau_0)$ we obtain for I_0 the final formula [178]:

$$I_0(\infty,x,y,\eta,\zeta,\varphi) = \frac{A(x,y)S\zeta\mu(\zeta,\tau_0)}{1-\bar{A}\tilde{C}(\tau_0)}\;, \tag{5.33}$$

where

$$\mu(\eta,\tau_0) = e^{-\tau_0/\eta} + 2\int_0^1 \tilde{\sigma}(\eta,\eta')\eta'\,d\eta'\;, \tag{5.34}$$

$$\tilde{C}(\tau_0) = 4\int_0^1 \eta\,d\eta \int_0^1 \tilde{\varrho}_0(\eta,\eta')\eta'\,d\eta'\;. \tag{5.35}$$

Coefficients of brightness $\tilde{\varrho}_0$ and $\tilde{\sigma}_0$ are zero harmonics in the expansion of total coefficients of reflection $\tilde{\varrho}$ and transmission $\tilde{\sigma}$ in Fourier series in azimuth φ. For an optically vertically-homogeneous atmosphere we have $\tilde{\varrho} = \varrho$, $\tilde{\sigma} = \sigma$. In this case instead of (5.33) for I_0 we obtain an expression:

$$I_0 = \frac{A(x,y)S\zeta\mu(\zeta,\tau_0)}{1-\bar{A}C(\tau_0)} . \qquad (5.36)$$

Because the effect of the vertical optical inhomogeneity of an aerosol atmosphere on the field of reflected radiation in the visible spectrum (400 – 800 nm) is weak (see Table 3.13), we shall confine ourselves to the case of a vertically-homogeneous atmosphere with the horizontally-inhomogeneous reflecting base, where the albedo $A(x,\theta_0)$ depends only on the coordinate and solar zenith angle θ_0.

We use the sun azimuth $\varphi_0 = 0°$. Then the spectral brightness I_s of the orthotropic horizontally-inhomogeneous surface, observed from space at an angle θ to the normal and azimuth φ, at a point with the x coordinate is [178]:

$$I_s(x,\theta,\theta_0,\varphi) = I_0(\tau_0, x-\Delta x,\theta)e^{-\tau_0 \sec\theta} + I(\tau_0, x-\Delta x,\theta,\theta_0,\varphi) , \qquad (5.37)$$

where I_h is the brightness of atmospheric haze; $\Delta x = \Delta z \operatorname{tg}\theta \cos\varphi$ the shift of the x horizontal coordinate caused by slant viewing of the surface from above. Apparent physical considerations suggest that the atmospheric haze brightness I_h in (5.37) is a sum of atmospheric brightness I_{atm} at $A = 0$ and of the brightness of the horizontally-inhomogeneous reflecting base I_0 diffusively attenuated towards θ and φ. In this case, the following expression holds for the atmospheric haze brightness:

$$I_h(x-\Delta x,\theta,\theta_0,\varphi,\tau_0) = S\zeta\varrho(\eta,\zeta,\varphi,\tau_0) + \int_{-\infty}^{\infty} I_0(x')Q_{sc}(x-x')dx' , \qquad (5.38)$$

where Q_{sc} is a function of scattering for an atmospheric layer which determines the diffuse transfer of photons reflected from the surface at the point with the current coordinate x′ towards the atmospheric top. Comparing (5.38) with the respective expression for haze brightness I_h in the 1-D case (horizontally homogeneous) (4.14), we are able to normalize the diffuse scattring function $Q_{sc}(x)$ [178]:

$$\int_{-\infty}^{\infty} Q_{sc}(x-x')dx' = T_\sigma . \qquad (5.39)$$

Introducing the Fourier transformation of functions Q_{sc} and I_0 by the horizontal coordinate x and using the respective theorem of summing up the functions, we obtain, instead of (5.38) [178].

$$I_h = S\zeta\left[\varrho(\eta,\zeta,\varphi,\tau_0) + \frac{\mu(\tau_0\zeta)}{1-\bar{A}C(\tau_0)}\int_{-\infty}^{\infty} A(\nu)Q_{sc}e^{-i\nu(x-\Delta x)}d\nu\right] , \qquad (5.40)$$

where ν is the spatial frequency in the units of inverse length. Formula (5.40) implies that the phase-frequency shifts in a space-derived image are either absent or very small. These conditions are satisfied in space-based surveys of natural formations in nadir or near nadir ($\theta = 0°$). Comparing (5.40) with the respective expressions for two extreme asymptotic cases (extended object $\nu \to 0$ and point object ν), we obtain [178]:

$$Q_{sc}(\nu \to 0) \to T_\sigma(\eta, \tau_0) = 2\int_0^1 \sigma_0(\eta, \zeta)\zeta d\zeta \; ; \quad Q_{sc}(\nu \to \infty) \to 0 \; , \quad \nu \geqslant \nu_{thr} \tag{5.41}$$

where ν_{thr} is an asymptotic (threshold) value of a high spatial frequency ν of a horizontally-inhomogeneous surface, from which the brightnesses of all high frequencies are reduced by the atmosphere, following Buger's law. Expanding the phase function $Q(\nu)$ near zero frequency $\nu = 0$ in Taylor series and using its evenness $Q_{sc}(\nu) = Q_{sc}(-\nu)$ [136], we approximate the Q_{sc} in the interval $[0, \nu_{thr}]$ by a sum of even polynomials [178] with respect to the variable ν:

$$Q_{sc}(\nu) = T_\sigma(\eta, \tau_0) + a_2\nu^2 + a_4\nu^4 + a_6\nu^6 + \dots \; . \tag{5.42}$$

Using (5.42), instead of (5.40), we find:

$$I_h = S\zeta\left\{\varrho(\eta, \zeta, \varphi, \tau_0) + \frac{\mu(\zeta, \tau_0)}{1 - AC}\left[A(x - \Delta x)T_\sigma(\eta, \tau_0) + \sum_{n=1}^{\infty} a_{2n}(-1)^n \frac{d^{2n}}{dx^{2n}} A(x - \Delta x)\right]\right\} \; . \tag{5.43}$$

Since the sign-changing series (5.42) converges, its sum must not exceed the first term of the series. Thus, instead of (5.42), we have:

$$I_h \leqslant S\zeta\left\{\varrho(\eta, \zeta, \varphi, \tau_0) + \frac{\mu(\zeta, \tau_0)}{1 + \bar{A}C} \times \left[A(x - \Delta x)T_\sigma(\eta, \tau_0) - a_2 \frac{d^2}{dx^2} A(x - \Delta x)\right]\right\} \; . \tag{5.44}$$

Formula (5.44) shows that in estimating the horizontal diffusion of photons from above it is enough to approximate the scattering function Q_{sc} in the interval $[0, \nu_{thr}]$ by the parabola:

$$Q_{sc}(\nu) = -\frac{T_\sigma}{\nu_{thr}^2}\nu^2 + T_\sigma \; , \quad a_2 = -\frac{T_\sigma}{\nu^2} \; . \tag{5.45}$$

Thus, choosing I_h as a principal component of atmospheric haze brightness, a term corresponding to a uniform exponential attenuation of surface brightness for all frequencies ν above the threshold value ν_{thr}, we have, instead of (5.37):

$$I_s = S\zeta\left[\varrho(\eta, \zeta, \varphi, \tau_0) + \frac{\mu(\zeta, \tau_0)\mu(\eta, \tau_0)A(x - \Delta x)}{1 - \bar{A}C(\tau_0)}\right] + \Delta\tilde{I}_h \; . \tag{5.46}$$

In this approximation an additional brightness of haze $\Delta\tilde{I}_h$ in (5.46), determined by non-uniform attenuation by frequency of the reflecting basal brightness, is [178]:

$$\Delta \tilde{I}_h = \frac{\mu(\zeta,\tau_0) S\zeta}{1-\bar{A}C(\tau_0)} \cdot \frac{T_\sigma(\eta,\tau_0)}{\nu^2} \cdot \frac{d^2 A(x-\Delta x)}{dx^2} \; . \tag{5.47}$$

Let us select now in I_h a component obtained on the assumption that the atmospheric haze brightness forms an average radiation field for an average albedo $\bar{A}$. Since in this case, the atmospheric haze uniformly distorts the brightnesses of all frequencies of the reflecting base, we have, instead of (5.46):

$$I_s = S\zeta \left\{ \left[\varrho(\eta,\zeta,\varphi,\tau_0) + \frac{\bar{A}\mu(\zeta,\tau_0) T_\sigma(\eta,\tau_0)}{1-\bar{A}C(\tau_0)} \right] + \frac{A(x-\Delta x)\mu(\zeta,\tau_0) T_\sigma(\eta,\tau_0)}{1-\bar{A}C(\tau_0)} \right\} + \Delta I_h \; . \tag{5.48}$$

In this approximation an additional haze brightness in (5.48) is:

$$\Delta I_h = \frac{\mu(\zeta,\tau_0) T_\sigma(\eta,\tau_0)}{1-\bar{A}C(\tau_0)} \left[(A-\bar{A}) + \frac{1}{\nu^2} \cdot \frac{\partial^2 A(x-\Delta x)}{\partial x^2} \right] \; . \tag{5.49}$$

Fixing in (5.46 – 5.47) the spatial frequency ν, we set a harmonic distribution of albedo values along the x horizontal:

$$A(x,\nu) = A_0 + A_1 \cos x\nu \; . \tag{5.50}$$

Then, for each frequency maximum contribution of the horizontal diffusion of photons into the brightness of a horizontally-inhomogeneous surface can be estimated as a function of ν_{thr}, average albedo $\bar{A}$ and amplitude of its variation A_1. The ratio of the contrast of brightnesses for cosinusoidal model (5.50) at the atmospheric top level to the contrast of the brightness of this model at the atmospheric base level is equal to the FCCh of the atmosphere [70]. In the calculations of an additional brightness I_h, according to (5.47), the ν_{thr} value has been taken as 3.5 [101, 135 – 137]. In Table 5.2 results are given of calculations of the FCCh of an aerosol atmosphere (Elterman 1968) based on calculated ratio of contrasts of the cosinusoidal model (5.50) observed from above and beneath. Calculations were made for $\lambda = 550$ nm and atmospheric phase function $x(\gamma)$ (see Table 1.3) using an advanced method of spherical harmonics.

Table 5.2. Calculations of the frequency-contrast characteristic (FCCh)

ν, 1/km	0.1	1	3	4	5	50	100	10000
		$A_0 = 0.2$;	$A_1 = 0.2$					
$FCCh = \frac{K_0}{K_{00}}$	2.110	1.615	1.080	1.075	1.070	1.070	1.070	1.070
		$A_0 = 0$;	$A_1 = 0.3$					
	2.739	2.410	2.053	2.050	2.050	2.050	2.050	2.050

Table 5.3. Monochromatic estimates of an additional brightness of atmospheric haze with respect to S as a function of spatical frequency ν (summer: Elterman-1968 closed aerosol model); $A_0 = 0.2$; $A_1 = 0.1$; $\eta = 1$; $\zeta = 1$; $\nu_{thr} = 3.5$)

ν, 1/km	λ, nm				
	400	500	600	700	800
0.2	0.000113	0.000072	0.000066	0.000048	0.000035
0.4	0.000453	0.000291	0.000265	0.000195	0.000142
0.5	0.000707	0.000291	0.000265	0.000193	0.000142
0.6	0.00102	0.000655	0.000600	0.000434	0.000321
0.8	0.00181	0.00117	0.00107	0.000771	0.000570
1.0	0.00283	0.00182	0.00167	0.00121	0.00892
1.5	0.00637	0.00410	0.00375	0.00271	0.00200
2.0	0.01132	0.007288	0.006672	0.00480	0.00368
2.5	0.01769	0.01139	0.01042	0.00754	0.00557
3.0	0.02548	0.01639	0.01501	0.01085	0.00803
3.5	0.03468	0.02232	0.02043	0.01477	0.01093

Table 5.4. Monochromatic estimates of an additional brightness of atmospheric haze with respect to S as a function of spatial frequency ν (winter: Elterman-1968 closed aerosol model); $A_0 = 0.8$; $A_1 = 0.1$; $\eta = 1$; $\zeta = 1$; $\nu_{thr} = 3.5$

ν, 1/km	λ, nm				
	400	500	600	700	800
0.2	0.000152	0.000082	0.000064	0.000052	0.000039
0.4	0.000608	0.000329	0.000257	0.000209	0.000145
0.5	0.000950	0.000515	0.000402	0.000329	0.000245
0.6	0.00137	0.000938	0.000579	0.000474	0.000353
0.8	0.00243	0.00132	0.00103	0.000843	0.000627
1.0	0.00382	0.00206	0.00161	0.001318	0.000981
1.5	0.00855	0.00463	0.00362	0.00296	0.00221
2.0	0.01521	0.00824	0.00644	0.00527	0.00392
2.5	0.02376	0.01287	0.01006	0.00824	0.00613
3.0	0.03422	0.01854	0.01449	0.01186	0.00883
3.5	0.04654	0.02523	0.01972	0.01605	0.01202

The Table 5.2 data confirm the correct choice of the value $\nu_{thr} \sim 3.5$, which, probably depends weakly on the optical state of the atmosphere. Values of amplitudes of A_0 and A_1 were chosen for moderate summer and winter conditions: $A_0 = 0.2$ and $A_1 = 0.1$ (summer); $A_0 = 0.8$ and $A_1 = 0.1$ (winter). Tables 5.3 and 5.4 list, in S units, the respective spectral estimates of additional haze brightness I_h as a function of spatial frequency ν.

For the interval $0 < \nu \leq \nu_{thr}$ Tables 5.5 and 5.6 list the estimates of maximum contribution $\hat{\delta}$ (%) of horizontal diffusion of photons to total field of brightness as a function of wavelength:

Table 5.5. Maximum contribution (%) of horizontal diffusion of photons to total brightness of the medium as a function of wavelength and spatial frequency ν (summer: Elterman-1968 closed aerosol model); $A_0 = 0.2$; $A_1 = 0.1$; $\eta = 1$; $\zeta = 1$; $\nu_{thr} = 3.5$

ν, 1/km	λ, nm			
	500	600	700	800
0.2	0.03	0.02	0.02	0.01
0.4	0.12	0.10	0.07	0.05
0.5	0.12	0.10	0.07	0.05
0.6	0.26	0.24	0.15	0.12
0.8	0.47	0.43	0.27	0.22
1.0	0.73	0.67	0.72	0.34
1.5	0.64	1.50	0.95	0.77
2.0	2.92	2.66	1.68	1.42
2.5	4.56	4.16	2.64	4.16
3.0	6.56	5.99	3.79	3.11
3.5	8.93	8.16	5.16	4.24

Table 5.6. Maximum contribution (%) of horizontal diffusion of photons to total brightness of the medium as a function of wavelength and spatial frequency ν (winter: Elterman-1968 closed aerosol model); $A_0 = 0.8$; $A_1 = 0.1$; $\eta = 1$; $\zeta = 1$; $\nu_{thr} = 3.5$

ν, 1/km	λ, nm			
	500	600	700	800
0.2	0.01	0.008	0.006	0.004
0.4	0.04	0.03	0.02	0.017
0.5	0.06	0.05	0.04	0.029
0.6	0.11	0.97	0.05	0.044
0.8	0.16	0.13	0.10	0.073
1.0	0.24	0.20	0.15	0.11
1.5	0.55	0.45	0.34	0.26
2.0	0.97	0.79	0.61	0.46
2.5	1.52	1.24	0.95	0.72
3.0	2.19	1.79	1.36	1.04
3.5	2.98	2.49	1.84	1.41

$$\hat{\delta} = \Delta \tilde{I}_h / I \ . \tag{5.51}$$

Here, the contrasts K_0 and K_{00} of the brightnesses of the system "object-background" at the levels of atmospheric top and bottom have been introduced:

$$K_0 = \frac{I_0^{max} - I_0^{min}}{I_0^{max} + I_0^{min}}$$

$$K_{00} = \frac{I_\infty^{max} - I_\infty^{min}}{I_\infty^{max} + I_\infty^{min}} \ . \tag{5.52}$$

Using (5.46 – 5.49), we obtain for transfer functions Π and $\mathscr{P}$ the following expressions [178]:

$$\Pi = I_0/\mathcal{I} = \frac{A(x-\Delta x)\mu(\zeta,\tau_0)}{[1 - AC(\tau_0)\varrho(\eta,\zeta,\varphi,\tau_0) + A(x-\Delta x)\mu(\eta,\tau_0) - } \ldots \rightarrow$$

$$\ldots \rightarrow \frac{}{- \mu(\zeta,\tau_0)T_\sigma(\eta,\tau_0)\dfrac{\nu^2}{\nu_{thr}^2}(A - \bar{A})} , \qquad (5.53)$$

$$\mathscr{P} = K_0/K_{00} = \frac{\varrho(\eta,\zeta,\varphi,\tau_0)[1 - \bar{A}C(\tau_0)] + \bar{A}\mu(\eta,\tau_0)\mu(\zeta,\tau_0)}{\bar{A}\mu(\zeta,\tau_0)[T_\sigma(\eta,\tau_0)\delta_*(\nu) + T_\sigma(\eta,\tau_0)]}$$

$$0 \leqslant \delta_* \leqslant -\frac{\nu^2}{\nu_{thr}^2} + 1 . \qquad (5.54)$$

The function δ_* in (5.53) determines the frequency inhomogeneity of the atmospheric reduction of the brightness contrasts of the horizontally inhomogeneous base.

Formulas (5.53) are a generalization of the respective asymptotic expressions obtained for a 1-D case for a small (point) source and horizontally-homogeneous surfaces [97]. So, for instance, at $\nu \rightarrow 0$ formula (4.18) follows from (5.53) for a transfer function for a 1-D infinitely extended surface. At $\nu > \nu_{thr}$ from (5.53 – 5.54) we obtain formulas for Π and $\mathscr{P}$ for a small (point) object surrounded by an infinitely extended background.

Formulas (5.53 – 5.54) also permit simple analytical generalizations for vertical inhomogeneity of such optical parameters of the atmosphere as phase function $x(\gamma)$ and probability of quantum survival Λ. Also, they can be generalized for polarized radiation scattered in an aerosol atmosphere and reflected from the surface. Figure 5.14 shows a function $\Delta = [\Pi(\nu_{thr}) - \Pi(0)]/\Pi(0)$ which characterizes, depending on wavelength λ, the maximum contribution of horizontal diffusion of photons to the spectral transfer function for the brightnesses of natural formations spectrophotometered from space in nadir

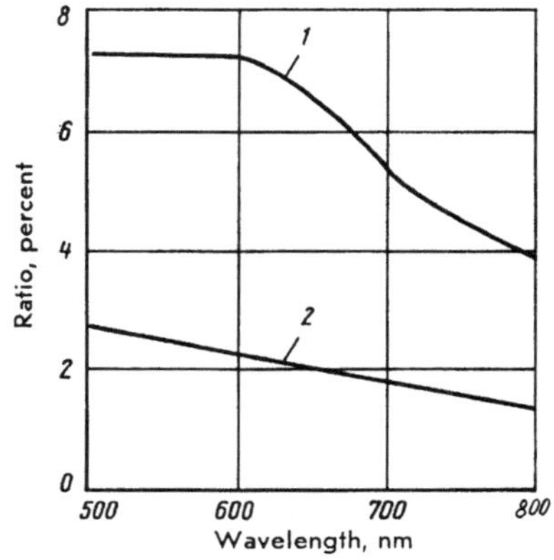

Fig. 5.14. Maximum contribution of the horizontal diffusion of photons into the spectral transfer function of the atmosphere for the brightnesses of natural formations spectrophotometered from space: (the Elterman-1968 closed aerosol model): $\nu_{thr} = 3.5$; $\theta_0 = 0°$; $\theta = 0°$; $A(x) = A_0 + A_1 \cos x\nu$; *1* $A_0 = 0.2$; $A_1 = 0.1$; *2* $A_0 = 0.8$; $A_1 = 0.1$

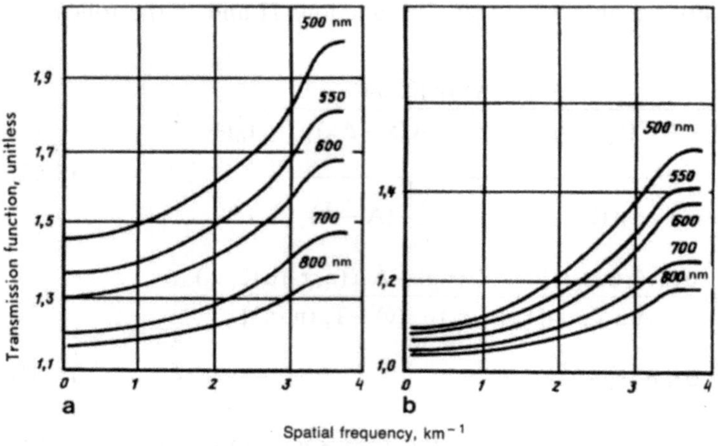

Fig. 5.15 a, b. The frequency dependence of the atmospheric transfer function for the monochromatic contrasts of the brightnesses of the consinusoidal model in surveying the Earth from space (the Elterman-1968) closed aerosol model): $\nu_{thr} = 3.5$; $\theta_0 = \theta = 0°$; $A(x, \nu) = A_0 + A_1 \cos x\nu$; **a** $A_0 = 0.2$; $A_1 = 0.1$; **b** $A_0 = 0.8$; $A_1 = 0.1$

($\theta = 0°$) with the sun in zenith ($\theta_0 = 0°$). It follows from Fig. 5.14 that, on average, over the spectrum for the Elterman-1968 aerosol model, for $\lambda = 500-800$ nm, nadir viewing $\theta = 0°$ and the sun in zenith $\theta_0 = 0°$ this contribution constitutes about 6%. When $A_0 = 0.8$ and $A_1 = 0.1$, it is about 2%. With increasing optical thickness τ_0 or increasing amplitude of albedo variations up to $A = 0.2$, the contribution of horizontal diffusion also grows.

Figure 5.15 a, b illustrates the frequency dependence of the transfer function for brightness contrasts $\mathscr{P}(\nu)$ calculated for the wavelength interval $\lambda = 500-800$ nm for the same conditions of viewing, sun location, average albedo A_0 and its variation A_1, as in Fig. 5.14. Analysis of the data in Fig. 5.15 a, b shows that the horizontal diffusion of photons affects markedly (up to 25 – 30%) the transfer of the contrasts of brightnesses of surface natural formations with characteristic sizes of horizontal-inhomogeneities either above 300 m or below 1 km. Outside this interval, with an accuracy of not more than 5%, the attenuation of the contrasts of brightnesses of a horizontally-inhomogeneous surface depending on spatial frequency ν, can be neglected. An aerosol atmosphere suppresses the high-frequency details of space-derived images and spectra more than the low-frequency ones. Thus, in automated initial or secondary processing of space-derived spectrophotometric information, an additional correction of both visual data and results of optical instrumental measurements must be included, artificially suppressing low frequencies and raising high frequencies of recorded spectral images.

For atmospheric radiative correction of high-frequency and low-frequency details of space-derived images according to (5.36) and (5.46), it is necessary to extract the atmospheric haze brightness I_h (at $A = 0$) from the brightnesses of a space image in each point, and to divide the result by $\mu(\eta, \tau_0)$.

As for space images of average spatial resolution, to improve their quality, this diffusion can be considered using formulas (5.53 – 5.54). The need for this depends on a finite application of the obtained spectrophotometric information, accuracy of its recording, and the quality of initial processing on board the space carrier.

Atmospheric frequency filtration of optical signals, which form a radiative image of surface (above-water) objects, is of great importance for solution of the problems of recognition, classification and assessment of the state of natural formations by their reflection spectra measured from a space platform. A standard solution of the problem of atmospheric filtration of aerospace video information in the first approximation must be based on the use of linear and invariant properties of the atmosphere-objects-background system. With this approach, fluctuations in the Fourier-transformed optical signal at the output and input of the system are known to be related by a transfer function $\Pi(\nu)$. An approximate analytical expression for module $\hat{T}(\nu)$ of the normalized transfer function $\Pi(\nu)$, equal to the FCCh of the optical system, is based on polynomial approximation of the atmospheric frequency filter (5.42) and solutions of the boundary problem of the 1-D theory of radiation transfer (5.24). Using formula (5.53), we find the following parameterization $\hat{T}(\nu) = |\Pi(\nu)|/|\Pi(0)|$:

$$\hat{T}(\nu) = \frac{T_\sigma(\eta, \tau_0)}{\mu(\eta, \tau_0)} \delta_*(\nu) + \frac{T(\eta, \tau_0)}{\mu(\eta, \tau_0)} = \delta_*(\nu) + [1 - \delta_*(\nu)] \frac{T(\eta, \tau_0)}{\mu(\eta, \tau_0)} , \qquad (5.55)$$

where the $\delta_*(\nu)$ function is determined by (5.54). Thus, the FCCh of the optical system "atmosphere-object-background" does not depend on average surface albedo $\bar{A}$ and is determined, in the approximation considered, by a threshold spatial frequency ν_{thr} and by a ratio between direct and diffuse transparencies of the atmosphere.

The physics of formula (5.55) is totally determined by the supposition that in the interaction between solar radiation and real horizontally-inhomogeneous surface the singly reflected photons carry information on its true albedo $A(x, \nu)$, and those reflected repeatedly – about average albedo $\bar{A}(x, \nu) = A_0$. In addition, it is supposed that the phase-frequency shiftings in an aerospace video image are absent, i.e., the surface (above-water) objects are viewed in nadir ($\eta = 1$).

This introduces the 1-D theory of radiation transfer of an important supporting parameter – threshold frequency ν_{thr}. From this the brightnesses and brightness contrasts of natural formations, surface of above-water objects and backgrounds are reduced by the atmosphere independently of spatial frequency ν. Thus, it is possible, based on exact physical interpretation, to study the effect of horizontal diffusion of photons on the values of spatial-frequency transfer functions in conditions, optimal for space-based survey of the Earth in the visible spectrum ($\lambda = 400 - 800$ nm). It is enough here to consider only the use of a simpler, 1-D theory of radiation transfer. The conclusions drawn suggest that with atmospheric radiative correction applied to space video images and spectra, the low- and high-frequency details of the processed images do not depend on the horizontal diffusion of photons.

5.5 An Approximate Assessment of the Transfer Function for a Surface Formed by Two Horizontally Homogeneous Half-Planes

For remote sensing of the environment from space of great interest is the problem of spectrophotometry of a horizontally-inhomogeneous surface formed by two homogeneous half-planes with different reflective characteristics [94, 97, 220]. Since the problem here is not standard and requires a special theoretical study (Section 5.3), first we shall take two extreme cases which permit a consideration similar to that already mentioned, on the assumption that the optical properties of the atmosphere do not vary horizontally.

First, we shall study the areas far from the interface between two homogeneous media. We shall call $I_i(0, \tau_0, \eta, \zeta, \varphi)$ and $I_{oi}(\tau_0, \eta, \zeta, \varphi)$ $(i = 1, 2)$ their brightnesses, respectively, at the levels of atmospheric top and the surface. The value of index $i = 1, 2$ denotes the location of a spectrophotometered area with albedo A_i with respect to the interface (on the right-hand or on the left-hand side). Then, instead of (4.1) and (4.16), we have the following equation:

$$I_i = I_i(\tau_0, \eta, \zeta, \varphi) + I_{oi}(\zeta, \tau_0) T(\eta, \tau_0) \ , \tag{5.56}$$

where $I_{h,i}$ $(i = 1, 2)$ is the atmospheric haze brightness measured on both sides of the interface. Following (4.14) and equivalent writing of this relationship follows from (5.56):

$$I_i = S\zeta\varrho(\eta, \zeta, \varphi, \tau_0) + I_{oi}[T(\eta, T_0) + T_\sigma(\eta, \tau_0)] = S\zeta\varrho(\eta, \zeta, \varphi, \tau_0) + I_{oi}\mu(\eta, \tau_0) \ . \tag{5.57}$$

It is easily seen that, far from the interface, for atmospheric haze brightness $I_{h,i}$, surface brightness, and transfer functions $\Pi_i = I_{0,i}/I_i$ $(i = 1, 2)$ expressions obtained in Section 5.1 hold, but with the homogeneous surface albedo in (4.14), (4.11), and (4.18) changed for albedos of the right-hand or left-hand half-planes A_i $(i = 1, 2)$. In particular, for the transfer function Π_i at $\eta \sim 1$, $\zeta \sim 1$ and large albedos A_i an approximate expression is valid:

$$\Pi_i \sim \frac{1}{\mu(\eta, \tau_0)} = \frac{1}{T(\eta, \tau_0) + T_\sigma(\eta, \tau_0)} \ . \tag{5.58}$$

For the transfer function of brightness contrasts $\mathscr{P} = \dfrac{K_{oi}}{K_i}$ $(i = 1, 2)$, where:

$$K_i = \frac{|I_1 - I_2|}{I_i} \geqslant 0 \ , \quad K_{0,i} = \frac{|I_{01} - I_{02}|}{I_{oi}} \geqslant 0 \ , \tag{5.59}$$

we obtain an expression:

$$\mathscr{P}_i = 1 + \frac{\varrho(\eta, \zeta, \varphi, \tau_0) S \zeta}{I_{0i}(\zeta, \tau_0)\,\mu(\eta, \tau_0)} \ , \tag{5.60}$$

i.e., the transfer functions for brightnesses Π_i and for contrasts $\mathscr{P}_i$ are related as

$$\Pi_i \mathscr{P}_i = \frac{1}{\mu(\eta, \tau_0)} = \frac{1}{T(\eta, \tau_0) + T_\sigma(\eta, \tau_0)} \ . \tag{5.61}$$

Substituting (5.58) into (5.61), we find a simple approximate relationship:

$$\mathscr{P}_i \sim 1 \ , \tag{5.62}$$

which shows that under certain conditions of viewing geometry, solar irradiance and albedo value, brightness contrasts of the two homogeneous surfaces with a common boundary, when observed from space, varies little compared to their contrast at the surface level. Note, that in (5.59 – 5.68) the values of index $i = 1, 2$ must be fixed depending on condition (5.59) providing a positive value of contrasts.

Now we shall consider the problem of the assessment of the spectral transfer functions Π and $\mathscr{P}$ near the interface between two homogeneous media. In this transfer zone, the field of brightness of the medium is very complicated due to multiple reflection of radiation from both the half-planes, whereas far from the interface it depends only on either A_1 or A_2. For a model of horizontally homogeneous atmosphere one may assume intermixing of the fluxes which is reflected by the left- and right-hand homogeneous half-planes. In this case we can state that the brightnesses of hazes $I_{h,i}$ over the interface are equal to an average haze brightness, determined at an average albedo $A = 1/2(A_1 + A_2)$:

$$\bar{I}_0 = S\zeta \left[\varrho(\eta, \zeta, \varphi, \tau_0) + \frac{\bar{A}\,\mu(\zeta, \tau_0)\,T_\sigma(\eta, \tau_0)}{1 + \bar{A}\,C(\tau_0)} \right] \ . \tag{5.63}$$

The surface brightnesses $I_{0,i}$ near the interface are found by the sum of the brightness at an average albedo $A_i = \bar{A}$ and an addition, determined by a deviation of the albedo of either left-hand or right-hand half-plane, from an average value $\bar{A}$, at a single scattering of solar radiation from it [97]:

$$I_{oi} = \frac{A_i\,\mu(\zeta, \tau_0) S \zeta}{1 - \bar{A}\,C(\tau_0)} + \frac{(A_i - \bar{A})\,\mu(\zeta, \tau_0) S \zeta}{1 - \bar{A}\,C(\tau_0)} = \frac{A_i\,\mu(\zeta, \tau_0) S \zeta}{1 - \bar{A}\,C(\tau_0)} \ . \tag{5.64}$$

Formula (5.64) has a simple physical meaning apparent from relationship (4.13). It is assumed in its derivation that, near the interface, the brightness of an inhomogeneous surface with a single reflection of radiation, is determined by the albedo of a half-plane from which direct radiation is reflected at a given viewing angle. In this case, a diffuse illumination due to multiple reflection from both the half-planes is determined as the brightness of a horizontally-homogeneous surface at an average albedo $\bar{A} = 1/2\ (A_1 + A_2)$. Taking $I_{h1} = A_{h2} = A_h$ and $I_{0,i} = \bar{I}_{0,i}$ in (5.56 – 5.57), we find:

$$I_i = I_h(\nu, \zeta, \varphi, \tau_0) + I_{oi}(\zeta, \tau_0)T(\eta, \tau_0) \ ;$$

$$I_i = S\zeta\varrho(\eta, \zeta, \varphi, \tau_0) + \bar{I}_{oi}(\zeta, \tau_0)\left[T(\eta, \tau_0) + \frac{\bar{A}}{\bar{A}_i}T_\sigma(\eta, \tau_0)\right] , \quad (5.65)$$

where the brightnesses of haze $\bar{I}_h$ and surfaces $\bar{I}_{o,i}$ (i = 1, 2) are determined by formulas (5.63 – 5.64), with an average albedo $\bar{A} = 1/2(A_1 + A_2)$. Using relationships obtained in Section 4.1, and bearing in mind (5.63 – 5.64), we find the following expressions for the transfer functions Π_i:

$$\bar{\Pi}_i = \frac{A_i\mu(\zeta, \tau_0)}{[1 - AC(\tau_0)]\varrho(\eta, \zeta, \varphi, \tau_0) + \mu(\zeta, \tau_0)[\bar{A}T_\sigma + A_iT]} , \quad (5.66)$$

$$\bar{\Pi}_i \sim \frac{1}{T(\eta, \tau_0) + \dfrac{\bar{A}}{\bar{A}_1}T_\sigma(\eta, \tau_0)} . \quad (5.67)$$

Formulas (5.63 – 5.67) were obtained for an orthotropic horizontally-inhomogeneous surface. The non-orthotropic reflection is considered in a similar way to that of a horizontally-homogeneous surface.

Near the interface the contrast between haze brightnesses $K_h \sim 0$, the transfer function $\mathscr{P}_i (i = 1, 2)$ for the brightness contrasts are:

$$\bar{\mathscr{P}}_i = 1 + \frac{I(\eta, \zeta, \varphi, \tau_0)}{I_{0,i}(\zeta, \tau_0)T(\eta, \tau_0)} , \quad (5.68)$$

i.e.,

$$\bar{\mathscr{P}}_i = e^{\tau_0/\eta} - \frac{1}{\bar{\Pi}_i(\eta, \zeta, \varphi, \tau_0)} . \quad (5.69)$$

Using (5.66) and (5.67), we find a simple approximate formula for $\mathscr{P}_i$ [97]:

$$\bar{\mathscr{P}}_i \sim e^{\tau_0/\eta}\left[T(\eta, \tau_0) + T_\sigma(\eta, \tau_0)\frac{\bar{A}}{\bar{A}_i}\right] = 1 + T_\sigma(\eta, \tau_0)e^{\tau_0/\eta} . \quad (5.70)$$

To determine the surface (above-water) contrast, finally we obtain, instead of (5.70):

$$\begin{aligned}\bar{\mathscr{P}}_i &= e^{\tau_0/\eta}\left[T(\eta, \tau_0) + \frac{\bar{A}}{\bar{A}_i}T_\sigma(\eta, \tau_0)\right] \\ &= e^{\tau_0/\eta}[T(\eta, \tau_0) + \tfrac{1}{2}(2 - K_{0,i})T_\sigma(\eta, \tau_0)] \\ &= 1 + \tfrac{1}{2}(2 - K_{oi})T_\sigma(\eta, \tau_0)e^{\tau_0/\eta} . \end{aligned} \quad (5.71)$$

Apart from these two extreme cases (far from and near the interface between two homogeneous media), an intermediate case is often observed in practice – spectrophotometry of the transition zone at some distance from the interface. In this case the following approximate semiempirical formula of brightnesses [31, 97] can be used:

$$I_i = S\zeta\left[\varrho(\eta,\zeta,\varphi,\tau_0) + \frac{\alpha_i\,\mu(\zeta,\tau_0)\,T_\sigma(\eta,\tau_0)}{1-\bar{A}C(\tau_0)} + \frac{\beta_i\,\mu(\zeta,\tau_0)\,T(\eta,\tau_0)}{1-\bar{A}C(\tau_0)}\right], \tag{5.72}$$

where the approximating coefficients $A_{i,\min} \leqslant \alpha_i \leqslant A_{i,\max}$ and $A_{i,\min} \leqslant \beta_i \leqslant A_{i,\max}$ have a meaning in terms of albedo and are found from comparison of the right-hand part of relationship (5.72) with the values of brightnesses I_i obtained experimentally. This intermediate case is considered in detail, based on analysis of the Soyus-9 spectrophotometric data obtained for the land-water interface [31].

The formulas for brightnesses, brightness contrasts and the respective transfer functions just obtained are easily generalized, within the applied physical assumptions of a horizontally inhomogeneous surface, albedo $A(x,\nu)$, where x is the horizontal coordinate, and ν the spatial frequency equal to the number of periods of the harmonic function per unit distance by horizons.

Instead of (5.64), (5.66) and (5.67) for the brightness of a horizontally-inhomogeneous surface and the respective transfer function at each fixed spatial frequency ν the following expressions are valid:

$$\bar{I}_0(x,\nu) = \frac{A(x,\nu)\,\mu(\zeta,\tau_0)\,S\zeta}{1-\bar{A}C(\tau_0)} = \frac{\bar{A}\,\mu(\zeta,\tau_0)\,S\zeta}{1-\bar{A}C(\tau_0)} + \frac{[A(x,\nu)-\bar{A}]\,\mu(\zeta,\tau_0)\,S\zeta}{1-\bar{A}C(\tau_0)}, \tag{5.73}$$

$$\bar{\Pi}(x,\nu) = \frac{A(x,\nu)\,\mu(\zeta,\tau_0)}{[1-\bar{A}C(\tau_0)]\,\varrho(\eta,\zeta,\varphi,\tau_0) + \mu(\zeta,\tau_0)\,[\bar{A}\,T_\sigma(\eta,\tau_0) + A(x,\nu)\,T]}, \tag{5.74}$$

$$\bar{\Pi}(x,\nu) \sim \left[T(\eta,\tau_0) + \frac{\bar{A}}{A(x,\nu)}\,T_\sigma(\eta,\tau_0)\right]^{-1}. \tag{5.75}$$

As in the case of the interface between two homogeneous media, when introducing formulas (5.73 – 5.75), it is assumed that an average brightness of atmospheric haze I_h is independent of the horizontal coordinate and uniformly reduces the brightness of an inhomogeneous surface at all spatial frequencies ν. The haze brightness I_h is assessed by expression (5.63), where $A = \bar{A}$ is an average albedo of an inhomogeneous surface:

$$\bar{A} = \frac{A_{\max}(x,\nu) + A_{\min}(x,\nu)}{2}. \tag{5.76}$$

The brightness of a horizontally-inhomogeneous surface in the given approximation in each point with the coordinate is determined at a single reflection for true albedo $A(x,\nu)$ and an "illumination" due to multiple reflection of photons from the surface, for an average albedo $\bar{A}$.

Similarly to the expressions for the transfer function of the contrasts $\mathscr{P}(x, \nu)$ at a fixed frequency ν already obtained, we find the following relationships:

$$\bar{\mathscr{P}}(x, \nu) = 1 + \frac{\bar{I}}{I_{0,\max}(x, \nu) T(\eta, \tau_0)} , \tag{5.77}$$

$$\bar{\mathscr{P}}(x, \nu) = e^{\tau_0/\eta} \frac{1}{\Pi_{\max}(x, \nu)} , \tag{5.78}$$

$$\bar{\mathscr{P}}(x, \nu) \approx e^{\tau_0 \eta} \left[T(\eta, \tau_0) + \frac{\bar{A}}{A_{\max}(x, \nu)} T_\sigma(\eta, \tau_0) \right] . \tag{5.79}$$

Using the definition of the surface (above-water) contrast of brightnesses K_0 of the given horizontally-inhomogeneous surface at a fixed spatial frequency ν:

$$K_0 = \frac{I_{0,\max}(x, \nu) - I_{0,\min}(x, \nu)}{I_{0,\max}(x, \nu)} \geq 0 \tag{5.80}$$

we find the final expression for $\bar{\mathscr{P}}(x, \nu)$ from (5.79):

$$\bar{\mathscr{P}}(x, \nu) \approx e^{\tau_0/\eta} [T(\eta, \tau_0) + \tfrac{1}{2}(2 - K_0) T_\sigma(\eta, \tau_0)] . \tag{5.81}$$

The surface (above-water) contrast K_0 in (5.80) is assessed by the formula

$$K_0 = \frac{A_{\max}(x, \nu) - A_{\min}(x, \nu)}{A_{\max}(x, \nu)} > 0 . \tag{5.82}$$

These formulas describe the effect of a scattering and absorbing atmospheric layer between a space-borne optical instrument and a surface (above-water) background, without consideration of horizontal diffusion of solar photons due to a spatially inhomogeneous distribution of the surface reflective characteristics by the x coordinate. A horizontally-inhomogeneous surface where the albedo varies after the harmonic law [174, 178] was considered. However, in real conditions of survey of natural formations at $\eta = 1$ (nadir) and high sun elevations over the horizon ($\zeta = 1$) from space, the horizontal diffusion of photons in an aerosol terrestrial atmosphere at $\tau_0 = 0.3 - 0.6$ cannot be very large for a horizontally-homogeneous atmosphere, from the physical point of view. Therefore, the first approximation suggests that the atmospheric haze uniformly covers the surface (above-water) background or object of different spatial extent. In this case the atmosphere must filter out high spatial frequencies ($\nu \to \infty$) uniformly, reducing the brightness contrasts for low frequencies ($\nu \to 0$).

5.6 A Comparison of Experimental and Theoretical Estimates of the Spectral Transfer Functions

To find the constituents of the spectral transfer function, the synchronous spectrophotometry from satellite and aircraft of a horizontally-inhomogeneous surface (background) is important. This may be either far from or near to the interface between its two homogeneous sites, with substantially different reflecting properties (for instance, the land-sea interface) [68, 97]. Suppose, the spectral brightnesses of two horizontally-homogeneous adjacent media at the atmospheric top level I_i and surface level $I_{o,i} (i = 1, 2)$ have been estimated from combined satellite and aircraft experiments. If the brightnesses are measured near the interface, when $I_{h1} \simeq I_{h2} = \bar{I}_h$, then, using a system of two linear algebraic equations (5.56), we find the unknown functions T and $\bar{T}_h$ [97]:

$$\bar{T}_h = \frac{\bar{I}_{01} I_2 - \bar{I}_{02} I_1}{\bar{I}_{01} - \bar{I}_{02}} ; \tag{5.83}$$

$$T = \frac{I_1 - I_2}{\bar{I}_{01} - \bar{I}_{02}} . \tag{5.84}$$

Using definitions of transfer functions $\Pi_i = I_{oi}/I_i$ and $\mathscr{P}_i = K_{oi}/K_i$, given in the last section, we find, using relationships (5.83–5.84):

$$\bar{\Pi}_i = \frac{\bar{I}_{01} (\bar{I}_{01} - \bar{I}_{02})}{\bar{I}_{01} I_2 - \bar{I}_{02} I_1 + \bar{I}_{01} (I_1 - I_2)} ; \tag{5.85}$$

$$\mathscr{P}_i = \frac{\bar{I}_{01} I_2 - \bar{I}_{02} I_1 + \bar{I}_{01} (I_1 - I_2)}{\bar{I}_{oi} (I_1 - I_2)} . \tag{5.86}$$

If the survey is made far from the interface, an assessment of transfer functions Π_i and $\mathscr{P}_i$ is strictly reduced to that just given (under condition of horizontal-homogeneity of the atmosphere). For this purpose the system of linear algebraic equations (5.57) must only be used to present the brightnesses $I_i (\eta, \zeta, \varphi, \tau_0)$ $(i = 1, 2)$, since, in contrast to (5.56), this system is definite. The solution for this system regarding the unknown functions $S\zeta\varrho$ and $\mu = T + T_\sigma$ has formally the form (5.83–5.84). The transfer functions Π_i and $\mathscr{P}_i$ $(i = 1, 2)$ are found with formulas (5.85–5.86). It must be emphasized, however, that in contrast to the last experiment, here the sums of direct and diffuse atmospheric transparencies are estimated, as well as haze brightness (without considering "illumination" by the surface, $A = 0$). This can differ substantially from the brightness of real haze $(A \neq 0)$. To estimate the brightness of real haze I_h, additional, independent estimations of the atmospheric transparency $T_\lambda (\eta, \tau_0)$, in the sub-satellite zone [28] must be made.

Figure 5.16 illustrates the spectral change of transfer functions Π_λ for the brightness contrasts of the surfaces "desert-water-surface" and "desert-

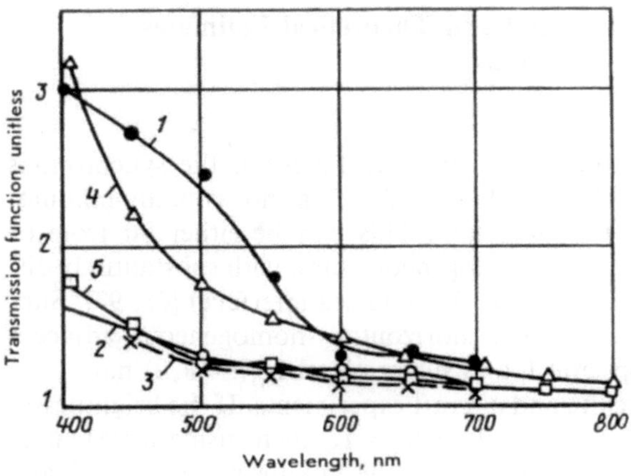

Fig. 5.16. The transfer functions for the brightness contrasts between two homogeneous surfaces: *1* desert-water surface; Soyuz-9 data; *2* cloudiness-desert; Soyuz-7 data; *3* cloudiness-desert; Soyuz-9 data; *4* sand-sea; calculated data: $A_1 = 0.24$; $A_2 = 0.06$; $\theta_0 = 40°$; *5* sand-cloudiness; calculated data: $A_1 = 0.78$; $A_2 = 0.24$; $\theta_0 = 40°$

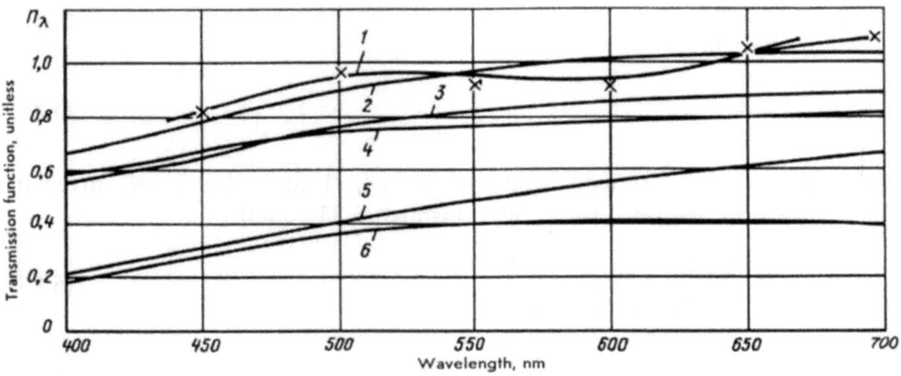

Fig. 5.17. Comparison of experimental and theoretical values of the atmospheric transfer functions near the interface between homogeneous media: *1* Soyuz-9; aircraft (2.7 km); *2* calculated data: $A_1 = 0.24$; $A_2 = 0.06$; sand near sea surface; *3* calculated data: $A_1 = 0.24$; $A_2 = 0.78$; sand near the edge of continuous cloudiness; *4* calculated data: $A_1 = 0.24$; homogeneous sand surface; *5* calculated data: $A_1 = 0.06$; $A_2 = 0.24$; sea near sand surface; *6* calculated data: $A_1 = 0.06$; homogeneous water surface. Calculated data are given for the Elterman-1968 model; $x_1 = 0.63$

cloudiness", drawn from results of measurements from the Soyuz-9 satellite, from a low-flying aircraft and near the Earth's surface [31].

Figure 5.17 compares theoretical and experimental estimates of transfer functions $\Pi_{\lambda,i}$ near the sand-water interface. In this case the theoretical basis for such comparisons is a supposition that there is a transition zone near the interface. Over this "pumping" of photons from the surface with large albedo (land) to the surface with low albedo (water) and mixing of diffuse radiative fluxes in the atmosphere, causes the brightnesses of atmospheric hazes $I_{h,i}$ ($i = 1, 2$) over water and land to be almost equal ($I_h = I_{h1} = I_{h2}$). The compared empirical curves have been obtained from Soyuz-9 and flying laboratory,

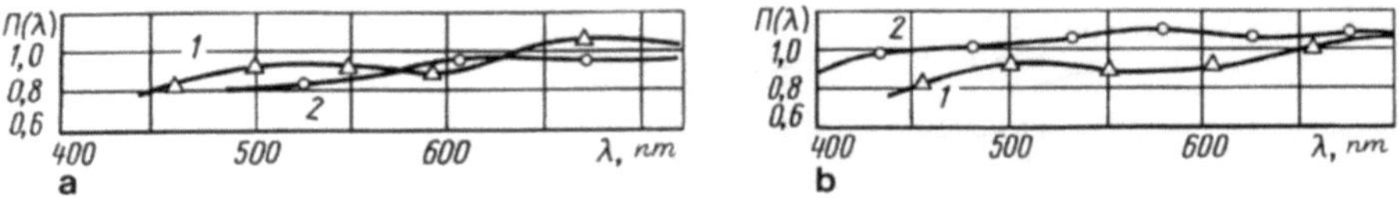

Fig. 5.18 a, b. Comparison of experimental and theoretical values of the spectral transfer function near the interface between two horizontally homogeneous media: **a** sand near the edge of continuous cloudiness; *1* experiment (Soyuz-9-aircraft); *2* calculations (3-term phase function). **b** Sand near water surface; *1* experiment (Soyuz-9-aircraft); *2* calculations (three-term phase function)

and theoretical $\Pi_{\lambda,i}$ values have been calculated with formula (5.66) based on the data from [97] and Tables 4.3–4.4 (Fig. 5.18). For the sand-sea interface, a good agreement of experimental and theoretical data is observed (Fig. 5.17, curves 1 and 2). This testifies to the correctness of the supposition about the existence of the transition zone and the equality of haze brightnesses near the interface of those two horizontally-homogeneous media. This supposition has served as the basis for the approximate formula (5.66).

In addition to the comparison of theoretical and experimental estimates of the $\Pi_{\lambda i}$ transfer function, an interesting study concerns the effect on its value of the spatial inhomogeneity of the surface spectrophotometered from space. The character of variations of the transfer function here is determined by the respective variations in the brightnesses I_{0i} and I_{hi} approaching or moving away from the interface. For instance, near the interface, according to (5.64) and (5.65), the atmospheric haze brightness I_h over a low-albedo surface, as grows the brightness of the surface itself I_0, compared to the homogeneous case, because of an adjacent high-albedo surface. For the high-albedo surfaces the brightnesses of atmospheric haze I_h and of the surface itself decrease, by constrast with the homogeneous surface.

Figure 5.19 illustrates several successive photometric sections of the spectral brightness of water surfaces and of a lake shore in South-West Afghanistan from the Soyuz-9 MS data. The position of the spectrograph's slit, relative to the surface to be spectrophotometered, is shown on the left-hand side of the figure. Horizontal lines mark the positions of photometric sections relative to the interface. From the Fig. 5.19 data one can trace the transformation of the spectral brightness of a horizontally-inhomogeneous surface in the transition zone "water surface-shore": approaching the shoreline the spectral brightness of water surface grows, and the brightness of the land surface decreases, becoming average over the interface (curve 10). The upper (1) and lower (2) curves are spectra of horizontally-homogeneous surfaces. The respective calculated data are given as dotted curves. Analysis of Fig. 5.19 allows each wavelength to be used to estimate characteristic lengths of the transition inhomogeneous zone over the interface. An approximate solution for this problem has been given in [143].

As an example of the mutual effect of two homogeneous media with different albedos on their brightnesses near to and far from the interface, we use theoretical estimates of the spectral brightnesses of atmospheric haze, sand

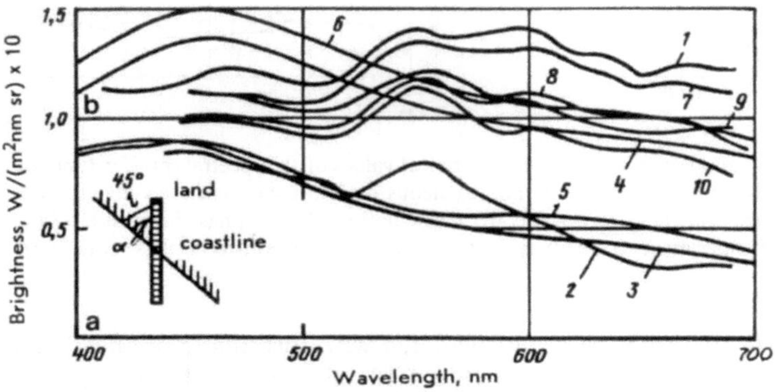

Fig. 5.19 a, b. Spectral brightnesses of the surface near the interface between two media ($\theta_0 = 33.8°$). **a** Position of the projection of the slit and photometric sections onto the spectrophotometered surface; **b** spectral brightnesses. *1, 7* sand surface; experiment (different distance from the interface); *2* water surface; *3* water surface; calculations: $\theta_0 = 33°$; *4* sand near the water surface; calculations: $\theta_0 = 33°$; *5* water surface near the sand shore; calculations: $\theta_0 = 33°$; *6* sand far from the water surface; calculations: $\theta_0 = 39°$; *8–10* the interface

and cloud surfaces when observed from space. Spectral brightnesses in extreme cases (near to and far from the interface) were calculated with the following approximate formulas:

$$I_{hi} = S\zeta \left[\frac{A_l \mu(\zeta, \tau_0) T_\sigma(\eta, \tau_0)}{1 - \bar{A} C(\tau_0)} + \varrho(\eta, \zeta, \varphi, \tau_0) \right] ; \tag{5.87}$$

$$I_i = S\zeta \left[\frac{A_i \mu(\zeta, \tau_0) \mu(\eta, \tau_0)}{1 - \bar{A} C(\tau_0)} + \varrho(\eta, \zeta, \varphi, \tau_0) \right] . \tag{5.88}$$

The effect of the atmosphere on space-derived spectra of natural formations can be assessed more accurately by calculating the atmospheric haze brightness I_h with formulas (4.5–4.6), where the values of the Π transfer function are taken from direct comparison of satellite and aircraft (surface) experiments. However, for this purpose, the atmospheric transparency $T_\lambda(\eta, \tau_0)$ over the spectrophotometered areas of the surface in the region of the combined experiment must be estimated independently. Since direct measurements of atmospheric transparency have not been made during the Soyuz-7 and 9 combined sub-satellite experiments, the only possibility of obtaining an experimental assessment of the haze brightness I_h and atmospheric transparency T is to solve the system of Eqs. (5.56) or (5.57).

A comparison of theoretical and experimental estimates of $I_{h\lambda}$ and T_λ obtained for the sand-sea interface have been made in [31, 145]. The spectral data of synchronous aircraft and Soyuz-9 satellite measurements have been taken as experiments. Theoretical curves of I_h have been calculated with (5.63), in which the atmospheric haze brightnesses near the interface between two homogeneous surfaces are assumed to be equal. A comparison of experimental and calculated curves shows their consistency in the given wavelength interval.

Now we compare the experimental and theoretical curves of the transfer functions for the brightness contrasts of two horizontally-homogeneous half-planes with different albedos A_i $(i = 1, 2)$.

Figure 5.16, apart from experimental data, illustrates theoretical estimates of $\mathcal{P}_{i\lambda}$ calculated with (4.10) and (5.60) for $K_h = 0$. According to both experimental and theoretical data, at low values of $A_{\lambda i}$, contrasting surfaces (for instance, "sand-sea"), the transfer of contrasts to the atmospheric top level is substantially worse, compared to the case of high albedos (clouds-sand). However, with increasing wavelength λ this difference is much reduced due to the lowered scattering ability of the atmosphere.

For the cloudiness-sand system, the theoretical and experimental estimates of $\mathcal{P}_{i\lambda}$ agree well through the whole wavelength interval in question, which is determined by high albedos of clouds and sand in the visible. A great difference is observed in the case of the sand-water system in the interval $\Delta\lambda = 450-550$ nm, where the reducing effect of atmospheric haze smoothing the contrasts of low-albedo surfaces is the strongest. With increasing wavelength ($\lambda > 500$ nm) the consistency between experimental and theoretical curves becomes much better.

5.7 The Effect of the Atmospheric Radiative Factors on the Detection of the Small-Sized Optical Inhomogeneities of the Terrestrial Surface

The solution of this important practical problem is very significant for the methods of optimal planning, instrumental and program provision of aerospace optical experiments connected with the problem of remote sensing and identifying of small-sized, weak-contrast surface or above-water optical inhomogeneities of different origin (plankton, oil spots, internal waves in the oceans, etc.). Theoretical modeling of the effect of the atmospheric radiative factors on the spectral brightnesses and brightness contrasts of such inhomogeneities is not standard and requires special consideration of the horizontal diffusion of photons in the system "atmosphere-object-background" [135–137, 182]. However, if the optical properties of the atmosphere over a small (point) object and its background are assumed to be equal and unchanged in the horizontal direction, simple approximate formulas can be obtained for haze brightnesses I_h, of object I_{ob}, of background I_B from general relationships (3.2), and formulas for transfer functions Π_{ob}, Π_B, p_{ob}, and p_B assessed from (3.3–3.5).

These approximate formulas, as for those for interfaces (Section 5.5) are based on the supposed equality of atmospheric haze brightnesses over an extended background and a small object. This enables presentation of the brightness of a small-sized surface (above-water) object spectrophotometered from space or aircraft as the brightness of an extended background and an addition to it, due to difference in the reflective properties of an object from its respective background properties.

From the physical point of view, this approach assumes that the effect of multiple reflection of photons from a small-sized object, compared to a similar effect for its background, can be neglected. Thus, the brightness of a single small object (Fig. 4.1) spectrophotometered from space, is determined by single reflection of solar radiation from this object and by diffuse illumination due to multiple reflection of photons from its background.

Bearing in mind these suppositions and using relationships already obtained, we find the following expressions for the brightnesses $I_{h,B}$, $I_{h,ob}$, $I_{o,ob}$, $I_{o,B}$, I_{ob} and I_B in the case of orthotropic reflective characteristics of a small object and its extended background [97]:

$$I_{h,B} = I_{h,ob} = S\zeta\left[\varrho(\eta,\zeta,\varphi,\tau_0) + \frac{A_B\mu(\zeta,\tau_0)T_\sigma(\eta,\tau_0)}{1-AC(\tau_0)}\right], \tag{5.89}$$

$$\begin{cases} I_{o,ob} = \dfrac{A_B\mu(\zeta,\tau_0)S\zeta}{1+A_BC(\tau_0)} + \dfrac{(A_{ob}-A_B)\mu(\zeta,\tau_0)S\zeta}{1-A_BC(\tau_0)} = \dfrac{A_{ob}\mu(\zeta,\tau_0)S\zeta}{1-A_BC(\tau_0)} \\ I_{o,b} = \dfrac{A_B\mu(\zeta,\tau_0)S\zeta}{1-A_BC(\tau_0)}, \end{cases} \tag{5.90}$$

$$\begin{cases} I_{ob} = S\zeta\left\{\left[\varrho(\eta,\zeta,\varphi,\tau_0) + \dfrac{A\mu(\zeta,\tau_0)T_\sigma(\eta,\tau_0)}{1-AC(\tau_0)}\right] + \dfrac{A_{ob}\mu(\zeta,\tau_0)T(\eta,\tau_0)}{1-A_BC(\tau_0)}\right\} \\ I_B = S\zeta\left[\varrho(\eta,\zeta,\varphi,\tau_0) + \dfrac{A_B\mu(\zeta,\tau_0)\mu(\eta,\tau_0)}{1-A_BC(\tau_0)}\right], \end{cases} \tag{5.91}$$

where A_B and A_{ob} are, respectively, the spectral albedo of background and surface (above-water) object.

The transfer function for the brightness of extended background Π_B is estimated from (4.18) at $A = A_B$:

$$\Pi_B = \frac{A_B\mu(\zeta,\tau_0)}{[1-A_BC(\tau_0)]\varrho(\eta,\zeta,\varphi,\tau_0) + A_B\mu(\zeta,\tau_0)\mu(\eta,\tau_0)}. \tag{5.92}$$

The transfer function for the brightness of a small (point) object, considering (5.89–5.91), is [97]:

$$\Pi_{ob} = \frac{A_{ob}\mu(\zeta,\tau_0)}{[1-A_BC(\tau_0)]\varrho(\eta,\zeta,\varphi,\tau_0) + \mu(\zeta,\tau_0)[A_BT_\sigma(\eta,\tau_0) + A_{ob}T(\eta,\tau_0)]} \tag{5.93}$$

Bearing in mind an inequality similar to (4.19), with survey in nadir ($\eta = 1$), high sun location over the horizon ($\zeta \sim 1$) and high albedos of an object and background ($A_{ob} \gtrsim 0.5$, $A_B \gtrsim 0.5$), we obtain from (5.92–5.93) the following simple relationships:

$$\Pi_B \simeq \frac{1}{\mu(\eta,\tau_0)} = \frac{1}{T(\eta,\tau_0)+T_\sigma(\eta,\tau_0)} , \tag{5.94}$$

$$\Pi_{ob} \simeq \frac{1}{T(\eta,\tau_0)+\dfrac{A_B}{A_{ob}}T_\sigma(\eta,\tau_0)} . \tag{5.95}$$

With these assumptions made, the brightness contrast over an object and background is nearly zero ($K_h \simeq 0$). Therefore, it follows from (3.3 – 3.5) and (3.11) that the transfer functions for brightness contrasts $\mathcal{P}_{ob}$ and $\mathcal{P}_B$ are simply related to the respective transfer functions for brightness Π_{ob} and Π_B [97]:

$$\begin{aligned} \mathcal{P}_B &= e^{+\tau_0/\eta}\Pi_B^{-1} , \quad A_B \geqslant A_{ob} , \\ \mathcal{P}_{ob} &= e^{+\tau_0/\eta}\Pi_{ob}^{-1} , \quad A_{ob} \geqslant A_B . \end{aligned} \tag{5.96}$$

Bearing in mind (5.96), we find from (5.92) and (5.93):

$$\mathcal{P}_B = \frac{[1-A_B C(\tau_0)]\,\varrho(\eta,\zeta,\varphi,\tau_0)+A_B\mu(\zeta,\tau_0)\mu(\eta,\tau_0)}{A_B\mu(\zeta,\tau_0)}e^{+\tau_0/\eta}$$
$$A_B \geqslant A_{ob} , \tag{5.97}$$

$$\mathcal{P}_{ob} = \frac{[1-A_B C(\tau_0)]\,\varrho(\eta,\zeta,\varphi,\tau_0)+\mu(\zeta,\tau_0)[A_{ob}T(\eta,\tau_0)+A_B T_\sigma(\eta,\tau_0)]}{A_{ob}\mu(\zeta,\tau_0)}e^{+\tau_0/\eta}$$
$$A_{ob} \geqslant A_B . \tag{5.98}$$

Using approximate expressions (5.94 – 5.95) for $\eta \to 1$ and $\zeta \to 1$ for large albedos of an object and background, remembering (5.96), we obtain [97]:

$$\begin{aligned} \mathcal{P}_B &\simeq e^{+\tau_0/\eta}\mu(\eta,\tau_0) = 1+\frac{T_\sigma(\eta,\tau_0)}{T(\eta,\tau_0)} , \quad A_B \geqslant A_{ob} , \\ \mathcal{P}_{ob} &\simeq e^{+\tau_0/\eta}\left[T(\eta,\tau_0)+\frac{A_B}{A_{ob}}T_\sigma(\eta,\tau_0)\right] , \quad A_{ob} \geqslant A_B . \end{aligned} \tag{5.99}$$

In the presence of objects and backgrounds weakly reflecting the incident radiation (i.e., water surface), the brightness of atmospheric haze greatly exceeds the brightness of the background at the atmospheric base level. Here, from (4.21):

$$\mathcal{P}_B = \frac{\varrho(\eta,\zeta,\varphi,\tau_0)[1-A_B C(\tau_0)]}{A_B\mu(\zeta,\tau_0)T(\eta,\tau_0)} , \quad A_B \geqslant A_{ob} , \tag{5.100}$$

$$\mathcal{P}_{ob} = \frac{\varrho(\eta,\zeta,\varphi,\tau_0)[1-A_B C(\tau_0)]}{A_{ob}\mu(\zeta,\tau_0)T(\eta,\tau_0)} , \quad A_{ob} \geqslant A_B . \tag{5.101}$$

Assuming that the surface contrast K_0 of the brightnesses of a small object and its extended background is determined, according to (3.10), when $I_{o,ob}$ $I_{o,B}$, we find from (5.99):

$$\mathscr{P}_{ob} \simeq 1 + (1 - K_0) \frac{T_\sigma(\eta, \tau_0)}{T(\eta, \tau_0)}, \quad A_0 \geqslant A . \tag{5.102}$$

Formula (5.102) makes it possible to estimate minimum spectral contrasts K_0 of the brightnesses of surface objects and their extended artificial (natural) backgrounds. Also to solve the problem of remote detection of the optical inhomogeneity of the terrestrial surface from an air- and space-borne platform, utilizing the transforming effect of the atmosphere and threshold characteristics of the on-board optical instruments.

Relationships (5.89 – 5.91) and formulas (5.92 – 5.93) hold for orthotropic reflective characteristics of small objects and their extended backgrounds. However, in conditions of real space-based survey of the Earth's surface (especially, with snow, ice or sea backgrounds the reflective characteristics of surface (above-water) objects and backgrounds can be non-orthotropic. In this case in finding expressions for $I_{o,B}$, I_{ob}, Π_B, Π_{ob}, $\mathscr{P}_B$, and $\mathscr{P}_{ob}$, we use an approximation applied already for a non-orthotropic horizontally inhomogeneous surface (background) at the atmospheric base level. Assuming the photons reflected singly from a small object and from its background to be characterized by the respective non-orthotropic coefficients of reflection r_{ob} and r_B. In addition, assuming radiation reflected repeatedly in the object-background system to be determined only by the albedo of an extended background A_B, instead of (5.89 – 5.91) we find the approximate expressions:

$$I_{h,B} = I_{h,ob} = S\zeta \left[\varrho(\eta, \zeta, \varphi, \tau_0) + \frac{r_B(\eta, \zeta, \varphi)\,\mu(\zeta, \tau_0)\,T_\sigma(\eta, \tau_0)}{1 - A_B C(\tau_0)} \right], \tag{5.103}$$

$$\begin{cases} I_{o,ob} = \dfrac{r_B(\eta, \zeta, \varphi)\,\mu(\zeta, \tau_0) S\zeta}{1 - A_B C(\tau_0)} \\[2ex] I_{o,B} = \dfrac{r_B(\eta, \zeta, \varphi)\,\mu(\zeta, \tau_0) S\zeta}{1 - A_B C(\tau_0)}, \end{cases} \tag{5.104}$$

$$\begin{cases} I_{ob} = S\zeta \left[\varrho(\eta, \zeta, \varphi, \tau_0) + \dfrac{r_B(\eta, \zeta, \varphi)\,\mu(\zeta, \tau_0)\,T_\sigma(\eta, \tau_0)}{1 - A_B C(\tau_0)} \right. \\[2ex] \qquad \left. + \dfrac{r_{ob}(\eta, \zeta, \varphi)\,\mu(\zeta, \tau_0)\,T(\eta, \tau_0)}{1 - A_b C(\tau_0)} \right. \\[2ex] I_B = S\zeta \left[\varrho(\eta, \zeta, \varphi, \tau_0) + \dfrac{r_B(\eta, \zeta, \varphi)\,\mu(\zeta, \tau_0)}{1 - A_B C(\tau_0)} \right]. \end{cases} \tag{5.105}$$

For the transfer functions Π_B and Π_{ob} for non-orthotropic reflective properties of the object and its background the following approximate expressions hold:

$$\Pi_B = \frac{r_B(\eta, \zeta, \varphi)\,\mu(\zeta, \tau_0)}{[1 - A_B C(\tau_0)]\,\varrho(\eta, \zeta, \varphi, \tau_0) + r_B(\eta, \zeta, \varphi)\,\mu(\zeta, \tau_0)\,\mu(\eta, \tau_0)}, \tag{5.106}$$

$$\Pi_{ob} = \frac{r_{ob}(\eta,\zeta,\varphi)\,\mu(\zeta,\tau_0)}{[1-A_B C(\tau)]\,\varrho(\eta,\zeta,\varphi,\tau_0)+\mu(\zeta,\tau_0)[r_{ob}T(\eta,\tau_0)+r_B T_\sigma(\eta,\tau_0)]}\,. \tag{5.107}$$

A similar re-writing of formulas (5.96 – 5.102) is made by taking the coefficients of reflection r_B and r_{ob}, instead of albedos A_B and A_{ob} only for the terms characterized by single reflection of solar radiation, from the object and its background.

5.8 The Formation of Reflection Spectra Near the Interface Between Two Horizontally Homogeneous Media

To solve this problem approximately, we use the results of Soyuz-9 nadir spectrophotometry of the sand shore-sea interface ($\theta = 0°$) in July 1970 [23]. In processing the spectrograms, measurements were made of reflected radiation from terrestrial surface sites whose minimum area was determined by the width of the RSS-2 slit ($\sim 0.13°$) and by threshold vertical resolution ($\sim 3'$). Here spectra of $450 \times 450\,\mathrm{km}^2$ surface sites have been analyzed.

Figure 5.20 illustrates monochromatic brightnesses for $\lambda = 400 - 690$ nm of 19 surfaces sites (near the sea-sand shore interface) located successively through the vertical slit of the spectrograph and "cut out" by this slit from the total area ($10 \times 0.45\,\mathrm{km}^2$) of the spectrophotometered band. If, for each wavelength in the considered spectral interval $\Delta\lambda$, we estimate the distance L from the interface, from which the monochromatic brightnesses of the medium vary little horizontally, we discover how the characteristic lengths L of the optical inhomogeneity zone depend on wavelength λ near the sea-sand shore interface (Fig. 5.21). Figure 5.21 shows the characteristic lengths ($l = 4.55$ and $l_1 = 2.54$ km) of this zone for the integral (photographic) brightness of inhomogeneous surface near the sand shore-sea interface from the Soyuz-9 data. Note, that in this case the estimates of both integral (photographic) brightness and monochromatic brightnesses for sand exceed the estimates for water surface. This shows that the optically calm (placid) sea surface is more homogeneous than the desert sand surface. Note, that in estimating the linear sites of the inhomogeneities zone in Figs. 5.20 – 5.21, there is an additional effect of linear "blurring" of spectra (~ 0.8 km) due to the MS moving in the near-terrestrial orbit.

An estimation of characteristic lengths of the zone of optical inhomogeneities, for monochromatic and integral brightness contrasts of the spectrophotometered sites of water and sand surfaces, near their interface is also of interest [29].

Figure 5.22 shows how the value of monochromatic contrast $K_\lambda = (I_{\lambda\,max} - I_{\lambda\,min})/I_{max}$ depends on the distance L for the brightnesses of the sites of sand and water surfaces located through the vertical slit of the spectrograph

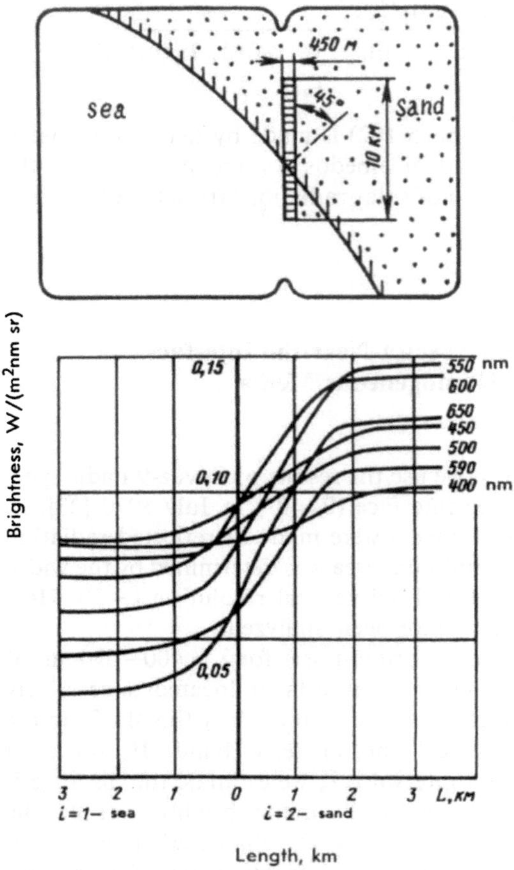

Fig. 5.20. Monochromatic brightnesses of sand and water surfaces with their interface spectrometered from space (Soyuz-9; $H_{orb} = 200$ km; $\theta_0 = 33.8°$; $\theta = 0°$)

at successively equal distances from their interface. Curves $K_\lambda(L)$ have been drawn from the data of Fig. 5.22, but for a distance counted vertically on both sides of the interface, i.e., at an angle of 45° to the slit of the spectrograph. Figure 5.23 shows these characteristic lengths of the optical inhomogeneity zone for the monochromatic contrast of brightnesses of sand and sea surfaces in the wavelength interval: $\lambda = 400-700$ nm. Also, this figure gives an estimate of the characteristic lengths $l = 2.56$ km for integral (photographic) contrast of brightnesses of the sand and sea surfaces from the Soyuz-9 data (Point Begdash). A comparison of Fig. 5.23 and Figs. 5.19–5.22 shows that characteristic lengths of the zone of optical inhomogeneity for brightness contrasts are almost equal to those of these brightnesses. The lengths of this zone, as for brightnesses, are somewhat overestimated, since the effect of the respective linear shift during the orbital moving of the MS has not been considered.

A similar problem has been considered in [298] for simulating the procedure of the radiation transfer near the interface of two half-planes with dif-

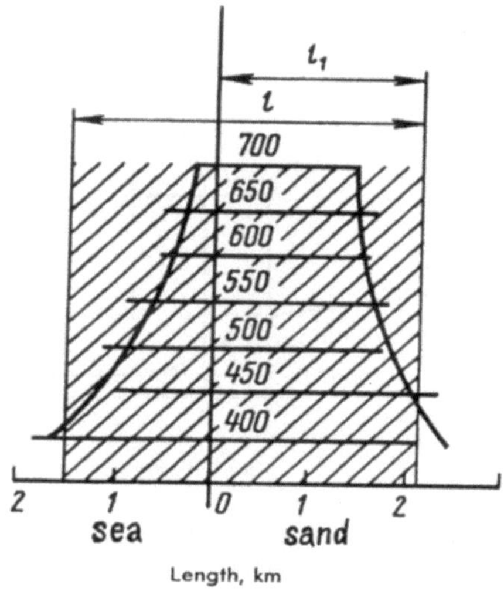

Fig. 5.21. Linear sizes of the zone of optical inhomogeneity for the monochromatic and integral (photographic) brightnesses near the sand-water interface (Soyuz-9; $H_{orb} = 200$ km; $\theta_0 = 33.8°$; $\theta = 0°$)

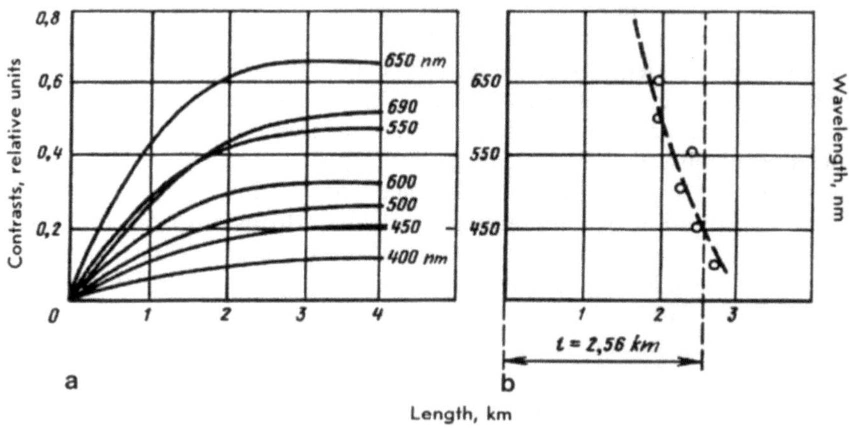

Fig. 5.22 a, b. Optical characteristics of the environment near the sand-water interface (Soyuz-9; $H_{orb} = 200$ km; $\theta_0 = 33.8°$; $\theta = 0°$). **a** The dependence of the monochromatic brightness contrast K_λ on distance L; **b** characteristic linear sizes of the zone of optical inhomogeneity for the monochromatic contrast of the brightnesses near the sand-water interface

ferent surface albedos (for instance, the interface near the seashore). The calculations of the field of diffuse radiation are based on a two-flux approximation and on an exact real phase function for the first-order scattering. This solution permits one to calculate the brightness on both sides of the interface, to assess the transfer function of the atmosphere and to analyze the transition to an asymptotic regime upon moving away from the land-sea interface. Comparison of results of approximate calculations with almost accurate Monte-

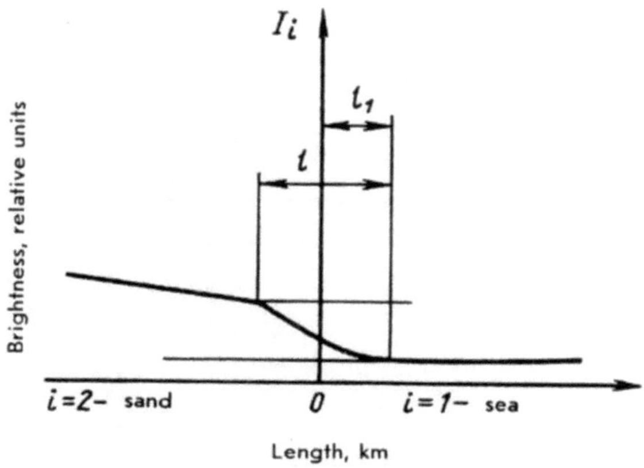

Fig. 5.23. The dependence of the characteristic linear sizes of the zone of optical inhomogeneity near the sand-sea interface

Carlo simulations revealed a good agreement and showed that the applied technique is useful for any surface albedo, if the aerosol optical thickness is less than 1. The adequacy of the technique has been confirmed by comparison with Landsat-1 data for the African Atlantic ocean coastline (80 m spatial resolution), including simultaneous surface measurements of the transparency and aerosol optical thickness of the atmosphere. Analysis of the results of calculations has shown that, at a distance from the interface, the dependence of the contribution of illumination (horizontal diffusion) from the adjacent higher-albedo site (land) to the formation of the low-albedo system (ocean) is insensitive to the features of the vertical profile of aerosol concentration. However, it varies with variations in the average height of the aerosol layer. This is not dependent on the aerosol size distribution, its optical thickness and average surface albedo.

To characterize the effect of the interface on the field of brightness of the system a parameter x_L has been introduced in [350]. This determines the distance to the interface at which, the horizontal inhomogeneity – induced contribution to the brightness of the system, halves. Calculations of this parameter have shown that it is almost independent of the average albedos of half-planes. With increasing optical thickness of the atmosphere the x_L decreases, reaching minimum values at $\tau_0 \sim 1$, and then increases. In the case of the Elterman-1968 aerosol model minimum values of x_L constitute about 0.7 km, and maximum values reach about 1.2 km.

In the approximation of the first-order scattering in [361] estimates have been obtained of the effect of horizontal-inhomogeneity of the surface on its brightness when it is observed from space. A solution has been considered for the initial 3-D transfer equation for nadir observations and the chess-like structure of the surface. Firm calculations, modeling the Landsat data, have been made for horizontal visibilities 3, 10 and 40 km and wavelengths 0.55,

0.65, 0.75, and 0.95 μm. The site under study is in the center of the grid consisting of 19×19×160-m-side squares. The following conditions were modeled: a field of wheat surrounded by a dark background, by other fields of wheat or by grass cover. The calculations have shown that brightness variations in the transition from the second to the first case can reach 25% (towards increasing brightness, since the grass albedo exceeds that of wheat). The effect of the background on the site under observation is maximum when the albedo of this site is low. So, for instance, the brightness of soil surrounded by grass cover reaches 85%. Study of the effect of atmospheric haze with loam surrounded by different backgrounds (the same loam, grass and black surface) has led to the conclusion that the brightness grows greatly with decreasing visibility (the haze intensifies).

A technique has been proposed [245] for atmospheric correction of remote sensing data of the optical characteristics of an inhomogeneous surface in the wavelength interval 0.35–4 μm. It is based on the consideration of three components responsible for the formation of the brightness of the system "object-background-atmosphere": (1) direct solar radiation reflected by the observed site of an inhomogeneous surface (and transmitted by the atmosphere); (2) radiation scattered back by the atmosphere; (3) "illumination" due to adjacent sites of an inhomogeneous surface (the contribution of this component depends on the size of the site under study). Based on the approximate analytical technique, estimates have been obtained of the enumerated components for different atmospheric models for atmospheric visibility 5 and 25 km. Comparison with results of exact calculations has shown that the atmospheric correction error does not exceed 5%.

Calculations have been made [387] with different optical models of an aerosol atmosphere in the approximation of single scattering and with Monte-Carlo simulations to assess the limits of applicability of the approximation of horizontal-homogeneity of the surface. The reflection by the surface is assumed to be isotropic and the atmosphere horizontally-homogeneous. The size distribution of aerosol is assumed to be after Yungs and the substantial index of refraction to be 1.50. Calculations were made for $\lambda = 450$ and 850 nm and nadir direction $\theta = 0°$. Two versions of atmospheric optical thickness τ_{aer} were considered: 0.2801 and 0.9305 (450 nm); 0.1550 and 0.5151 (850 nm).

The Rayleigh optical thickness τ_{mol} constitutes 0.2157 and 0.0163, respectively. The transfer function of the atmosphere (TFA) is defined as a quantity characterizing the contribution of a site of background, at a distance L, from the given to the space-derived spectral brightness coefficient (SBC). Analysis of calculation results has shown that the single-scattering approximation is acceptable at $\lambda = 850$ nm. The SBC is largely determined by aerosol optical thickness τ_{aer} which, along with Rayleigh optical thickness τ_{mol}, determines the weights characterizing the contribution of the given site and background to the signal to be measured. With extreme aerosol conditions excluded, the atmospheric transfer function can reliably be estimated with an average aerosol model. Typical horizontal scales of the effect of the contribution of the background constitute 1 km for aerosol scattering and 10 km for molecular scattering. With homogeneous sites of intermediate sizes of several hundred meters (when the asymptotic cases of a small or large object are unacceptable), the horizontal gradients of SBC turn out to be so small that, they permit the use of simple formulas to estimate the principal contribution of the TFA.

Since the present space-borne scanning systems enable one to reach a high spatial resolution (10 – 100 m), but their zone of scanning is confined to angles close to nadir (8° in the case of Landsat), the repeatability of satellite data is very rare (18 days for Landsat). The frequency of observations can be increased by broadening the zone of scanning to about 50°, but in this case the distorting effect of the atmosphere grows considerably. In [243] the problem of the retrieval of the brightness of a 5×5-km site with orthotropic reflection (A = 0.05) surrounded by a brighter surface with albedo A = 0.30, depending on solar irradiance ($\theta_0 = 0-75°$), wavelength, and the position of the surveyed site relative to the sub-satellite point has been considered. A ratio of the observed brightness of the surface-atmosphere system to the brightness of the given site of the surface (q) is reduced by atmospheric thickness (ignoring illumination and multiple scattering). The variability has been assessed using three atmospheric models: without aerosol; with moderate and with very large aerosol content. Calculations made for wavelengths corresponding to the channels of the Landsat scanning telephotometer (λ = 555, 655, 752, and 867 nm), revealed a strong dependence of q on wavelength λ, solar zenith angle and position of the site with respect to the sub-satellite point. As a rule, the largest contribution to the formation of brightness to be measured is made by atmospheric haze, especially at large viewing angles of the size relative to the sub-satellite point. The effect of illumination from the brighter environment is traced, at the distance at least L = 2 km from the edge of the site, and at viewing angles $\theta = 42-43°$ it propagates over the entire site. In this case, where there is a very turbid aerosol atmosphere, $q > 320$ at $\lambda = 555$ nm and $\theta_0 = 0.75°$. The contribution of illumination is strongly reduced with growing wavelength. However, the horizontal size of the area where the effect of illumination is strong and almost independent of wavelength.

The effect of variations in the aerosol particles size distribution, connected with meteorological conditions on the visibility of surface (above-water) objects as a function of spectral range $\Delta\lambda$, has been considered in [243]. Since the size distribution and optical thickness of aerosol are, as a rule, sufficiently homogeneous in the lower 5-km atmospheric layer, a deterioration of the quality of aerospace images can be determined not only by turbidity but also by inhomogeneous length of the run of photons in the process of multiple scattering.

The effect of aerosol scattering has been considered not only from the viewpoint of persistent (independent of spatial frequency) background and signal attenuation at low spatial frequencies, but also in connection with the poorer quality of aero-space images due to the effect of other frequency intervals. This problem will be considered in detail in the next chapter, associated with the problem of atmospheric radiative correction of space-derived optical information.

Chapter 6 Radiative Correction of the Space-Derived Images of the Earth Surface

6.1 Setting the Problem of Atmospheric Radiative Correction

Techniques for the transformation of air- and space-derived images based on the mathematical apparatus of statistical solutions are now widely used. On the one hand, these transformations are aimed at the removal of distortions from space-derived information about the state of natural objects, and on the other hand, they improve the quality of the multispectral survey from both aircraft and satellites. Very often, however, transformations of images brightnesses are based on pure mathematical formalism, which does not consider the principal laws of formation of the outgoing radiation field. Transformations based on preliminary modeling of the processes of interaction of electromagnetic emission with the atmosphere and Earth's surface are a qualitatively different way of removing atmospheric radiative distortions [347]. Here, the principal energy balance relationships and the related functions of conversion of radiation into the data recorded remotely by sensors are included. This approach to the transformation of spectral brightnesses obtained from multispectral air- and space-based survey will be used next.

Due to the low accuracy of retrieval of space-derived optical parameters of the atmosphere and surface needed for radiative correction, it is necessary to use similar information obtained in other ways [342]. For instance, microphysical parameters of atmospheric aerosol can be obtained with actinometric measurements from networks, aerosondes, meteorological rockets, and the atmospheric optical thickness from laser sounding data, etc.

Figure 6.1 exemplifies the spectral change of nadir brightnesses of three types of natural objects at the Earth surface level and at the satellite's orbiting level. The values in Fig. 6.1 have been obtained from data on the spectral albedos of these objects considering the calculation scheme of radiation transfer in the surface-atmosphere system previously described. The curves in Fig. 6.1 have been obtained by averaging spectral histograms of the intensity of the nadir-outgoing radiation. The same figure gives the limits of the atmospheric effect variability for vegetation. The variability means the differences in the intensity of radiation outgoing from the atmospheric top for Rayleigh atmosphere without aerosol and for real atmosphere with continental aerosol. The variability of the atmospheric effect for spectral intervals with $\lambda < 0.8$ μm is caused by the combined effect of aerosol and molecular scatter-

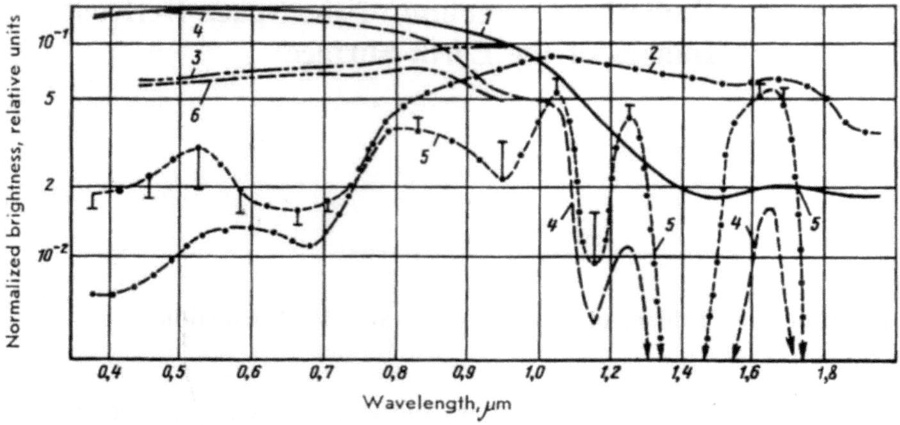

Fig. 6.1. Spectral brightnesses of three types of natural formations: *1* snow cover; *2* vegetation cover (grass); *3* bare soil (light-colored sand); *4, 5, 6* spectral images of these objects recorded in nadir from satellite. The solar zenith angle cosine is 0.6; the limits are shown of the variability of the atmospheric effect for vegetation cover

ing; in relative transparency windows by the extent of aerosol absorption and scattering of radiation; in the absorption bands for gaseous substances by the concentration of optically active radiation absorbers.

As follows from Fig. 6.1, in the wavelength interval 0.5 – 0.6 μm the atmospheric effect varies widely: variations of the brightnesses of vegetation due to the atmospheric effect can reach 200% and more. Large variations of the atmospheric effect occur in the region of absorption bands for atmospheric gases. However, in the spectral interval 0.7 – 0.8 μm, the atmospheric effect variation is weak. Here, apparently, the statistical mean parameters of the atmosphere are enough to introduce radiative correction to air-and space-derived images.

The technical means of processing the data of remote optical sensing are now being developed towards presentation of information as images, with subsequent recognition of images and an assessment of the state of natural objects. In image processing it is necessary to consider the distorting effect of the atmosphere, i.e., the construction of the optical transfer function relating the recorded images of natural objects to their properties at the surface level. Radiative correction is a conversion of video images, which eliminates radiative distortions due to the interaction of radiation with the atmosphere and surface. An algorithm of 2-D filtration of atmospheric interference in reflected radiation considers the principal factors of the formation of the outgoing optical radiation field: the angular conditions of survey, non-orthotropic reflection, horizontal-inhomogeneity of natural formations, etc. The filtering is based on the solution of multi-dimensional and one-dimensional boundary problems of radiation transfer in the Earth's surface – atmosphere system; see (5.1) through (5.3).

The algorithm of filtration (see Chapter 5) can be based, however, on the use of solutions for simpler 1-D boundary problems (5.24–5.26) and (5.27–5.29). The assumption that, photons singly reflected from the surface, are considered for real albedo and those multiply reflected for average albedo, is used as zero approximation in assessing the brightness of natural formations.

Practically, the radiative correction of air- and space-derived images is carried out in two stages: for the mean constituent of the brightness of the processed image I; and for linear deviations of brightness I′ from its average. The physics of the first stage of correction consists in calculating the atmospheric haze brightness for an average surface albedo $\bar{A}$, i.e., in the assessment of the transformation of radiation multiply scattered in the atmosphere and re-reflected by the surface [84]. That of the second stage in the assessment of radiative distortions in the atmosphere due to the spatial distribution of horizontal-inhomogeneities of the atmosphere and surface. An algorithm of radiative correction at the second stage implies a linear relationship between fluctuations of brightness and deviations of albedo from an average.

For different states of the atmosphere and different natural objects the solution for the boundary problem (3.14–3.16) for the mean constituent of brightness is written as:

$$\bar{I}_{ij} = \bar{I}_{ij}(\eta^+, \varphi, \zeta, \bar{A}_j) \tag{6.1}$$

where (η^+, φ) are, respectively, cosine of zenith viewing angle and azimuth; ζ – the solar zenith angle; the i index characterizes the model of the atmospheric vertical stratification used in calculations; $\bar{A}_j$ – the average albedo of the image or of its separately processed fragment; j – the type of anisotropy of surface reflection characterizing the difference between a real surface and an idealized Lambertian surface.

To use the direct problem solution in processing the images, the values $\bar{I}_{ij} = \bar{I}_{ij}(\eta^+, \varphi, \zeta, \bar{A}_j)$ for zenith viewing angles below 60° and individual magnitudes of average albedo (0–1) have been approximated with the help of Chebyshev's second-order orthogonal polynomials $\hat{T}_k$ [83]:

$$\begin{aligned}\bar{I}_{ij}(\eta^+, \varphi, \zeta, \bar{A}_j) \approx\ & a_1 + a_2\hat{T}_1(\bar{A}_j) + a_3\hat{T}_1(\zeta) + a_4 T_1(\eta^+) + a_5\hat{T}_1(\varphi)\\ & + a_6\hat{T}_2(\bar{A}_j) + a_7\hat{T}_1(\zeta)\hat{T}_1(\bar{A}_j) + a_8\hat{T}_1(\bar{A}_j)\hat{T}_1(\eta^+)\\ & + a_9\hat{T}_1(\bar{A}_j)\hat{T}_1(\varphi) + a_{10}\hat{T}_2(\zeta) + a_{11}\hat{T}_1(\zeta)\hat{T}_1(\eta^+)\\ & + a_{12}\hat{T}_1(\zeta)\hat{T}_1(\varphi) + a_{13}\hat{T}_2(\eta^+) + a_{14}\hat{T}_1(\eta^+)\hat{T}_1(\varphi)\\ & + a_{15}\hat{T}_2(\varphi)\ ,\end{aligned} \tag{6.2}$$

where $a_k (k = 1, \ldots, 15)$ is the coefficient of approximation:

$$\hat{T}_0 = 1\ ;\quad \hat{T}_1(z) = 2z - 1\ ;\quad \hat{T}_2(z) = 2(2z-1)^2 - 1\ .$$

The solution of the 3-D boundary problem (5.1 – 5.3) results in establishing a relationship between surface albedo variations $\tilde{A}_j(x,y)$ and a varying constituent of brightness $\hat{I}_{ij}(x,y,\eta^+,\varphi,\zeta)$:

$$\hat{I}_{ij}(x,y,\eta^+,\varphi,\zeta,\bar{A}_j) = \int_{-\infty}^{\infty}\int Q_i(x-\zeta',y-\eta,\eta^+,\varphi,\zeta)\tilde{A}_j(\zeta',\eta)d\zeta' d\eta \ , \quad (6.3)$$

where (x,y) are the Cartesian coordinates; Q_i the scattering function equal to the Fourier transformation of the SFCh of the linear system of the optical radiation transfer in the atmosphere.

The Q_i function is easily approximated by a family of Gauss curves:

$$Q_i(x,y,\eta^+,\varphi,\zeta) = f_i(\eta^+,\varphi,\zeta)\delta(x-\tilde{x},y-\tilde{y}) + S_i(\eta^+,\varphi,\zeta)l e^{-l(x^2+y^2)} \ , \quad (6.4)$$

where f_i is the function which characterizes the direct reduction of reflected radiation in the atmosphere (it depends on τ_i, x_i and on surface irradiance E); δ – the delta-function; (x,y) the current coordinates of individual elements of the image; S_i the function which determines the contribution of multiple scattering; l the numerical factor.

In processing the images, the angular coordinates (η^+,φ,ζ) are known for each pixel, from the conditions of survey. The factor of reflection anisotropy for the given objects is connected with the angular conditions of survey and must be known a-priori, for instance, based on preliminary division of the territories under study or referencing of remote measurements against this factor.

Thus, the first stage of radiative correction to images is reduced to a pixel to pixel retrieval of $\bar{A}_j$ using (6.4) for the known sets of approximation coefficients a_k, depending on the survey conditions and characterizing the atmospheric transfer function for the mean constituent of the image brightness. At the second stage of correction using (6.3) and (6.4) the spatial distribution of the varying constituent of albedo $\tilde{A}_j(x,y)$, is retrieved by solving the Fredholm 2-order integral equation:

$$\tilde{A}_j(x,y) = E_i(\eta^+,\varphi,\zeta)[\tilde{I}_{ij}(x,y) - \lambda_i \iint_{\tilde{D}} \tilde{I}_{ij}(\zeta',\eta') \times \exp\{-l[(x-\zeta')^2+(y-\eta')^2]\}d\zeta' d\eta'] \ , \quad (6.5)$$

where $E_i = 1/f_i$, $\lambda_i = S_i E_i$. The size of the area $\tilde{D}$ is determined by an efficient area of mutual effect of natural objects: the size of this area depends on the quantity of horizontal diffusion of photons and on the angular resolution of instruments, decreasing as the angle of the field of view grows. Both the stages of atmospheric correction of images can be realized on a special complex of images on a real time scale. The procedure of the 2-D filtration of images consists in: (1) selecting the mean coefficient of brightness and linear fluctuations of brightness; (2) solving algebraic 2-order equations; (3) accomplishing the fast Fourier transformation; (4) calculating the convolution integral.

Digitizing the multispectral space-derived information [211] makes it possible to realize the four stages of processing enumerated special processors and analog arrangements [155].

6.2 The Status of the Problem of Radiative Correction of Multispectral Space-Derived Images

The data of multispectral space-based survey of the Earth's surface are distorted due to scattering and absorption of electromagnetic radiation by molecules of atmospheric gases and by atmospheric aerosol particles [294]. The 2-D filtration of atmospheric distortions implies the construction of a correcting filter based on the solution of the boundary problems of the radiation transfer theory, mathematically describing the process of radiation transformation in the surface-atmosphere system. In addition to the need to solve the boundary problems of radiation transfer (5.1 – 5.3), the solution of the problem of radiative correction requires additional measurements described next.

The setting and solving of these boundary problems have been described in section 5.1. The actual physical and mathematical content of the atmospheric transfer function is more complicated. A unique reduction of space brightnesses to the surface level cannot be done without knowledge of the optical thickness of the atmosphere, its phase function, concrete conditions of objects' irradiation, viewing angles, non-orthotropic reflection of radiation by different objects, horizontal-inhomogeneity of some fragments of a space-derived image, etc. The atmospheric transfer function turns out, thereby, to be a function of transformation of the data of space-based survey into the parameters of natural objects, constant relative to the illumination conditions, viewing, etc. In this case, the problem of the determination of the atmospheric transfer function, must include numerous factors affecting the processes of the formation of the outgoing radiation field. Note, that a considerable part of these factors can be considered efficiently in analytical calculation schemes for radiation transfer, despite substantial mathematical difficulties in solving the respective boundary problems [302, 303]. Nevertheless, although statistical methods of solving similar problems require much computer time, numerical methods seem the most suitable practically.

Thus, the problem of radiative correction of the data of multispectral space-based survey has two principal features. First, determined by mathematical difficulty of an adequate description of the interaction of radiation with the surface-atmosphere system. Secondly by the necessity of real-time radiative correction of large volumes of information. In the first case, upon overcoming the related mathematical difficulties in solving the boundary problems of the radiation transfer theory, the angular and spectral distributions of the outgoing radiation flux intensity are calculated as a function of viewing angle, sun elevation, albedo or coefficient of the Earth's surface brightness, factor of reflection anisotropy, horizontal inhomogeneity and other characteristics of the surface-atmosphere system. In the second case, the solution is reduced to an approximation of these values by a system of orthogonal polynomials, aimed at presenting the atmospheric transfer function as approximation coefficients.

One more important aspect of the problem of radiative correction is the necessity to use additional a priori information for its solution [304]. If the problems connected with the solution of inverse problems of atmospheric optics are not considered, two questions arise: what must be measured additionally to introduce atmospheric correction and how must these additional measurements be realized? One of the principal optical characteristics of the atmosphere is its optical thickness which depends on atmospheric turbidity, spectral and vertical distribution of aerosol particles, composition and humidity, and other factors. The optical thickness can be measured from the Earth surface from reduced direct solar radiation; however, satellite data are most promising for atmospheric correction of space-derived images. Studies are establishing correlations between atmospheric optical thickness and its phase function, to solve the problem of radiative correction of satellite-derived information [114, 200, 217].

Thus, at present the problem of atmospheric correction of space-derived information can be solved with the use of complex optical atmospheric models in practice. However, in the process of processing the data of space-based multispectral survey the parameters of these models at the moment of survey must be specified. The setting of the problem of radiative correction of space-derive multi-spectral images, techniques for its practical solution and connection with the problem of recognition of the images of natural objects have been described in [84] using a mathematical model from [85]. The statistical processing of initial and corrected images has been made [80] for both photographic and scanner measuring systems. The status of the problem of radiative correction has been described in [77], including the solution of direct and inverse boundary problems in the radiation transfer theory, obtaining and systematization of the initial data on scattering and absorption of radiation by the gaseous and aerosol atmospheric substances, the results of processing multispectral space-derived images, etc.

To correct aerospace video information, it is necessary to know the optical parameters of the atmosphere. However, direct measurements of these parameters are usually impossible during the recording of images. In this case, atmospheric distortions of video data can be corrected only partially. Besides, use of climatological information is hindered by large variations in the state of aerosol particles. Therefore, methods of atmospheric correction based on obtaining atmospheric parameters directly, from data of the image itself, are interesting [111]. The choice of the atmospheric correction technique depends on the problem which is solved using data of remote sensing and recording instruments. For instance, in contrast to low-resolution images (where the size of an element is more than 1 km), for which a supposition of surface homogeneity is acceptable, the high-resolution images (where the size of an element is less than 100 m) require consideration of surface inhomogeneity [297]. Atmospheric correction of video information can be based either on calculations of radiation transfer in the surface-atmosphere system, or on an estimation of radiation intensity transformed by the atmosphere, directly from data of multichannel video images. An application of accurate techniques for calculations of radiation transfer is hindered by a limited set of available input parameters of the atmosphere and surface. However, a precise modeling of radiation fields makes it possible to check and specify approximate calculation schemes and to elucidate basic parameters of the surface-atmosphere system, which have a prevailing effect on the process of radiation transfer. In approximate approaches to calculations of radiation transfer, some simplified assumptions are made about the properties of the atmosphere and surface and a small set of basic parameters is used [323]. Approximate methods for calculating the radiation transfer make it possible to specify easily the input parameters to satisfy some criteria of the quality of correction of aerospace information connected with the special character of the problem to be solved [378].

An approximate analytical expression for radiance can be obtained in different ways. The first approach is based on obtaining an approximate analytical expression for the function of sources in the right-hand part of the integrated differential equation of transfer with the lower boundary condition for the homogeneous Lambertian reflecting surface. The second is based on consideration of the processes of interaction of radiation with the atmosphere and reflecting surface and on obtaining expressions for radiances without solving the transfer equation. In the third case, an accurate solution of the transfer equation obtained for the case of a homogeneous Lambertian surface, is used to derive an approximate expression for radiation intensity for an inhomogeneous non-Lambertian reflecting surface [239].

Publications [391, 409] recommend, for calculations of the atmospheric haze intensity (without reflecting surface), approximation of the atmospheric phase function by two delta-functions, since the aerosol phase function has a strongly stretched peak in the straight direction of scattering. For the background haze intensity (radiation, which, at least, once interacted with the surface), an assumption is made that it does not depend on the direction of propagation within the lower and upper hemispheres (for the downward and upward radiation) and is expressed through the fluxes (downward and upward). This presentation makes it possible to move from the initial integrodifferential transfer equation to a system of two differential equations for the downward and upward fluxes (separately, for the atmospheric haze and for the background haze). Using approximate analytical expressions for the fluxes, we obtain an approximate expression for the source function. The atmosphere is considered to be homogeneous, plane-parallel, and the surface-homogeneous Lambertian.

Examples of using this approach to correct remote sensing data have been given in [214]. The spectral optical thickness of the atmosphere is estimated there by comparing results of theoretical calculations of radiance at the atmospheric upper boundaries with intensities recorded over pure lakes. The reflective characteristics of pure lakes are assumed to be known.

In contrast to this approach [391, 409], it has been assumed in [296] that both the atmospheric haze intensity and the background haze intensity do not depend on the direction of radiation propagation (within the lower and upper hemispheres) and are expressed through the downward and upward fluxes.

A comparison has been made between results of calculations from obtained approximate expressions with results of Monte-Carlo simulations. It appeared that the difference constituted less than 13% for the optical thickness $\tau_0 = 0.4 + \tau_R$ for wavelengths between 0.4 – 0.8 μm and with different values of single scattering albedo, sun elevation and surface albedo. An analytical solution, with the help of the least squares method was completed for a better agreement with results of Monte-Carlo simulations. As a result, maximum difference was reduced to 6%.

Use of an approximate solution for the transfer equation obtained by V. V. Sobolev has been considered in [272, 413]. This solution is based on averaging the radiance over angular coordinates. A two-term phase function is used, obtained by expansion in Legendre-polynomials, and radiance is represented respectively as a sum of two first terms of the Fourier series for azumuth φ. The source function of this method is expressed not only through diffused vertical fluxes, but also through horizontal diffused flux and average diffused radiance. The analytical approximate solution obtained with this approach, is used for atmospheric correction of remote sensing data of the Adriatic Sea from different altitudes, with the help of the airborne module multichannel scanner MMS. To take into account the height of survey, the atmosphere is divided into two homogeneous layers, one lying below the height of survey, the other above it. Since variations of the phase function due to variations of particles size distributions are due to more than wavelength variations, the aerosol phase function is assumed to be spectrally independent in the visible and near IR spectral regions. Since it is assumed that the aerosol size distribution is independent of the height, the total phase function for each layer is a weighted sum of Rayleigh and aerosol phase functions with the weights determined by Rayleigh and aerosol optical thicknesses of these layers. In calculations, the absorption by ozone and by water vapour is considered. The effect of radiation reflected from the sea surface outside the instant field of view and then diffusively transported in the viewing direction, is ignored in view of its small value. Direct and diffused radiation reflected by sea surface in the viewing direction is modeled using non-orthotropic reflection, wind speed and the angle between the external normal to elementary areas modeling the rough sea surface and the vertical. During the experiment, ground measurements were made of the atmospheric spectral optical thickness $\tau(\lambda, \infty)$. It is assumed that the intensity of the volume emission of water in the near IR spectrum is zero. Then, from radiance recorded by the scanner, with the use of successive approximation, the optical thickness $\tau(\lambda, h)$ of the atmospheric layer between the surface and the height of survey in the near IR was estimated. Then, the parameter $M_A(h) = \tau(\lambda, h)/\tau(\lambda, \infty)$ is estimated and $\tau(\lambda, h)$ is determined for other wavelengths. With the help of estimates obtained of the spectral optical thickness $\tau(\lambda, h)$, atmospheric correction of aircraft data was made. It was shown that values of intensities of the volume emission of water, obtained from data of the two-height survey (1.5 and 2.9 km) over the same region of the sea, are very close.

The second approach to obtaining approximate expressions was considered by Otterman et al. [350]. They propose an approximate formula to calculate the coefficient of spectral brightness of the surface-atmosphere system a_s, for the case of small atmospheric optical thickness. This formula has been obtained in the single scattering approximation. Radiation intensity recorded in nadir by the satellite system is represented as a sum of three items:

$$L_n = L_{nr} + L_{nd} + L_{na} \ .$$

The first item L_{nr} is the radiance reflected from the viewed surface object with albedo r and attenuated by the atmosphere (exp $(-Q)$), where $Q = R+M+W$ is the vertical optical thickness of the atmosphere, R is Rayleigh and M, the aerosol optical thickness determined by absorption. The object is considered to be illuminated by singly scattered solar radiation directed to the object, radiation reflected from the background surrounding the object, and subsequently scattered back by the atmosphere towards the surface. For the intensity L_{nd} of direct solar radiation singly scattered by the atmosphere in the viewing direction, either an exact expression or approximate one is used.

The third item L_{na} is the radiance reflected from a homogeneous background with albedo a and singly scattered upward over the viewed object (an expression can be obtained for the inhomogeneous surface using the model considered). In deriving an expression for L_{na}, an attenuation of radiation reflected from the background before the moment of scattering and after it, is not considered. It is considered in the following way. An optical path for radiation reflected from the background to the point of scattering and farther to the sensor, is longer than that during the vertical movement. However, in the case of reflection from the background close to an object, these optical paths can be considered to be equal, and the attenuation can be calculated using a factor exp $(-\theta)$.

Using this model, expressions for L_{na} have been obtained for the inhomogeneous atmosphere, with vertical distribution of scatterers. Here, the atmospheric scattering infinitely thin layers are considered. First, for the vertically homogeneous atmosphere, the reduction of radiation is ignored, which makes it possible to summarize the scatterings of different layers, and consider the vertical distribution of the scatterers. When the vertical distribution of the absorbers is different from that of the scatterers, the reduction is calculated in the following way. If the absorbers are above the scattering particles (for instance, ozone), then the factor of attenuation due to absorption is exp $(-W)$. When the absorbers (water vapor) are in the surface layer of the atmosphere, below the scattering particles, then the exponential factor of attenuation depends on the distance between the object and the given element of the background, which must be considered when calculating L_{na}.

An approximate expression has been given in [350] for the spectral brightness coefficient a_s, in which only linear terms relative to optical thickness have been left:

$$a_s = r\{1-[(B+W)(1/\mu_0+1)-2rB]\}+(a-r)(F+2rB)+\frac{g(\mu_0)}{2\mu_0}B \ ,$$

where $\mu_0 = \cos\theta_0$; θ_0 is the solar zenith angle; F and B are optical thicknesses determined by direct and backscattering; $g(\mu_0)$ is the backscattering anisotropy from solar zenith angle; $g(\mu_0)B/2\pi Q$ is a combined phase function for aerosols and Rayleigh particles for a scattering angle ($\pi - \arccos\mu_0$). The term $(a-r)F$ characterizes the reflection from the background and the subsequent single scattering in zenith; the term $2(a-r)B$ corresponds to the object's irradiance due to the reflection from the background and subsequent scattering by the atmosphere onto the object. Calculations have shown that the first term is an order of magnitude larger than the second one. It has been shown in [350] that the effect of the background can be local. For instance, where large aerosol particles are concentrated at high altitudes (1 or 2 km), the principal effect of the background is confined to several hundred meters from an object. On the other hand, the background can affect at large distances, too, with Rayleigh scattering or small aerosol particles located at high altitudes.

For the quantitative assessment of the transforming effect of the atmosphere, in the interpretation of remote sounding data, calculations have been in [349] of the quantity $a_s - a_o$ for the four channels of the Landsat MSS (multispectral scanning system). This difference is presented as a sum of the contributions of the background atmosphere $(a_s - a_o)_m$ and a variable component $(a_s - a_o)_v$. The background atmosphere characterizes the effect of Rayleigh scattering as well as molecular absorption by ozone, oxygen and water vapor (with the total content of the latter not more than 5 mm of precipitated water). The variable component takes into account the contributions of water vapor (more than 5 mm) and aerosol. The representative aerosol optical thickness of the atmosphere at $\lambda = 0.55$ µm is taken to be 0.1. For the turbid atmosphere it is 0.2. The aerosol is assumed to be purely scattering. The total content of water vapor is taken to be 20 mm (representative) and 35 mm (high). Calculations have shown that the difference $(a_s - a_o)_v$ for channels 5 (0.6 – 0.7 µm) and 6 (0.7 – 0.8 µm) is determined only by aerosol scattering, whereas for channel 7 (0.8 – 1.1 µm) the contributions of both aerosol scattering and molecular absorption can be substantial. The dependence of $(a_s - a_o)$ on a_o is very important. In this connection, the following cases should be mentioned in the estimation of the atmospheric transfer function:

(1) a small value of a_o, ($a_o < 0.1$) i.e., water basins and thick vegetation, when the effect of backscattering prevails; (2) a moderate value, $0.1 < a_o < 0.35$, with the prevailing contribution of attenuation by atmospheric thickness; (3) a high value, $a_o > 0.35$ (this case, with the snow cover excluded, is always characteristic of channels 6 and 7). Various values of the difference $(a_s - a_o)$ are given in [349]. The parameters B and W can be estimated either from ground spectral measurements of direct solar and scattered radiation or from satellite-measured radiances over dark water objects larger than 1 km, to reduce the effect of the background.

Let us consider the third approach to obtain approximate expressions for radiances. The effects of the atmosphere and surface inhomogeneity on the data of remote sounding are closely connected. Therefore, it is interesting to assess their relative contribution to the formation of the upward radiation at the atmospheric top [386]. For a homogeneous Lambertian surface, an expression for the coefficients of reflection for the surface-atmosphere system is:

$$\varrho^*(\mu_0, \mu, \varphi) = \varrho_a(\mu_0, \mu, \varphi) + \frac{\varrho}{1 - \varrho S} \times [\exp(-\tau/\mu_0) + E(\mu_0)][\exp(-\tau/\mu) + E(\mu)] \ ,$$

where $\theta_0 = \arccos \mu_0$; $\theta = \arccos \mu$ are solar zenith angle and viewing angle; φ azimuth, ϱ_a the reflection coefficient of the atmosphere, ϱ surface albedo, $(E(\mu_0) + \exp(-\tau/\mu_0)]$ the ratio of the atmospheric base irradiance to the irradiance of its top from above, S the atmospheric spherical albedo. Based on this formula, an approximate expression is derived for the reflection coefficient ϱ^* which takes into account both surface inhomogeneity and non-Lambertian reflection $\varrho = \varrho(M, \mu_0, \mu, \varphi)$ in point M:

$$\varrho^* \simeq \varrho_a + \frac{1}{1 - \langle\varrho\rangle S}(A\varrho + B\bar{\varrho} + C\langle\varrho\rangle) \ ,$$

where:

$$A(\tau, \mu_0, \mu) = \exp(-\tau/\mu)\exp(-\tau/\mu_0) \ ; \quad B(\tau, \mu_0, \mu) = \exp(-\tau/\mu)E(\mu_0) \ ;$$

$$C(\tau, \mu_0, \mu) = E(\mu)[\exp(-\tau/\mu_0) + E(\mu_0)] \ .$$

The term $A\varrho$ characterizes direct solar radiation reflected by an object and directly attenuated by the atmosphere in the viewing direction. The term $B\bar{\varrho}$ characterizes the downward diffused radiation reflected by an object and directly attenuated by the atmosphere. The coefficient $\bar{\varrho}(M, \mu, \mu_0, \varphi)$ describes the reflection of the downward diffused radiation and includes the bidirectional reflection by an object. The term $C\langle\varrho\rangle$ characterizes the radiation reflected by the background and diffused by the atmosphere in the viewing direction. The reflection coefficient $\langle\varrho(M, \mu_0, \mu, \varphi)\rangle$ describes the reflection of direct and diffused downward radiation by the background elements, including non-orthotropic reflection and inhomogeneous background. Calculation of radiation repeatedly reflected by the surface using the coefficient $(1 - \langle\varrho\rangle S)^{-1}$ is not quite correct but may be used since the contribution of this coefficient is not so significant.

Such an approximate presentation of the reflection coefficient does not give a complete separation of the effects of the atmosphere and surface on the formation of the upward radiation, since $\bar{\varrho}$ and $\langle\varrho\rangle$ depend on the properties of the atmosphere. However, the effect of atmospheric characteristics on $\bar{\varrho}$ and $\langle\varrho\rangle$ is shown to be much weaker than on their relative contribution to the total coefficient of reflection.

Calculations of functions A, B and C for $\mu = 1$ have shown that the relative contributions of the terms with reflection coefficients $\varrho, \bar{\varrho}$ and $\langle\varrho\rangle$ are of the same order, except for the cases of sufficiently transparent atmosphere (visibility $V = 30$ km) and near IR spectral region ($\lambda = 850$ nm).

Based on the model of radiation calculation previously mentioned, the effect of the isotropic background inhomogeneity has been considered in [387]. In this case, for nadir angles of observation in polar coordinates (r, ψ), the coefficient $\langle \varrho(M) \rangle$ is written as:

$$\langle \varrho(M) \rangle = \int_0^{2\pi} \int_0^{\infty} \varrho(r, \psi) r \varrho(r) \, dr \, d\psi \ ,$$

where $r\varrho(r) = r e(r, \psi, +1)$ and $e(r, \psi, \mu)$ determines the contribution of an element of isotropic surface with coordinates (r, ψ) to diffused transmission $E(\mu)$. An integral function $F(r) = \int_0^r 2\pi r' \varrho(r') dr'$ determines the relative contribution to $\langle \varrho(\mu) \rangle$ of the surface elements at a distance not more than r from an object (the origin of coordinates). A comparison of the results of calculations of the function $F(r)$ in the single-scattering approximation and using Monte-Carlo simulations has shown their good agreement for large wavelengths (850 nm). For short wavelengths (450 nm), the single-scattering approximation leads to a systematic underestimation of $F(r)$ because multiple severe scattering by aerosol particles was not considered. The behavior of functions $F(r)$ and $r\varrho(r)$ is largely determined by the ratio of optical thicknesses τ_R/τ_A and by the relative vertical profile of aerosol density and, to a lesser extent, by the form of aerosol phase function.

The following presentation of the function of the background effect is suggested:

$$F(r) = [\alpha_A \tau_A F_A(r) + \alpha_R \tau_R F_R(r)]/[\alpha_A \tau_A + \alpha_R \tau_R] \ .$$

This kind of function is confirmed by performed calculations and by a simplified expression for $E(\tau, \mu) \simeq \exp(-(\tau_R + \tau_A)/\mu)(\alpha_R \tau_R + \alpha_A \tau_A)$. It shows that the molecular and aerosol constituents of the background have respective weights $\alpha_R \tau_R$ and $\alpha_A \tau_A$. Therefore, in calculations of the function of the background effect $F(r)$, it is sufficient to assess $F_A(r)$ for an average model of aerosol. Typical horizontal scales of the effect of the background contribution constitutes 1 km for the aerosol and 10 km for the molecular scattering. With homogeneous areas of intermediate sizes of an order of several hundred meters, the horizontal gradients ϱ turn out to be so small that they enable one to use simple formulas to assess the principal contribution of the atmospheric transfer function.

The case of a homogeneous anisotropic surface has been analyzed in [388]. Here the term $C\langle \varrho \rangle$ can be written as a sum of two items: $\bar{\varrho}'(\tau, \mu, \mu_0, \varphi) \times \exp(-\tau/\mu_0) E(\mu) + \bar{\bar{\varrho}}(\tau, \mu_0, \mu, \varphi) E(\mu) E(\mu_0)$. The term with the coefficient $\bar{\varrho}'$ describes anisotropic reflection of direct solar radiation by the background and its diffused transfer in the viewing direction, and the term with $\bar{\bar{\varrho}}$ describes the reflection of diffused radiation by the surface and its diffused transfer in the viewing direction. Using the mutuality principle, we can show that $\bar{\varrho} = \bar{\varrho}'$. Calculations of $\bar{\varrho}$ and $\bar{\bar{\varrho}}$ were made for several models of the atmosphere. Experimental values were taken for $\varrho(\mu_0, \mu, \varphi)$. Radiances were calculated using the method of successive approximations. Calculations made for different geometries of viewing and illumination have shown that the bidirectional character of reflection in the coefficients $\bar{\varrho}$ and $\bar{\bar{\varrho}}$ is partly preserved and can be written as:

$$\bar{\varrho}(\mu_0, \mu, \varphi) \simeq a\,\varrho(\mu_0, \mu, \varphi) + b \ ,$$

$$\bar{\bar{\varrho}}(\mu_0, \mu, \varphi) \simeq a'\,\varrho(\mu_0, \mu, \varphi) + b' \ ,$$

where a, b, a′, b′ are regression parameters. They are insensitive to changes in geometric parameters and depend weakly on the reflection law ϱ. On the whole, these parameters are determined by the field of diffused radiation, i.e., by the atmospheric parameters. Thus, the atmosphere and the reflective effects have been separated very accurately.

The preservation of reflection anisotropy in $\bar{\varrho}$ and $\bar{\bar{\varrho}}$ is explained, generally, by great forward scattering by the aerosol components of the atmosphere. Due to this scattering, the directivity of solar radiation is preserved in the downward diffused radiation, which is confirmed both by calculation results and by published experimental data. Therefore, the intensity of diffused downward radiation at the surface level is presented in terms of δ-function as $L_a^{\downarrow}(\mu_0, \mu, \varphi) = \mu_0 E(\mu_0)(A\,\delta(\mu_0, \mu) + B)$, where coefficients A and B ($A + B = 1$) result from an approximation of results of accurate calculations of intensity, with different values of μ_0, by this function (using the least squares technique). Hence, we obtain the expressions for $\bar{\varrho}$ and $\bar{\bar{\varrho}}$ through ϱ. It follows from calculations that $a \simeq A$, i.e., a corresponds to preserved directivity of solar radiation in the downward diffused radiation flux. The relationship $a' \simeq A^2$ is assessed in the same way; here the power 2 is obtained due to the dual effect of the forward scattering in the expression for $\bar{\bar{\varrho}}$.

In order to consider the effect of surface inhomogeneity on the surface-atmosphere system radiance, a generalization has been made in [333] of the two-flux approach to calculate the radiation transfer for a homogeneous Lambertian surface [296]. An approximate analytical solution was obtained for the equation of radiation transfer in the purely scattering atmosphere, for an underlying surface presented as two homogeneous half-planes with different albedos and straight linear interface. The cases of homogeneous and exponential distribution of the aerosol layer are considered. Analytical dependences of radiances at the atmospheric top on the distance to the interface make it possible to calculate the function of the line blurring. The inverse Fourier cosine-transformation gives the atmospheric transfer function for the frequency-contrast characteristic (FCCh), which describes the effect of the atmosphere on radiation transfer from the Earth's surface to the atmospheric top in zenith, including radiation transfer from brighter objects to darker ones. There is a negligible dependence of the atmospheric transfer function on surface albedo, which distinguishes it from the classical definition of the transfer function. This dependence is explained by a non-linear "feedback" between the atmosphere and the surface. Two parameters, A_c and $x_{1/2}$ are introduced to characterize the extent of the atmospheric effect. The parameter $x_{1/2}$ determines the distance from the boundary, at which the horizontal inhomogeneity-induced contribution to the system's radiance gets halved. Calculations show that it is, practically, independent of average albedos of the half-planes. With increasing τ, the parameter $x_{1/2}$ decreases, reaching minima at $\tau \approx 1$, and then it grows. For the Elterman aerosol model, minimum values constitute about 0.7 km and maximum values (at small τ and at $\tau \approx 5$) reach about 1.2 km. The values obtained can be used to assess the optical thickness and scale height of the aerosol layer from the distribution of the system radiance recorded on the satellite.

Based on the dependence of the atmospheric-top radiance on the distance to the interface between two half-planes obtained in [333], an iterative algorithm has been proposed in [334] to retrieve the albedo of the surface with any 1-D spatial distribution (and not only of a step-like type). To realize this algorithm, it is necessary to know the aerosol optical thickness and other atmospheric parameters, as well as calibration parameters of the recording instruments. The technique was tested for sensitivity to the errors in the input parameters of the model. It was shown that for a 20% uncertainty in the optical thickness, the algorithm had a convergence within 3 – 5 iterations. The algorithm was tested for the values of optical thickness below unity. The processing of Landsat-1 MSS data was exemplified.

The authors of [298] suggest an improvement for the technique for an approximate calculation of the surface-atmosphere system radiance in nadir in the near UV, visible and near IR spectral regions in the presence of an inhomogeneous surface. In contrast to [333], a two-flux approximation [296] is used to describe the contribution of multiple scattering, and single scattering is considered precisely. Besides, the atmospheric absorption is included. The vertical profiles of both scatterers and absorbers can be arbitrary. Comparison of results of calculations of the transfer function using approximate formulas from [333, 298] and Monte-Carlo simulations, has shown a very good agreement for the formulas from [298], in contrast to those from [333]. The adequacy of the technique was also confirmed by comparing with Landsat-1 MSS data, for which the data on ground measurements of atmospheric transparency and aerosol optical thickness were available. Analysis of calculation results has shown that, the dependence of the contribution of side light from a higher-albedo site (land) to the formation of the low-albedo system radiance (ocean) on the distance to the interface is, for practical purposes, insensitive to the special features of the vertical profile of concentration. However, it does alter with varying average height of the aerosol layer.

This dependence is not that sensitive to the aerosol particles' size distribution or to optical thickness and surface albedo. This makes it possible to retrieve the scale height for aerosols from data of nadir measurements of the dependence of the system's brightness on the distance to the interface.

A laboratory simulation has been considered of the formation of the radiation field in the atmosphere over a horizontally inhomogeneous surface [335]. The atmosphere was simulated by latex balls suspended in water (with their density varied), and the surface by half-planes painted black and white. A comparison of the results of calculations based on an approximate solution of the 2-D transfer equation already considered, with the observational data, revealed a satisfactory agreement. This testifies to a possibility of using approximate analytical calculations in real conditions, to assess the effect of radiation reflection from the background. The contribution of radiation from an adjacent site has been somewhat overestimated. This is explained by an insufficient consideration of multiple scattering.

Results from the consideration of the effect of the inhomogeneous surface [333] have formed the basis for an estimation of the albedo and the atmospheric characteristics from satellite data [299]. To retrieve these quantities, it is necessary to have two extended objects with different albedos (sea-land) on an interface image. The retrieval procedure is based on the approximate analytical solution of the transfer equation, and it can be used for the wavelength interval 0.4 – 0.8 μm, solar zenith angles 10° – 60°, albedo 0.03 – 0.5, and optical thicknesses 0.2 – 2. Studies have shown that this procedure is very sensitive to single scattering albedo and to optical thickness. This enables one to obtain reliable estimates of these parameters, confirmed by the results of the Landsat MSS data processing.

A technique has been considered in [300] to retrieve the aerosol absorption from satellite-derived images of both the transparent and turbid atmosphere over the horizontally-inhomogeneous surface [299]. Note is taken here, that a certain value of surface albedo, at which the shortwave radiation is independent of the atmospheric optical thickness (with the constant phase function), corresponds to any fixed value of the single scattering albedo W_0. Thus, there is a unique dependence of W_0 on a "critical" surface albedo value, with the given sun elevations and aerosol size distributions, which determine the phase function. The equality of radiances for the transparent and turbid atmosphere serves as the criterion for the retrieval of the critical albedo. Numerical experiments have shown that W_0 can be retrieved in the wavelength intervals 500 – 600 nm and 600 – 700 nm, with accuracies $\Delta W_0 = \pm 0.02$ at $W_0 > 0.8$, and $\Delta W_0 = \pm 0.4$ at $0.6 < W_0 < 0.8$. A comparison of results of real-time data processing and control laboratory experiments suggested that the values retrieved are systematically overestimated by 0.007, on average, with the MSD 0.03. If $W_0 > 0.9$, then the MSD decreases to 0.02. The proposed technique can be used in global monitoring of W_0 averaged over the atmospheric thickness.

The authors of [262] consider the problem of the relative effect of aerosol scattering and absorption on remote sensing data. Based on the spherical harmonics technique, estimates have been obtained of individual contributions of aerosol scattering and absorption in the visible and near IR spectral regions. Principal parameters determining the atmospheric transfer function are τ and W_0. Calculations have shown that the effect of aerosol absorption substantially shows itself when the surface albedo is high. A detailed analysis has been made of the results of calculations of the critical albedo, when the increase of atmospheric haze brightness is compensated by the increasingly attenuated atmospheric radiation. Values have been obtained of the critical albedo on the viewing geometry W_0 and other factors. Analysis of the results has shown that a given variation of the aerosol optical thickness can always be compared to an equivalent change in W_0, at a fixed surface albedo.

A technique has been considered in [246] to solve a 3-D radiation transfer equation in the vertically inhomogeneous atmosphere, with an inhomogeneous non-Lambertian surface. This technique is based on the 2-D Fourier transformation of the transfer equation and on the obtaining of solutions for each of the Fourier components of the radiation field, with the use of the Gauss-Zeidel iterative approach. The 3-D intensity field is retrieved using the inverse Fourier transformation. The problem of the principal advantage of the separation of radiation into direct solar and scattered radiation is discussed. The advantage is connected with the fact that the atmosphere is a filter which suppresses high special frequences determining the horizontal inhomogeneity of the surface. Therefore, the retrieval of the scattered radiation field requires, as a rule, consideration of many fewer Fourier-components than in the case of direct solar radiation, when the sophisticated spatial variability of the surface must be evaluated. The introduction of a spatial class of functions has made it possible to analyze the combined effect of the geometrical and optical structures of the atmosphere on the smoothing of the details of the inhomogeneous surface, which is manifested in the formation of the scattered radiation field. Use of the Fourier-transformation reduces the problem for each of the components to a 1-D case, which permits use of mini-computers. The approach considered permits a simultaneous calculation of the radiance field at different altitudes. The reliability of the technique has been tested by comparing with Monte-Carlo simulations [247]. Calculations have shown that in measurements from space over a bright surface (albedo 0.6), the effect of an adjacent dark surface (albedo zero) can be noticed only at a distance of 10 km from the interface.

An analysis has been performed in [248] of the effect of the side diffusion of photons on the identifiability of characteristic surface details, depending on viewing conditions (nadir or slant viewing) and atmospheric optical properties (optical thickness, phase function).

It is shown that using only primary scattering, variations in the aerosol particles size distribution substantially affect the blurring due to scattering. This effect is observed in the center of the zone of blurring 60 – 500 m wide. In the interval of zenith angles larger than 40° the blurring function becomes asymmetric. In addition, for non-nadir angles, the zone of diffused blurring broadens [248].

Now let us consider techniques for atmospheric correction of the data of remote sensing of the ocean in the visible and near IR spectra. Paper [267] deals with the technique for atmospheric correction of the data obtained with Nimbus-7 CZCS which has five channels in the visible and near IR (443, 520, 650, 670, and 750 nm) and one channel in the thermal IR interval (10.5 – 12.5 μm). The scanner has total FOV 78° and spatial resolution 825 m in nadir at a flight altitude of 955 km.

The satellite is in the sun-synchronous orbit, and local time in the subsatellite point is close to noon. A possibility may exist to incline the scanning plane in order to exclude the light spots of direct solar radiation.

Data from this satellite have been used to estimate the concentration of the phytoplankton pigment connected with the volume emission of water. To obtain volume radiances from a general scanner-recorded signal, it is necessary to exclude the contributions of the atmosphere and of surface reflection, which are at least, an order of magnitude larger than volume radiance. The atmospheric correction technique is based on the following suppositions [265, 267]. The intensity of radiation scattered by the atmosphere and reflected by the flat sea surface can be presented as a sum of two terms, one of which appears due to Rayleigh scattering, the other due to aerosol scattering in the atmosphere. This supposition holds not only for small optical thicknesses, when a single-scattering approximation is valid, but also for large optical thicknesses. The specular reflection of diffused downward radiation and of direct solar radiation (then diffusively scattered by the atmosphere into the instant FOV of the sensor) is considered with the help of the Fresnel coefficient of reflection.

Calculations in the single-scattering approximation and precise calculations of radiation transfer have shown that for solar zenith angles less than 40°, the principal contribution (96%) into the Rayleigh component of radiance is made by atmospheric haze (i.e., radiation never reflected from the surface), which is explained by the Rayleigh phase function symmetry. Therefore, for viewing angles outside the solar light spot, the Rayleigh component can accurately be estimated without considering sea surface waves.

In contrast to this, the aerosol phase function is strongly anisotropic in direct scattering and, therefore, the contribution of the atmospheric aerosol haze to the total aerosol intensity is rarely more than 60% – 70%.

An assessment of the aerosol scattering intensity is based on that [266] for pure water with a concentration of phytoplankton pigment $C < 0.25\ \mathrm{mg/m^3}$, the normalized value of the water volume emission for wavelengths 520 and 550 nm is known and is constant when C varies, and for 670 nm it can be considered zero. Choosing the areas of pure water in the image, and calculating Rayleigh scattering intensities for wavelengths 520, 550 and 670 nm, it is possible to assess the ratios S (520, 670), S (550, 670) of aerosol scattering intensities at wavelengths 520 and 550 nm with the aerosol scattering intensity at 670 nm. The quantity S (443, 670) is determined using an extrapolation procedure. These ratios are used to correct other fragments of images, on the assumption that the volume radiance is equal to zero and for $C > 0.25\ \mathrm{mg/m^3}$ at $\lambda = 670$ nm. The S ratios are considered to remain constant for a given type of aerosol where there is a horizontally inhomogeneous aerosol and they are determined by the ratio of aerosol optical thicknesses for respective wavelengths. This is connected with the fact that, the intensity of aerosol optical thickness and the aerosol phase function and the aerosol constituent of single-scattering albedo, exhibit a weak wavelength dependence in the interval 443 – 670 nm.

A comparison of in-situ measured data and estimates from satellite corrected data has shown that in the interval 0.08 – 1.5 $\mathrm{mg/m^3}$, the error in phytoplankton pigment concentration estimation constitues about 30 – 40%. The error in the estimation of the sea surface volume radiance is about 10%. Note should be taken that this method for atmospheric correction does not require any ground measurements.

The possibility was substantiated in [406] of separating the contributions of the Rayleigh and aerosol constituents of radiation scattering in the atmosphere to estimate the chlorophyll concentrations in water. The contribution of these constituents may be estimated from satellite measurements in a channel where radiation reflection from water surface can be neglected (for instance, in the red and near IR spectral regions). It is shown that the effect of aerosol scattering can linearly be extrapolated from one wavelength to another. Natural variability, of the aerosol optical thickness of the atmosphere, has led to an uncertainty in the chlorophyll concentration estimated from the data of regressive analysis of experimental differences of water surface reflectivities, at two wavelengths. This also affects chlorophyll content from measurements in the Atlantic Ocean, where the concentrations ranged between 0.2 and 0.8 mg/m^3 and in coastal regions of the North Sea with concentrations exceeding 3 mg/m^3. In the regions far from the coastline this uncertainty in concentrations is estimated by a factor of $1.5-3$, depending on extreme cases of atmospheric turbidity [406].

In the approaches considered in [267, 406] the contribution of the atmospheric haze brightness was assessed on the supposition that the Rayleigh component of brightness is independent of aerosol, and the aerosol haze brightness does not depend on Rayleigh scattering.

In the single-scattering approximation, more accurate calculations of the atmospheric haze brightness have been made, unconnected with the assumptions mentioned above [377]. Use of a new algorithm for atmospheric correction, with the Nimbus-7 CZCS data as an example, has shown that it eliminates the previous underestimation (25%) of the ratio of the ocean-surface upward radiances at wavelengths 443 and 550 nm, which characterize the chlorophyll concentration in the surface layer of the ocean.

Based on numerical solution of the transfer equation, using the successive approximations technique, for several atmospheric models, the problem is investigated of the effect of various input parameters on the sea surface. These include atmosphere system brightness, of great importance in an estimation of the phytoplankton pigment concentration from remote sensing data [364]. The results of calculations show that for $\lambda = 0.443$ μm the contribution of emission from the ocean, to total radiance recorded by the satellite without phytoplankton, constitutes 5%. With phytoplankton concentration 1 mg/m^3 (10 mg/m^3), the contribution decreases to 4% (3%) due to radiation absorption by phytoplankton. At other wavelengths, corresponding to CZCS channels, relative variations of recorded radiance, with varying phytoplankton concentration, are still smaller. On the other hand, characteristic real-time variations of atmospheric transparency and of the extent of foam-covered ocean areas cause variations in recorded radiance from 1% to 10% [294, 364]. Therefore, an exact consideration of the distorting effects of the atmosphere and a precise calibration of the sensors are needed. For instance, if a 1% accuracy of correction of total satellite-recorded radiance is required, then the aerosol optical thickness must be known within 0.11 at $\lambda = 0.44\ \mu m$ for sun elevations $15-20°$, and within ~ 0.04 for sun elevations more than 30°. The accuracy of the aerosol optical thickness ~ 0.04 means a 10% accuracy for total optical thickness of aerosol and molecules at $\lambda = 0.44\ \mu m$, with a moderate turbidity of the atmosphere. For $\lambda = 0.75\ \mu m$, and all sun elevations, the aerosol optical thickness should be known within 0.003.

In publications [311, 365], problems have been discussed connected with revealing the most favourable conditions for retrieving the atmospheric optical thickness, from satellite data on the outgoing shortwave radiance over the clear-sky water surface. An analysis was based on numerical modeling data. A quasi-linear correlation between measured and aerosol optical thickness is also confirmed by the observational results. Conditions are most favorable when the aerosol optical thickness exhibits the prevailing effect on the outgoing radiation formation.

Results are given in [295] of retrieving the aerosol optical thickness of the Saharan aerosol layer over the Atlantic Ocean from the satellite scannning radiometer data in the visible spectrum. Favorable clear-sky viewing angles have been chosen, outside the sun path zone. In this case, the retrieval errors do not exceed ±20%.

Based on theoretical calculations, a linear dependence was obtained between the upward diffused radiance and the atmospheric optical thickness [215]. It was shown that variations in optical thickness are connected with varying aerosol particles' size distribution. Optical thickness increases approximately linearly when the concentration of large particles grows. Thus, the inhomogeneity of the aerosol size distribution may affect the upward radiance. The spectral optical thickness was estimated from comparison of results of solution of the transfer equation and Nimbus-7 CZCS data.

Model calculations were based on the aerosol size distribution measured $1-2$ hours before space-based survey. The results obtained point to a strong dependence of optical thickness on wavelength and to the variability of the ratio between optical thicknesses at two wavelengths. The supposition about the constancy of the ratio has been made in [267] to solve the problem of atmospheric correction. Atmospheric correction of the images of mountains is a more complicated problem [320, 378]. To describe the interaction of radiation with the atmosphere and mountains, it is necessary to consider the following effects [378]. First, the thickness of the atmospheric layer between the sensor and an object depends on the height of this object above sea level. Second, variations in the orientation of the elements of the mountain surface cause changes in their illumination. Third, the elements of the mountain features can either be shaded by other features or self-shaded. The proposed model of the formation of the upward emission is based on the following simplifying suppositions. The atmosphere is considered to be plane-parallel, horizontally homogeneous and vertically inhomogeneous. Multiple scattering and absorption of radiation in the atmosphere are considered through experimentally determined coefficients of the model. The features are assumed to be illuminated only by radiation from the upper hemisphere and the effect of the background is neglected. Anisotropic distribution of the sky brightness is replaced by an equivalent homogeneous distribution. The reflection from the elements of the mountain surface follows the Lambert law. The altitudinal dependence of the optical thickness, illumination and intensity of the atmospheric haze is prescribed as independent exponential functions. To assess the altitudinal distribution and orientation of the surface elements, a digital model of the image was constructed. The altitudinal dependence of the atmospheric haze is determined using the following procedure.

For each altitude, minimal radiances were determined from an initial image. Thus, we obtained an experimental vertical profile of the atmospheric haze on the assumption that the minimum value corresponds to a shaded object, and the effect of albedo can be neglected. The vertical profile of minimum values is approximated by an exponential function in order to assess the atmospheric haze. Here an average range of altitudes should be borne in mind, since at high altitudes, a high-albedo snow cover is observed, and at lower altitudes there may be light spots from water (river).

A model from [391, 409] has been used to assess the altitudinal dependence of irradiance. Three criteria were introduced to assess the quality of atmospheric correction: (1) objects with equal true albedos and different orientations should have close calculated albedos; (2) a homogeneous area (within which the orientation and type of cover are constant); (3) calculated albedos should be positive and not exceed unity.

Three results are given of the correction of an image of a mountain massif obtained with Landsat-1 MSS. One of them is made using a theoretical altitudinal dependence of the atmospheric haze, calculated with model [391, 409], and the other two using experimental altitudinal dependence of the haze intensity. The correction made with experimental dependence meets the quality better criteria. Despite the shortcomings of the considered model of radiation transformation in the mountains-atmosphere system, the procedure of atmospheric correction has certain advantages. To realize it, a minimum of input data is needed, and the input parameters can be varied due to realization simplicity, to meet the quality criteria of atmospheric image correction. Besides, it should be mentioned that it is easier to create a digital model of the image than to perform synchronous measurements of atmospheric parameters. The correction makes it possible to obtain information about shaded features.

A technique to assess atmospheric transmission with the use of multi-angle sensing has been considered in [249]. In order to reduce the effect of the atmosphere and of the instrumental noise, the image is filtered based on Fourier transformations [246, 247]. The technique requires knowledge of a phase function and of coefficients of bidirectional reflection of the surface. The technique makes it possible to estimate the transparency with an accuracy up to 10%, enough to solve some applied problems.

An estimation of parameters for atmospheric correction of images with the use of the multi-angle approach, has also been considered in [383]. The authors mention that the difference for optical paths due to the scanning angle

($\pm 30°$) is not enough to determine the atmospheric parameters with sufficient accuracy.

An assessment of the atmospheric haze, using the covariance matrix technique, is another approach to the determination of atmospheric parameters from remote sensing data [385]. This method is based on the correlation between the data of all the four Landsat MSS spectral channels, simultaneously and not in pairs. This method does not require additional information and uses the digital survey data. In addition, this method does not require the presence of low-reflectivity objects (pure lakes) or weakly illuminated objects (shaded by clouds or due to the scene topography) on the image. To achieve this, areas of hilly landscape, homogeneous in their reflective properties must be on images, which are used as test areas to determine the atmospheric haze spectral intensities. This correction technique has been used in processing space-derived images in various types of scene and various atmospheric conditions. The technique has given satisfactory results for all areas.

In connection with the method for determining the atmospheric optical thickness with the help of dark objects (mentioned previously in consideration of the first approximate solution of the transfer equation), it should be noted that the hazes have been calculated independently, in different spectral channels. An advanced technique has been proposed in [232] to assess the haze from dark objects, which involves the relative model of the spectral change of atmospheric scattering. The model of spectral change is chosen based on the data of one of the channels. Calculations reveal substantial differences between the cases when the spectral dependence of the atmospheric haze is considered and when it is neglected.

Atmospheric parameters have been estimated in [221] from reference objects. The urban features (roofs, take-off and landing runways, etc.) were proposed as reference reflectors.

The annual variation of their reflection coefficients is assumed to be weak. A large number of such features, with different reflection coefficients have been analyzed. In the visible and near IR, the surface albedo can be estimated as a sum of albedos for Landsat MSS channels 4 and 7, with respective weights.

Apart from water bodies, vegetation, with reflective characteristics already known, has been investigated to determine atmospheric parameters [301]. A comparison was made of measured optical thickness and calculated from radiances over the water body, forest, maize and pasture. The best agreement has been obtained for the data for water bodies and non-nadir observations of forest, which is connected with low reflectivities of these surfaces. In addition, problems have been considered, directly connected with the correction procedure itself.

The atmospheric effect on the classification of Landsat data has been described in [336]. Multi-spectral images of Brazil obtained from Landsat and digitized have been used to consider the atmospheric effect and for subsequent classification of features. The scattering properties of the atmosphere were calculated with a model suggested in [330]. A plane-parallel model of the atmosphere with Lambertian reflection was used to correct for the atmospheric effect on images. The angular conditions of survey (viewing angles from nadir to 10°, viewing azimuths, sun location) were calculated for images following a special program. Spectral albedos of the objects on images have been taken from literature. Atmospheric correction consisted in solving the direct problem of atmospheric optics using enumerated initial data and available data of radiosondes.

The exemplified initial and ocrrected Landsat images of channels 4 (spectral interval 0.5 – 0.6 μm) and 7 (0.8 – 1.1 μm) have shown intensified contrasts of images for both channels resulting from atmospheric correction and a 9% increase of the accuracy of classification of the sugar-cane plantations.

Optimal conditions of viewing have been determined in [310] based on calculations of radiation transfer in the atmosphere at a given sun elevation and with the phase function known. For geostationary satellites the optimal conditions are found to be complicated because of the need to consider variations in viewing direction, sun elevation and azimuth of the viewed point at the moment of survey. From measurements at 0.7 μm the directions of the upward and downward radiation have been chosen in which the radiation variability due to varying atmospheric parameters reached a minimum. The related calculations were made for marine aerosol and that of the Sahara desert. The result is that for these types of aerosol the atmospheric turbidity can be estimated to an accuracy of better than ±25% for cloudless conditions, if the accuracy of radiometric measurements is ignored. Errors in calculations of the upward radiation flux due to uncertain knowledge of initial data and a limited accuracy of calculation technique have been tabulated:

Affecting factor	Error (%)
Neglected foam cover	±6
Type of aerosol	±4
Spectral density of the extra atmospheric solar radiation	±4
Air pressure near the Earth's surface	±0.6
Ozone content	±0.4
Aerosol absorption	±2
Vertical stratification of aerosol	±2
Asphericity of aerosol particles	±4
Accuracy of calculation technique	±0.5
Neglected polarization	±2

A mathematical model of radiation transfer to study the atmospheric effect on variations of the brightnesses of individual surface objects ("test targets") has been described in [394]. The model considers the scattering and absorbing properties of principal atmospheric components (gases and aerosols) with Lambertian reflection of radiation from the surface. An analytical method for solving the radiation transfer equation in a plane-parallel inhomogeneous anisotropically scattering atmosphere has been described.

The atmospheric effect on the space-derived images of the Earth has been illustrated in [273]. The procedure for cluster-analysis with available measurements at the test areas of the Earth's surface enables one to classify space-derived images. However, due to a variety of data obtained for different objects at different times, under varying atmospheric conditions, this procedure is difficult to realize. Even under constant atmospheric conditions the distorting effect of the atmosphere causes individual image pixels to be dependent on those adjacent. This latter effect is particularly important in studies of small-scale parts of space-derived images. These features are typical of the Landsat-2 images of Japan. The effective radius of the "adjacent" effect was 300 m. A linear connection of satellite-recorded brightnesses with surface albedo has been used: since the albedos of test areas are known, the albedo of the entire image can be retrieved using this connection. For initial and corrected images standard procedures of cluster-analysis was used and a classification by the criterion of minimum Euclidean distance between the clusters.

The following variations in the results of classification of corrected images have been marked: the number of pixels belonging to each of the four classes has changed; a fine structure of objects has developed on corrected images (inside the "forest" class there appeared imbeddings of the "water" class, which has not been observed before the correction). Comparison of data with a detailed map of the given region has confirmed the existence of these small hydrological features over the forested territories. The results of classification suggest the conclusion that, in the first approximation, the data of a standard atmosphere can be used to consider the atmospheric effect. It is stated that knowledge of the optical thickness of a real atmosphere will raise the accuracy of classification.

Let us consider principal theoretical and practical aspects of the solution for the problem of atmospheric correction of the Landsat scanner images.

A "tasselled cap" is a characteristic geometric feature of the graphical presentation of the spectral change of the phenological development of agricultural plants from MSS Landsat scanner data [289]. When considering the atmospheric effect from these data it is important to separate the effects of atmospheric haze and the mutual "adjacent" effects of image pixels. A model of radiative transfer was used, to reduce the surface measurement data, to a conditional atmospheric top level with several levels of atmospheric haze brightness. Channel 0.8 – 1.1 μm of MSS 7 is divided into separate parts, in accordance with different effects of the water fraction of the atmosphere on radiation transfer. With the help of the radiation transfer model for four gradations of haze brightness and four gradations of atmospheric precipitation calculations were made of the curves of the "tasselled cap" brightness, dominant green and yellow colors, and of the same curves without any dominant color. The first of the enumerated factors is regulated by surface conditions and by atmospheric haze; the dominant green color by the same two factors and by precipitation; the dominant yellow color by atmospheric haze; the fourth factor of the model by precipitation, the latter two factors depending weakly on the state of the surface: they are used to adjust the first two factors to varying conditions of soil and vegetation and to reduce their sensitivity to atmospheric haze and precipitation.

To estimate the contribution of radiation from adjacent objects to total quantity of radiation field, the Landsat data were analyzed with the help of a regression model [240]. Radiation singly scattered by the surface increases due to two additional components, one of which is determined by the brightness of atmospheric haze, the other by radiation reflected multiply from the surface. The regression model made it possible to consider both subsidiary components within certain concentric circumferences whose maximum radius does not exceed 500 m, from the Landsat data.

Now we consider the geometric and atmospheric distortions of images obtained from the NOAA AVHRR data [223]. The AVHRR gives a better perspective than the Landsat MSS (multi-spectral scanner) to monitor the phenological development of agricultural crops. The shortcomings of the MSS in solving this problem, are explained by the fact that its 80-m spatial resolution results in enormous volumes of current information. Also, at the same time, its $+6°$ angle of scanning from nadir confining the viewed band to 185×185 km, prevents a daily analysis of large regions of crops because of a 18-day period of the satellite's rotation. Therefore, the AVHRR, despite a poorer spatial resolution (about 1 km in nadir) at an angle of scanning $\pm 56°$ from nadir gives a better perspective, since it makes it possible to perform a daily analysis of the given territory. At the same time, the MSS data are useful in specifying daily results of rapid analysis of the AVHRR data, since the AVHRR channels 1 and 2 are most informative from the vegetation viewpoint, close in location and sensitivity to the MSS channels 5 and 7. With the help of AVHRR, it was found that substantial variations in measurements data cannot be explained by variations in vegetation cover. For instance, in channel 1 there is no effect of the 0.68 μm characteristic absorption band for chlorophyll; in channel 2 the reflection does not increase with growing vegetation mass, etc. These effects are explained by the effects of atmospheric haze and viewing geometry, which is confirmed by high coefficients of correlation between the data of channels 1 and 2 (varied for 6 test areas from 0.3 to 0.66), which cannot be explained by reflection of radiation only from vegetation cover. Three procedures were applied to introduce corrections for these effects: from satellite-derived information for sufficiently transparent lakes; from the curve of linear regression of satellite measurements and of scanning angle, by reducing the results of measurements over test areas to nadir angle; and finally a combined procedure of an account of scanning angle, solar irradiance geometry for test areas, and of atmospheric haze with the use of current variations in atmospheric conditions. The shortcoming of the first correction procedure is in neglecting the variations in sun elevation and atmospheric aerosol properties, as well as the non-Lambertian reflection whose effect can reach 10% – 30% with the viewing angles varying within $\pm 20°$. In the second procedure variations of the viewing angle are considered; in the third procedure the entire scene is analyzed. The first procedure resulted in a decrease of the coefficient of correlation of channels for six test areas to 0.01 – 0.14; in the second case the coefficients of correlation were close to those obtained after the first procedure; in the third case they increased to 0.39 for two areas, and for the remaining areas they remained at the level 0.01 – 0.13. As a result, variations in satellite data have substantially decreased over time, especially when the third of the enumerated procedures has been used.

Usually, in problems considering the effect of atmospheric haze in processing of the data of satellite measurements, the effect of diffused reflection of light in sea water is neglected. Hence, there is a possibility to retrieve the aerosol optical thickness of the atmosphere, which serves as the basis for sensing the phytoplankton with the Nimbus-7 coastal zone color scanner (CZCS) at a wavelength near 670 nm [410]. For areas with a high content of phytoplankton and suspensions in water, measurements are recommended at wavelengths near 745, 880 and 1060 nm. The retrieval possibilities are fulfilled, however, when measurements are made over sufficiently unpolluted water. For the continental lakes an empirical correlation of optical thickness with the Landsat channel-7 data (800 – 1100 nm) was found to be unexpectedly low, which can be explained by two reasons: (1) the effect of diffuse reflection from the turbid water; (2) the effect of the coastal zone.

To analyze the effect of the first factor, laboratory measurements were made of the diffused reflection by water surfaces with different suspension of contents in the wavelength interval 400 – 1600 nm. The composition of suspensions in pure water varied, covering the real variability of suspension concentrations from units to hundreds of mg per litre. A 5% variability of real deviations of the aerosol optical thickness of the atmosphere from averages (deviations constitute 0.033 to 0.4 at $\lambda = 700$ nm) was considered to be the level, of which the effect of variations in the content of suspensions on diffused reflection, can be neglected. From laboratory measurements average coefficients of reflection almost doubled in the wavelength interval 440 to 670 nm, then almost halved up to $\lambda = 1240$ nm, with the assumptions made for variations in the suspension content reaching 1.5 – 2.

The effect of the atmosphere on the measurements with the thematic mapper (TM) has been studied with two models of radiation transfer [298]: in calculations of reflection by the Earth-atmosphere system a doubling method was used; and for the TM thermal channel (10.4 – 12.5 μm) – a model of the transmission function. In the first case the effect of atmospheric aerosol, water vapor content and sun elevation on the system's reflectivity was analyzed; in the second case the effect of water vapor variations on the measurement results was assessed. A comparison of the channels of TM and MSS has shown that variations in the aerosol content lead to substantial variations in reflection for both types of instruments. Also, variations in the water vapor content lead to a much weaker effect of the atmosphere on the TM 4-channel measurements (0.76 – 0.90 μm) compared to those of MSS channel 7. Note, that there is difference between the TM spatial resolution ($30 \times 30\ m^2$) and that of MSS ($79 \times 79\ m^2$) and in the number of levels of the image quantizing (256 for TM against 64 for MSS), which must both be considered during the atmospheric correction of satellite images.

Now we shall consider the role of meteorological satellites in remote sensing of crops [33].

The program of monitoring the development and resources of agricultural production with the help of air- and space-borne AgRISTARS includes the use of Landsat information to develop models of the crop yield forecast. The U.S. system of operative satellites aimed at solving these problems consists of two geostationary satellites GOES over 75° and 135 °W, two NOAA polar-orbit satellites. The GOES satellites are equipped with the visible and IR scanning instruments VISSR which ensure the transmission of images every 30 min with a spatial resolution of about 1 km at the sub-satellite point in channel (0.55 – 0.7 μm) and about 8 km in channel (10.5 – 12.5 μm). The principal objective of the GOES system is to provide weather forecasts with the information needed about the development of atmospheric processes. An additional goal is to retrieve the paths of air-surface winds from the motion of cloud formations, sea surface temperature, and to draw maps of snow cover propagation.

The NOAA polar-orbit satellites, in constrast to GOES, operate in orbits which ensure global-scale coverage with measurements four times a day (03 : 30, 07 : 30, 15 : 30, and 19 : 30 local time). They carry TOVS instruments for operative vertical sounding of the atmosphere permitting a retrieval of the vertical profiles of temperature and humidity, and AVHRR to retrieve sea surface temperature and to obtain images. The TOVS includes the HIRS/2 spectrometer in 20 intervals of IR spectral region, MSU 5-channel microwave radiometer in the 5.5-mm oxygen band and the SSU radiometer to retrieve the atmospheric-top temperature. The AVHRR 5-channel radiometer (0.55 – 0.75, 0.7 – 1.2, 3.7 – 3.9, 10 – 11, and 11 – 12 μm) provides images with nadir resolution of 1 km. In addition, for agricultural meteorology purposes, knowledge of the amount of precipitation is needed for the period of the crops' growth 24 hours beforehand and with spatial resolution 1/3°. Geostationary satellites provide a continuous monitoring of the source of precipitation, and polar-orbit satellites make it possible to analyze precipitation over the regions with thin meteological networks. Also, for precipitation forecasts, information from the surface meteological network has largely been used.

An important characteristic for the development of the models of yield forecasts is the minimum and maximum values of day-time temperature. Instead of the difficult to assess characteristic of crops' temperature which exhibits considerable correlation with the phenological phases of development, the models of the yield forecast usually use air temperature near the soil surface. This characteristic too, exhibits a very strong spatial variability. The space-borne TOVS instruments, including a 20-channel IR spectrometer, 5-channel microwave radiation and the radiometer of thermal sounding of the stratosphere make it possible to estimate minimum temperature in the early morning and maximum temperature after noon. But these temperatures retrieved from satellite data contain very large errors due to the effect of atmospheric water vapor, aerosol and cloudiness: standard deviations of retrieved and temperatures measured in situ reach 2K in the case of partial cloudiness, and where cloudiness is complete, when microwave measurements can be used, they reach 3K.

The incident solar radiation flux is the next parameter needed for the yield forecast, evapotranspiration of vegetation and soil moisture. Snow cover is one more important parameter needed to assess the growth of winter wheat in northern regions. The NOAA-7 data make it possible to draw maps of snow cover distribution in polar stereographic projection. The vegetation index equal to the ratio of Landsat channels 7 and 5 (NOAA AVHRR channels 2 and 1 measuring daily as against a 18-day period of Landsat rotation) is an important factor in assessing the state of vegetation cover, though this spectral combination of measurements is useful in a classification of vegetation. An increase of the vegetation index from 0.1 to 0.5 characterizes the growth of the plant biomass and its decrease is caused by atmospheric attenuation, cloudiness, diffuse reflection, etc.

6.3 An Approximation of the Dependences of Outgoing Radiation Intensities by Orthogonal Polynomials

The polynomial approximation of the dependences of outgoing radiation intensity on the input data of the boundary problem solution makes it possible to use the calculation results for a given atmospheric stratification on a real time scale.

The values of outgoing radiation intensities at the atmospheric top $I = I(A, \zeta, \eta^+)$, which can be identified with those measured from space, are subject to errors resulting from calculation technique errors. The values are of a strongly oscillating function in character. Oscillations are caused by the behavior of high-order Legendre polynomials. As has been found, for $\eta^+ \geqslant 0.5$ at fixed values of A and ζ the intensity as a function η^+ can well be approximated by a second-order polynomial of η^+.

Smoothing the oscillations of curve $I = I(\eta^+)$ obtained at fixed A and ζ values has been made in two ways: (1) a simple averaging of oscillations so that the resulting averaged curve lies practically in the middle of oscillations; (2) smoothing made with an approximation by a second-order polynomial of one variable η^+ after the monotonic approximation method which, for the case of a strongly oscillating function, gives results close to mean-square approximation [160]. The values of thus-smoothed functions for all considered intervals of variations in A and ζ were taken as initial for an approximation of a function $I(A, \zeta, \eta^+)$. The results of smoothing made in these two ways are shown in Fig. 6.2.

Let S be a multitude of points $\{\bar{x}_j\}_{j=1}^{N}$ where $\bar{x}_j = (\bar{A}_j, \zeta_j, \eta_j^+)$ and N is the total number of points chosen for approximation. We shall call $\max_{\bar{x} \in S} I(\bar{x})$ the deviation of the function, and:

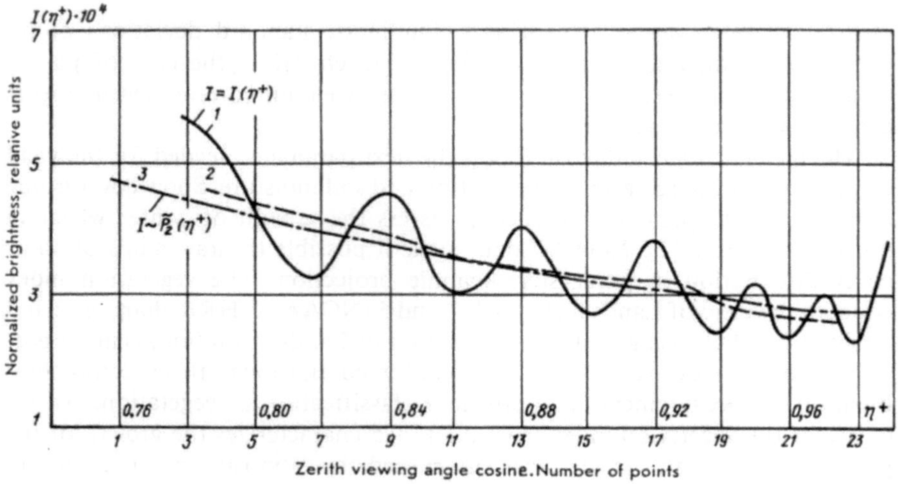

Fig. 6.2. Smoothing of curves $I = I(\eta^+)$ using the averaging technique and the Chebyshev uniform approximation technique: *1* calculated curve; *2* smoothed by simple averaging; *3* smoothed by the 2-power polynomial

$$E_n(I) = \min_{P_n} \max_{\bar{x} \in S} |I(\bar{x}) - P_n(\bar{x})|$$

the least deviation and the best approximation of a continuous function $I(\bar{x})$ in polynomial S by n-order polynomials [160]. We shall determine $P_n(\bar{x})$ as:

$$P_n(\bar{x}) = \sum_{j=1}^{n} a_j \bar{\varphi}_j(\bar{x}) \ , \qquad (6.6)$$

where $\bar{\varphi}_j(\bar{x})(j = \overline{1, n})$ is a linearly independent system of functions.

The polynomial $P_n^*(\bar{x})$ in the form (6.6) is needed which satisfies a condition:

$$\max_S |I(\bar{x}) - P_n^*(\bar{x})| - \min_{P_n} \max_S |I(\bar{x}) - P_n| \leqslant \varepsilon$$

for a given small $\varepsilon > 0$.

We shall choose a algorithm with an iterative process of constructing a best-approximation polynomial as an algorithm for constructing a polynomial of monotonic approximation. Proceeding from an initial polynomial $P_{n,0}$, we construct a succession of polynomials $P_{n,1}; P_{n,2}; \ldots; P_{n,m}$ so that:

$$\max_S |I - P_{n,0}| > \max_S |I - P_{n,1}| > \ldots > \max_S |I - P_{n,n}| > \ldots \ .$$

An approaching polynomial for the $(k+1)$-th iteration is required in the form:

$$P_{n,k+1} = P_{nk} + t_k \theta_{n,k} \ ,$$

where

$$\theta_{n,k} = \sum_{j=1}^{n} q_{jk} \varphi_j(\bar{x}) \ .$$

Coefficients q_{jk} are found from solution of the following problem of linear programming (PLP):

$$1 \leqslant \theta_{n,k}(\bar{x}) \underset{\min\alpha}{\operatorname{sign}} [I(\bar{x}) - P_{n,k}(\bar{x})] \leqslant \alpha$$

for multiple points of maximum deviation

$$S(\varepsilon) = \{\bar{x} \in S \mid I(\bar{x}) - P_{n,k}(\bar{x})| \geqslant \max_S |I(\bar{x}) - P_{n,k}| - \varepsilon\} \ .$$

By the introduction of fictitious variables, the PLP with limitations-in-equalities is reduced to the PLP with limitations-equalities.

Based on a matrix of constraints and a minimizing function, a simplex-table is compiled, where the sought-for coefficients q_{nk} $(i = \overline{1,n})$ of the correction polynomial θ_{nk} are used as free variables. The problem can be solved by a double simplex-method, since the coefficients of a minimizing function are not negative. The processing of the simplex table is made with the use of a modified Jordan exclusion, using the procedures that eliminate the cycling. We find a correction factor t_k by solving the problem of minimization:

$$\min_t \max_S |I - (P_{n,k} + t\theta_{n,k})|$$

using the "broken Fourier" method.

The sought-for polynomial is considered found if the difference between iterations becomes less than a given accuracy ε:

$$E_n(I) - E_{n+1}(I) < \varepsilon \ .$$

As an approximating system of functions we suggest for consideration generalized polynomials of three variables of an order below n:

$$P_n(\bar{x}) = \sum_{0 \leqslant i_1 + i_2 + i_3 \leqslant n} a_{j(i_1, i_2, i_3)} \hat{T}_{i_1}(A) \hat{T}_{i_2}(\zeta) \hat{T}_{i_3}(\eta^+) \qquad (6.7)$$

where $\hat{T}_i$ represents Chebyshev's combined polynomials, orthogonal on a length [0,1] for which the following recurrent relationships are valid:

$$\hat{T}_0 = 1 \ , \quad \hat{T}_1(x) = 2x - 1$$

$$\hat{T}_{n+1}(x) = 2(2x-1)\hat{T}_n(x) - \hat{T}_{n-1}(x) \ .$$

A regularization of the sets $j(i_1, i_2, i_3)$ is made in increased j numbers in the following way:

$$j = j(i_1, i_2, i_3) = 1 + \sum_{p=1}^{3} C_{3-p}^{4-p} + (i_p + \ldots + i_3) \ .$$

The writing of approximating polynomials in the form (6.7), compared to other known forms, is advantageous, because the number of iterations in this form decreases by about one fourth and thus, the computer time decreases.

For a real mid-latitude summertime atmosphere Table 6.1 gives some of the areas where an approximation $I = I(A, \zeta, \eta^+)$ is made by orthogonal polynomials. Calculations have shown that the value 10^{-6} for the $[\varepsilon]$ quantity, determining the way out of the iteration process, is enough for the areas given in Table 6.1 so that maximum absolute error in approximation is within an uncertainty of the technique for calculations of $I(A, \zeta, \eta^+)$. The differences $I(\bar{x}_j) - P_2(\bar{x}_j)$ $(n = 1, N)$ for some numbers are positive, for others – negative.

The BESM-6 computer time for different areas shown in Table 6.1 varies from 3 to 10 min. An approximation of the function of three variables $I = I(A, \zeta, \eta^+)$ by generalized four-order polynomials of the form (6.7) increases the computer time for $0.6 \leqslant \eta^+ \leqslant 0.99$ to 30 min at $E = 10^{-6}$, maximum error in approximation, compared to that by generalized 2-order polynomials which decreases slightly.

Table 6.2 lists the coefficients of the 2-order generalized polynomials approximating the outgoing radiation intensity as a function of three variables: albedo A, viewing angle η^+, and solar zenith angle ζ for the four versions of $\eta^+_{\text{init.}}$ and η^+_{final}, setting the interval of η^+ variations and shown in Table 6.1. For this case we write an approximating polynomial in an extended form:

$$\begin{aligned} I(A,\zeta,\eta) \sim P_2(A,\zeta,\eta) &= a_1 + a_2\hat{T}_1(A) \\ &+ a_3\hat{T}_1(\zeta) + a_4\hat{T}_1(\eta) + a_5\hat{T}_2(A) + a_6\hat{T}_1(A)\hat{T}_1(\zeta) \\ &+ a_7\hat{T}_1(A)\hat{T}_1(\eta) + a_8\hat{T}_2(\zeta) + a_9\hat{T}_1(\zeta)\hat{T}_1(\eta) + a_{10}\hat{T}_2(\eta) \ . \end{aligned}$$

The studies performed suggest the following conclusions:

1. The quantity $\varepsilon = 10^{-6}$, requiring minimum computer time for an approximation of the results of radiation transfer calculations, is enough for the accuracy of approximation not to exceed the accuracy of the calculation technique using the order of calculation approximation 71 for $\eta^+ \geqslant 0.5$.
2. The accuracy of approximation of the results by the 2-order polynomial does not exceed the accuracy of the technique for radiation calculations.
3. Expression (6.7) is the best presentation of the approximating polynomials.

The polynomial dependences for $I = I(A, \zeta, \eta^+)$ obtained with the method just described make it possible to estimate surface albedo $A_{\text{calc.}}$ from space-derived intensity $I_{\text{obs.}}$ at fixed solar zenith angles and viewing angles η^+ which will differ from true albedo value A_{true}.

The difference is connected both with an erroneous setting of the input data, with the errors in the calculation technique for the outgoing radiation intensity and an approximation technique. Therefore, surface measurements of reference sites are needed to estimate A_{true}. By finding discrepancies between $A_{\text{calc.}}$ and A_{true} for reference sites one can estimate the A for other sites of a space-derived image (with other angular values).

Discrepancies will serve as the characteristics to show on which side of $A_{\text{calc.}}$ one must place A_{true} over the whole image, provided that the atmosphere is the same (both horizontally and vertically).

Table 6.1. Some versions of calculations of the outgoing radiation intensity square approximation as a function of three variables (A, ζ, η^+) by orthogonal polynomials for different areas in A, ζ, η^+

Number of versions	Albedo A			ζ			η			Time of counting min	Maximum absolute error (%)
	Beginning	End	Step	Beginning	End	Step	Beginning	End	Step		
1	0.0	0.2	0.1	0.6	1.0	0.1	0.83	0.98	0.01	2.0	7.1
2	0.0	0.2	0.1	0.7	1.0	0.1	0.83	0.98	0.01	1.3	2.0
3	0.0	0.2	0.1	0.7	1.0	0.1	0.79	0.98	0.01	1.9	2.1
4	0.0	0.2	0.1	0.8	1.0	0.1	0.83	0.98	0.01	1.5	2.8
5	0.0	0.2	0.1	0.9	1.0	0.1	0.83	0.98	0.01	0.8	3.5
6	0.0	0.2	0.1	0.8	1.0	0.1	0.59	0.98	0.01	3.1	7.9
7	0.3	0.5	0.1	0.7	1.0	0.1	0.69	0.98	0.01	2.4	2.3
8	0.0	0.2	0.1	0.5	1.0	0.1	0.83	0.98	0.01	3.1	7.1

Table 6.2. Results of approximation of dependences $I = I(A, \zeta, \eta^+)$ by generalized second-order polynomials: $a_i (i = \overline{1, 10})$ are the coefficients of approximating polynomials

a_i	Number of version							
	1	2	3	4	5	6	7	8
a_1	9.251×10^{-3}	8.908×10^{-3}	9.128×10^{-3}	1.022×10^{-2}	1.313×10^{-2}	8.255×10^{-3}	8.289×10^{-3}	3.239×10^{-3}
a_2	3.931×10^{-3}	3.990×10^{-3}	3.967×10^{-3}	3.984×10^{-2}	4.133×10^{-3}	3.197×10^{-3}	3.697×10^{-3}	4.812×10^{-4}
a_3	1.952×10^{-3}	5.562×10^{-4}	1.722×10^{-3}	-1.492×10^{-3}	-3.090×10^{-3}	2.621×10^{-3}	3.067×10^{-3}	6.390×10^{-3}
a_4	-4.773×10^{-3}	-3.707×10^{-3}	-4.808×10^{-3}	-5.152×10^{-3}	-9.646×10^{-3}	-2.954×10^{-3}	-4.233×10^{-3}	-2.357×10^{-3}
a_5	3.019×10^{-4}	3.411×10^{-4}	3.395×10^{-4}	3.428×10^{-4}	3.814×10^{-4}	3.169×10^{-4}	3.873×10^{-4}	-7.839×10^{-4}
a_6	4.805×10^{-3}	4.863×10^{-3}	4.856×10^{-3}	4.886×10^{-3}	4.821×10^{-3}	4.635×10^{-3}	5.217×10^{-3}	6.070×10^{-3}
a_7	6.953×10^{-4}	7.002×10^{-4}	7.209×10^{-4}	6.929×10^{-4}	7.220×10^{-4}	9.477×10^{-4}	8.504×10^{-4}	-5.905×10^{-4}
a_8	2.462×10^{-4}	3.405×10^{-4}	1.110×10^{-4}	3.328×10^{-4}	0	3.264×10^{-4}	-8.856×10^{-6}	-4.258×10^{-4}
a_9	3.043×10^{-3}	4.388×10^{-3}	3.864×10^{-3}	6.827×10^{-3}	9.992×10^{-3}	1.587×10^{-3}	2.932×10^{-3}	9.230×10^{-4}
a_{10}	9.496×10^{-4}	2.772×10^{-4}	7.509×10^{-4}	6.282×10^{-5}	5.440×10^{-4}	7.944×10^{-4}	8.357×10^{-4}	2.995×10^{-4}

The uniqueness of the determination of $A_{calc.}$ from the outgoing radiation intensity at given angles of survey and solar zenith angles has been proved in [181]. Therefore, the signs of discrepancies characterize the direction of deviations of $A_{calc.}$ from A_{true} on the assumption that the initial data of the considering boundary problem of radiation transfer are known very accurately. Therefore, the atmospheric parameters affecting the radiation transfer must be estimated with an accuracy not less than the accuracy of the suggested calculation technique.

The effect of non-orthotropicity of radiation reflection by natural formations [158] constricts the possibilities of simplified algorithms to consider the distorting effect of the atmosphere. The non-orthotropic reflection results in the fact that in natural conditions, even with similar optical characteristics of the atmosphere, its transfer functions for the same natural object can differ, since the brightnesses of this object may alter at different viewing angles. The result is that, in the recognition of the images of natural objects from the spectral change of their brightnesses, the same object can be assigned to different classes.

When considering the non-orthotropicity in processing the space-derived information, it is worthwhile using the results of preliminary modeling of the outgoing radiation formation processes referring to the azimuthal dependence. Here one can obtain values for the spectral and angular distribution of the outgoing radiation intensity I on albedo A, cosine of zenith viewing angle $\theta(\eta^+ = \cos\theta)$, cosine of solar zenith angle $\zeta = \cos\theta_0$ and the difference between the viewing and the sun azimuths φ.

The factor of reflection anisotropy is contained in each of the four arguments of the function:

$$I_i = I_i(A, \zeta, \eta^+, \varphi)$$

where i is the index of the type of anisotropy.

The values $I = I(A, \zeta, \eta^+, \varphi)$ for homogeneous extended objects with non-orthotropic law of reflection have been obtained from the square approximation of the outgoing radiation intensity I on albedo A, cosine of solar zenith angle ζ, viewing angle η^+ and azimuth by orthogonal polynomials using the monotonic approximation technique already described:

$$I \approx P_2(A, \zeta, \eta^+, \varphi) = \sum_{0 \leqslant i_1 + i_2 + i_3 \leqslant 2} a_j(i_1, i_2, i_3, i_4)\hat{T}_{i_1}(A)\hat{T}_{i_2}(\zeta)\hat{T}_{i_3}(\eta^+)\hat{T}_{i_4}(\varphi) \ , \qquad (6.8)$$

where $\hat{T}_{ik}(k = \overline{1,4})$ represents combined i_k-order polynomials by Chebyshev orthogonal on a length [0, 1]. The terms of expansion can be regulated in increased numbers in the following way:

$$j = j(i_1, i_2, i_3, i_4) = 1 + \sum_{p=1}^{4} C_{4-p}^{5-p} + (i_p + \ldots + i_4) \ .$$

Table 6.3 gives different intervals of albedo variations for which the square approximation of the outgoing radiation intensity has been made, with

Table 6.3. Some versions of calculations of the square approximation of the outgoing radiation intensity orthogonal polynomials as a function of A, ζ, η^{+} for $0.6 \leqslant \zeta \leqslant 0.7$; $5° \leqslant \theta \leqslant 45°$; $0° \leqslant \varphi \leqslant 360°$ (versions 1 through 9 from the data of the left-hand part of Table 6.4; 10 through 13 from the right-hand part)

Number of version	Albedo A		Maximum absolute error (%)	Time of counting (min)
	Beginning of the interval	End of the interval		
1	0.01	1.0	50	29
2	0.1	1.0	26	25
3	0.3	1.0	13	10
4	0.01	0.3	20	7
5	0.01	0.2	18	4
6	0.01	0.1	9	3
7	0.1	0.3	8	6
8	0.3	0.5	6	7
9	0.5	1.0	6	10
10	0.01	0.1	9	3
11	0.1	0.3	8	4
12	0.3	0.5	5	3
13	0.5	1.0	5	6

Table 6.4. Experimental values of indicatrices of snow cover brightness measured of heights H = 8000 and 200 m under two conditions of survey

φ, degree	θ, degree									
	H = 8000 m					H = 200 m				
	5	15	25	35	45	5	15	25	35	45
0	0.66	0.66	0.66	0.68	0.74	0.82	0.81	0.84	0.88	0.94
30	0.66	0.68	0.71	0.74	0.81	0.79	0.79	0.80	0.81	0.86
60	0.68	0.70	0.68	0.66	0.72	0.77	0.77	0.80	0.80	0.82
90	0.68	0.68	0.68	0.71	0.73	0.77	0.79	0.77	0.77	0.79
120	0.70	0.70	0.68	0.68	0.71	0.79	0.79	0.77	0.79	0.79
150	0.68	0.66	0.70	0.72	0.74	0.80	0.80	0.79	0.79	0.79
180	0.70	0.68	0.66	0.70	0.72	0.80	0.79	0.80	0.82	0.82
210	0.68	0.64	0.68	0.70	0.75	0.80	0.79	0.77	0.79	0.80
240	0.70	0.68	0.68	0.68	0.70	0.79	0.79	0.79	0.81	0.81
270	0.68	0.68	0.68	0.70	0.70	0.79	0.79	0.79	0.79	0.80
300	0.64	0.64	0.66	0.66	0.68	0.79	0.79	0.80	0.81	0.84
330	0.64	0.64	0.68	0.72	0.77	0.77	0.77	0.77	0.77	0.80

$0.6 \leqslant \eta^+ \leqslant 0.7$; $5° \leqslant \theta \leqslant 45°$; $0° \leqslant \varphi \leqslant 360°$. The approximation was based on experimental phase functions [6] (Table 6.4).

Table 6.5 presents respective coefficients of approximation $a_j (j = \overline{1, 15})$ for each interval. The coefficients for orthogonal polynomials of the φ variable are an order and more below the coefficients for orthogonal polynomials of ζ, η^+ and A in an expression for intensity $I = I(A, \zeta, \eta^+, \varphi)$. Thus, in this case we have a weak dependence of the outgoing radiation intensity on the viewing azimuth φ.

If the accuracy of the approximation method is within the accuracy limits of the technique for calculating the radiation transfer in the system "surface-atmosphere", then an estimate of albedo A can be obtained uniquely from a set of approximation coefficients for the recorded outgoing radiation $I_{rec.}$ at fixed values of ζ, η^+ and φ using (6.8).

To ensure that the approximation error did not exceed the error in the technique for obtaining dependencies $I = I(A, \zeta, \eta^+, \varphi)$ the interval [0, 1] for albedo was divided into the following four intervals (Table 6.3): [0.01; 0.1], [0.1; 0.3], [0.3; 0.5], [0.5; 0.1].

Proceeding from the uniqueness of the estimation of albedo from satellite-recorded outgoing radiation intensity and from estimates of the accuracies of techniques for calculation and approximation, the true set of approximation coefficients of the four intervals will be that for which, at a given $I_{calc.}$ and fixed values of ζ, η^+, φ the A value is found in the interval:

$$\alpha - \frac{\alpha}{10} \leqslant A \leqslant \beta + \frac{\beta}{10}, \qquad (6.9)$$

where α and β are, respectively, the left- and right-hand ends of the interval of division by an albedo corresponding to a set of approximation coefficients (Tables 6.3, 6.5). The determination of the true set of approximation coefficients is made by successive examination of the four intervals and by checking the conditions (6.9) for them.

The radiative correction is reduced to retrieving the albedos of natural objects from absolute brightnesses of space-derived images where the angular conditions of survey (ζ, η^+, φ) are known. For each pixel of image, it is necessary to know the type of reflection anisotropy which can be defined, based on either, preliminary areal division of the natural features to be studied or, some statistical mean material about the reflective characteristics of these objects.

6.4 Sensitivity of the Calculation Scheme Considering the Atmospheric Haze Brightness to Variations in the Initial Data

The presentation of the atmospheric haze brightness as coefficients of approximation of the angular distributions of radiation intensity by Chebyshev's orthogonal polynomials (Sect. 6.3) for certain optical models of the atmosphere

Table 6.5. Results of the square approximation of the outgoing radiation intensity $I = I(A, \zeta, \eta^+, \varphi)$ by orthogonal polynomials; $[a_j (j = \overline{1, 15})]$ are coefficients of the approximating polynomial

a_j	Number of version 1	2	3	4	5	6	7
a_1	8.095×10^{-3}	8.431×10^{-3}	8.139×10^{-3}	7.127×10^{-3}	7.096×10^{-3}	6.704×10^{-3}	7.281×10^{-3}
a_2	6.263×10^{-3}	5.872×10^{-3}	6.558×10^{-3}	5.851×10^{-3}	6.138×10^{-3}	5.592×10^{-3}	6.153×10^{-3}
a_3	6.626×10^{-3}	6.241×10^{-3}	6.125×10^{-3}	5.709×10^{-3}	5.175×10^{-3}	5.001×10^{-3}	5.618×10^{-3}
a_4	-4.441×10^{-3}	-4.416×10^{-3}	-4.016×10^{-3}	-2.460×10^{-3}	-2.185×10^{-3}	-2.004×10^{-3}	-2.812×10^{-3}
a_5	-1.960×10^{-4}	-1.868×10^{-4}	-1.886×10^{-4}	-7.093×10^{-3}	-5.886×10^{-5}	-4.562×10^{-5}	-6.574×10^{-5}
a_6	1.738×10^{-4}	1.879×10^{-4}	2.343×10^{-4}	9.882×10^{-5}	1.629×10^{-4}	0.000	2.682×10^{-5}
a_7	5.103×10^{-3}	5.718×10^{-3}	5.802×10^{-3}	5.055×10^{-3}	5.095×10^{-3}	4.799×10^{-3}	5.340×10^{-3}
a_8	-1.527×10^{-3}	-1.384×10^{-3}	-1.790×10^{-3}	-1.191×10^{-3}	-1.248×10^{-3}	-1.177×10^{-3}	-1.781×10^{-3}
a_9	-1.998×10^{-5}	-1.364×10^{-5}	-1.912×10^{-5}	-1.049×10^{-5}	-2.047×10^{-5}	-1.991×10^{-5}	-6.656×10^{-6}
a_{10}	0.000	0.000	0.000	0.000	0.000	0.000	0.000
a_{11}	-1.537×10^{-3}	-1.049×10^{-3}	-1.117×10^{-3}	-5.639×10^{-4}	-5.257×10^{-4}	-7.016×10^{-6}	-2.753×10^{-4}
a_{12}	1.954×10^{-5}	-3.550×10^{-5}	-2.156×10^{-5}	-6.000×10^{-6}	-6.701×10^{-6}	-8.790×10^{-5}	-1.515×10^{-5}
a_{13}	9.261×10^{-4}	1.214×10^{-3}	1.232×10^{-3}	4.195×10^{-4}	3.051×10^{-4}	1.925×10^{-4}	4.195×10^{-4}
a_{14}	1.323×10^{-4}	-4.224×10^{-5}	-4.338×10^{-5}	-1.368×10^{-5}	-8.820×10^{-6}	-4.698×10^{-6}	-1.301×10^{-5}
a_{15}	6.784×10^{-6}	1.179×10^{-5}	1.191×10^{-5}	3.771×10^{-6}	2.670×10^{-6}	1.585×10^{-6}	3.771×10^{-6}

Table 6.5 (continued)

a_j	Number of version					
	8	9	10	11	12	13
a_1	7.681×10^{-3}	7.911×10^{-3}	5.679×10^{-3}	6.122×10^{-3}	6.569×10^{-3}	7.823×10^{-3}
a_2	5.554×10^{-3}	5.965×10^{-3}	4.669×10^{-3}	4.796×10^{-3}	4.947×10^{-3}	4.505×10^{-3}
a_3	5.501×10^{-3}	6.883×10^{-3}	4.907×10^{-3}	5.186×10^{-3}	5.052×10^{-3}	5.120×10^{-3}
a_4	-3.179×10^{-3}	-3.638×10^{-3}	-9.228×10^{-4}	-1.268×10^{-3}	-1.894×10^{-3}	-2.971×10^{-3}
a_5	-1.098×10^{-4}	-1.851×10^{-4}	-4.696×10^{-5}	-9.367×10^{-5}	-1.420×10^{-4}	-2.682×10^{-4}
a_6	1.194×10^{-4}	2.982×10^{-4}	0.000	7.208×10^{-5}	-2.220×10^{-5}	2.630×10^{-4}
a_7	5.041×10^{-3}	5.988×10^{-3}	4.805×10^{-3}	4.918×10^{-3}	5.063×10^{-3}	5.882×10^{-3}
a_8	-8.596×10^{-4}	-1.838×10^{-3}	2.449×10^{-4}	-7.708×10^{-5}	-3.020×10^{-4}	-2.842×10^{-4}
a_9	6.134×10^{-7}	-2.140×10^{-5}	-1.244×10^{-5}	-1.715×10^{-5}	-1.274×10^{-5}	-1.011×10^{-5}
a_{10}	0.000	0.000	0.000	0.000	0.000	0.000
a_{11}	-4.571×10^{-4}	-2.009×10^{-3}	1.257×10^{-4}	-4.348×10^{-5}	1.082×10^{-4}	-2.116×10^{-4}
a_{12}	-5.347×10^{-3}	-2.829×10^{-5}	-3.491×10^{-6}	-9.052×10^{-6}	-5.258×10^{-6}	5.839×10^{-4}
a_{13}	6.527×10^{-4}	1.246×10^{-3}	9.732×10^{-5}	2.030×10^{-4}	3.121×10^{-4}	5.839×10^{-4}
a_{14}	-2.128×10^{-5}	-4.491×10^{-5}	9.173×10^{-6}	2.126×10^{-5}	3.373×10^{-5}	6.801×10^{-5}
a_{15}	6.018×10^{-6}	1.201×10^{-5}	1.394×10^{-6}	3.285×10^{-6}	5.235×10^{-6}	1.207×10^{-5}

has been used in [84] for atmospheric correction of data for the photographic systems, in [80] for the scanner systems ignoring non-orthotropic reflection, and in [77] including this factor. Since a real optical state of the atmosphere is assessed with certain errors, a question arises: what are the variations necessary in the initial data, for the calculation scheme to retrieve surface albedo, to be sensitive to them? In general, analysis of the sensitivity of the calculation model to variations in atmospheric parameters requires the use of the concept "the function of the information content" [122]. This considers the effect of variations in the functionals, recorded by air- and space-borne instruments, on variations in the atmospheric optical characteristics. The latter occur not only because of uncertain optical parameters of the atmosphere but also from imprecise angular conditions of survey, spectral solar constant, and for lack of precise knowledge of the function of the sensitivity of the receivers and filters of measuring instruments, etc.

In solving the problem of radiative correction of space-derived images one faces two categories of difficulties: firstly, solution of complicated boundary problems of the radiation transfer theory serving as the basis for the problem considered; secondly, the necessity to process large real-time information volumes. Two questions arise whether it is possible to simplify the calculation formulas of the problem in order to accelerate the information processing and how far the radiation transfer problem can be simplified without loss of accuracy. How does an image change due to measurement errors? How important is it to know the actual state of the atmosphere or in radiative correction is it enough to consider to a statistical mean atmosphere? How do the angular conditions of survey affect the results of image processing? These are the problems, whose solution characterizes the sensitivity of the calculation scheme of radiation transfer and makes it possible to estimate possible errors in processing space-derived images.

This section is aimed at assessing the relationships between the initial and corrected brightnesses of natural objects on digitized images. A consideration of the contribution of atmospheric haze (the initial stage of radiative correction) is reduced to retrieving the albedo of natural objects from their space-derived images. The respective functions of transformation of the data of survey from space have been obtained from solution of 1-D boundary problems of the radiation transfer theory (3.14 – 3.16) with an approximation of the resulting dependences by orthogonal polynomials (Section 6.3).

Note, that initial images have been quantized into 256 conditional gradations of brightness corresponding to the transition from dark objects (zero gradation – cloud formations on negatives) to light objects (gradation 255 – water surface). As a result of radiative correction, each pixel of the image achieved the respective surface cover albedo (within 0 – 1) and further – the same 256 gradations of brightness which now characterize the corrected images.

The transformation of images consists in solution of the square equation (6.2) relating the albedo of individual pixels to recorded brightnesses under the known angular conditions of survey, known characteristics of measuring instruments, and known sets of coefficients of the atmospheric effect approximation.

Let us consider the effect of errors in setting the initial data of the calculation scheme, on the results of the digitized information processing, as applied to optical-electronic scanning systems of observation.

The errors in processing can conditionally be divided into three groups: measurement errors (instrumental noise and channel noise, errors in calibration of images, different energetic characteristics of the sun and of the on-board reference source); inadequate atmosphere model (including methodical errors of the calculation technique and errors in the approximation of calculation results); cases when the angular conditions of survey are not taken into account in the scheme of radiative correction (deviations of viewing angles from nadir, difference in the azimuths of viewing of of the sun for individual pixels, inaccurate knowledge of the factor of reflection non-orthotropicity, etc.).

The brightnesses of corrected images I are related to the brightnesses of initial images I_0 in the following way:

$$I = f(I_0, b_0, \Pi, \varphi)$$

where the form of function f is known from solution of the radiation transfer problems with a subsequent approximation of calculation results. The b_0 parameter characterizes the instrumental distortions of information (depends on the brightness of a reference source for information calibration $\tilde{B}$, on the function of spectral sensitivity of receivers and filters $\tau_{r,f}$, as well as on solar radiation energetics in a given measuring channel: the S-quantity relating calculations to measurements); the Π parameter characterizes the atmospheric transfer function; the φ parameter is determined by the angular coordinates of survey.

If the atmosphere did not affect the Earth's surface brightness, the dependence of I on I_0 would be a bisector (dashed-lines in Figs. 6.3–6.5), i.e., space-derived images upon considering atmospheric haze would not change. Curve 1 in Figs. 6.3–6.5 describes the connection of I with I_0 for a Rayleigh atmosphere with an optical thickness of $\tau_0 = 0.1$.

Curve 1 in Fig. 6.3 corresponds to average functions of instrumental distortions $b_{0,\text{aver.}} = B\tau_{r,f}/S \approx 0.00007$. Curves 2 and 3 correspond to a scatter of the b_0 parameter for space-borne low-resolution scanning instruments with reference to $b_{0,\text{aver.}}$ (30% on both sides – such measurement errors have been obtained from assessing real data on the photometric wedge of the scanner's reverse).

It is seen from Fig. 6.3 that measurement errors and noise in channels lead to mainly erroneous corrections in case of dark objects (large gradations of brightness on negative images). In particular, the brightness gradations exceeding 22 have an upper limit of 255 due to radiative correction.

Curves 1 and 2 in Fig. 6.4 have been obtained with varying parameters by the transition from a Rayleigh weakly turbid atmosphere (curve 1) to a turbid aerosol atmospheres (curve 2 corresponding to total optical thickness of the at-

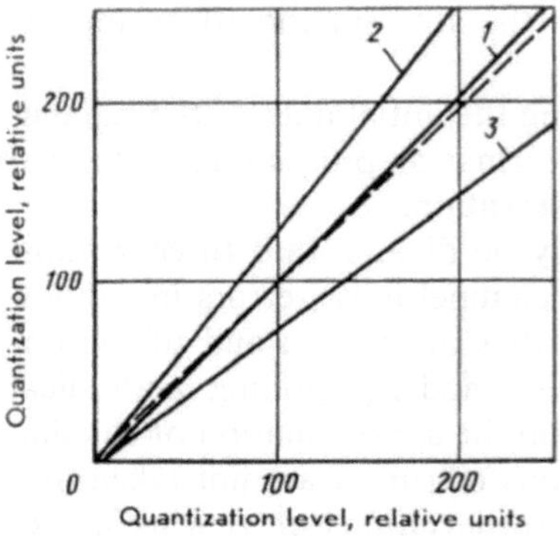

Fig. 6.3

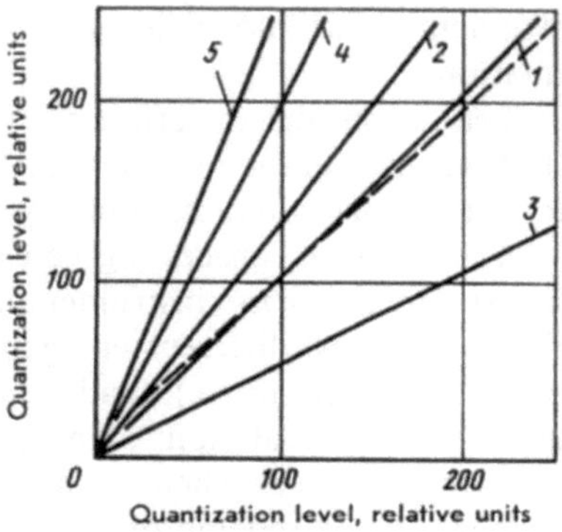

Fig. 6.4

Fig. 6.3. The effect of radiometric distortions on the conversion of the space-derived image brightnesses under standard survey conditions and in the transparent atmosphere: *1* without measurement errors; *2, 3* extreme values for measurement errors

Fig. 6.4. The effect of the atmospheric transfer function on the conversion of the space-derived image brightnesses: *1* transparent atmosphere; *2* turbid atmosphere

mosphere 0.4), with fixed $b_{0,\text{aver.}}$ and φ parameters. In the latter case, standard conditions are presumed: nadir viewing, the sun in zenith, Lambertian reflection. Errors due to substituting the real aerosol atmosphere (curve 2) with an idealized Rayleigh atmosphere (curve 1) for the curves in Fig. 6.4 can reach 60%.

In accordance with the physics of the procedure considered, of inverse filtration of images at high gradations of quantizing the brightnesses, the I exceeds the I_0; at low gradations the effect is the opposite, both for a weakly turbid atmosphere and a strongly turbid one.

As seen from the data of Fig. 6.4, gradations above 125 are, actually, the noise viewed from the problem of atmospheric filtration: for real aerosol atmosphere all these gradations are limited from above (curve 2). In this respect, it is better to quantize the range of image brightnesses by a smaller number of gradations than 256 (for instance, by 16).

Figure 6.5 shows three kinds of curves. First, the position of curve 3 relative to curve 1 characterizes the maximum effect of variations of the angular conditions of survey: zenith angles 0°–50°, viewing azimuths 0°–360°. Second, for the same weakly turbid atmosphere, the position of curve 2 relative to curve 1, characterizes the maximum effect of real non-Lambertian reflection. Third, curves 4 and 5 characterize the same regularities as curves 2 and 3, but for the strongly turbid atmosphere.

It follows, from the data of Fig. 6.5, that, neglecting the angular conditions of survey within given zenith angles and azimuths, can lead to 70% errors in the results of image processing. Neglecting the factor of reflection non-orthotropicity can give 25% errors. For the turbid atmosphere the effect of the

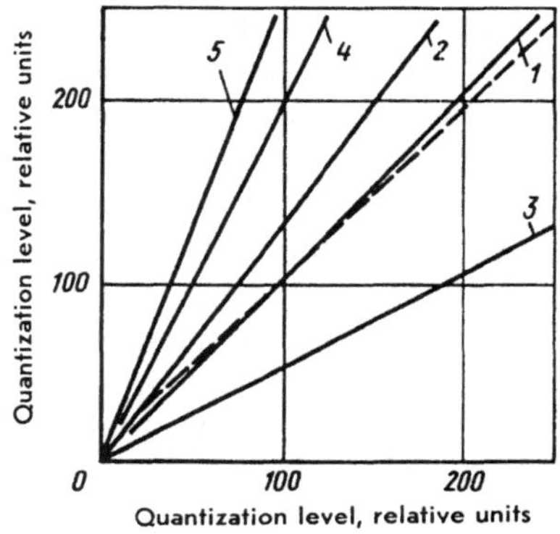

Fig. 6.5. The effect of angular conditions of survey on the conversion of the space-derived image brightnesses: *1, 4* under standard conditions, for the transparent and turbid atmosphere, respectively; *2* maximum effect of non-orthotropicity for the transparent atmosphere; *3, 5* maximum effect of the angular conditions of survey for the transparent and turbid atmosphere

angular conditions of survey and of the factor of non-orthotropicity is not important: in the transition from curve 1 to curve 4, a larger part of the haze-induced distortion is eliminated, so that the effect of the angular coordinates variations (curves 4 and 5) is less important here than the effect of variations in the state of the atmosphere.

The laws mentioned, corresponding to the curves in Fig. 6.5, are explained in the following way. In the transparent atmosphere a "limb darkening" is observed, i.e., with increased zenith viewing angle and decreased sun elevation the objects darken (their brightness decreases). On the contrary, on negative images (abscissa axis) their brightness grows; with haze brightness eliminated this additional enhancement is liquidated (ordinate axis, curve 3). The presence of the specular constituent of reflection leads to a decreasing brightness in viewing directions close to the specular reflection peak, and the transformation eliminates this increase (curve 2). For the turbid atmosphere the "limb darkening" takes place, i.e., the transition from curve 5 to curve 4 is similar to the transition from curve 2 to curve 1. This is the physics of the established procedure of image filtration.

Note, in conclusion, that the calculation scheme involving the contribution of atmospheric haze on space-derived images is sensitive in different ways to variations in these three types of initial data. Differences in sensitivity are caused by variations in physical processes of radiation transformation in the atmosphere and in the measuring instruments. As has been shown, the largest errors may occur, firstly, by neglecting the angular conditions of survey and the real state of the atmosphere at the moment of survey. Note, however, that the first group of factors (angular coordinates of measurements) has a great effect only at zenith angles exceeding 30°, but the second group (atmospheric state) effects are constant. The calculation scheme is sensitive to a fixed accuracy of radiometric measurements, but it seems that a 3–5% accuracy of absolute measurements is enough to solve the problem. The errors in assessing the angular conditions of survey and their effect on the brightness of natural objects can be minimized, too. The principal problem consists in the determination of the actual atmospheres state at the moment of survey, to eliminate the uncertainties caused by the difference between the actual state of the atmosphere and its statistical mean. The errors can reach 60%. Assessing the actual atmospheric state makes it possible to reduce these errors.

Now let us consider a more general case: how are initial and retrieved brightnesses related, when individual coefficients of approximation vary widely for the chosen "optimal" parameter $b_{0,\text{aver.}} = 0.00007$?

Here the word "optimal" means the least difference between the transformed intensities (gradations of quantization) and the initial ones.

The respective values, similar to those in Figs. 6.3–6.5, are shown in Fig. 6.6, with successive doubling of each of 15 coefficients of approximation within each of the four intervals of albedo. The initial coefficients correspond to a mid-latitude statistical mean atmospheric model [331]. It is seen from Fig. 6.6 that the calculation scheme is most sensitive to the first coefficients of approximation (especially to the first two which most transform the curves considered). The effect of coefficients with square terms of approximation (starting from coefficient 7) is small: with doubled coefficients from 7 to 15, practically, there is no variation in the way the corrected brightnesses depend

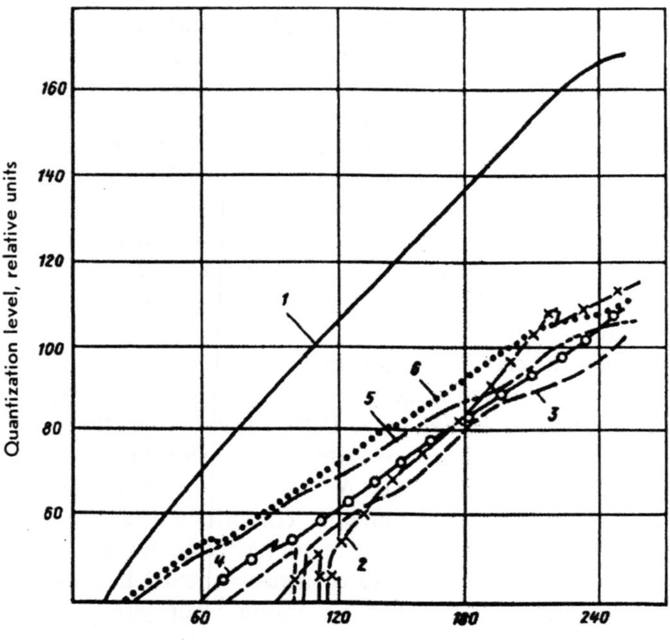

Fig. 6.6. Variations of the gradations of image quantization with doubled coefficients of approximation: *1* for the initial atmospheric model; *2* variations of the first four coefficients of approximation for each of the albedo intervals; *3* variations of the second coefficients; *4* variations of the third coefficients; *5* variations from the 4th to the 6th coefficients; *6* variations from the 7th to the 15th coefficients

on the initial ones. Thus, the first 6 terms of this approximation must be known very accurately. In general, a doubling of each of the coefficients of approximation halves the transformed brightnesses, on average. However, a "fine structure" of the curves can be observed in Fig. 6.6 due to the method of finding the roots of the square equation with respect to surface albedo. Individual bulges of curves generally do not break the monotonic course of the dependence of transformed brightnesses on the initial ones. This makes it possible to tabulate these controls to avoid the need to solve the square equation (6.2) in processing space-derived information.

Analysis of the sensitivity of the calculation scheme for the surface albedo retrieval from satellite-derived images makes it possible to tabulate the respective functions of transformation for a given optical state of the atmosphere. The errors in the transition from the coefficients of approximation to the tabulated functions of transformation are easily estimated in each firm case (for the given state of the atmosphere characterized by a certain haze brightness due to multiple scattering of solar radiation on certain ensembles of polydisperse aerosol particles).

6.5 Atmospheric Filter for Spatial Frequencies

The procedure of filtering the distortions caused by natural horizontal inhomogeneity of the Earth surface is based on the presentation of brightness $I(z,x,y,\theta,\varphi)$ and surface albedo $A(x,y)$ as a sum of constant and varying constitutens:

$$I(z,x,y,\theta,\varphi) = I_{\bar{A}}(z,\theta,\varphi) + \tilde{I}(z,x,y,\theta,\varphi) \ ,$$

$$A(x,y) = \bar{A} + \tilde{A}(x,y) \ .$$

The quantities $A(x,y)$ and $I(z,x,y,\theta,\varphi)$ are considered to be realizations of stationary random processes. In [77], in an approximation of single reflecting of the varying constituent from the Earth surface, the problem of calculating the regime of the light in the atmosphere is reduced to a successive estimation of the quantities $I_{\bar{A}}(z,\theta,\varphi)$ and $\tilde{I}(z,x,y,\theta,\varphi)$ from solutions of respective boundary problems of the theory of radiation transfer. Such an approach enables one to divide an algorithm of radiative correction into two stages.

First, using a solution for the 1-D boundary problem (3.14 – 3.16) a functional dependence is established between $I_{\bar{A}}(z,\theta,\varphi)$ and $\bar{A}$, which enables estimation of average surface albedo from space-derived average brightness of the light field. The use of this variable ensures the extraction of the atmospheric haze brightness from total intensity of the light field. Thereby, average brightness of image is corrected and its contrast intensifies. Then, the problem of retrieving the distribution of the surface albedo inhomogeneities is solved. For this purpose a relationship between $\tilde{A}(x,y)$ and the fluctuating constituent of space-measured[1] brightness is used [82]:

$$\tilde{I}(z,x,y,\theta,\varphi) = \iint_{-\infty}^{\infty} Q(z,x-\xi,y-\eta,\theta,\varphi)\tilde{A}(\xi',\eta')d\xi'd\eta' \ . \tag{6.10}$$

We shall suppose now that the first stage of correction has been realized and consider a procedure of calculation of a function $\tilde{A}(x,y)$ from (6.10). A function $Q(z,x,y,\theta,\varphi)$, being a core term of integral equation (6.10), is related to the spatial-frequency characteristic of the atmosphere by Fourier transformation:

$$Q(z,x,y,\theta,\varphi) = \frac{1}{(2\pi)^2}\iint_{-\infty}^{\infty} \psi(z,p_x,p_y,\theta,\varphi)e^{-i(xp_x+yp_y)}dp_x dp_y \ . \tag{6.11}$$

Considering the atmosphere as a linear filter of spatial frequencies we estimate the contribution of the horizontally inhomogeneous constituent of brightness, to total outgoing radiation field, for a real system of nadir observations from space. To solve this problem, we have the following initial expression [137]:

$$I(H,\vec{O},\vec{S}) = I_{\bar{A}}(H,\vec{S}) + E\left\{\tilde{A}(-\vec{r}_1)e^{-\tau_0/\eta} + \iint_{-\infty}^{\infty} Q(H_1-\vec{r}_1-\vec{\xi}')\tilde{A}(\vec{\xi}')d\vec{\xi}'\right\} \ . \tag{6.12}$$

Let us write a general expression to calculate the field of light at the receiver's aperture. Suppose, the scanner's objective has a round aperture with the FOV angle $2q$. Then the flux received is calculated with the formula:

$$\varkappa = \int_0^{2\pi}\int_0^q J(H,\vec{O},\vec{S})\eta' d\vec{S}' \tag{6.13}$$

and is proportional to the irradiance of an image pixel corresponding to a resolving element on a scene of radius $\bar{r}$:

$$\bar{r} = \varkappa \operatorname{tg} q \ . \tag{6.14}$$

The average constituent of the brightness of the field of light at the atmospheric top limited by an orthotropic surface, varies slightly within a narrow solid angle:

$$\Delta\omega = \int_0^{2\pi}\int_0^q \eta' d\vec{S}' = \frac{\pi}{2}(1-\cos 2q) < \overline{\Delta\omega} \ . \tag{6.15}$$

In addition, the function of point blurring hardly varies within $\overline{\Delta\omega}$. Therefore, taking (6.12) into account, we have the following approximate equality, instead of (6.13):

$$\varkappa \simeq I_{\bar{A}}(H,\vec{S}_H)\Delta\omega + Ee^{-\tau_0}\int_0^{2\pi}\int_0^q \tilde{A}(-\vec{r}_1)d\vec{S} + \varkappa_r \ , \tag{6.16}$$

[1] Measurements are intended to be made with an ideal optical system; the effect of the real optical system will be considered in Section 6.7.

where the quantity

$$\varkappa_r = E \int_0^{2\pi} \int_0^{q} \iint_{-\infty}^{\infty} Q(H, |\vec{r}+\vec{\xi}'|, \vec{S}_H) \tilde{A}(\vec{\xi}') d\vec{\xi}' \eta d\vec{S} , \tag{6.17}$$

is a flux of scattered radition falling on the aperture, determined by surface albedo inhomogeneity.

Taking:

$$\varkappa_{\bar{r}} = E \int_0^{2\pi} \int_0^{q} \iint_{|\vec{r}_\perp + \vec{\xi}| \le \bar{r}} Q(H, |\vec{r}+\vec{\xi}'|, \vec{S}_H) \tilde{A}(\vec{\xi}') d\vec{\xi}' \eta d\vec{S} , \tag{6.18}$$

we call:

$$\varkappa_s[\tilde{A}(\vec{\xi})\bar{r}] = \varkappa_r - \varkappa_{\bar{r}} = E \int_0^{2\pi} \int_0^{q} \iint_{|\vec{r}+\vec{\xi}| \geqslant r} Q(H, \vec{r}+\vec{\xi}', \vec{S}_H) \tilde{A}(\vec{\xi}') d\vec{\xi}' \eta d\vec{S} \tag{6.19}$$

a side additional illumination.

The side additional illumination (6.19) is a scattered radiation flux reached by an instrument sensor because of surface albedo inhomogeneities outside the FOV of the optical observational system. The following estimate can be obtained from (6.19):

$$|\varkappa_s[\tilde{A}(\vec{\xi}), \bar{r}]| \leqslant \bar{\varkappa}(\bar{r}) = |\Delta A| E \tilde{A}_0 \int_{\Delta\omega} \eta \int_{|\vec{r}+\vec{\xi}| \geqslant \bar{r}} \frac{e^{-|\vec{r}_1 + \vec{\xi}'|/l}}{2\pi l |\vec{r}_1 + \vec{\xi}'|} d\vec{\xi}' d\vec{S} \tag{6.20}$$

where $\tilde{A}_0$ is an average varying constituent of albedo for a given scene. At $\bar{r} \to 0$ (an ideal image) we find from (6.20):

$$\lim_{\bar{r} \to 0} \left[\frac{\varkappa(\bar{r})}{\Delta\omega} \right] = |\Delta A| E \tilde{A}_0 . \tag{6.21}$$

Formula (6.21) estimates maximum contribution of the horizontally inhomogeneous constituent of scattered light to the brightness of a fixed point on the photograph.

The contribution of side additional illumination to the pixel irradiance is characterized by a relative uncertainty:

$$\bar{\delta}(\bar{r}) = \frac{\varkappa(\bar{r})}{\varkappa} . \tag{6.22}$$

Maximum relative uncertainty is calculated with the formula:

$$\bar{\delta}(0) = \frac{E |\Delta A| \tilde{A}_0}{I_{\tilde{A}}(H, \vec{S}_H) + E[\tilde{A}(0) e^{-\tau_0} + |\Delta A| \tilde{A}_0]} . \tag{6.23}$$

In another extreme case $\lim_{\bar{r} \to \infty} \bar{\delta}(\bar{r}) = 0$, i.e., at large solid angles of the scattered radiation averaging, the horizontally inhomogeneous constituent of scattered radiation determined by reflection from the outlying areas of landscape, does not affect the average brightness of a fixed image pixel.

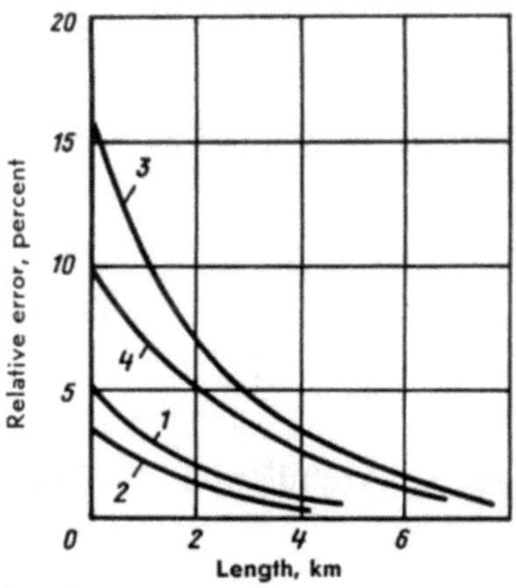

Fig. 6.7

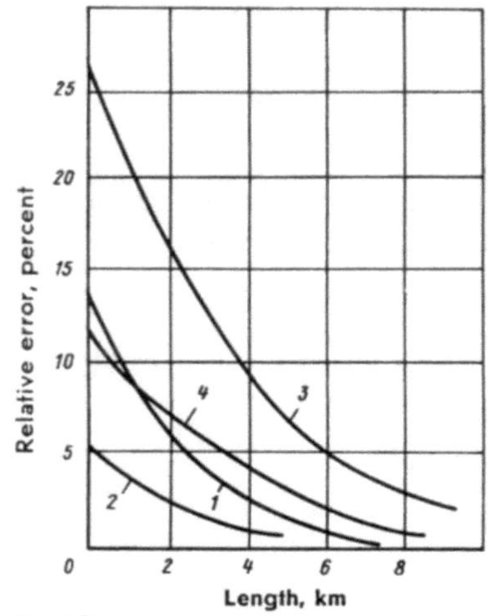

Fig. 6.8

Fig. 6.7. The relative error curves characterizing the contribution of side light to the illumination of a scanner image pixel. A light-coloured object; $\tau = 0$; $\eta = 0.866$. *1* – $\bar{A} = 0.2$, *2* – $\bar{A} = 0.6\}$ $\tau_0 = 0.2$; $x_1 = 0.796$. *3* – $\bar{A} = 0.2$, *4* – $\bar{A} = 0.6\}$ $\tau_0 = 0.6$; $x_1 = 1.7$

Fig. 6.8. The relative error curves characterizing the contribution of side light to the illumination of a scanner image pixel. A dark object; $\tau = 0$; $\eta = 0.866$. Curves *1–4* correspond to Fig. 6.7

Figures 6.7, 6.8 show the curves $\bar{\delta}(\bar{r})$ calculated with (6.22) at $\bar{A} = 0.2, 0.6$, and $\tau_0 = 0.2, 0.6$. The graphs of functions $\bar{\delta}(\bar{r})$ in Figs. 6.7, 6.8 correspond to the varying constituent of albedo at $\tilde{A}(\vec{\xi}') = |\Delta A| = 0.15$; $|\vec{\xi}'| \leqslant \bar{r}$ is a light object and $\tilde{A}(\vec{\xi}') = -\Delta A = -0.15$; $|\vec{\xi}'| \leqslant r$ is a dark object. The curves indicate the contribution of the additional side light source into the irradiance of the pixel, depending on the size of the FOV of an optical observation system. Consider, for instance, curve 3 in Fig. 6.7. Let $\bar{r} = 0.5$ km, then $\bar{\delta}(0.5) = 13\%$. For $\bar{r} = 3$ km we obtain similarly $\bar{\delta}(3) = 5\%$.

Taking 5% conditionally for relative uncertainty in space-based measurements of the light field brightness, the additional side light can be neglected in some cases (curves 1 and 2 in Fig. 6.7). The given conditions show that the effect of the side additional light on the image structure grows with increased contrast of brightnesses between object and background. These regularities are explained by increasing horizontal diffusion of photons with growing multiple scattering in a medium and growing amplitude of spatial variations in the Earth's surface brightness coefficients. Comparison of the curves in Figs. 6.7, 6.8 suggests that the side light most affects the images of weakly reflecting surface areas surrounded by strongly reflecting areas. This results from horizontal diffusion of photons from a strongly reflecting landscape towards a weakly reflecting object.

Thus, in image formation, the role of horizontal fluxes of scattered radiation determined by inhomogeneous albedos of the outlying landscape areas,

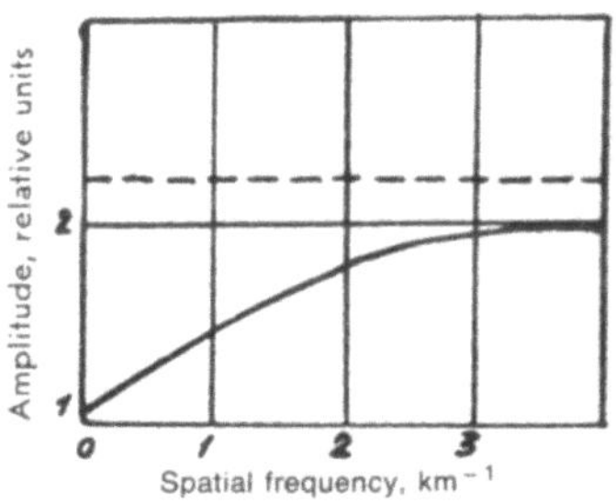

Fig. 6.9

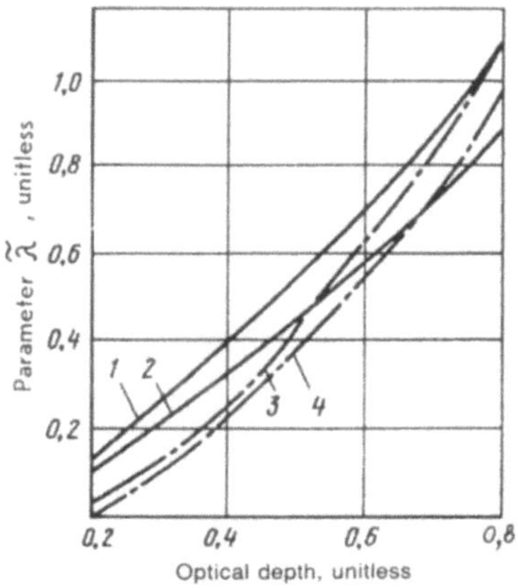

Fig. 6.10

Fig. 6.9. The spatial-frequency characteristic of the reconstructing filter

Fig. 6.10. The dependence of the parameter on the optical thickness of the environment: $\zeta = 1$; $\tau = 0$. *1* $\bar{A} = 0.2$; $x_1 = 1.19$; *2* $\bar{A} = 0.6$; $x_1 = 1.19$; *3* $\bar{A} = 0.2$; $x_1 = 1.894$; *4* $\bar{A} = 0.6$; $x_1 = 1.894$

grows with the narrowing of the directional diagram of the receiving optical system. The side light affects a photographic image most. Conversely, the structure of an image obtained with a wide-FOV scanner is independent of the side light and is distorted only by the photons reflected from surface areas lying within the FOV of the optical observation system.

6.6 Retrieving the Spatial Structure of an Ideal Image

Since the surface albedo is presented as a sum of the constant and varying constituents, the retrieval problem is solved in two stages. First, using a solution for a 1-D boundary problem, a functional dependence is established between the quantities $I_{\bar{A}}(z,S)$ and $\bar{A}$ which enables estimation of an average surface albedo from space-derived average brightness of the field of light. Also, retrieved value of $\bar{A}$ will be considered known. In the second stage the distribution of surface albedo inhomogeneities is retrieved.

For this purpose we shall use an integral equation relating $\tilde{A}(\vec{r}_\perp)$ and $\tilde{I}(z,\vec{r}_\perp,\vec{S})$. This equation can be written as:

$$E\tilde{A}(\vec{r}_\perp) = \frac{1}{(2\pi)^2}\iint\limits_{-\infty}^{\infty} \hat{\tilde{Z}}(z,\vec{p},\vec{S})\hat{\tilde{I}}(z,\vec{p},\vec{S})e^{i(\vec{p},\vec{r}_\perp)}d\vec{p}\ . \tag{6.24}$$

We shall call the function

$$\hat{\tilde{Z}}(z,\vec{p},\vec{S}) = \psi^{-1}(z,\vec{p},\vec{S}) = \frac{1}{\Lambda^{-1}+\chi(z,\vec{p},\vec{S})} \tag{6.25}$$

the spatial-frequency characteristic of the retrieving filter.

The physics of solution (6.24) is clear from analysis of (6.25). Figure 6.9 shows a section of the amplitude-frequency characteristic of the retrieving factor:

$$\tilde{R}(z,\vec{p},\vec{S}) = |\hat{\tilde{Z}}(z,\vec{p},\vec{S})| = \frac{1}{\sqrt{\Lambda^{-2}+\tilde{A}^2(z,\vec{p},\vec{S})+2\Lambda^{-1}\tilde{A}(z,\vec{p},\vec{S})\cos\tilde{\varphi}(\vec{p})}} \tag{6.26}$$

by a plane $p_y = 0$ in nadir observations. The asymptote (dashed line) lies at the level $\tilde{R}_\infty = \Lambda$.

It follows from Fig. 6.9 that, the spatial spectrum if the image is retrieved at the expense of partial suppressing of low spatial frequencies.

The solution (6.24) holds for scattering layers of a finite optical thickness and indicates mainly a possibility of retrieving an optical image passing through a very turbid atmospheric layer. Practically, at large values of $\tau(H)-\tau(z)$ the procedure of calculation of integral (6.24) becomes incorrect, since $\lim_{|\vec{p}|\to\infty}\chi(z,\vec{p},\vec{S}) = 0$. With the non-ideal character of the input optical system considered, this solution is incorrect for the layers of any optical thickness and is formally written as:

$$E\tilde{A}(\vec{r}_\perp) = \frac{1}{(2\pi)^2}\iint_{-\infty}^{\infty} \frac{\hat{\tilde{Z}}(z,\vec{p},\vec{S})}{\Pi(\vec{p})}\hat{\tilde{I}}(z,\vec{p},\vec{S})\,e^{-i(\vec{p},\vec{r}_\perp+\vec{\bar{r}}_\perp)}d\vec{p}\ , \tag{6.27}$$

where $\Pi(\vec{p})$ is the spatial-frequency characteristic of the optical system of observation.

The integral (6.27) can be calculated correctly if the limits of integrations are confined to a region in which a function is very different from zero, thereby neglecting, the super-high spatial image frequencies. A strict solution can be obtained using a regularization which will be discussed below (Sect. 6.7).

Formula (6.24) shows that the solution of this inverse problem is reduced to the calculation of the atmospheric SFCh and to a fast 2-D Fourier transformation. Technically, formula (6.24) can easily be realized by means of optical information processing. In the absence of an analog elementary base the formula can be realized by digital modeling. In this case it is more suitable to use a solution considered below.

The $E\tilde{A}(\vec{r}_\perp)$ function is equal to the following integral of convolution:

$$E\tilde{A}(\vec{r}_\perp) = \int_{-\infty}^{\infty} \hat{\tilde{Z}}(z,\vec{r}_\perp+\vec{\bar{r}}_\perp-\vec{\zeta}',\vec{S})\tilde{I}(z,\vec{\zeta},\vec{S})d\vec{\zeta}\ , \tag{6.28}$$

where

$$\hat{\tilde{Z}}(z,\vec{r}_\perp,\vec{S}) = \Lambda\left[\delta(\vec{r}_\perp)\sum_{n=1}^{\infty}(-\tilde{\lambda})^n\tilde{Q}_n(z,\vec{r}_\perp,\vec{S})\right]\ , \tag{6.29}$$

where $\tilde{Q}_n$ is a succession of the blurring functions connected with an expansion of the Q scattering function in series in multiple collisions with parameter $\tilde{\lambda} < 1$ (Fig. 6.10). In the N-th approximation we obtain:

$$E\tilde{A}_N(\vec{r}_\perp) = \Lambda \left\{ \tilde{I}(z, \vec{r}_\perp + \vec{\tilde{r}}_\perp, \vec{S}) + \sum_{n=1}^{N} (-\tilde{\lambda})^n \iint_{-\infty}^{\infty} \tilde{Q}_n(z, \vec{r}_\perp + \vec{\tilde{r}}_\perp - \vec{\zeta}, \vec{S}) \tilde{I}(z, \vec{\zeta}, \vec{S}) d\vec{\zeta} \right\} . \quad (6.30)$$

An absolute error in solution (6.30) constitutes:

$$\delta = E\,|\tilde{A}_N(\vec{r}_\perp) - A(\vec{r}_\perp)| \leqslant E\,\Delta I \Lambda \tilde{\lambda}^N , \quad (6.31)$$

where $\Delta I = \max |\tilde{I}(\vec{r}_\perp)|$.

In the first approximation the solution appears as:

$$E\tilde{A}(\vec{r}_\perp) = \Lambda \left[\tilde{I}(z, \vec{r}_\perp + \vec{\tilde{r}}_\perp, \vec{S}) - \tilde{\lambda} \iint_{-\infty}^{\infty} Q_1(z, \vec{r}_\perp + \vec{r}_\perp - \vec{\zeta}', \vec{S}) \times \tilde{I}(z, \vec{\zeta}', \vec{S}) d\vec{\zeta}' \right] . \quad (6.32)$$

The difference in solutions (6.28) and (6.32) from (6.24) occurs, because to obtain them, there is no need for the 2-D Fourier transformation of the measured field of brightness. However, expression (6.28) holds only for optically thin layers, since the $\tilde{\lambda}$ parameter grows exponentially with increasing $H - z$.

At small $\tilde{\lambda}$ the first approximation (6.32) is enough. In this case calculations of albedo variations is reduced to calculations of the convolution integral in the right-hand part of (6.32).

So, the varying constituent of space-recorded upward radiation brightness at the receiving point in the direction $\vec{S}$ can be described with an expression:

$$\tilde{I}(H, \vec{O}, \vec{S}) = E\Lambda^{-1} \left\{ \tilde{A}(-\vec{\tilde{r}}_\perp) + \tilde{\lambda} \iint_{-\infty}^{\infty} \tilde{Q}(H, |\vec{\tilde{r}}_\perp + \vec{\zeta}'|, \vec{S}_H) \tilde{A}(\vec{\zeta}') d\vec{\zeta}' \right\} , \quad (6.33)$$

where an approximate substitution is used:

$$\tilde{Q}(H, |\vec{\tilde{r}}_\perp + \vec{\zeta}'|, \vec{S}) \approx \tilde{Q}(H, |\vec{\tilde{r}}_\perp + \vec{\zeta}'|, \vec{S}_H) .$$

For an ideal photographic image, the $\tilde{I}(H, \vec{O}, \vec{S})$ value is equal to the brightness of the respective point on the image, and for a scanner image to the average brightness of respective resolving element.

We rewrite (6.23) for the coordinates of the plane of a photographic or scanner image $\vec{x} = \{x', y'\}$. Taking:

$$\tilde{A}(-\vec{\tilde{r}}_\perp) = q'(\vec{x}) , \quad \tilde{I}(H, \vec{O}, \vec{S}) \equiv \varphi'(\vec{x}) ,$$

$$\tilde{Q}(H, |\vec{\tilde{r}}_\perp + \vec{\zeta}'|, \vec{S}_H) \equiv Q'(\vec{x} - \vec{\zeta}')$$

and confining infinite limits of integration in (6.33) to the limits of an image $[-\bar{M}, \bar{M}] \times [-\bar{M}, \bar{M}]$, we obtain:

$$\varphi'(\vec{x}) = E\tilde{\Lambda}^{-1}\left\{q'(\vec{x}) + \tilde{\lambda}' \int_{-\bar{M}}^{\bar{M}} \int_{-\bar{M}}^{\bar{M}} Q'(\vec{x}-\vec{\zeta}')q'(\vec{\zeta}')d\vec{\zeta}'\right\} \tag{6.34}$$

where:

$$\tilde{\Lambda} = \Lambda|_\eta = \frac{\tilde{f}}{\sqrt{|\vec{x}'|^2+\tilde{f}^2}}, \quad \tilde{\lambda}' = \lambda|_\eta = \frac{\tilde{f}}{\sqrt{|\vec{x}'|^2+\tilde{f}^2}},$$

$\tilde{f}$ is the focus distance of the input objective.

The solution for integral Eq. (6.34) with respect to $q'(\vec{x})$ is written similarly to (6.32):

$$Eq'^{(1)}(\vec{x}') = \tilde{\Lambda}[\varphi'(\vec{x}') - \tilde{\lambda}'\tilde{F}(\vec{x}')] \tag{6.35}$$

$$\tilde{F}(\vec{x}') = \int_{\bar{M}\times\bar{M}} Q(\vec{x}'-\vec{\zeta}')\varphi'(\vec{\zeta}')d\vec{\zeta}' . \tag{6.36}$$

The convolution integral (6.36) should be understood as a term compensating the effect of scattered radiation due to inhomogeneous surface albedo. Thus, formula (6.36) expresses the means of compensation of the varying constituent of scattered radiation in the algorithm of radiative correction of space-derived images. The formula shows that, in processing a fragment of image, the sum of adjacent pixels with weights proportional to respective values of the blurring function, must be extracted from the value of the varying constituent of the brightness of a fixed pixel.

It follows from (6.36) that if the true brightness of a fixed pixel of the retrieved image is close to zero, the errors of retrieval due to neglecting the convolution integral (6.36), may be very high. Since the radius of an area of real radiative connection between natural objects under the influence of atmospheric haze decreases to some efficient quantity $\bar{\bar{R}}_\beta$, it is enough to carry out an integration in (6.36) within a circle of radius $\bar{\bar{R}}'_\beta = \bar{\bar{R}}_\beta\tilde{f}/H$. The final form of the algorithm is:

$$Eq'(\vec{x}') = \tilde{\Lambda}\left[\varphi'(\vec{x}') - \frac{\tilde{\lambda}'1}{2\pi}\int_{|\vec{x}'-\vec{\zeta}'|\leqslant\bar{\bar{R}}'_\beta} \frac{\varphi'(\vec{\zeta}')d\vec{\zeta}'}{(l^2+|\vec{x}'-\vec{\zeta}'|^2)^{3/2}}\right. . \tag{6.37}$$

For the fragments of the central part of image for $\Delta A = 0.15$ we have $\bar{R}_{0.05} = 1.2$ km. Hence, with a spatial resolution of $2\bar{r} = 200$ m the number of resolving elements in the region of efficient radiative connection $N = \bar{R}_{0.05}/\bar{r} = 12$. Since $N \sim 1/\bar{r}$, a technical realization of a discrete analog to the procedure (6.37) is greatly hindered at a high scanner's resolution. Therefore, in general, a selection of the interference caused by the horizontally inhomogeneous constituent of scattered radiation, should be made with the use of optical-electronic systems functioning independently of the image discretization step.

A total image is retrieved in accordance with the formula:

$$U(\vec{x}') = E[\bar{A} + q'^{(1)}(\vec{x}')] . \tag{6.38}$$

Formulas (6.37) and (6.38) express the essence of radiative correction of a space-derived image, with an average albedo $\bar{A}$ known.

Let us consider a case of large $\bar{A}$ – such that the resolving element of a scene is compatible with an area $\bar{\bar{D}}_\beta = 2\bar{\bar{R}}_\beta$. According to (6.32), there is the formula:

$$E\tilde{A}(-\vec{r}_\perp) = \Lambda\left[\tilde{I}(H,\vec{O},\vec{S}) - \tilde{\lambda}\int_{-\infty}^{\infty} Q(H,|\vec{r}_\perp + \vec{\zeta}'|,\vec{S})\tilde{I}(H,\vec{\zeta}',\vec{S})\,d\vec{\zeta}'\right] . \quad (6.39)$$

It is easily seen that (6.35) and (6.39) are equivalent. We multiply the left- and right-hand parts of (6.32) by η and integrate in η within a solid angle $\Delta\omega_\eta$, whose axis coincides with direction η. As a result, we obtain a formula to calculate the varying constituent of the irradiance of the resolving element:

$$\tilde{E} = E' - \Delta E' , \quad (6.40)$$

where

$$E' = \Lambda' E_{meas.} \; ; \quad \Delta E' = \lambda'\Lambda' E_{calc.} \; ; \quad E_{meas.} = \int_{\Delta\omega_\eta} \tilde{I}(H,\vec{O},\vec{S})\eta\, d\vec{S} ,$$

is the measured varying constituent of the given element; and

$$E_{calc.} = \frac{1}{2\pi}\int_{\Delta\omega_\eta}\int_{-\infty}^{\infty}\frac{\tilde{I}(H,\vec{\zeta}',\vec{S})\,d\vec{\zeta}'}{(1^2 + |\vec{\bar{r}}_\perp + \vec{\zeta}'|^2)^{3/2}}\eta\, d\vec{S} , \quad (6.41)$$

is the calculated correction for irradiance.

Thus, at large $\bar{A}$ the compensation algorithm is reduced to an element-by-element processing of the whole image and expressed by formulas (6.40), (6.41), once the full irradiance is given by the sum $\bar{E} + \tilde{E}$.

Now we shall consider the difference between the initial and filtered (as a result of the procedure of 2-D filtration described) images. The measure of difference between them is a ratio of mean square deviation $\tilde{\sigma} = \langle\varphi'(\vec{x}')\varphi'(\vec{x})\rangle^{1/2}$ to mathematical expectance $I_{\bar{A}} \approx I_{\bar{A}}(H,\vec{S}_H)$:

$$K = \tilde{\sigma}/J_{\bar{A}} . \quad (6.42)$$

Usually, the K quantity is found from the digital matrix of a real image. However, the use of a real image to study the effect of the atmosphere on its contrast is not necessary. These variations can be assessed from the analytically assumed images.

Let us consider Eq. (6.35). To simplify further samplings, we substitute, in this equation, an integration over the image area with an integration in infinite limits:

$$E q'(\vec{x}) = \tilde{\Lambda}\left[\varphi'(\vec{x}) - \tilde{\lambda}'\iint_{-\infty}^{\infty} Q'(\vec{x} - \vec{\zeta}')\varphi'(\vec{\zeta}')\,d\vec{\zeta}\right] . \quad (6.43)$$

Suppose, $E q'(\vec{x})$ is a stationary random process with a correlation function $K(\Delta\vec{x})$. Using (6.34) and (6.43) it can easily be shown that from the stationarity of the process $\varphi'(\vec{x})$ there follows the stationarity of the process $E q'(\vec{x})$, and vice versa. We shall call $\tilde{K}(\Delta\vec{x})$ the correlation function of the latter.

On our assumptions, an equality follows from (6.43):

$$K(\Delta\vec{x}) = \tilde{\Lambda}^2 \left\{ \tilde{K}(\Delta\vec{x}) + \tilde{\lambda}^2 \iiiint\limits_{-\infty}^{\infty} \tilde{K}(\Delta\vec{x}-\vec{\zeta}')G(\vec{\zeta}')d\vec{\zeta}' - \tilde{\lambda}' \iint\limits_{-\infty}^{\infty} \left[Q'(\vec{\zeta}')\tilde{K}(\Delta\vec{x}+\vec{\zeta}') + \tilde{K}(\Delta\vec{x}-\vec{\zeta}')Q'(\vec{\zeta}') \right] d\vec{\zeta}' \right\} , \quad (6.44)$$

where

$$G(\vec{\zeta}') = \iint\limits_{-\infty}^{\infty} Q'(\vec{U})Q'(\vec{U}+\vec{\zeta}')d\vec{U} . \quad (6.45)$$

The integral in (6.45) is taken in elementary functions, if an exponential approximation of the function of point blurring is used. In this case:

$$G(\vec{\zeta}) = \frac{1}{\pi} \frac{1}{(\sqrt{4l^2 + |\vec{\zeta}'|^2})^3} .$$

We take the correlation function of the process $\varphi'(\vec{x})$ as:

$$\tilde{K}(\Delta\vec{x}) = \tilde{\sigma}^2 e^{-\tilde{\beta}|\Delta\vec{x}|} .$$

Then:

$$K = (\Delta\vec{x}) = \tilde{\sigma}^2\ \tilde{\Lambda}^2 \left\{ e^{-\tilde{\beta}|\Delta\vec{x}|} + \frac{l\tilde{\lambda}^2}{\pi} \iint\limits_{-\infty}^{\infty} \frac{e^{-\tilde{\beta}|\Delta\vec{x}-\vec{\zeta}'|}}{(\sqrt{4l^2 + |\vec{\zeta}'|^2})^3} d\vec{\zeta}' - \frac{\tilde{\lambda}l}{\pi} \iint\limits_{-\infty}^{\infty} e^{-\tilde{\beta}|\Delta\vec{x}-\vec{\zeta}'|} \left[\frac{1}{(\sqrt{l^2 + |\vec{\zeta}'|^2})^3} - \frac{\tilde{\lambda}}{(\sqrt{4l^2 + |\vec{\zeta}'|^2})^3} \right] d\vec{\zeta}' \right\} . \quad (6.46)$$

Formula (6.46) shows that the correlation function of the retrieved image has a smaller half-width than the initial one. This is confirmed by smoothing the spatial distributions of brightness, by the atmosphere: in a more turbid optical image the correlation of the resolving elements intensifies. Taking $\Delta\vec{x} = 0$, we find from (6.46) a dispersion correlation between the initial and retrieved images:

$$\sigma^2 = \tilde{\sigma}^2 \tilde{\Lambda}^2 \left\{ 1 - 2l\tilde{\lambda} \int\limits_0^{\infty} \varrho e^{-\tilde{\beta}\varrho} \left[\frac{1}{\sqrt{l^2+\varrho^2}} - \frac{\lambda}{\sqrt{4l^2+\varrho^2}} \right] d\varrho \right\} . \quad (6.47)$$

Here, a factor $\tilde{\Lambda}^2$ compensates an attenuation of variations in radiance in the atmosphere after the Buger law. The integral in the right-hand part of (6.47) compensates an attenuation of variations due to horizontal fluxes of scattered radiation.

If, the $\tilde{\beta}$ parameter characterizing the size of surface albedo in homogeneities is such that, the energy spectrum of an optical image on a real scale of spatial frequencies falls on a working part of the AFCh, from $p_{thr.}$ to 3 km^{-1}, then the contrast of an optical image depends on the SFCh of the atmosphere. At $\tilde{\beta} \to \infty$ this dependence vanishes, since the integral in the right-hand part of (6.47) in this case becomes zero. At $\tilde{\beta} \to 0$ the effect of SFCh reaches its maximum:

$$\sigma^2|_{\tilde{\beta}=0} = \bar{\sigma}^2 = \tilde{\sigma}^2 \tilde{\Lambda}^2 (1 + \tilde{\lambda})^2 . \quad (6.48)$$

Table 6.6. Changes in the quality of image due to radiative correction (%)

τ_0	0.2		0.6	
$\bar{A}$	0.2	0.6	0.2	0.6
δ, %	68	52	236	204
δ_1, %	49	31	138	105
$\Delta\delta$, %	19	21	98	99

We shall introduce a relative uncertainty appearing in the comparison of the initial and retrieved images:

$$\delta = \frac{K-\tilde{K}}{\tilde{K}} \cdot 100\% \ , \tag{6.49}$$

where:

$$K = \sigma/|(E\bar{A}) \ , \quad \tilde{K} = \tilde{\sigma}/I_{\bar{A}} \ . \tag{6.50}$$

Substituting (6.50) into (6.49) we obtain:

$$\delta = \left[\frac{J_{\bar{A}}(1+\tilde{\lambda})}{\bar{A}\bar{E}\tilde{\Lambda}^{-1}} - 1\right] \cdot 100\% \ . \tag{6.51}$$

With the AFCh of scattered radiation neglected the formula (6.51) becomes:

$$\delta_1 = \left[\frac{I_{\bar{A}}}{\bar{A}\bar{E}\tilde{\Lambda}^{-1}} - 1\right] \cdot 100\% \ . \tag{6.52}$$

The difference between δ and δ_1 characterizes variations in the image quality resulting from considering the AFCh of the atmosphere in retrieving the spatial structure of the initial image:

$$\Delta\delta = \delta_1 - \delta = \frac{\tilde{\lambda} I_{\bar{A}}}{\bar{A}E\tilde{\Lambda}^{-1}} \ . \tag{6.53}$$

The results of calculations of δ, δ_1 and $\Delta\delta$ parameters for different values of the average surface albedo and the atmospheric optical thickness in the case of the nadir-viewing optical system and $\sigma^2 = \overline{\sigma^2}$ are given in Table 6.6.

It is clear from this table that, the principal way of improving the image quality, is to extract the atmospheric haze brightness from the brightness distribution over an aerospace image, with a given average albedo value.

6.7 A Regularized Solution for Inverse Problem

The problems of considering the atmospheric and instrumental distortions of the outgoing radiation, have features of their own, determined by interaction of respective procedures of image filtration in retrieving the radiative characteristics of the Earth surface.

The physics of these interrelationships is that substantial instrumental noise leads to roughening of the radiation transfer models. Vice versa, considerable errors in setting the parameters of the transfer models cause the need for higher measurement accuracies.

Optical images of the surface, fixed by a passive observational system, contain spatial-frequency distortions determined by the atmospheric effect and the transfer function of an optical receiver. The problem of retrieving a true image is reduced to the filtering of space-derived image, with the help of a linear spatial-frequency characteristic:

$$Z = [\tilde{\psi} Q]^{-1} = \Pi^{-1} \, . \tag{6.54}$$

Here, $\tilde{\psi}$ is the SFCh of the surface-atmosphere system [136]; Q is the optical transfer function of the optical system connected with the pulse transfer function [199], i.e., with the reaction of the measuring system to an input pulse forcing, and defined as a Fourier image of the pulse transfer function.

For an ideal optical system, with the use of the mathematical apparatus of the linear systems theory [199], the following relationships between the characteristics of the system measured by radiation intensity fluctuations, and required variations in surface albedo can easily be obtained:

$$\tilde{A}(x,y) = \frac{1}{(2\pi)^2} \iint\limits_{-\infty}^{\infty} \tilde{I}(z,\zeta,\eta,\vec{S}) \iint\limits_{-\infty}^{\infty} \frac{e^{-i[(\zeta-x)p_x+(\eta-y)p_y]-i\,\mathrm{arctg}\,[(\mathrm{Im}\,\tilde{\psi})/(\mathrm{Re}\,\tilde{\psi})]}}{|\tilde{\psi}(z,p_x,p_y,\vec{S})|} \, dp \, d\vec{\zeta} \, . \tag{6.55}$$

Integrating in (6.55) is a correct operation, since $\tilde{\psi}(z,p_x,p_y,\vec{S})$ at all $\vec{p} = \{p_x, p_y\}$ does not become zero. This is because there always is direct, surface-reflected radiation determining other than zero asymptotics at $|\vec{p}| \to \infty$ of the function $\tilde{\psi}(z,p_x,p_y,\vec{S})$. It is easily seen that ignoring the non-ideal character of the optical system leads to incorrect solution of (6.55). Really, in this case in (6.55) the function $\tilde{\psi}(z,\vec{p},\vec{S})$ should be changed by $\Pi = \tilde{\psi} Q$ which at $|\vec{p}| \to \infty$ is zero (finite function), and the integral in this expression is diverged. For the solution of (6.55) to be correct, infinite limits of integration are usually substituted by a region where the Π function differs greatly from zero. In this retrieving procedure, the high-frequency part of the spectrum of the true image, is lost.

Since the Q function monotonically decreases with increased spatial frequency $Q(p_x, p_y)$, the retrieval operation is reduced to raising higher frequencies and suppressing lower spatial frequencies. However, this operation is incorrect, too [38, 193], since there are errors in measuring the intensity $I(z,x,S)$ and, besides, the functions Q and $\tilde{\psi}$ are known only with some uncertainty. The necessity, to solve this problem regarding measurement errors, is explained by concern for correct retrieving of the high-frequency part of the spectrum, containing a considerable volume of information about natural objects.

Thus, to solve the problem of retrieving the object's spectrum from its space-derived values the solution (6.55) must be regularized using the techniques known [193].

Note some features of this problem determining a definite choice of the regularization procedure. First, availability of only one space-derived image of a given landscape confines the applicability of the statistical regularization technique [193] usually used in studies of natural objects. Therefore, $A(x, y)$ may be considered a deterministic function. Second, since the linear filtration theory is a suitable apparatus, to choose an algorithm of regularization we use the theory developed in [193].

Moving along the line on axis x, the scanner converts the input video signal into 1-D analog (N lines altoghether):

$$U(t) = \sum_{n=1}^{N} \iint_{-\infty}^{\infty} Q_{sc.}[v(t-nT)-x', ut-y'] I_{ent.}(z, x', y') dx' dy' \, , \tag{6.56}$$

where v – is the rate of scanning, u – the velocity of the satellite orbiting; T – the time of passing one line.

Simply assuming that the same analog signal enters an image reproducer, we write:

$$I_{exit}(x, y) = \sum_{m=1}^{M} \int_{-\infty}^{\infty} U(t) Q_{repr}[x - v(t - mT), y - ut] dt \, . \tag{6.57}$$

Combining formulas (6.55) and (6.57), we obtain an expression for the transfer function of the optical device:

$$I_{exit}(x, y) = \iint_{-\infty}^{\infty} I_{entr}(x', y') Q(x - x'; y - y') dx' dy' \, , \tag{6.58}$$

$$Q(x-x', y-y') = \sum_{n=1}^{N} \sum_{m=1}^{M} \int_{-\infty}^{\infty} Q_{sc}[v(t-nT)-x', ut-y'] \times Q_{repr}[x - v(t - mT), y - ut] dt \, . \tag{6.59}$$

A fixed space-derived image contains distortions due to fluctuating noise of instruments and of the atmosphere. The second group of errors affecting the accuracy and also leading to incorrect retrieving, appears because the calculated transfer function $\Pi(p_x, p_y)$ differs from the real one. A finite regularizing algorithm must be developed taking account of both sources of errors. However, from the methodical point of view, it is worthwhile first to consider only measurement errors.

An approximate solution for the sought-for problem has been given in [38]:

$$Z_\delta^\alpha(x, y) = \iint_{-\infty}^{\infty} \Pi^\alpha(x - \zeta', y - \eta') U_\delta(\zeta', \eta') d\zeta' d\eta' \, , \tag{6.60}$$

$$\Pi^\alpha(x - \zeta', y - \eta') = \frac{1}{(2\pi)^2} \iint_{-\infty}^{\infty} \frac{\tilde{\Pi}^*(p_x, p_y) e^{i[p_x(x-\zeta') + p_y(y-\eta')]} dp_x dp_y}{\tilde{\Pi}^*(p_x, p_y)\tilde{\Pi}(p_x, p_y) + \alpha[1 + (p_x^2 + p_y^2)^2]^2} \tag{6.61}$$

where $U_\delta(\zeta',\eta')$ is the fixed space-derived image: α – the parameter of regularization; the tilda sign points to the Fourier transformation of the core of equation Π^α; an asterix marks a complex conjugation. The finding of an optimal regularizing parameter $\alpha_{opt.}$ is difficult because it is necessary to have a-priori information about noise introduced into $U(x,y)$ and $\Pi(x-\zeta',y-\eta')$. But one can obtain the upper limit of an approximate estimate of $\alpha^*_{opt.}$ for $\alpha_{opt.}$ using an accurate formula for a 1-D problem from [193]:

$$\alpha_{opt} = B^{2/(2n+3)} \left\{ \frac{2\sigma^2\delta t}{\left(\dfrac{dz}{dt}\right)^2} \cdot \frac{1+\cos\dfrac{\pi}{n+1}}{\sin\dfrac{\pi}{2n+2}\left(\sin\dfrac{\pi}{2n+2}+\sin\dfrac{3\pi}{2n+2}\right)} \right\}^{(2n+2)/(2n+3)}, \tag{6.62}$$

where $\sigma^2\delta t$ is the spectral power of noise introduced into $U(t)$ and $\tilde{\Pi}(p) \sim B/p^n$. The order of $\alpha_{opt.}$ is preserved in a multidimensional case too.

Optimal values of the α parameter can be estimated from the following discrete scheme based on calculations of discrepancy:

$$\alpha_{S+1} = C\delta_S^2 \quad (0<C<1) \, ,$$

where $Z^0(\zeta',\eta')$ is the solution for the required problem at $\alpha_0 = \alpha^*_{opt.}$. A discrepancy δ_S which characterizes the effect of regularization, should be coordinated with an accuracy of the input data prescription:

$$\delta_S = \mathrm{mod}\left\{ \iint_{-\infty}^{\infty} \Pi(x-\zeta',y-\eta')[Z_S(\zeta',\eta')-Z_{S-1}(\zeta',\eta')]\,d\zeta' d\eta' \right\}$$

since any object has finite sizes, the estimates of $\Pi(x-\zeta',y-\eta')$ are always concentrated in a final region, therefore, the limits in (6.60) and (6.61) will be finite.

The program is suggested to be developed following the scheme:

1. A unit for $\alpha^*_{opt.}$ calculations.
2. A unit for $\Pi^\alpha(x-\zeta',y-\eta')$ calculations.
3. A unit for $Z^\alpha_\delta(x,y)$ calculations.
4. A unit to calculate the discrepancy δ_S.

The enumerated transformations of multi-spectral space-derived images can be realized on special processors on a real time scale. The problem of 2-D filtration of space derived surface images is solved thereby. The next stage of processing is an assessment of parameters of the physical state of surface objects from their corrected images.

6.8 Results of Radiative Correction of the Digitized Photographic and Scanner Information

The need for absolute measurements aimed at radiative correction of space-derived images makes severe demands on satellite-borne instruments. In this respect the MKF-6 instruments are of special interest which permit an easy conversion of measurement results into absolute energy units.

Figure 6.11 shows results of correction of multizonal images obtained experimentally with MKF-6. Principal distortions were corrected in two stages.

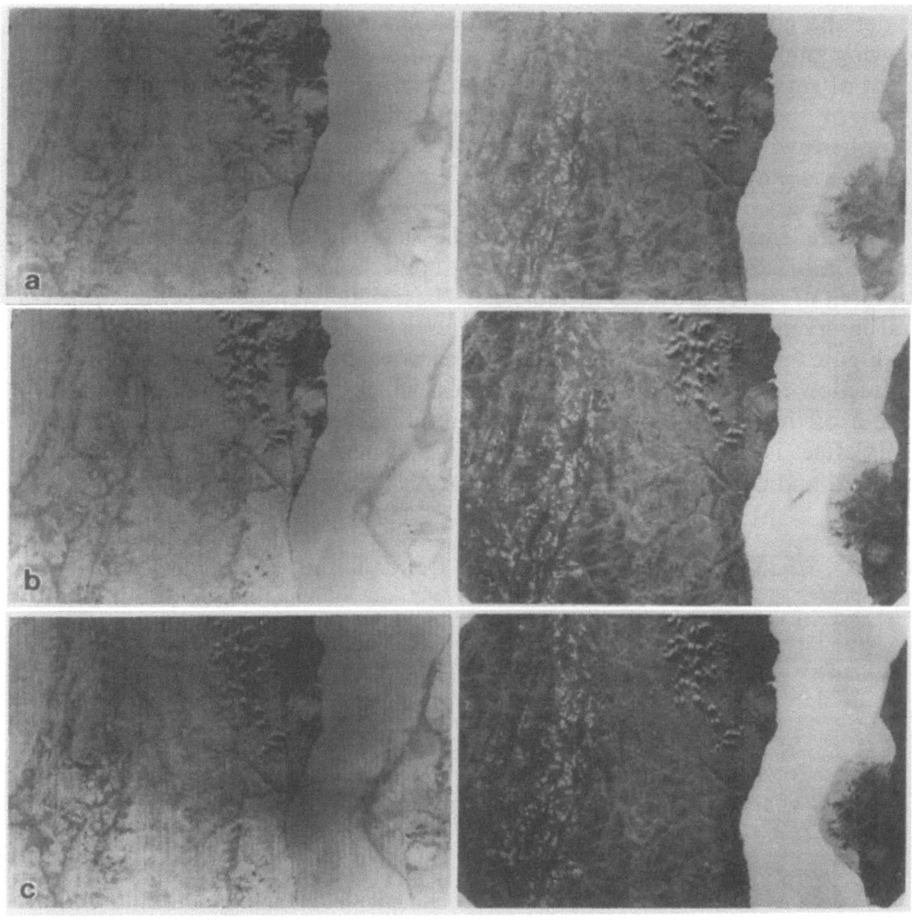

Fig. 6.11 a–c. Results of radiative correction of space-derived images obtained with the MKF-6, channels 1 and 5: **a** initial images; **b** after radiative correction based on the IFA atmospheric model; **c** after radiative correction based on the LSU atmospheric model

First, distortions caused by the inhomogeneous distribution of irradiance over the film field (radiometric correction) were removed; then distortions due to the interaction of radiation with optically active atmospheric components (radiative correction) were corrected. When accomplishing radiometric correction, the function of instrumental distortions was considered. Then, based on the photometric wedge with ten gradations of optical density, the image was presented in absolute energy units. A linear interpolation was made between the nodes of the tabulated functions of instrumental distortions and the fields of wedge brightnesses.

For radiative correction, results of modeling the processes of the formation of the outgoing radiation field (item 2) were used as well as results of approximation of the obtained dependences by orthogonal polynomials (Sect. 6.3) under the given conditions of observation (solar angle, latitude of location, viewing angles). Lacking measurements of atmospheric parameters at the moment of survey, it was decided to use results of preliminary modeling for two types of the atmosphere: an LSU (Leningrad State University) model atmosphere with a strong aerosol effect [153] and an IAP (Institute for Atmospheric Physics, USSR Academy of Sciences) model atmosphere with a weaker effect of aerosol [163].

Coefficients of approximation for the outgoing radiation intensity as a function of albedo for the given viewing angles and solar zenith angle at the moment of survey are listed in Table 6.7. These data refer to two (1 and 5) spectral intervals of MKF-6.

Figure 6.11 shows the images before and after their processing. Radiometric correction resulted in increased brightness of objects at a distance from the center of the image. Radiative correction intensified the contrasts of images. This effect is illustrated by comparison of dispersion of the brightnesses of original and corrected images (Table 6.8). As follows from the Table 6.8 data,

Table 6.7. Coefficients of approximation a_j ($j = 1, 2, \ldots, 10$) for the distorting effect of the atmosphere

j	First channel		Fifth channel	
	LSU model	IAP model	LSU model	IAP model
1	5.997×10^{-3}	5.177×10^{-3}	8.312×10^{-3}	8.343×10^{-3}
2	3.726×10^{-3}	4.439×10^{-3}	3.667×10^{-3}	3.723×10^{-3}
3	4.701×10^{-3}	4.829×10^{-3}	3.091×10^{-3}	3.114×10^{-3}
4	-1.012×10^{-3}	-1.742×10^{-4}	-4.239×10^{-3}	-4.308×10^{-3}
5	2.935×10^{-4}	1.603×10^{-4}	3.887×10^{-4}	3.879×10^{-4}
6	5.413×10^{-3}	5.325×10^{-3}	5.212×10^{-3}	5.154×10^{-3}
7	6.347×10^{-4}	2.249×10^{-4}	8.511×10^{-4}	8.517×10^{-3}
8	0	0	8.950×10^{-6}	9.005×10^{-6}
9	7.790×10^{-4}	4.926×10^{-4}	2.955×10^{-3}	1.989×10^{-3}
10	5.726×10^{-5}	9.375×10^{-6}	8.323×10^{-4}	7.675×10^{-4}

Table 6.8. Average values of digital equivalents to the optical density of image $\bar{I}_s$ and brightness dispersion I_s in MKF-6 channels 1 and 5 before and after correction

Image	I_s		I_s	
	Channel 1	Channel 5	Channel 1	Channel 5
Original	66	107	49	57
After correction (LSU model)	37	151	59	65
After correction (IAP model)	48	132	55	60

the dispersion of image brightnesses (hence, contrast) increased after correction by factors 1.45 and 1.15 (for channels 1 and 5, respectively).

An average value of post-correction image brightness changed, too. For channel 1 the brightness halved upon correction, and for channel 5 average brightness variability was negligible – by a factor of 1.1. These results agree with numerical modeling data (see Fig. 6.1). The atmospheric effect for channels 1 and 5 is substantially different: for images in channel 5 radiative correction from data on the statistical mean atmosphere seems to be enough; it is insufficient for measurements in channel 1.

Radiometric and radiative corrections to the Meteor MSU-L scanner information are not as simple as for the MKF-6 data. This is because MSU-L, in contrast to MKF-6, does not exhibit, originally, any metric properties. Nevertheless, the presence in the MSU-L scanning radiometer reverse of the half-tone and electric wedge, together with results of bench tests of these instruments, makes it possible to convert the brightnesses of image pixels from relative values into absolute ones, which is necessary for realization of radiative correction.

The radiometric correction of the MSU-L data involves elimination of distortions in information caused by the instruments themselves and in the connection channels for signals between the satellite and its recording at the surface processing complex.

The radiometric correction makes it possible to establish a unique consistency between calculated radiation transfer and data on satellite-recorded outgoing radiation. As a result of radiometric correction, the brightnesses of pixels are expressed in terms of the densities of reference-plate blackening on board the satellite. With the help of the known brightness of a reference source and the transmission functions of light filters, the field of the images is converted into absolute energy units for radiative correction to be made.

Figure 6.12 shows results of radiative correction for MSU-L channels 1 and 4. The radiometrically corrected images serve as the basis for radiative correction. Radiometric correction results in the conversion of each brightness image pixel into absolute energy units. Radiative correction reduces brightnesses recorded by MSU-L. Radiative correction was based on results in Section 6.3

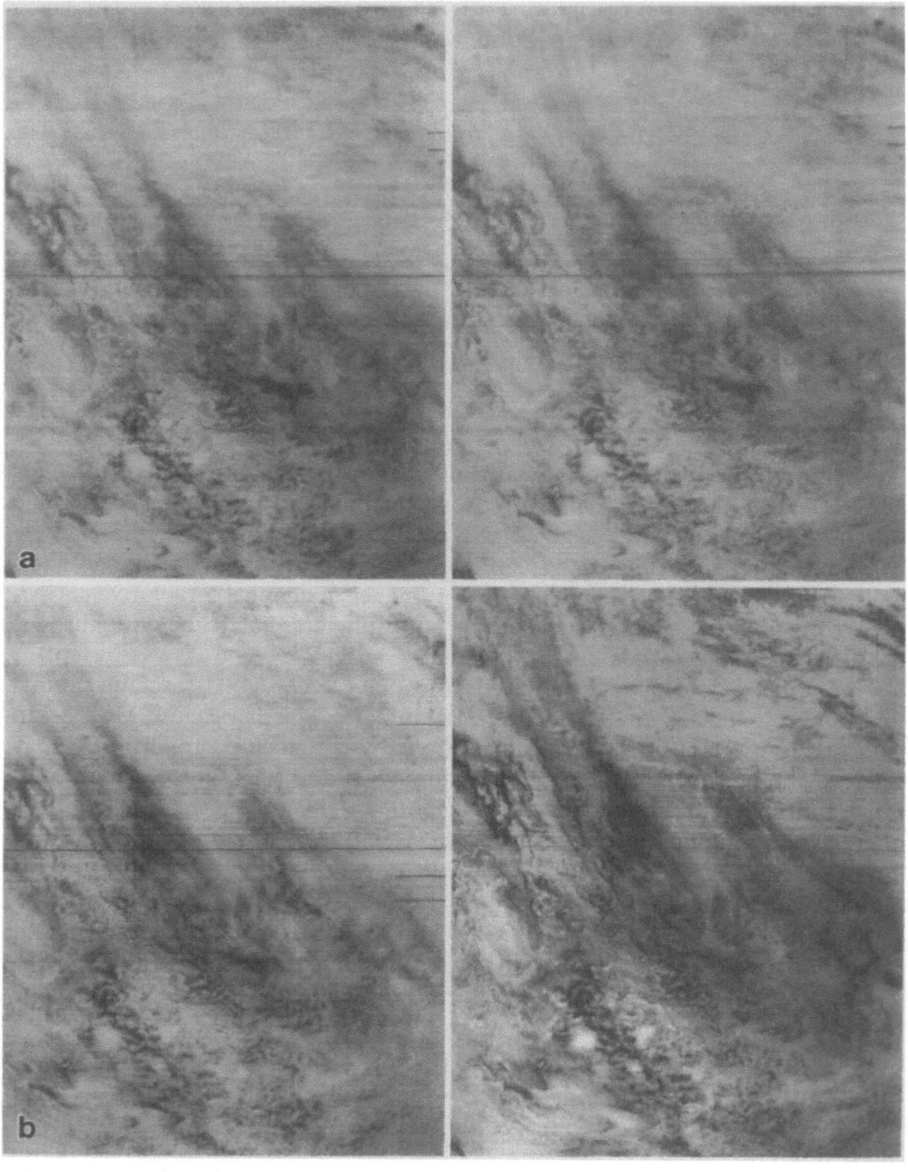

Fig. 6.12 a, b. Results of radiative correction of the MSU-L data (channels 1 and 4): **a** after radiometric correction; **b** after radiative correction

Table 6.9. The $\bar{I}_s$ averages and their dispersions I_s for meteor channels 1 and 4

Image	$\bar{I}_s$		$\tilde{I}_s$	
	Channel 1	Channel 4	Channel 1	Channel 4
Original	169	102	38	56
After correction				
LSU model	119	118	66	84
IAP model	135	120	63	57

presented as approximation coefficients (see Table 6.5) as well as on the discretizing and quantizing of images (Sect. 6.4).

Results of radiative correction shown in Fig. 6.12 refer to the mid-latitude atmosphere loaded with continental aerosol. As follows from Fig. 6.12, radiative correction has raised the contrast of images. This conclusion has been verified by calculations: a dispersion of the brightness of the image after radiative correction in channel 1 almost doubled, and in channel 4 it increased by a factor of 1.5 (Table 6.9) compared to the dispersion of the image not corrected radiatively (upon radiometric correction).

With consideration of the atmospheric effect the average brightness of image has decreased in channel 1, and in channel 4 its increases were negligible, in total accordance with results of preliminary modeling of the processes of formation of the outgoing radiation field.

The effect of the atmosphere in MSU-L channels 1 and 4 is quite different. In channel 1 this effect is reduced so that the brightnesses of dark natural objects (with low albedo) increase, and those of light-colored objects (with high albedo) decrease. As a result, the contrasts between different natural formations become smoothed. This is explained by substantial scattering of radiation in a given spectral region by the gaseous and aerosol substances of the atmosphere. In channel 4 the effect of the atmosphere is manifested through a general decrease of brightnesses of both light- and dark-colored objects, although contrasts between various objects decrease negligibly. This results from radiation absorption by atmospheric components (mainly, water vapor). Radiative correction eliminates these distortions.

The Fig. 6.12 and Table 6.9 data show that radiative correction has intensified the representing properties of the MSU-L images (their contrasts have intensified). The final conclusion about an increased information content of corrected images is determined by the identification of spectral images of natural objects, before correction and after it. Nevertheless, these results point to the possibility of radiative correction based on the described calculation scheme of radiation transfer in the surface-atmosphere system. Results of radiative correction confirm the necessity of combined sub-satellite experiments to specify atmospheric parameters at the moment of survey.

6.9 Comparison of Results of Surface Albedo Estimation from Satellites and Aircraft

The solution when considering the transforming effect of the atmosphere is reduced to the construction of atmospheric transfer function based on these or those techniques for solving the boundary problems of the theory of radiation transfer (Section 3.4). The effect of the atmosphere on the spectral brightnesses and contrasts of natural formations has been considered in [105]. Analysis has also been made, of the results from the combined sub-satellite optical experiment. Publication [77] contains calculated results for the atmospheric transfer function for some models of atmospheric aerosols as a function of viewing geometry and solar illumination of the Earth's surface.

This section is dedicated to a comparison of surface reflective characteristics measured synchronously from the MGO flying laboratory and from the Meteor satellite in November 1978. An air-borne complex has made it possible to measure spectral albedo and phase function at some altitudes in the atmosphere at different sun elevations. The digitized data of multizonal space-borne low-resolution instruments were conversed into spectral albedo with the help of an algorithm for the radiative correction of images. Radiative correction included: calculations of radiation transfer in the surface-atmosphere system for the Meteor spectral channels; obtaining of the angular distributions of radiation as functions of angular coordinates and reflective characteristics; and an approximation of calculated data by orthogonal polynomials. The atmospheric transfer function is a set of approximation coefficients for atmospheric effects which involve real processes of interaction of radiation with the gaseous and aerosol components of the atmosphere (see Section 6.3).

In accomplishing radiative correction with a specialized processing complex, signal distortions in the satellite system (radiometric correction) and distortions caused by transformation of reflected radiation in the atmosphere were investigated. Radiometric correction was based on the wedge acquired at the ground receiving stations along with space-borne scanning radiometer data. In the process of radiative correction, the albedos were retrieved for each spectral channel, viewing angles and solar zenith angles were known, the sets of approximation coefficients for atmospheric effects obtained from solutions of the radiation transfer boundary problems were used and recorded on magnetic discs in a specialized processing complex.

The air-borne measuring complex provided a unique solution for the inverse problem of retrieving the reflective characteristics for the areas observed from aircraft. The flight routes, the schemes of measurements and respective results of aircraft measurements have been given in the Proceedings of the Main Geophysical Observatory (1982, issue 462).

In constructing the atmospheric transfer function, a model of complex reflection index for atmospheric aerosol particles [330], and a model of aerosol particles size distribution [153], have been used. The scattering and absorbing properties of aerosol (cross-sections of interaction and phase functions) were

Table 6.10. Vertical distributions of the macrosections of interaction (volume coefficients of attenuation) in km^{-1} for two channels of MSU-S and two atmospheric models (model 1 – from data of LSU model of aerosol measurements; model 2 – Rayleigh atmosphere without aerosol

z, km	Channel 1		Channel 4	
	1	2	1	2
0 – 1	1.7746 – 01	1.1453 – 02	1.5851 – 01	3.6936 – 02
1 – 2	8.1209 – 02	1.0412 – 02	8.9075 – 02	2.8045 – 02
2 – 3	3.4065 – 02	9.2718 – 03	4.1091 – 02	1.9599 – 02
3 – 4	2.4892 – 02	8.1606 – 03	2.7968 – 02	1.3203 – 02
4 – 5	2.1169 – 02	7.1268 – 03	2.1284 – 02	8.6010 – 03
6 – 6	1.5087 – 02	6.1876 – 03	1.4007 – 02	5.6537 – 03
6 – 7	1.1559 – 02	5.3450 – 03	9.5259 – 03	3.7642 – 03
7 – 8	8.4578 – 03	4.5931 – 03	5.9413 – 03	2.5293 – 03
8 – 9	7.3339 – 03	3.9382 – 03	4.6807 – 03	1.7327 – 03
9 – 10	8.8600 – 03	3.3615 – 03	4.1881 – 03	1.2094 – 03
10 – 11	4.0897 – 03	2.8711 – 03	1.6769 – 03	8.5122 – 03
11 – 12	2.9839 – 03	2.4467 – 03	9.2804 – 04	6.1504 – 04
12 – 13	2.4691 – 03	2.0899 – 03	7.0989 – 04	4.6126 – 04
13 – 14	2.0896 – 03	1.8750 – 03	5.9709 – 04	3.6137 – 04
14 – 15	1.9190 – 03	1.5852 – 03	6.0935 – 04	2.9165 – 04
15 – 16	8.9511 – 03	1.3907 – 03	8.0135 – 03	2.4055 – 04
15 – 17	9.4596 – 03	1.2409 – 03	8.6644 – 03	2.0136 – 04
17 – 18	1.0704 – 02	1.1284 – 03	1.0035 – 02	1.7001 – 04
18 – 19	1.3066 – 02	1.0451 – 03	1.2534 – 02	1.4470 – 04
19 – 20	1.2190 – 02	9.7070 – 04	1.1687 – 02	1.2372 – 04
20 – 21	3.4833 – 03	9.0195 – 04	2.7569 – 03	1.0625 – 04
21 – 22	1.5847 – 03	8.3592 – 04	8.5860 – 04	9.1427 – 05
22 – 23	2.0829 – 03	7.6177 – 04	1.4291 – 03	7.9094 – 05
23 – 24	1.3206 – 03	6.8561 – 04	6.9940 – 04	6.8435 – 05
24 – 25	1.1656 – 03	6.1743 – 04	5.9642 – 04	5.9375 – 05
25 – 26	5.6456 – 04	4.6877 – 04	1.3456 – 04	4.2258 – 05
30 – 35	3.6935 – 04	2.8838 – 04	1.0507 – 04	2.7534 – 05
35 – 40	2.2300 – 04	1.5502 – 04	8.1919 – 05	1.7192 – 05
40 – 45	1.4273 – 04	8.0787 – 05	6.8947 – 05	1.0484 – 05
45 – 50	8.8278 – 05	4.0772 – 05	4.0922 – 05	6.2792 – 06
50 – 70	4.0496 – 05	1.8503 – 05	2.3865 – 05	3.1970 – 06
70 – 100	9.9342 – 06	8.1747 – 06	3.2522 – 06	1.5988 – 06
0 – 100	4.7370 – 01	9.6023 – 02	4.4085 – 01	1.2577 – 01

Note: The last three symbols denote the sign and order of respective numbers

calculated for these models based on Mie theory [75]. To assess the aerosol effect radiative correction was made for an idealized Rayleigh atmosphere without aerosol. The vertical distributions of total cross-sections of interaction of radiation with principal optically active components for the models of the atmosphere just mentioned are given in Table 6.10.

The optical thickness of the applied atmospheric models was calculated by multiplying Table 6.10 data by the geometric thickness of the respective atmospheric layers.

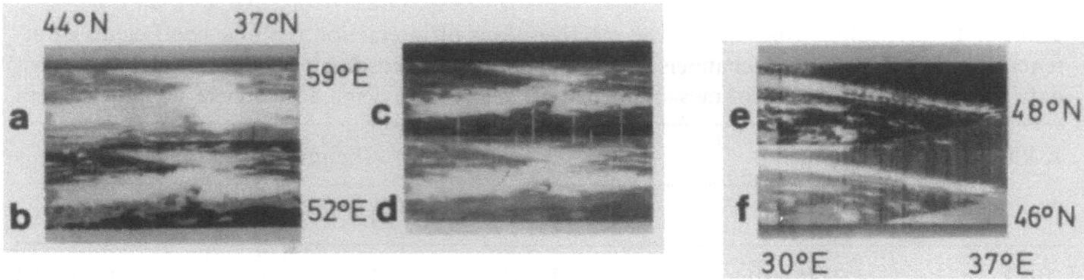

Fig. 6.13 a–f. A fragment of space images obtained with MSU-L, channel 1, and results of their radiative correction: **a** initial images of desert; **b** after correction considering reflection, from aircraft data; **c, d** after correction using aerosol models 1 and 2; **e** initial images of agricultural fields; **f** after radiative correction (model 2). Images **a–d** correspond to 37–44 °N, 52–59 °E; images **e–f** to 46–48 °N, 30–37 °E

Figure 6.13 shows fragments of original space-derived images (200×600 pixels) of the Meteor MSU-L channel 1 as well as results of corrections of these images. Fragments of images in Fig. 6.13 e, f approximately correspond to nadir viewing angles of MSU-L. Fragments of Fig. 6.13 a–d, however, correspond to slant viewing angles 40–50°. Therefore, the scales of images in Fig. 6.13 a–d and Fig. 6.13 e, f are different: in the first case the illustrated fragments cover an area 600×600 km, in the second case – 600×200 km. A visual analysis of images in Fig. 6.13 shows an increase of their contrasts after radiative correction. Results of statistical processing of the digital equivalents to images in Fig. 6.13 for channels 1 and 4 are given in Table 6.11.

The Table 6.11 data show a decrease of average brightness of image in channel 1 after correction (the atmospheric haze brightness is eliminated) and its increase in channel 4 (radiation absorbed in water vapor bands). A considerable increase of the brightness dispersion in channel 1 is a manifestation of the visually observed increase of image contrasts in Fig. 6.13. In channel 4 the growth of dispersion is less substantial.

The non-orthotropic reflection of radiation was considered by extrapolation of the air-derived anisotropy factor onto part of the image not covered by clouds. The azimuth of viewing from space of a test area of desert relative to the sun azimuth is close to 90°, and the aircraft measurement data showed the presence of the specular component at the moment of survey. This circumstance results in total brightness of areas not covered with clouds decreasing in satellite channel 1, and a dispersion in this channel increased. Under other conditions of survey it is probable that, the result of considering the reflection anisotropy could have differed. Table 6.12 gives results of comparison of the albedos of surface areas measured with air-borne instruments and obtained from radiative correction of multizonal space-derived information. As follows from the Table 6.12 data, differences in spectral albedos of a desert territory do not exceed 15% after radiative correction. Differences for agricultural sites in satellite channel 4 exceed 50%. These differences could be explained by low accuracy of radiometric correction of satellite information as

Table 6.11. Average brightnesses of images $\bar{I}_s$ and their dispersions σI_s for original images ($\bar{I}_H s, \sigma_H s$); upon radiative correction after models 1 ($\bar{I}_1 s, \sigma_1 s$) and 2 ($\bar{I}_2 s, \sigma_2 s$) of atmospheric aerosol, with reference to reflection anisotropy ($\bar{I}_3 s, \sigma_3 s$). Nominator, desert; denominator, agricultural field

Channels	$\bar{I}^S_H$	$\bar{I}^S_3$	$\bar{I}^S_2$	$\bar{I}^S_3$
1	*181*	*153*	*128*	*109*
	149	136	108	–
4	*94*	*101*	*115*	*116*
	73	75	82	–
1	*120*	*144*	*184*	*196*
	162	156	203	–
4	*67*	*75*	*80*	*88*
	52	59	74	–

Table 6.12. Surface albedo values from aircraft measurements over a desert territory (A_{200} – 200-m height) and in an agricultural zone (A_{3000} – 3-km height)

λ, µm	A_{200}	A_H	A_1	A_2	A_3	$\Delta\lambda$, µm
0.509	0.34	0.45	0.39	0.36	0.35	0.5 – 0.6
0.701	0.40	0.46	0.39	0.38	0.35	0.6 – 0.7
0.96	0.44	0.47	0.46	0.42	0.37	0.7 – 0.8
0.99	0.44	0.40	0.42	0.43	0.43	0.8 – 1.0
	A_{3000}					
0.99	0.28	0.15	0.16	0.17	–	0.8 – 1.0

Note: A_H is the albedo of the surface-atmosphere system from data of satellite MSU-L instruments; A_1 and A_2 are surface albedos after radiative correction with the use of aerosol models 1 and 2; A_3 – using reflection anisotropy; λ – the wave length; $\Delta\lambda$ – the working spectral interval for satellite-borne measuring instruments

well as by averaged radiation in channel 4 over a wide frequency band, in contrast to aircraft monochromatic measurements. However, agreement of results of aircraft and satellite measurements for homogeneous surface covers (of desert type) testifies that differences are due to neglecting the horizontal inhomogeneity of the Earth's surface.

An averaging of radiation by measuring instruments over a certain area and the neglecting of the mutual effect by natural formations can lead to disagreement between measurement results.

Some differences in the characteristics of radiation reflection (see Table 6.12) are explained both by the model of atmospheric aerosol and by neglecting reflection anisotropy. Therefore, when assessing the informative characteristics of air- and space-borne instruments and planning combined experiments with them as the basis, the optical properties of the atmosphere at the moment of experiments, have to be specified.

Also, it follows from Table 6.12 data that neglecting the atmospheric effects leads to a difference exceeding 40% between the data of aircraft and satellite measurements. Using the sets of coefficients of approximation of the atmospheric effect for the model, of Rayleigh atmosphere this error decreases to 15%. Use of the LSU aerosol model for radiative correction reduces this error below 10%. Thus, a description of the atmospheric transfer function in terms of optical models of atmospheric aerosol enables one to improve the quality of interpretation of satellite information. However, these models must be specified at the moment of survey.

6.10 Thematic Interpretation of Air- and Space-Derived Digitized Video Information

In the development of communication and management, the processing and interpretation of air- and space-based surveys have not gained importance, as one of the spheres of application of satellite information techniques to the solution of scientific and applied problems. Video information is one of the potential forms of presentation of remote sensing data on magnetic carriers of interactive (display) computers. A digital presentation of video information favors the solution of various thematic problems based on remote sensing data. Thematic interpretation is aimed at algorithms for interactive analysis of digital data onto the structure of dialogue (man – computer) processing systems with subsequent thematic mapping of the parameters characterizing the state of objects under study [79].

Separation of interdisciplinary processing from thematic interpretation is often discussed. Interdisciplinary processing implies the "normalization" of data: geographic referencing, geometric correction, conversion of data into a certain map projection, correction of radiometric (photometric) distortions, etc. At this stage of processing (it is often called "bulk processing") it is enough to have a set of special processors performing certain standard operations. It is quite different with thematic interpretation or processing in general: here a subsequent "intellectualization" of processing is needed, with due regard to the laws of formation of the outgoing radiation fields; to understanding the fine structure of spectral data and the laws of development (spatial and temporal profiles) of the objects under study, etc. The problem is that at the stage of thematic interpretation specialized mathematical transformations are needed. But they are not enough: knowledge of physical laws of the fields recorded in satellite-derived images and spectra is necessary. The related interpretation techniques are based on the principle of recognition of objects and on the assessment of their state. So far, there has not been any agreed terminology with respect to the separation between these two stages of thematic interpretation of air- and space-derived images. For instance, based on recognition of vegetation cover (agricultural crops, forests) from multispectral images, sometimes an assessment of the state (good, bad, satisfactory, etc.) is made.

Traditional visual decoding (an identification of objects with the human eye) has made it possible for an interpreter to assess the state of targets. The subsequent visual-instrumental processing has led to an interactive analysis of images and spectra obtained from aircraft and satellites. The development of the highest level thematic interpretation – with the use of "artificial intelligence" – is in prospect.

In this section, we shall describe an experience of improving the imaging properties of remote sensing data, as one of the stages in the transition from traditional methods of visual decoding to thematic interpretation. Methods for recognition of natural objects from their multispectral images are now most widely used. In this connection the emphasis is placed on the problem of atmospheric correction of video information as one of the stages of improving the imaging properties, on the one hand, and raising the accuracy of recognition, on the other hand.

6.10.1 Improving the Properties of Images

Specialized computers for thematic interpretation are usually combined into one interactive system involving special processors and universal computers. There are numerous subsidiary instruments including arrangements for the input of additional graphic or digital information, alphabetic-digital printing machines, digitizers, graphers, colour and half-tone displays, etc. Interactive schemes of processing are now being developed to ensure: the possiblity of interaction with a data set in the dialogue regime, i.e., obtaining of images and an access to a-priori data by request, memorizing the intermediate and final results of processing and classification;

- possibility of visualization, i.e., representing the digital equivalent as a black-white, half-tone or false-colour image, simultaneous viewing of several films or channels of survey, the use of the cursor and a capability of recording its trace;
- fast operating due to time-optimal algorithms, fast-input displays, easy to in operate cursors and minimum delay in the presentation of the processing results.

It is possible to feed the multispectral air- and space-derived video information directly on communication channels. In this case a multispectral videofile (a matrix of digital equivalents to a multispectral image) is formed on magnetic tape or disc with the help of a special input program packet. Before feeding the information, a continuous analogue signal is discretized along the scanning line (discretization in the direction normal to scanning lines is made by the scanner's raster) and the signal is quantized into 256 conditional gradations, by the analogue-digital converter.

All the conversions of the multispectral videofile accomplished in the process of thematic interpretation can be divided conditionally into two large classes: brightness and geometry. Independent of the class, the conversions may be both intrachannel and interchannel. Principal conversions are:

- intrachannel element-by-element,
- intrachannel local,
- interchannel element-by-element.

Geometric conversions consist of changing the geometry of objects' images, the brightness properties of images varying negligibly. Various equations of geometric conversions serve as the law for changing the element indices.

The videofile is subject to similar geometric conversions during transformation into the map projection, but the transformation equations are much more complicated.

The class of brightness conversions covers a variety of arithmetic, algebraic and trigonometric operations with the matrix elements of digital equivalents to multispectral images.

The brightness conversions are usually aimed at:

1. Changing the imaging properties of the survey material prior to visual interpretation, for instance, underlining the contours, synthesizing, changing the dynamic range of brightnesses;
2. Filtration, i.e., eliminating one or other constituent of the recorded signal (for instance, elimination of various interferences);
3. Compressing, i.e., reducing the volume of processed data, and decreasing the number of internal fluctuations of a homogeneous object, without information content loss;
4. Presenting results of processing (for instance, probability analysis) in the desired form.

The underlying contour conversion is based on obtaining a gradient image by choosing the recorded radiance gradients between the elements of the original matrix of digital equivalents. The gradient is estimated by a sliding window, contouring each element with subsequent comparison of the function obtained with that given beforehand (a threshold value of gradient).

By superimposing the obtained image gradient on the original we obtain a new image with underlined contours.

The synthesizing may be both intrachannel and interchannel. As a result, we obtain a new image with the characteristics very different from those of the original. For instance, such operations as summation, extraction, raising to a power, using a logarithm, using exponential functions, etc. are widely used in thematic interpretation of the materials of survey and are aimed at selecting a group of objects on an image with similar spectral features. Also, these operations, especially power and exponential, can be useful in the fine-structure analysis of homogeneous objects.

A special place among the brightness conversions of the class "synthesizing" is occupied by color images, the most interesting being results of technique for additive mixing of digital brightness equivalents from several channels. With color displays or color photoregistrators this operation is quite easy to perform.

Digitized spectral images of natural objects are contoured at all levels of the assumed range of quantization. However, in several cases the interpreter is interested in a particular object (for instance, water surface image), whose spectral range of brightness is much narower than the range of quantization. In this case the procedures of changing the dynamic range of brightnesses for the original digital equivalent, such as: photometric cuttings with subsequent linear scaling of the brightness fields; equalization; hyperequalization, can be useful.

Linear scaling of the brightness fields, interesting for the interpreter of an assumed range of quantization, is accomplished using the formula of linear expansion of the dynamic range of brightness for the chosen class of objects.

Equalization is an attempt at an optimal presentation of image in the assumed range of quantization: since the human visual system is more sensitive to half-tone gradients than to their absolute quantities, images with brightnesses uniformly distributed over the entire dynamic range are an optimal form of presenting video information. In accordance with this, the areas of the dynamic range with numerous image details are divided into smaller ones, and the remaining areas of the range into larger areas. The conversion is accomplished so that the digital image equivalent with an arbitrary histogram be transformed into a digital equivalent with the histogram of uniform distribution of brightness. However, this method does not consider some important features of the human eye. In the visual system the effect is proportional to a logarithm of the incident light intensity. The conversion leading to a uniform distribution of the brightness logarithms was called "hyperequalization". This conversion gives a hyperbolic histogram.

From the thematic interpretation viewpoint, the problem of filtering the digital image equivalents is usually considered as a method enabling exclusion from the image of some constituents of the recorded signal caused by interference. This problem will be considered next, together with the problem of atmospheric correction of video information as one of the ways of image filtration.

Materials of air- and space-based survey transmitted to the processing system via communication channels, in some cases, cannot be processed in real-time because of imperfect computation systems. Information must be compressed without information loss in digital image equivalents. The problem of information content of remote sensing data is closely connected with the problem of recognition of natural objects from their multispectral images.

6.10.2 Principles of Recognition of Images

An image is a description of an object (for instance, by sight in visual decoding, mathematical in visual-instrumental processing of images, etc.). A class of images is a category characterized by some properties common for every elements. Recognition of images is an attribution of the current data to a certain class by choosing significant indications or properties from a mass of insignificant details. Solution of the problem of recognition presumes the existence of classification system (the problem of classification of objects) and inventory of indications (the problem of identification of objects by certain indications), as well as some decision-making rule: about attributing selected objects to one or other classes by certain features. In crop recognition from their multispectral images the targets are winter wheat, maize, sugar-beet, spring barley, etc.

The indications by which the chosen classes can be described are as follows: height, crop thickness, projective vegetation cover, phytomass volume, etc. It is a question of selecting classes for the state of enumerated objects (good, satisfactory, bad, etc.). The task is to choose the most informative indications (both quantitative and qualitative) and to describe each class in language with sufficient indications. A description of classes usually contains information: are or are not these (or other) indications inherent to a given class? (logical recognition); what are the laws of distribution of indications by their significance (statistical recognition)? Here we consider in detail the more widely used statistical recognition. If the survey revealed some indications of an unknown object to be recognized, a posteriori data must be compared with an a priori description of all classes of objects in the indications language. This comparison makes it possible to attribute an unknown object to this or that class, i.e., to recognize it.

The problem of description of the classes of objects in an indication language has no unique solution [39]. Depending on the volume of a priori information, there are the following methods for the recognition: direct processing of initial data (without learning); methods for learning; and methods for self-learning. In the first case the initial information on the target is enough to divide a variety of objects into classes $\Omega_1, \ldots, \Omega_m$, to compose an inventory of indications $x_1, \ldots, x_N$, and, on the basis of direct processing of initial data, to describe each class of objects in an indication language. In the second case, the initial a priori information is enough so that, in accordance with the chosen indication for classification, to divide a multitude of objects into classes $\Omega_1, \ldots, \Omega_m$ and to compose an inventory of indications $x_1, \ldots, x_N$, but not enough for the classes to be described in an indication language. Here, a learning succession or learning sampling must take place, and a system of recognition must be constructed by a scheme "with a teacher" (see later). In the third case, the initial a priori information is enough to compose an inventory of indications $x_1, \ldots, x_N$ but not enough to classify the objects. These are self-learning systems where the initial information is either absent or less than in the two other cases.

The systems or recognition "without learning" are constructed, if the kind and parameters of conditional probability densities $f_i(x_1, \ldots, x_N)$, where $i = 1, \ldots, m$, are known a priori, as well as probabilities $P(\Omega_i)$ of occurrence of classes and payment matrix (loss function) $||C||$, whose elements, with a given description of classes $\{f_i(x), P(\Omega_i)\}$, $\vec{x} = \{x_1, \ldots, x_N\}$, characterize the risk (mathematical expectation of penalty) for correct and erroneous solutions. These systems function when the volume of a priori information makes it possible, by direct processing of initial data, to detemine respective functional dependences with an accuracy ensuring the solution of the problem of recognition with a pre-set probability of erroneous solutions.

In the systems "with learning" a priori probabilities $P(\Omega_i)$ and probability densities $f_i(x_1, \ldots, x_N)$ are unknown. In this case based on the presentation to the system of recognition of a learning sample with indicated respective classes, hyper-surfaces are constructed in a multidimensional indication space which divide this space (in the best way, in a sense) into regions D_i corresponding to classes Ω_i, $i = 1, \ldots, m$.

In the "self-learning" systems functions $P(\Omega_i)$ and $f_i\ (x_1, \ldots, x_N)$ are unknown, too, but on the basis of measurement results, the objects are classified so that substantial maxima of a function $f(\vec{x}) = \sum_{i=1}^{m} P(\Omega_i) f_i(\vec{x})$ correspond to the "centers" of classes in the indication space.

Let us consider further the problem of classification "without learning" based on the methods for cluster analysis [47]. We proceed from the fact that we have m lines of a space-derived image, each of them containing n pixels. Each pixel is the k-dimensional vector of observations $(k = 1, 2, \ldots, K)$.

This problem consists of a representation of a given succession of multispectral data as a certain number of classes, each being characterized either by statistical homogeneity or by a certain similarity of spectral characteristics. The problem is whether the sampled multispectral data belong to one population. Here a testing is made, whether a given small number M (usually less than or equal to 6) of sampled data really forms a homogeneous population. Each sample is a K-dimensional vector.

If M samples are defined through $I_{i,k}\,(i = 1, 2, \ldots, M; k = 1, 2, \ldots, K)$, one can calculate the vector of averages $\bar{I}_{1,k}$ and the square of the distance between the sample and average vector:

$$\bar{I}_{1,k} = \frac{1}{M} \sum_{i=1}^{M} I_{i,k} \tag{6.63}$$

$$\Delta I_i^2 = \sum_{k=1}^{K} (I_{i,k} - \bar{I}_{1,k})^2 \ . \tag{6.64}$$

Now we shall estimate the maximum of these M quantities (ΔI_{max}^2) and compare it with the square of the vector module of averages $(|\bar{I}_{1,k}|^2)$. If:

$$\frac{\Delta I_{max}^2}{|\bar{I}_{1,k}|^2} \leqslant T \ , \tag{6.65}$$

where T is the some given threshold (as we see, here some type of "learning" is mentioned, hence, everything is arbitrary), then it is possible to combine these M samples into a certain population. Here $|\bar{I}_{1,k}|$ is the length of average vector. If condition (6.65) is not fulfilled, the whole process is repeated for the next sample until the first population is obtained.

Let W be the number of populations thus obtained. For each i-th population the number of samples m_i, averages $\bar{I}_{i,k}$ and other statistical parameters (for instance, dispersions, mean-square deviations, etc.) are known. A major objective of the cluster-analysis procedure is to reduce the number of populations obtained to some pre-set number W_{max} (again, there is a certain amount of "learning", which is unavoidable: a division into classes in real conditions has a goal-oriented character). The possibility appears, thereby, to combine two most similar populations with the help of the most often used measure of similarity – Euclidian distance-between averages of these populations. Then, averages and dispersions are calculated for a new population from populations i_1 and i_2. If the numbers of samples in original populations are m_{i_1} and m_{i_2}, then an average for these populations is found from the expression:

$$\bar{I}_{i,k} = \frac{1}{m_{i_1} + m_{i_2}} (m_{i_1} \bar{I}_{i_{1,k}} + m_{i_2} \bar{I}_{i_{2,k}}) \ . \tag{6.66}$$

A similar expression can be written for dispersions and mean-square deviations.

The next stage of cluster-analysis consists of testing whether new samples belong to the established populations. This test consists of two stages. The first stage is based on the distribution χ^2, the second stage on the normal distribution. We shall denote them as χ^2-test and N-test.

The sampled dispersion vector of the i-th population with m_i accumulated samples is determined from the expression:

$$S^2_{m_{i,k}} = \frac{1}{m_i - 1} \sum_{i=1}^{m_i} (I_{i,k} - \bar{I}_{i,k})^2 \ . \tag{6.67}$$

This expression is usually used to obtain a random variable χ^2 with $m_i - 1$ degrees of freedom:

$$\chi^2(m_i - 1) = \frac{(m_i - 1) S^2_{m_{i,k}}}{\sigma^2_{i,k}} \ , \tag{6.68}$$

where $\sigma^2_{i,k}$ is an unknown expected dispersion vector, in contrast to a sampled dispersion vector $S^2_{m_i,k}$. The $\sigma^2_{i,k}$ parameter is constant, if the resolving elements to be classified are statistically independent and belong to one class. Under these two conditions, the function of distribution of χ^2 variable is known from statistical tables. In practice, these conditions are not fulfilled, therefore, it is ncessary to obtain related estimates. To estimate the $\sigma^2_{i,k}$ parameter, we can use the most probable $\chi^2(m_i - 1)_{M:P.}$ value:

$$\chi^2(m_i - 1)|_{M.P.} = m_i - 3 \ , \quad m_i \geqslant 4 \ . \tag{6.69}$$

From this expression we obtain the most probable dispersion:

$$\hat{S}^2_{m_{i,k}} = \frac{m_i - 3}{m_i - 1}\hat{\sigma}^2_{i,k} \; , \tag{6.70}$$

where $\hat{\sigma}^2_{i,k}$ is the most probable value of $\sigma^2_{i,k}$.

The $\hat{\sigma}^2_{i,k}$ value is estimated from the expression:

$$\hat{\sigma}^2_{i,k} = \frac{\sum_{m_i'=4}^{m_i} \frac{m_i' - 1}{m_i' - 3} S^2_{m_{i,k}}}{2 \sum_{m_i'=4}^{m_i} \sqrt{\frac{m_i' - 1}{m_i' - 3}} S_{m'_{i,k}}} \; . \tag{6.71}$$

as the least square deviation between the most probable dispersion and the current (actual) sampled dispersion, i.e., from the condition:

$$\frac{\partial}{\partial \hat{\sigma}_{i,k}} \sum_{m_i'=4}^{m_i} \left(\frac{S_{m'_{i,k}} - \frac{m_i' - 3}{m_i' - 1}\hat{\sigma}_{i,k}}{\frac{m_i' - 3}{m_i' - 1}\hat{\sigma}_{i,k}} \right) = 0 \; . \tag{6.72}$$

The current dispersion will oscillate about its most probable value. The measure of this oscillation is expressed, as usual, by the following confidence interval with a $(1-2p)\cdot 100\%$ confidence level:

$$\frac{\lambda_{1-p}(m_i - 1)}{\sqrt{m_i - 1}} \leqslant \max_k \left[\frac{S_{m_{i,k}}}{\hat{\sigma}_{i,k}} \right] \leqslant \frac{\chi_p(m_i - 1)}{\sqrt{m_i - 1}} \; , \tag{6.73}$$

where $\chi_{1-p}(m_i - 1)$ is a $(1-p)\cdot 100\%$ point of distribution of χ^2 with $(m_i - 1)$ degrees of freedom.

A new sample does not belong to the i-th population, if (6.73) is not fulfilled. Thus, we say that the image is classified with a confidence level determined from expression (6.73). However, each sample may satisfy this inequality for more than one population. Therefore, an N-test is needed for the final classification, which solves the ambiguities of the χ^2-test by checking the normalized values of a new sample $I_{j,k}$ based on current sampled averages and standard deviations. Those $I_{j,k}$ are checked which satisfy (6.73), with values

$$Y_{i,k} \neq \frac{I_{j,k} - \bar{I}_{i,k}}{S_{m_{i,k}}} \tag{6.74}$$

meeting this inequality for all i. Then, we find a minimum by i and refer the new sample to a population for which the $Y_{i,k}$ value is minimum.

In practice, the stage if image classification without learning is reduced to a division of the digital image equivalent into homogeneous sub-divisions with the help of the foregoing procedure of cluster-analysis. This division is based on constructing a multidimensional histogram with selection of oriented graphs, each of them (its apexes) corresponding to one cluster. An identification of thus obtained clusters is made with the use of materials of synchronous sub-satellite (surface, aircraft) measurements. Because of a limited number of such ("test") measurements, the identification is usually reduced to the problem of linear interpolation and extrapolation.

The system of recognition with learning and self-learning consists of the same stages of selecting homogeneous sub-divisions described above. In the process of learning, the recognition system faces some "references" (reference objects), each describing a spectral image of one of N-classes. The function of probability distribution density $f(\vec{x})$ is retrieved for each of these references based, for instance, on the multidimensional normal distribution law (the most often used supposition). At the same stage, based on some considerations, a priori probabilities $P(\Omega_i)$ of occurrence of each of the classes of objects should be prescribed, and then a function $f(x)$ is retrieved over the whole image. The N classes of interest are to be recognized. The remaining classes of objects are combined into the $(N+1)$ class. At the stage of classification with learning, the probabilities of their belonging to each of $N+1$ classes are calculated for each of the vectors. An image pixel is considered to belong to a class for which the obtained a posteriori probability is the greatest.

In statistical recognition of images the Bayes strategy is usually used, which ensures a minimum average risk (minimum of erroneous solutions). The a posteriori probability that in realization of the event $a_N = (x_1, x_2, \ldots, x_N)$ the object refers to the i-th class, is:

$$P(\Omega_i/a_N) = \frac{P(\Omega_i) f_i(x_1, \ldots, x_N)}{\sum\limits_{i=1}^{m} P(\Omega_i) f_i(x_1, \ldots, x_N)}. \tag{6.75}$$

If a priori probabilities of occurrence of the objects of respective classes are unknown, a mini-max criterion is usually used – a maximum possible value of average risk is minimized. In this case the Bayes strategy is used with such value of $P(\Omega_i)$ that an average risk of erroneous solution is maximum, i.e., this strategy gives a cautious but guaranteed value of an average risk. The a forementioned statistical separability of objects characterizes the accuracy of recongnition and is connected with the probability of making correct and erroneous decisions.

When not only a priori probability is unknown but also the payment matrix, the Neiman-Pierson criterion [39] is often used: a permissible (prescribed) value of conditional probability of the first-order error (probability of false alarm) is estimated and then the boundary between classes is determined by keeping to the one which contrives to reach a minimum of conditional probability of the second-order error (probability of missing the target).

As far as the essence of the learning procedure is concerned, here, together with a description of classes, it is necessary, to know the list of objects with an indicated class for each object. If $\omega_1, \ldots, \omega_l$ are the objects on this list, and $\Omega_1, \ldots, \Omega_m$ are classes, then, with available initial information, one can refer to this or that set of objects as belonging to certain classes. Since each object can be described in the indication language $x_1, \ldots, x_N$, and values of indications for an object ω_1 constitute $(x_1^{(1)}, \ldots, x_N^{(1)})$, and for an object $\omega_l(x_1^{(l)}, \ldots, x_N^{(l)})$, the learning succession or learning sample is:

$$\begin{aligned}&[(x_1^{(1)}, x_2^{(1)}, \ldots, x_N^{(1)}); \ldots; (x_1^{(r)}, x_2^{(r)}, \ldots, x_N^{(r)})] \,\varepsilon\, \Omega_1\\ &[(x_1^{(q)}, x_2^{(q)}, \ldots, x_N^{(q)}); \ldots; (x_1^{(l)}, x_2^{(l)}, \ldots, x_N^{(l)})] \,\varepsilon\, \Omega_m \;.\end{aligned} \tag{6.76}$$

As has been said above, a priori probabilities $P(\Omega_i)$ and conditional densities of distribution $f_i(x_1, \ldots, x_N)$ are unknown. The problem is reduced to constructing in the multidimensional indication space a hyper-surface dividing this space into areas D_i corresponding to classes Ω_i, $i = 1, \ldots, m$. Expressing the dividing function through:

$$\hat{y} = f(\vec{x}\,\vec{c}) \;, \tag{6.77}$$

where $\vec{c} = (c_1, \ldots, c_N)$ is an unknown vector of parameters of the state of the objects under study, $(\vec{x}\,\vec{c})$ the scalar product of vectors $\vec{x}$ and $\vec{c}$, we can determine some degree of deviation $f(\vec{x}\,\vec{c})$ from the indication function y determined for two classes as:

$$y = \begin{cases} -1 \;, & \text{if } \omega_d \,\varepsilon\, \Omega_1 \\ \;\;\,1 \;, & \text{if } \omega_d \,\varepsilon\, \Omega_2 \;, \quad d = 1, 2, \ldots, l \;. \end{cases} \tag{6.78}$$

If we choose the following functional (mathematical expectation of a function F from the $y - \hat{y}$ difference):

$$J(\vec{c}) = E\{F[y - f(\vec{x}\,\vec{c})]\} \tag{6.79}$$

as a degree characterizing the choice of the division function, the problem is reduced to an assessment of vector $\vec{c} = \vec{c}^{\,0}$, giving a minimum to functional (6.79). This problem is similar, to some extent, to a search for the most probable value of dispersion in expression (6.71) and consists of equalizing the gradient of functional $J(\vec{c})$ in $\vec{c}$ to zero with subsequent determination of an optimal value for $\vec{c} = \vec{c}^{\,0}$, at which the functional $J(\vec{c})$ reaches an extreme. In conditions of total a priori information, the density of distribution $p(\vec{x})$ is known, the vector $\vec{c} = (c_1, \ldots, c_N)$ is prescribed, the functional $Q(\vec{x}, \vec{c}) = F[y - f(\vec{x}\,\vec{c})]$ is known, and, therefore, a random functional is also known:

$$J(\vec{c}) = \int_X Q(\vec{x}\,\vec{c})\, p(\vec{x})\, d\vec{x} \;, \tag{6.80}$$

where X is the space of vector x. Therefore, $J(\vec{c})$ and its gradient in $\vec{c}$ can be expressed in explicit form, which makes it possible to estimate an optimal value for $\vec{c} = \vec{c}^{\,0}$. The lack of total a priori information in learning systems (the density of distribution and mathematical expectation E{F} are unknown), at the first stage of determining the vector $\vec{c}^{\,0}$ causes, use of a priori information contained in the learning sample, and at the second stage – the current a posteriori information. The usefulness (convolution) of the learning procedure is characterized by a limit:

$$\lim_{n \to \infty} E\{||\vec{c}(n) - \vec{c}^0||^2\} = 0 , \tag{6.81}$$

where $\vec{c}(n)$ is the estimate of the vector $\vec{c}$ at the n-th step for the discrete algorithm of learning and:

$$\lim_{t \to \infty} E\{||\vec{c}(t) - \vec{c}^0||^2\} = 0 \tag{6.82}$$

for the continuous algorithm of learning.

An experience of thematic interpretation of remote sensing data shows that, the development of the enumerated studies on recognition of the images of natural objects, takes place in the direction of constructing a "self-learning system" – a component of the model of artificial intellect. A major objective of these studies is the use of certain information characterizing the quality of data and techniques for their transformation. Another component of thematic interpretation is the development of certain procedures of mathematical filtration of images to select useful signals against the background of "noise constituents" of the scenes under analysis. Atmospheric correction of digital air- and space-derived video information is an example of such procedures.

6.10.3 Features of the Applied Method for Classification

Difficulties of reliable classification of objects from space-derived middle-resolution images are determined by: mutual overlapping of spectral reflectivities of various natural formations; by strong variability of indications by which the classes are selected; by a large number of modes on histograms of relative brightness frequencies; and by other reasons. The quantitative characteristics of these indications are usually found using methods of cluster-analysis [47]. These methods are based on the grouping of various objects by indications, which leads, in particular, to an appearance of the brightness of individual clusters or modes on the histograms. By selecting these modes, one can determine the boundaries separating them – threshold values of indications and, thereby, to classify natural objects [211].

The principle of modes' quantizing [165] is used in the classification of natural objects using the histogram technique. Here an object is individual sub-divisions of images. These are not necessarily objects are usually understood (soil, vegetation, water, etc.) but some combined descriptions of classes which contain usual objects with definite weights. It is this methodical description of objects which is needed in analysis of variations taking place in the images after their atmospheric correction. Only additional information about the region of survey, the phase of vegetation development, the type of soil, etc. makes it possible to compare the term "object" and real natural formations (see item 6.12.2).

Based on [165], we use the respective distribution percentiles as threshold levels of histograms, and derivatives $d\tilde{v}/d\tilde{D}$ as characteristic features $\tilde{x}$ in the pattern recognition technique used after running means of the histograms by a three-point filter "arithmetical mean". The distribution percentiles divide the region of variation of $\tilde{x}$ into intervals with equal probabilities of getting into them. Sharp changes of derivatives $d\tilde{v}/d\tilde{D}$ point to the existence of "suspicious" boundaries of different classes. It does not mean that these different classes do exist really. By successive sorting of these boundaries, under conditions where the needed criteria are satisfied, first two objects can be separated, then two others, with the previous two objects considered as one object, etc. The problem is that, using the method of percentiles, maximum closeness of the number of elements located on both sides of the threshold (dividing boundary) be reached. For this purpose the following matrix was calculated [39]:

$$L(\omega, \Omega_g) = \sqrt{\frac{1}{k} \sum_{s=1}^{k} \sum_{j=1}^{N} W_j (\tilde{x}_j - \tilde{x}_{gsj})^2} \; ,$$

where W_j is the weights of the applied indications $\tilde{x}_j$ of an object ω and indications $\tilde{x}_{gs}$ of class $\Omega_g, g = 1, 2$. Obtaining minimum distance between the given and other objects, one can draw the deciding boundary in the first approximation.

This measure of closeness has been specified in [165] based on calculations of areas occupied by sub-divisions ("objects"). These areas play the role of a priori probabilities $P_r(\Omega_1)$ and $P_r(\Omega_2)$ needed for the best division of classes. The coefficient of probability:

$$\lambda_0 = \frac{P_r(\Omega_1)(C_{12} - C_{11})}{P_r(\Omega_2)(C_{21} - C_{22})}$$

determines the deciding boundary specifying the best place for the threshold [165]. Here C_{ij} is elements of payment matrix characterizing at $i = j$ correct decisions, and at $i \neq j$ the errors of "false alarm" and "missing the target" [39, 195]. Elements C_{ij} – are determined by comparing the current values of measure L with its local extreme.

Table 6.13 exemplifies the values of L and λ_0 for some levels of quantization $\tilde{D}$ of a histogram in Fig. 6.14 near the deciding boundaries. Analysis of Table 6.13 data shows that the thresholds can be placed best in the proposed multi-stage search for boundaries between the following levels: 82–83, 109–110, 122–123, 131–132, 145–146, 156–157. This division of classes is marked by shading in Fig. 6.14. Near these levels not only the derivative $d\tilde{v}/d\tilde{D}$ changes but also the measurement L has a local minimum and λ_0 becomes maximum. In this example of the assumed existence of a threshold between gradations 86 and 92, where $d\tilde{v}/d\tilde{D}$ varies substantially, there is no deciding boundary, since here L and λ_0 are monotonic.

From the point of view of the technique of percentiles and of calculations of areas occupied by the sought-for sub-divisions, the probability of drawing

Table 6.13. The distribution of measure L and probability coefficient λ_0 in drawing the deciding boundaries on the Fig. 6.14 Histogram before atmospheric correction

$\tilde{D}$	L	λ_0	$\tilde{D}$	L	λ_0
78	0.095	0.128	128	0.176	0.077
90	0.064	0.132	130	0.122	0.091
82	0.029	0.142	132	0.098	0.109
84	0.052	0.139	134	0.154	0.086
86	0.045	0.131	142	0.282	0.156
88	0.037	0.127	144	0.173	0.201
90	0.026	0.124	146	0.153	0.223
92	0.022	0.120	148	0.239	0.190
108	0.015	0.109	152	0.097	0.088
110	0.009	0.125	154	0.075	0.110
112	0.030	0.118	156	0.036	0.122
114	0.081	0.113	158	0.041	0.096
120	0.094	0.096			
122	0.043	0.118			
124	0.066	0.114			
126	0.107	0.095			

the boundaries of points 1 – 4 in Fig. 6.14 is nearly equal. However, a larger number of elements in the transition to boundary 3 raises the risk of correct and erroneous dicisions expressed by the payment matrix c_{ij} and reduces, thereby, the similarity ratio in Table 6.13 in the transition to boundary 4 from boundary 3.

These considerations on the histogram classification of the middle-resolution scanner images are used in item 6.12.2 in the recognition of soil and vegetation covers.

6.10.4 Classification of Images Before and After Their Radiative Correction

Let us consider some problems of processing multispectral information from the Meteor-satellite optical-electronic scanner. The ratio-transmitted information was digitized (as a 2-D matrix of image pixels) by discretizing the continuous analogue signal from the satellite. We leave aside the technical side of the digitizing and possible conversions of digital equivalents to images (see [155, 159]). Note only, that the originals have been quantized into 256 conditional gradations of brightness, and these images recorded on magnetic tapes of a specialized processing complex have been subject to element by element

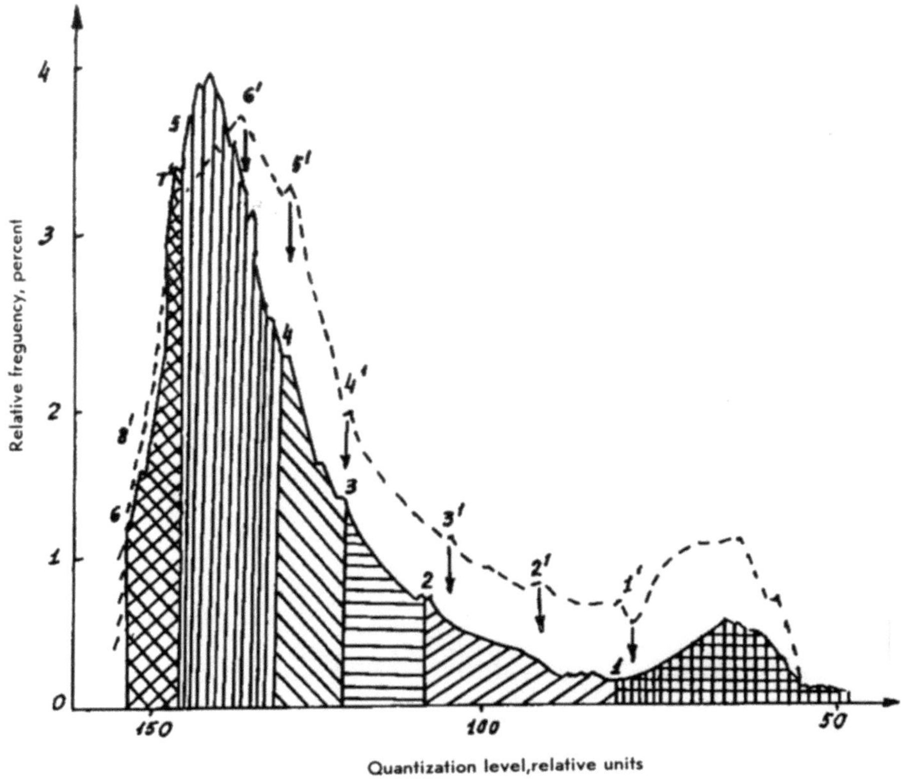

Fig. 6.14. Histogram of the distribution of the brightness of objects on a mid-resolution image before atmospheric correction (*dashed*) and after it. *Figures* point to the existence of the resolving boundaries of the histograms. *Arrows* correspond to the most probable boundaries of transition from one object to another after atmospheric correction

and areal transformations in accordance with the two stages of radiative correction discussed in Section 6.1. Results of transformation have been displayed. Simultaneously, both original and corrected images were represented through photorecorders as photographs. For correction purposes model 1 was used as shown in Table 6.10.

A fragment of the space-derived MSU-L low-resolution original image for channel 0.5 – 0.6 μm is shown in Fig. 6.15a. Figure 6.15b shows results of considering the contribution of atmospheric haze for a statistical mean mid-latitude model of the atmosphere (see Sect. 6.9). Comparison of Fig. 6.15a, b shows that, on the whole, the contrast of the corrected image increased: water objects became darker, individual agricultural fields showed themselves more clearly, being hardly seen in the original. Nevertheless, in some parts of the image the contrast decreases, for instance, the small-scale hydrological network is seen poorly. This can be explained by limiting individual brightnesses (main-

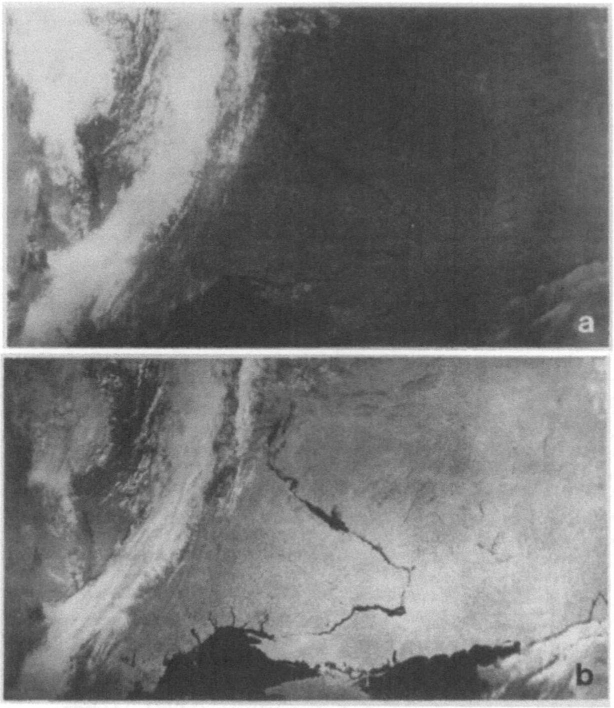

Fig. 6.15. **a** Initial MSU-L image (channel 0.5 – 0.6 μm); **b** the image after the first stage of radiative correction

ly for dark objects in positive images) at a certain level. These limitations are, apparently connected with the digital form of presentation of images as 256 conditional gradations of brightnesses and with impossibility of exceeding the upper threshold of brightnesses with this form of presentation, contrary to the real effect of the atmosphere. Probably, the alternative is quantizing an image into a lesser number of gradations.

The results shown in Fig. 6.16 a, b are similar to those in Fig. 6.15 a, b but correspond to the channel 0.6 – 0.7 μm. As follows from comparison of Fig. 6.16 a, b, consideration of the contribution of atmospheric haze in space-derived images leads to an increase of image contrast and to clearer sections in the image.

The Fig. 6.17 a, b data show the effect of the atmospheric aerosol model: Fig. 6.17 a corresponds to the IAP model of atmospheric aerosol and Fig. 6.17 b to the LSU aerosol model (see Sect. 6.8). Even a visual analysis of Fig. 6.17 a, b reveals differences in corrected images due to different optical properties of the atmosphere. The statistical analysis of images in Fig. 6.17 a, b confirms this conclusion (for instance, brightness dispersion varies by almost 1.5 from the data of Fig. 6.17 a, b).

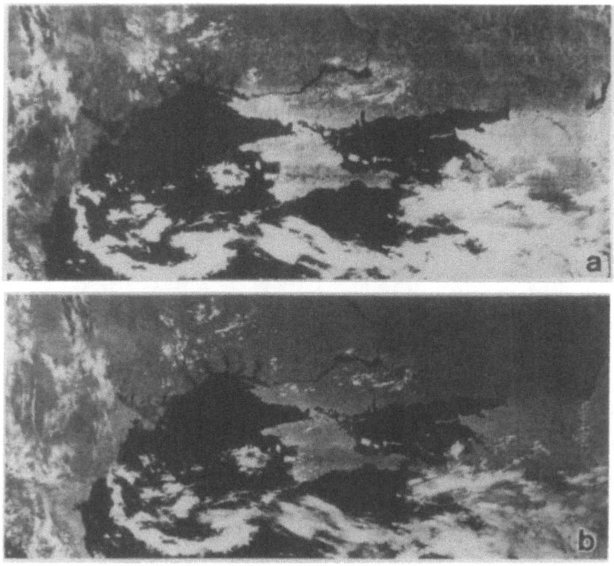

Fig. 6.16. **a** Initial MSU-L image (channel 0.6 – 0.7 μm); **b** the image after the first stage of radiative correction

Inclusion of non-orthotropic reflection in radiative correction of digital equivalents to images (Fig. 6.18) leads to still greater contrasts in individual image details. The original image was processed in the following way (Fig. 6.18 a). First, the objects were classified by individual modes of the histogram of the distribution of brightness. Based on this calculation procedure, four principal classes were selected – water bodies, agricultural fields forests and cloud formations. In accordance with Section 4.4, a certain type of non-orthotropic reflection, from specular to Lambertian, was attributed to each of these conditional classes. Then, with a given law of reflection, the contribution of atmospheric haze (Fig. 6.18) was considered. An appearance in Fig. 6.18 b of contrasting water objects, almost unseen in Fig. 6.18 a, is connected with the specular component of reflection. In particular, in considering the contribution of atmospheric haze, with supposed Lambertian reflection, these details were not observed.

Figure 6.19 shows the effect of the spatial-frequency characteristic in the problem of radiative correction of satellite image for the channel 0.5 – 0.6 μm. Figure 6.19 a corresponds to the first stage of radiative correction, Fig. 6.19 b – to the second stage. It is difficult to distinguish visually between Fig. 6.19 a and b, but the statistical processing of digital equivalents to these images shows negligible changes in the brightness dispersion (in Fig. 6.19 b the dispersion is 5% higher).

Figure 6.20 b, compared to Fig. 6.20 a, makes it possible to analyze the effect of all three types of distortions (including haze, non-orthotropic reflection and

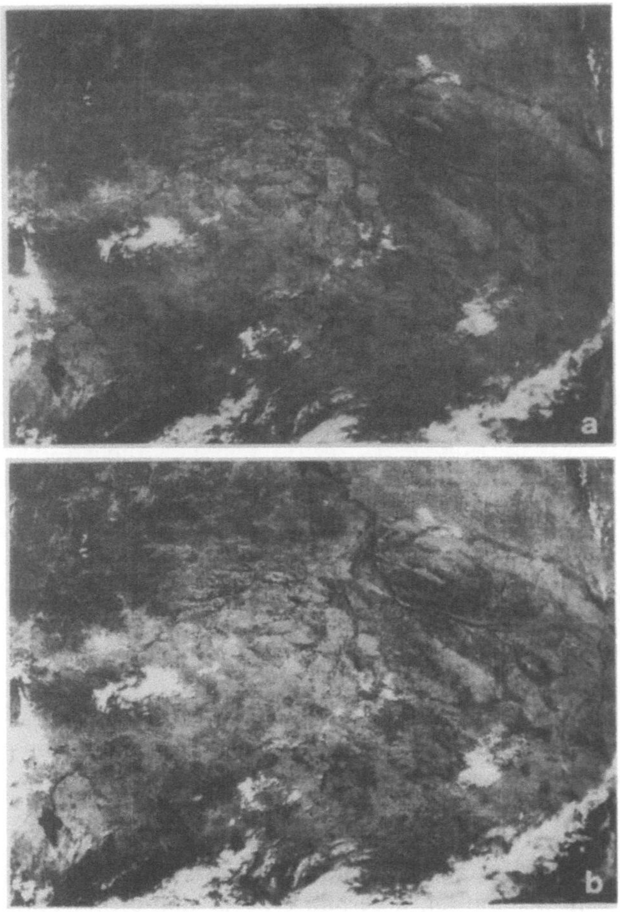

Fig. 6.17. a An image after the first stage of radiative correction using the IFA atmospheric model; **b** the same image corrected using the LSU model

horizontal inhomogeneities of the Earth's surface) in the problem of radiative correction. Some contrasting elements can easily be seen in the image due to radiative distortions in space-derived information.

Comparison of Figs. 6.15 – 6.20 suggests that for space-derived low-resolution (about 1.5 km) images inclusion of the first stage of radiative correction (the contribution of haze and non-orthotropic reflection) is more substantial than that of the second stage (the contribution of horizontal inhomogeneities of the Earth's surface). The effect of the second stage is strong for large optical thickness (about 1), but such images are very difficult to interpret both visually and with machines because the useful signal in a fixed image is weak.

Figures 6.21 and 6.22 show the results of two stages of radiative correction of space-derived mid-resolution information (MSU-M) with a resolving ele-

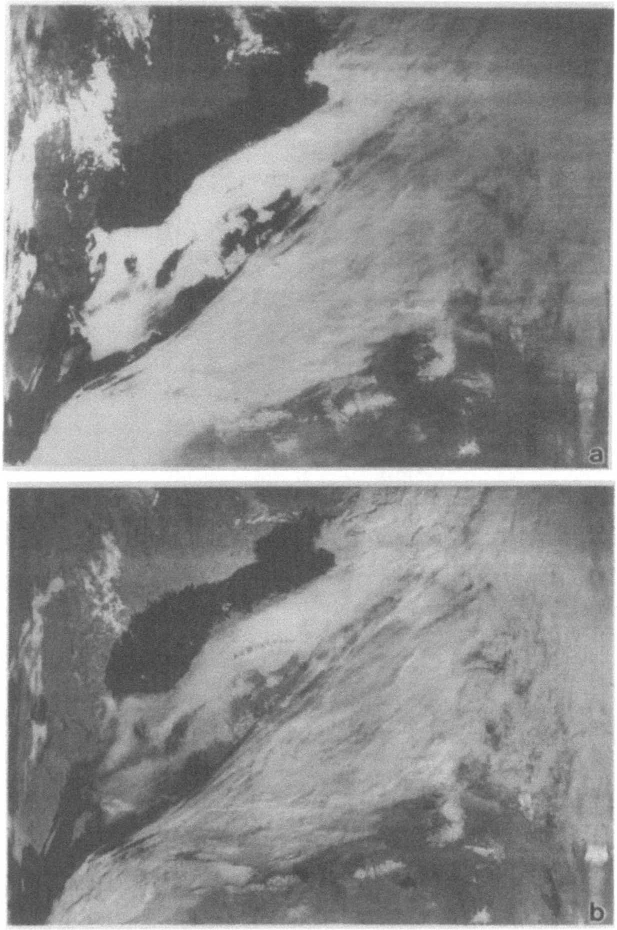

Fig. 6.18. **a** An initial image using MSU-L (channel 0.5 – 0.6 μm); **b** the image after the first stage of radiative correction with the image non-orthotropicity

ment of about $280\times280\ km^2$. Figure 6.21 shows the effect of atmospheric haze, whilst Fig. 6.22 shows the spatial-frequency characteristic of the surface-atmosphere system. Comparison of Fig. 6.21 a, b reveals slight variations in images, before and after correction. These changes are even smaller in Fig. 6.22 b compared to Fig. 6.22 a. As seen from Figs. 6.15 – 6.20, these variations in space-derived low-resolution images are more substantial. This can be explained by the fact that at a low resolution the viewing band of the area is large, so that the necessity grows to consider different angular conditions of survey, different types of non-orthotropic reflection, types of atmospheric states, etc. In addition, for middle resolution in a given period, in a given natural zone the contrasts of the original mid-resolution images grow substantially. Further

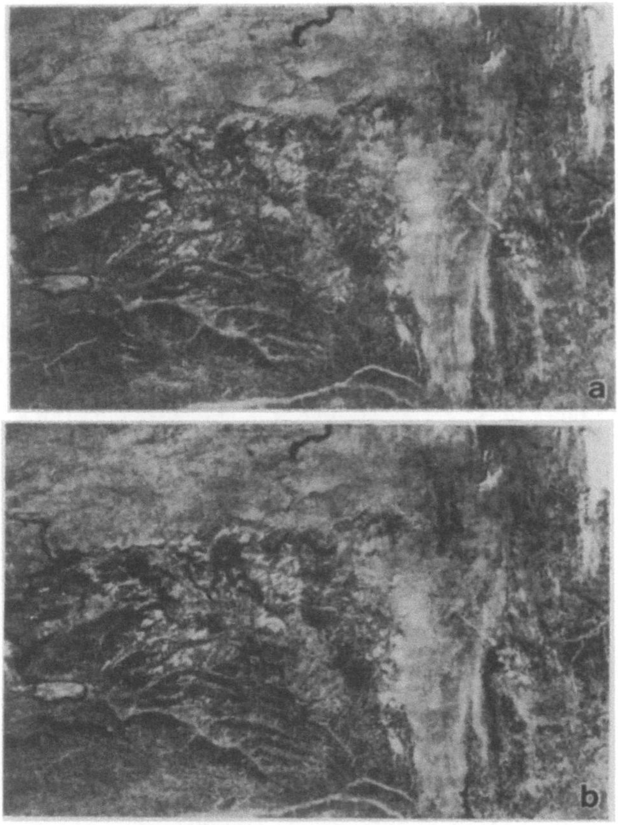

Fig. 6.19. **a** An initial MSU-L image (channel 0.5 – 0.6 μm); **b** the image after the second stage of radiative correction

increase of contrast, corresponding to the atmospheric effect, is confined to a value of initial dispersion of brightnesses. Therefore, mid-resolution images before and after radiative correction differ little.

Table 6.14 lists results of comparison of individual stages of atmospheric correction of Meteor low-resolution (MSU-L) and mid-resolution (MSU-M) images. These images correspond to the survey of the same area in the same season (but in different years). It follows from Table 6.14 that in solving the problem of atmospheric correction the factors determined by atmospheric haze brightness and non-orthotropic reflection, are the governing ones for both MSU-M and MSU-L. In both cases, the effect of horizontal inhomogeneity of the surface is less significant. It is probable that these conclusions depend on the area imaged, survey conditions and atmospheric turbidity. However, the fact that first two factors have a substantial (2 – 3-fold) effect over that of horizontal inhomogeneity, confirms the need for more accurate study of haze brightness and non-orthotropicity compared to horizontal diffusion of

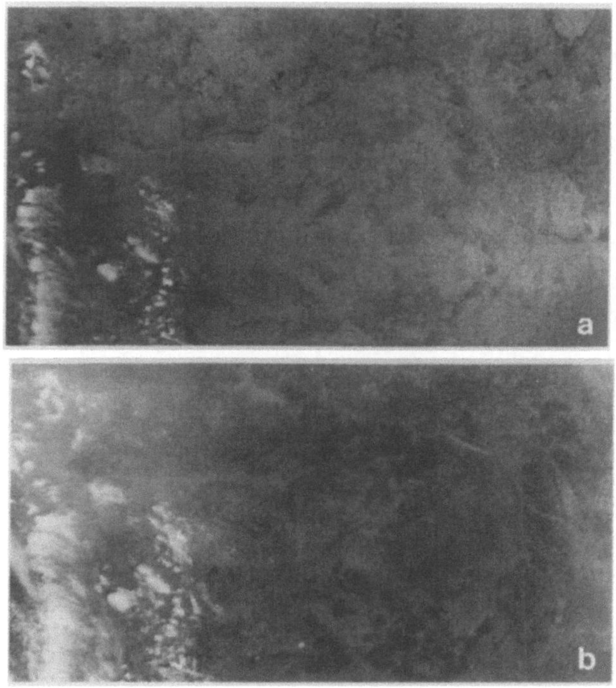

Fig. 6.20. a An initial MSU-L image (channel 0.5 – 0.6 μm); **b** the image after two stages of radiative correction

Table 6.14. Average values of $\bar{I}$ and brightness dispersions $\sigma_{\bar{I}}$ for low- (nominator) and middle- (denominator) resolution images and variations of averages and dispersions, as factors of the formation of the field of recorded radiation are being taken into account

Statistical characteristics	Original image	Atmospheric haze	Non-orthotropic reflection	Horizontal inhomogeneity	Sum of effects
$\bar{I}$	*164*	*122*	*130*	*151*	*118*
	149	108	114	140	99
$\sigma_{\bar{I}}$	*57*	*87*	*80*	*64*	*93*
	78	110	103	89	115
$\Delta\bar{I}$		*26*	*21*	*8*	*28*
		28	23	6	3
$\Delta\sigma_{\bar{I}}$		*53*	*40*	*12*	*63*
		41	32	14	47

photons in atmospheric correction of scanner images. This confirms the possibility of using the 1-D theory of radiation transfer, to solve the problem of compensation of distortions introduced by the atmosphere, into the scanner and photographic images of the Earth's surface of low and middle resolution.

Fig. 6.21. a An initial MSU-M image (channel 0.8 – 1.1 μm); **b** the image after the first stage of radiative correction

In conclusion, we shall discuss the procedure of the classification of objects in space-derived images. The solution of this problem is connected with mathematical problems of recognition of objects [40] based on statistical or logical rules of making decisions about attributing this or that image pixel to a certain class of objects. The authors used the histogram technique to classify objects. A histogram is a ratio of the number of elements of a given brightness to total number of brightness elements for a certain area (in this case an area fixed in a space-derived image or part of the image).

The presence on histograms of clearly manifested peaks (modes) or the so-called breaking points (change of the slope of the histogram) testifies to the presence of natural formations with different reflective properties, which serves as the basis for classification of objects in space-derived images. As follows from the histograms shown in Fig. 6.23 a, b, the radiative correction of low-resolution information caused individual modes which had been absent on the pre-corrected histogram. This testifies to an appearance of additional classes of objects which had been distorted by the atmosphere and had not been exposed in the originals. In particular, in the histograms of Fig. 6.23 a six basic classes of objects are clearly seen, corresponding to a different slope of histograms, but in Fig. 6.23 b only four classes, considered in analysis of Fig. 6.18.

Thus, the solution of the problem of radiative correction, together with the problem of classification of objects in images opens up wide prospects for

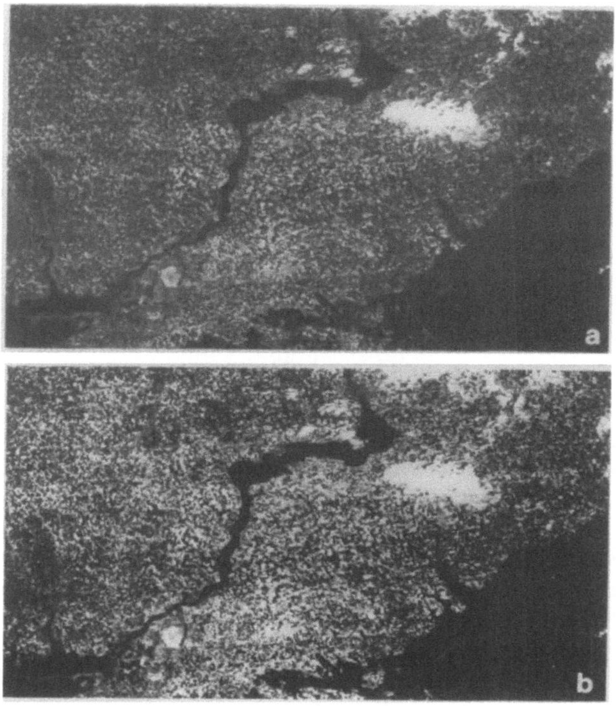

Fig. 6.22. **a** An initial MSU-M image (channel 0.8 – 1.1 μm); **b** the image after the second stage of radiative correction

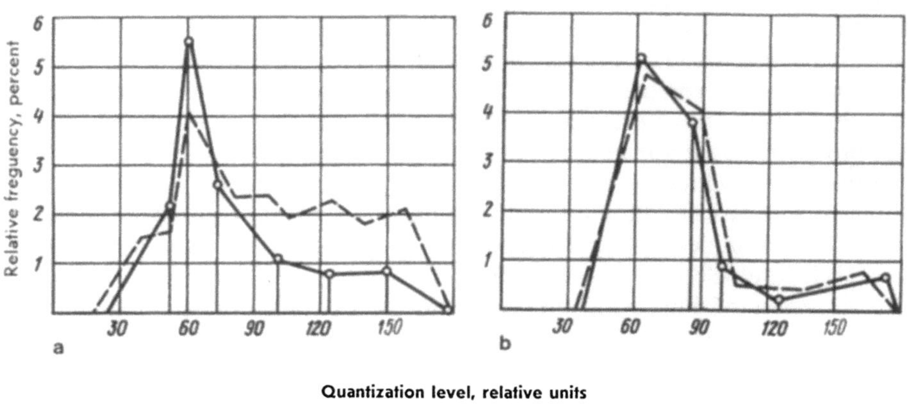

Fig. 6.23 a, b. Histograms of relative frequencies for different objects smoothed by running mean mechanism: **a** for Fig. 6.18; **b** for Fig. 6.22; *solid lines* are initial images, *dashed lines* are corrected images

automated processing and interpretation of the data of air- and space-based survey.

6.11 Atmospheric Correction as One of the Stages of Thematic Interpretation of Video Information

The problem of eliminating the distorting effect of the atmosphere (atmospheric correction) in air- and space-derived images is solved in two basic directions: theoretical, when the radiation transfer theory is developed to retrieve the brightnesses and albedos of the Earth's surface [314]; and technological, when one proceeds from space-derived video information. Here, the problem is reduced to the determination of certain functions of multispectral data conversion, both from the view point of pure mathematical considerations and with the use of transfer theory. Also, there are empirical approaches to the solution of this problem (see [386]): for instance, a presentation of the atmosphere as a multiplicative clutter [251] and the use of "black reflectors" (water reservoirs) to determine the atmospheric haze brightness with its subsequent extraction, the interchannel correlation method in which correlations between the channels of multispectral survey serve as the basis.

So far as the technique for thematic interpretation of remote sensing data is concerned, an abstract determination of the function of data conversion loses its importance, giving place to the aforementioned problem of recognition of objects. The atmospheric correction of video information should be approached from the viewpoint of solution of this problem. In this case, certain criteria of the qualities of images before and after their atmospheric correction are introduced. If the atmospheric correction does not improve images (i.e., an introduction of a certain measure of quality), it is worthwhile performing thematic interpretation of the initial data for the atmospheric top, without their additional transformation. In particular, atmospheric correction does not improve the quality of images when the atmosphere is very turbid [251]. The problem of finding the threshold of atmospheric turbidity, when respective transformations of data improve their quality, becomes urgent. This improvement is determined by increased probability of correct recognition of objects on their space-derived images [252, 253].

Examples of improving the accuracies of classification of satellite mid-resolution images upon their atmospheric correction, together with the spectrometric data of the same spatial resolution, have been given in [108]. The authors of [108] show that, based on the use of polynomial approximation of the outgoing radiation intensity, the accuracy of classification of crop fields rises due to retrieval of the mode in the histogram of the brightness distribution, distorted by an optically thin atmosphere. The authors of [108] exemplify an increased accuracy of recognition of crop fields due to decreased blurring of their boundaries by a transformation operator considering the morphological structure of the processed image. These examples illustrate some possibilities of technological solution of the problem under consideration.

In this case, the problem is to remove atmospheric distortions in processing multispectral data of high and middle resolution. The need for removing these distortions appears in the following cases:

- when it is necessary to estimate absolute radiation intensities (for instance, to determine surface temperature, evapotranspiration of vegetation, etc.);
- when it is necessary to spot relatively small differences in brightnesses (radiobrightness temperatures) against the background of considerable atmospheric turbidity (for instance, when measuring chlorophyll concentrations in the oceans, when eliminating the blurring of the boundaries of agricultural fields with the use of a morphostructural operator [17], etc.);
- when multitemporal data are analyzed (for instance, in studies of crops or of tidal effects on the coastal zones of the oceans).

6.11.1 A Comparative Analysis of Space Images and Spectra

The development of satellite systems of observations of the terrestrial atmosphere and surface is aimed at increasing the spatial resolution of space-based instruments, enhancing the number of spectral channels, using multiangular, trace, scanner, slant and other methods for remote sensing.[2] Both photographic systems of observation, ensuring images of high spatial resolution, and scanner systems promptly providing a large information volume about the Earth's surface – atmosphere system, and being developed. There are also special systems to study oceans, atmosphere cloudiness, radiation balance, with the use of both wide-angle receivers recording the outgoing radiation fluxes and narrow-angle scanning systems for sun-synchronous, geostationary satellites and satellite systems in precessing orbits. There is no need here to dwell upon various types of satellite measurements.

A major problem in practical use of satellite information consists of the automated obtaining and processing of data with interactive displays for processing and analyzing of measurement results.

This necessitates a combined analysis of data obtained by various types of instruments, revealing respective features of the data analyzed and finally, making decisions about classifying these data.

A special feature of photographic and optical-electronic scanner systems of measurements is that they allow presentation of information as images – two-dimensional matrices of digital equivalents of the results of measurements. The shortcomings of photographic information are connected with the difficulty to control errors in photoprocessing and with digitization of information using optical systems for scanning the photographs. The drawbacks of the scanning systems are determined by possible distortions of information

---

[2] This section was written with V. N. Dosov and L. A. Pakhomova.

transmitted via communication channels: these drawbacks can be eliminated by transmitting digitized signals, but then the information flux expands and must be compressed. In this connection there is a need for optimal presentation of results of scanner survey by discretizing and quantizing images.

A special feature of photographic and scanner information is the possibility of its visualization not only by photorecording but also, with digital presentation on magnetic carriers, by displaying it with subsequent interactive (dialogue) analysis of measurement results. Another feature of these types of information is the possibility of absolute calibration of photographic information by gradations of the brightness of photometric wedge imprinted on a photograph between survey sessions (photometric correction). Radiometric distortions of scanner information by the electric wedge of the scanner's reverse (radiometric correction) can also be considered. Along with the scanning systems, images can be obtained by non-scanning systems with the parallel read-out of signals ("sensor comb"), but such systems have only started to develop.

An alternative to photographic and scanning systems covering large areas either in narrow spectral intervals or in the entire short- or longwave region, are spectrometric systems in which high spectral resolutions are reached in non-scanning trace measurements. Possible combination of scanning and spectrometric measurements in one instrumental complex will lead to a substantial increases of information fluxes and will necessitate development of recording systems allowing compression of records (for instance, laser recorders).

Traditional spectrometric measurements made from aircraft have become widely used, because they permit visual or other control of the state of surface objects. However, satellite spectrometry techniques are not so popular because of low efficiency of the spectra of reflection and emission for the objects whose state is not known beforehand. In this connection, a combined referencing of information is needed, both from scanning systems and from trace spectrometric measurements to classify observed objects from scanner images and then to assign their reflection (emission) spectra to a definite class of objects. Multispectral techniques for recognition of space-derived measurement data, when images are obtained simultaneously in several spectral intervals, are now widely used.

Multispectral images of the Earth's surface form under the influence of solar radiation penetrating the atmospheric thickness; the interaction of direct solar and diffuse scattered radiation with natural formations; the transformation in the atmosphere of the Earth-reflected radiation; and finally, the conversion of radiation recorded by the optical system in satellite instruments, in communication channels, and when recording the received analogue signal at surface information-receiving stations. The processing of multi-spectral images to assess the state of natural objects must contain algorithms for conversion of brightness images of the objects into parameters of their state. Algorithms for assessing the distorting effect of the atmosphere and considering the distortions suffered by recorded radiation while the signal goes from satellite to surface stations are also needed.

Infrared images in which the atmosphere and surface are the source of eigen emission, are obtained in a similar way. In any case there is a need for comparative analysis of images and space-borne spectrometric measurements.

Within the programme "Bulgaria-1300" in 1981, the "Meteor-Priroda" satellite carried the trace 32-channel SMP-32 spectrometer, developed in Bulgaria [46]. It recorded reflection spectra of natural formations in the wavelength region 0.46–0.88 μm with a resolution of 12–14 mm. The spectrometer's FOV at the Earth's surface constitutes 280×280 m and is close to a mid-resolution scanner resolving element (MSU-M) carried by the same satellite. The MSU-M data (Earth surface images in the spectral intervals 0.5–0.7 μm and 0.7–1.0 μm) are widely used now in studies of natural resources.

One of the limitations of information content of the scanner data in recognition of natural objects, is connected with a small number of spectral intervals for space-based survey and low spectral resolution. In this connection, it is of interest, to assess the usefulness of additional information about the reflection spectra of natural objects within the MSU-M resolving element. This assessment can be made in combined analysis of data obtained simultaneously with SMP-32 spectrometer and with MSU-M. It appears that, the possibility of combined analysis of these data is closely connected with an element by element matching of the SMP-32 and MSU-M data. With almost equal geometric size of the fields of view of these sensors, the latter are independent instrumental complexes, and precise matching of SMP-32 measurements with certain image pixels from MSU-M has not been planned. Therefore, at the first stage of combined analysis of these data the need appears for the element by element referencing to consider their special presentation at surface receiving stations.

If it were possible to automate the process of combined referencing of the trace SMP-32 measurements to MSU-M images, further automation of the results of image analysis would be possible without using an interactive regime of processing, which always leads to certain subjective errors. Comparison of results of interactive classification of MSU-M images with results of classification of "referenced" SMP-32 data would be helpful, to analyze the number of object classes which can actually be selected with these two types of satellite instruments. Finally, the possibilities are still unclear, of using satellite spectrometric measurements to compile a data base on the spectra of natural formations which seems very promising. It may be that there is an urgent need for reducing the spectra to surface level, or perhaps the space-derived information can efficiently be processed with the use of space-derived spectra which must constitute the data base.

Each measurement with SMP-32 is a set of 32 numbers – spectral radiances at 32 wavelengths in the range 0.46–0.88 μm. Measurement data are transmitted from the satellite in a digital form, 128 gradations in each spectral interval of measurements being used to present spectral brightnesses (in relative units). The time for measuring each spectrum is coordinate with the velocity of the satellite to permit measurement of border surface sites spectrometered in the transition from one measurement to another. The surface-received data of SMP-32, after initial processing, are normalized – converted into absolute energy units and recorded on computer magnetic tapes for their further analysis and processing [49].

In the description of special features of SMP-32 data an important problem is the accuracy of measurements. The noise constituent of measurements can be estimated from dispersion of the spectrometer signal in the calibration regime. In this regime, a reference emitter is put into the spectrometer's FOV; with its assumed stability, variations of the signal are determined only by instrumental errors of measurements. Analysis of variations of the calibration signals suggests that the MSD value is within the interval of discretization. Thus, the process of discretization is a determining factor of measurement accuracy. The energetic ability of the spectrometer can be judged by spectral brightnesses of radiation corresponding to maximum signals. In the range of wave-numbers recorded by the spectrometer, 128 relative units of the signal correspond to spectral brightnesses within the interval 3.5×10^{-8} to 15×10^{-8} W/m^3 sr).

The MSU-M data are now presented, as a rule, as photographs with the help of photo recorders. For the computer interpretation of these data the MSU-M photo negatives can be digitized on computer magnetic tapes by an optical scanner. In our case, an optical scanner of a specialized processing complex was used to digitize photo images. The format of photo negatives and the possibility to choose a step of scanning in this system make it possible to obtain a digitized image with a discretization not exceeding the geometrical resolution of the mid-resolution scanner.

By reference to special features of the SMP-32 and MSU-M data the problem of their element by element referencing was reduced to the following. A line of elements was formed from a set of subsequent spectrometric measurements corresponding to a certain orbit of the satellite, each element representing the spectral brightness of measured radiation, averaged over the wavelength interval of one of the MSU-M channels.

Bearing in mind that the contrast of images of the mid-resolution scanner in the interval $0.7-1.0\,\mu$m is much higher than in channel $0.5-0.7\,\mu$m, the spectrometer brightnesses measured in the interval $0.7-0.88\,\mu$m were averaged. If $X(i,j)$, $i = \overline{1,M}$; $j = \overline{1,N}$ is the field of elements of an MSU-M digital image in the channel $0.7-1.0\,\mu$m, and $Y(k)$, $k = \overline{1,L}$ is a line of successively averaged spectral brightnesses measured by the spectrometer, the combined referencing of data is reduced to a search for succession of L elements from X coinciding to an accuracy of the scale factor with Y elements. Maximum correlation function was used as a measure of closeness of the succession L and the line Y.

The correlation function was determined by the position of the succession L with respect to the field X, by the step between the elements of L, as well as by the shift of Y with respect to L, which determines the principal features of the algorithm for combined referencing of the data.

In the computer-realized algorithm, this problem is solved by successive calculations of the correlation function taking possible locations of L and a

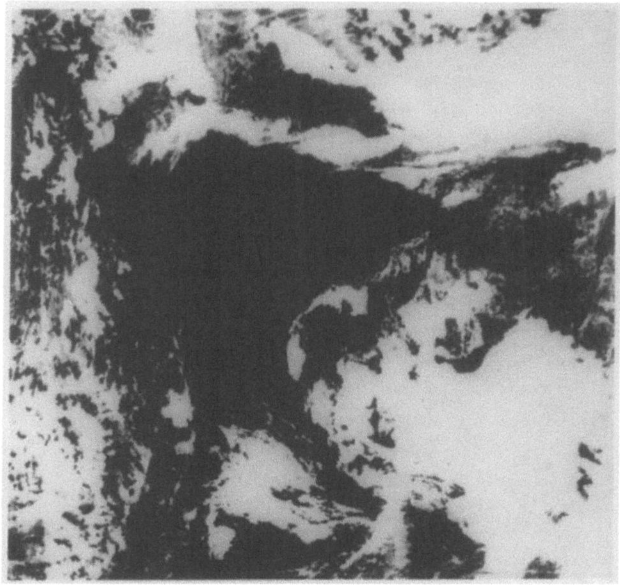

Fig. 6.24. Analyzed fraction of the mid-resolution scanner image

step inside L. Maximum correlation function obtained in our case constituted 0.93 for an orbit whose analysis will be discussed shortly. A fragment of this orbit is shown in Fig. 6.24. Bearing in mind the incomplete coincidence of the spectral interval of averaging in obtaining the line Y with the spectral interval of survey with MSU-M, and the non-linear transformation of X(i, j) in the process of photorecording, the obtained value of the correlation function may be considered to correspond to a sufficiently high level of combined referencing of the profiles of scanning lines to the trace data of the spectrometer. The fact that this condition is enough is confirmed by visual analysis of results of combined referencing on the display screen of a specialized complex of image processing.

Figure 6.25 shows a fragment of an MSU-M image in the channel 0.7 – 1.0 μm for which synchronous spectrometric data have been available. With the use of the aforementioned algorithm the spectra were referenced to the image in Fig. 6.25. To identify the spectra measured by SMP-32, the images were additionally classified.

The classification, was made in the interactive regime on display with the use of the histogram technique: relative frequencies of occurrence of these or other image densities in the analyzed fragment were calculated. The histogram of the distribution of the brightnesses of the image fragment in Fig. 6.25 a is shown in Fig. 6.25 b. Numerous peaks (modes) are observed resulting from a variety of objects in the fragment analyzed.

The characteristic breaking points on the histogram showing after calculating its running mean, suggested that it is possible to recognize 8 classes of

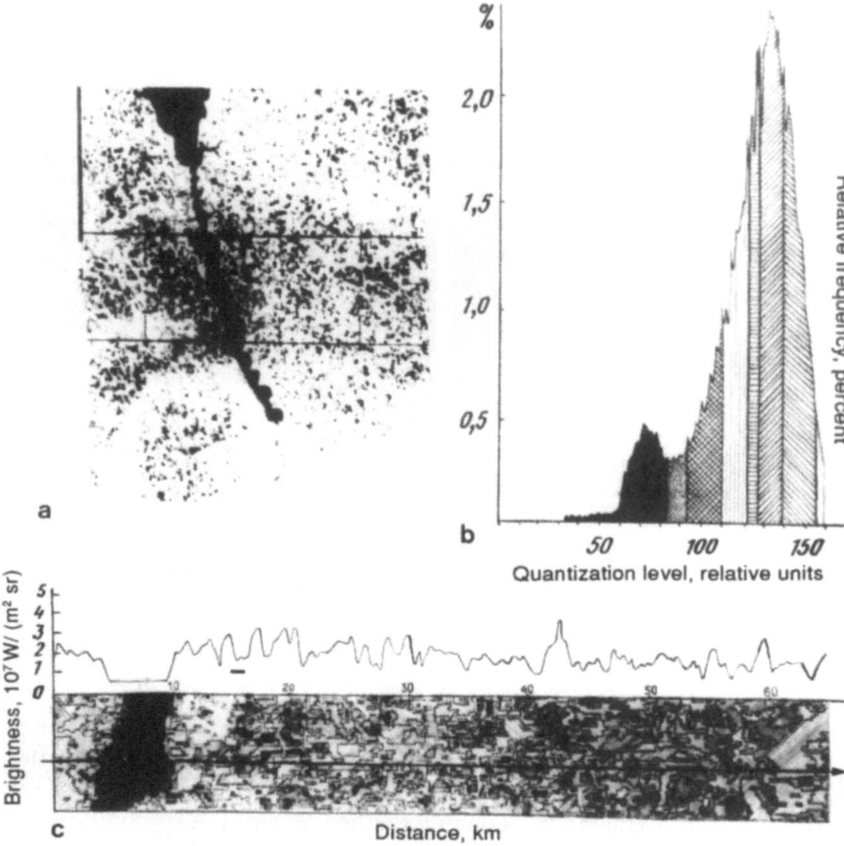

Fig. 6.25. **a** A fragment of Fig. 6.24, **b** histogram of the distribution of the brightnesses of this fragment and **c** referencing of spectral data of this fragment

objects with different brightnesses, according to the shading of the field of the histogram (Fig. 6.25 b).

Analysis of the 0.5 – 0.7 μm histogram of the same fragment has shown that the information in this channel does not improve the results of classification obtained from analysis of the longwave channel image. It appears that, the reason is insufficient contrasts of objects during survey, strong effect of atmospheric haze, and probably, errors in the photographic presentation of information. A priori information about the area of survey has made it possible to identify the selected classes of objects in the following way: water basins, three types of soil (moist, grey and light-coloured), three species of vegetation, and other objects. An example of referencing the spectrometric data to the fragment of image in Fig. 6.25 a is shown in Fig. 6.25 b; at the bottom is an enhanced site of fragment in Fig. 6.25 a classified in accordance with the histogram in Fig. 6.25 b, with the satellite's route shown; in the upper part is the course of the SMP-32 spectral data averaged over the interval 0.7 – 0.88 μm.

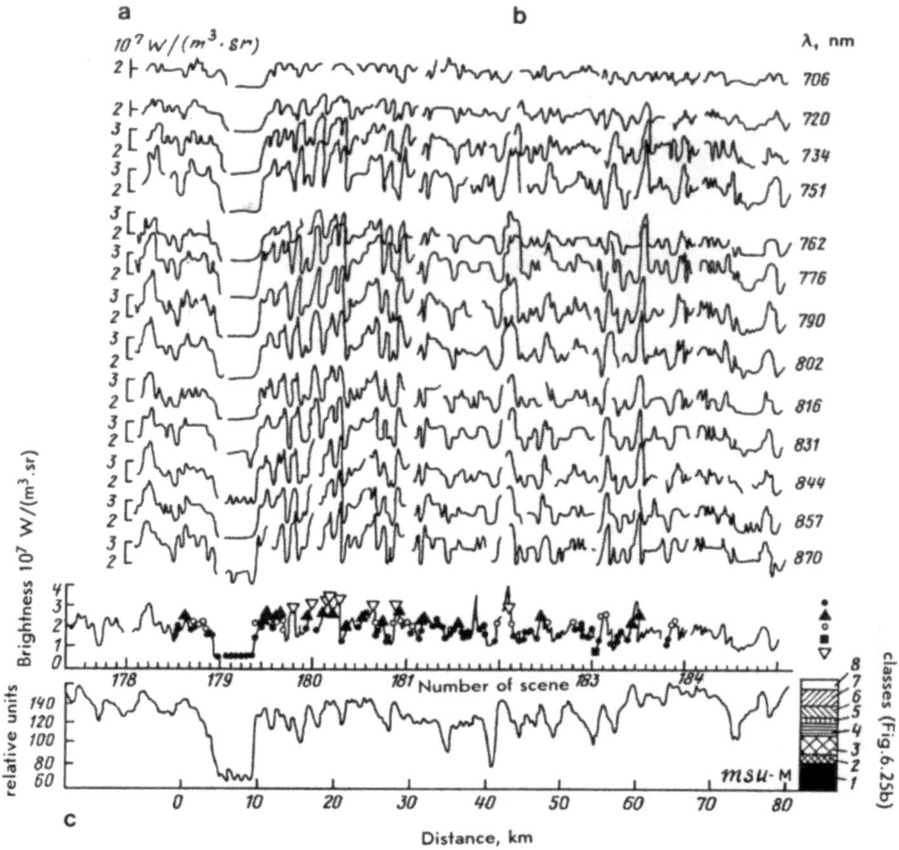

Fig. 6.26. a Spectral brightnesses in channels from 0.706 to 0.870 μm; **b** convolute spectra within the interval 0.7 – 0.88 μm; **c** spatially combined MSU-M signals

A qualitative analysis of spectrometric data referenced to a scanner image suggests certain conclusions about special features of spectral characteristics of individual natural objects recorded from space. Naturally, the similarity of conclusions is limited by the accuracy characteristics of spectral instruments and by the regions and conditions of survey.

Figure 6.26 shows the spatially combined MSU-M signals (Fig. 6.25) averaged (within the interval 0.7 – 0.88 μm) spectral brightnesses from SMP-32 spectrometer (Fig. 6.25 b), as well as non-averaged brightnesses in the channels centered between 0.706 and 0.870 μm (Fig. 6.25 c). It is of interest that in practically all the spectrometer channels, there is a synchronous variability of recorded spectral brightnesses in the transition from one natural formation to another. The variability of spectral brightness in narrow spectral intervals (12 – 14 nm) is practically identical to the variability of averaged spectral radiance or to the MSU-M signal.

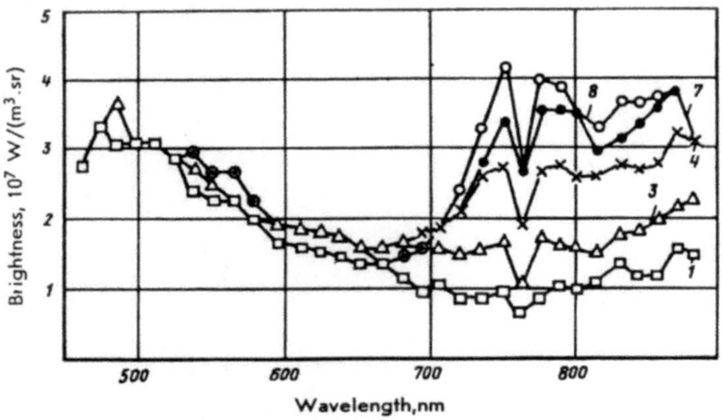

Fig. 6.27. SMP-32 absolute spectra corresponding to the classes of *dashed* objects in Fig. 6.25

The spectral brightness variabilities in the channels of the interval 751 – 870 nm are close in value. An exception is channels 762, 816, 831 nm affected by the absorbing components of the atmosphere. Starting from channel 734 nm and farther into the shortwave region, the variability of the amplitude of spectral brightnesses for different natural formations decreases. In channels of the interval 0.46 – 0.7 μm the spectral brightness is practically independent of natural formation type, especially when it refers to soil and vegetation cover.

Figure 6.27 shows the spectra measured with SMP-32 for respective natural formations selected with the use of the histograms in Fig. 6.27b. Differences in reflection spectra of various objects are observed, mainly in the longwave channels. The spectral brightnesses of vegetation (spectrum 7) increase with increased wavelength, and that of soil (spectra 3, 4) has a neutral course in the longwave region. Minimum spectral brightnesses are typical of water objects (spectrum 1). As Fig. 6.27 shows, the spectra in the shortwave region vary little. This could be explained by the effect of atmospheric haze, smoothing the differences equally in the spectral brightnesses of various objects. However, most likely it is connected with original sensitivity of spectrometric instruments, weakly responding to actual variability of spectra in this interval.

A preliminary conclusion from the aforementioned qualitative analysis of reflection spectra for some natural features is that in the conditions of the MSU-M measurements considered details of the type of spectral characteristics of objects do not substantially specify the information on the state of the object. To specify this conclusion, a sample of SMP-32 spectra within the image fragment in Fig. 6.25a was clustered. According to this procedure, a permissible number of the required classes K is taken successively, which in our case, varied between 3 and 25. For initial arbitrary distribution of spectra S_j by classes c_i, $i = \overline{1, K}$ elements were removed from one class and added to another by successive selecting of S_j so as to minimize the intraclass sum of squares for each of the classes c_i. The element (spectrum) S_m moved from class r to class j ($r \neq j; i, j \leqslant K$) in the case when:

$$\frac{n_r}{n_r - 1} \|\bar{S}_r - S_n\|^2 > \frac{n_j}{n_j - 1} \|\bar{S}_j - S_n\|^2 , \tag{6.83}$$

where n_r, n_j are the number of elements in classes c_r and c_n, and $\bar{S}_r$ and $\bar{S}_j$ are the centers of these classes. The Euclidean distance was used as a distance between elements. With each permutation of elements S_n new centers $\bar{S}_r$ and $\bar{S}_j$ of the class wewe calculated. The procedure stopped upon reaching such division into classes when for any n of succession S_j the condition (6.83) was not satisfied.

Since in the applied clustering procedure the number of the required clusters has been prescribed, it is important to specify the classes (and their number) which enable reliable assignment of the given objects to them by their spectral characteristics. For instance, at $K = 25$ the procedure revealed 19 clusters, with an exception of single measurements. However, starting from $K = 8$, when 6 centers are selected, the centers of four of them coinciding with MSU-M intervals 1, 3, 6 and 7 of signal changes (Fig. 6.25 b); two clusters with centers localized in intervals 4 and 5 are strongly overlapping. With a distance between classes of about 0.79 relative units, the radii of scattering of these classes constituted, respectively, 0.73 and 1.13. Similarly, at $K = 9$ the three clusters with centers localized in the intervals 4 and 5 are also overlapping. With distances between them, respectively, 0.8, 1.22 and 1.12, the radii of scattering constituted 0.63, 1.14 and 1.14. This served as the basis for the conclusion that in the given sample containing 170 spectral measurements only five types of natural formations are identified from spectral characteristics.

Bearing in mind that the number of spectra measured over moist soils (interval 2 of the MSU-M signal in Fig. 6.25 b) and over light-coloured vegetation (interval 8) is small, and over the objects characterized by MSU-M signals in intervals 5 and 4, an overlapping division into two classes is observed, it can be stated that in this case possibilities of recognition of the objects from multispectral measurements and from the MSU-M data are approximately equal. Attributing the SMP-32 multispectral measurements to the classes obtained is shown in Fig. 6.25 b. In all cases, the objects are uniquely assigned to a respective class both on the basis of the whole set of spectral measurements in 32 channels in the interval 0.4–0.88 μm and from one measurement over an object in the MSU-M interval 0.7–1.0 μm. There were no cases when, with equal average brightnesses in two measurements in the interval 0.7–0.88 μm, these measurements would be assigned to different classes.

With close or equal information content of MSU-M and SMP-32 data, from the viewpoint of the determination of the number of recognized objects, it is important to emphasize special features of the determinants in the classification procedure. For the MSU-M data the classifier, i.e., a division of the signal into an interval, each for this or another natural formation, has been obtained from a sample 256×256 elements, with the use of a priori information about the area of survey. From the MSU-M data obtained along the route of the satellite, it is possible to select regions of signals corresponding to only water surface and land. At the same time, using the route sample of multispectral measurements, one can locate the centers of classes and construct a classifier to identify soil and vegetation. As follows from Fig. 6.25, the type of object can be determined from the character of spectral dependence of the cluster's center.

Thus, space-derived spectrometric data, being combined with scanner images, may be useful in solving a number of methodological problems of remote sensing of natural objects.

6.11.2 The Stages of Atmospheric Correction of Video Information

Thus, the status of the problem of radiative correction of air- and space-based measurements is such that it may be possible to use theoretical methods, which serve as the basis for the solution of this problem, in thematic processing of crop images. The concept "radiative correction" is broader than a simple elimination of atmospheric effect, firstly not only an atmospheric layer between measuring instruments and the Earth's surface affects the results of remote sensing of natural objects greatly. Secondly a transformation of radiation by vegetation cover [85] manifested, initially, through non-orthotropic reflection is also relevant. The effect of the atmosphere and of non-orthotropic reflection is not a simple sum of distortions: they must be considered as much more complex.

Theoretically, it is not difficult to solve the boundary problem of radiation transfer in the atmosphere, though the problem of finding efficient algorithms to calculate radiation fields in the atmosphere remains urgent. At present, methods to solve this problem have been developed, both for the mean constituent of radiation intensity and for its fluctuating constituent [82], with the use of iterative techniques [136], advanced spherical harmonics techniques [77], Monte-Carlo simulations [54], etc.

However, the practical solution of the problem of radiative correction is hindered by the impossibility of real-time calculations in processing space-derived information. This emphasizes the urgency of approximating the results of calculations with an accuracy which does not exceed the accuracy of the techniques suggested. Radiative correction is made to raise the accuracy of thematic processing and, initially of recognition of spectral images of natural objects. Most of the problems of thematic processing and of analysis of space-derived information in the interests of agriculture are based on special procedures of classification [19]. In this connection, real possibilities of using the procedure of radiative correction, can be assessed in their combined use with the problems of classification and identification of objects.

Use of results of theoretical modeling of radiation fields in the atmosphere in processing air- and space-derived images is hindered by the fact that energy characteristics of radiation fields are calculated. This processing implies the transformation of the signals connected with the outgoing radiation energy through respective calibration quantities, i.e., for the transition from results of modeling to the energy of individual pixels of space-derived images it is necessary to know solar irradiance in a given spectral interval, functions of spectral sensitivity of receivers and filters, etc. Leaving behind the difficulty of assessing the state of the atmosphere at the moment of survey (which is a problem of its own requiring the development of techniques for solving incorrect inverse problems and empirical techniques with complex air- and space-based measurements) we emphasize the urgency of technological testing of the technique for atmospheric image correction.

To model the atmospheric distortions of images, values of recorded intensity $I = I(A, \zeta, \eta^+, \varphi)$ were used on surface albedo A, solar zenith angle cosine ζ, viewing zenith angle cosine $\eta^+ = \cos\theta$, and on the difference between viewing and solar azimuths φ (see Sect. 6.3). They have been found through an approximation by orthogonal polynomials of the dependence of the outgoing radiation intensity on the atmospheric state and angular conditions of survey. These were obtained from solution of the boundary problem of radiation transfer in the real stratified atmosphere over the terrestrial surface with non-orthotropic reflection. Note, that the boundary problem has been solved using an advanced technique of spherical harmonics and a system of specially chosen vectors (Sect. 6.3). Experimental values of the Earth's surface phase functions [83] were used as lower boundary conditions.

An approximation of results of solving the boundary problem for homogeneous extended objects has been made by orthogonal second-order polynomials using the uniform approximation technique (see Sect. 6.3):

$$I(A, \zeta, \eta^+, \varphi) = \sum_{0 \leqslant i_1 + i_2 + i_3 + i_4 \leqslant 2} a_j(i_1, i_2, i_3, i_4) \hat{T}_{i_1}(A) \times \hat{T}_{i_2}(|\zeta|) \hat{T}_{i_3}(\eta^+) \hat{T}_{i_4}(\varphi) , \tag{6.84}$$

where $\hat{T}_{i_k}(k = \overline{1,4})$ represents shifted polynomials by Chebyshev to the power i_k, orthogonal on a length [0, 1], for which the following recurrent relationships are valid:

$$\hat{T}_0 = 1, \hat{T}_1(x) = 2x - 1, \hat{T}_{n+1}(x) = 2(2x - 1)\hat{T}_n(x) - T_{n-1}(x) .$$

The terms of expansion were regulated in increasing numbers in the following way:

$$j = j(i_1, i_2, i_3, i_4) = 1 + \sum_{p=1}^{4} c^{5-p}_{4-p+(i_p+\ldots+i_4)} .$$

Coefficients of approximation $a_j(j = \overline{1,15})$ used to model atmospheric distortions and corresponding to two intervals of albedo: [0.01;0.1], [0.1;0.3], to angular conditions of survey: $0.6 \leqslant |\zeta| \leqslant 0.7$; $5° \leqslant \theta \leqslant 45°$; $0° \leqslant \varphi \leqslant 360°$, and to radiation wavelength $\lambda = 0.509\ \mu m$ has been given in [83].

In accordance with the model of the outgoing radiation formation (6.84), calculations were made of atmospheric distortions of radiation reflected from

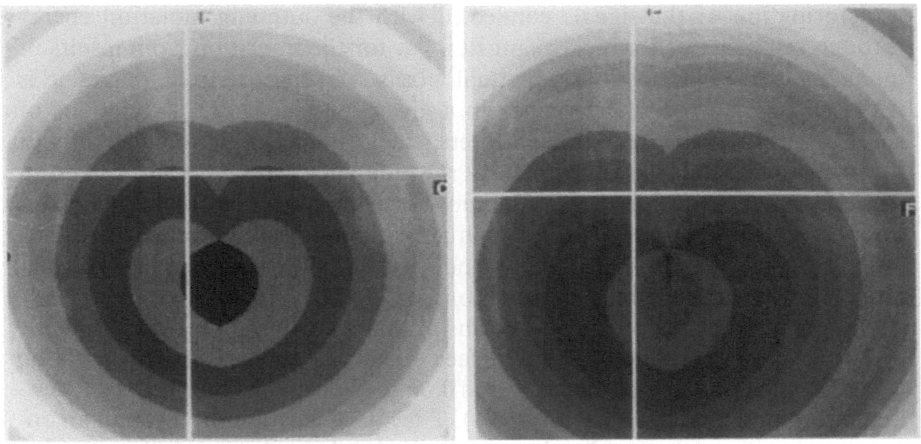

Fig. 6.28. a The atmosphere-distorted image of an homogeneous surface with albedos A = 0.03; **b** and A = 0.12. In Figs. 6.28, 6.30 and 6.32 the center of the image corresponds to nadir viewing. The upward direction is the direction to the sun in the image plane; solar zenith angle is $\cos\theta_0 = 0.6$

a homogeneous surface with albedo A = constant over its whole area. The results of calculations for A = 0.03 and A = 0.12 are shown as false-color images (256×256 pixels), respectively, in Fig. 6.28a, b. The chosen values of A correspond to the limits of the interval, in which real albedo values for crops, lie [108]. Figure 6.28 clearly shows the brightness levels of the dynamic range of images 0, 255, marked by different colors. These figures contain lines A, B, C, D, E, F to which the profiles of image brightnesses in Fig. 6.29 correspond. Comparison of images (Fig. 6.28) and their profiles (Fig. 6.29) reveals a consistency between numerical values of brightness and their color coding.

Centers of images correspond to Earth's surface viewing in nadir. For each image pixel the quantities $\eta^+ = \cos\theta$ and φ were estimated. To calculate η^+, linear sizes of an image pixel were prescribed and the height at which the radiation record is simulated, so that maximum viewing angle θ be less than 45°. The angle φ was measured from the Sun's direction in the image plane. In this way the photographic survey of surface objects was simulated. In a similar way a model scanner image can be obtained. For this purpose, the center of each line is considered the point of nadir viewing.

The estimates of intensity $I(A, \zeta, \eta^+, \varphi)$ calculated with formula (6.84) have an order of 10^{-4} to 10^{-2}. This is because the approximation coefficients have been calculated for absolute intensity normalized to solar constant measured at the atmospheric top. To present these quantities in whole numbers, a scale coefficient $K \sim 10^4$ has been chosen in the interval [0, 255] so that the estimates of $KI(A, \zeta, \eta^+, \varphi)$ calculated for any A, $\zeta = 0.6$, any φ and $\theta \in [5°, 45°]$ did not exceed the limits of the dynamic range of the image.

The coefficient K is a coefficient of linear transformation of absolute physical quantities, measured by recorders, into the dynamic range [0, 255] of digital presentation of the image.

The conversion of recorded light fluxes into a digital presentation can be both linear and non-linear. The latter is observed, for instance, in recording the images on photographic film, since the characteristic curve of photochemical process is non-linear. However, this non-linear transformation can be calculated based on the sensitometric wedge which is imprinted close to the recorded image. In the absence of the wedge, the recorded intensities can be calculated from results of measurements on test areas of the surface. Knowledge of this transformation is needed to solve the inverse problem – a retrieval of albedo from data of a digital image. First a transition is made from whole-number estimates in the dynamic range to absolute physical quantities of intensity and then the A parameter is found from (6.84) with I, ζ, η^+ and φ known.

Figures 6.28, 6.29 illustrate atmospheric distortions of the images of homogeneous terrestrial surfaces which exhibit a square dependence on η^+ and φ. Note, that on the edges the brightness values of the vertical profiles A, B, D, E increase. This has been caused by increased zenith viewing angle and hence by greater optical thickness of the atmosphere which determines a larger contribution of the diffuse scattered constituent to the formation of the outgoing radiation. For the same purpose, the brightnesses grow too, towards the edges of the profiles C and E. Large values of brightnesses of the first elements of the profiles A, B, D, F, compared to the last elements, and local maxima in the center of the profiles C, F are explained by the direction chosen to the Sun and by the specular type of surface reflection. As for the horizontal profiles at the bottom of the image, local maximum vanishes. The brightness of a model image of a homogeneous Earth surface with $A = 0.03$ (Fig. 6.28 a) varies towards its edge by 28% relative to the brightness near nadir, and with $A = 0.12$ (Fig. 6.28 b) by 23% for the given model of the formation of images and prescribed survey conditions. This means that with increasing albedo, the relative deviation of brightness decreases, though the number of gradations of the brightness grows from 11 (Fig. 6.28 a) to 21 (Fig. 6.28 b). This is explained in the following way. The albedo is a characteristic of the reflective properties of the Earth's surface. Therefore, at low albedos a major contribution to the formation of the outgoing radiation is made by atmospheric haze, i.e., solar radiation scattered in an atmospheric layer and not reflected from the surface. At $A = 0$ the outgoing radiation is only the atmospheric haze. As the albedo grows, the contribution of radiation reflected from the surface to the formation of the outgoing radiation increases, too. The effect of atmospheric haze on the image inhomogeneity thereby decreases although negligibly. Thus, results of calculations show that atmospheric distortions break the homogeneity of the Earth's surface.

The distortions of an image which models several agricultural fields with various crops happens in the following way. A model image (Fig. 6.30), where each field is a sum of real fluctuations of brightness and of average brightness proportional to the albedos of respective crops, is taken as an input one. The fields of wheat, barley, beets, and grazing thus formed are separated by shelter belts, whose profiles of brightness relative to the brightnesses of crops corre-

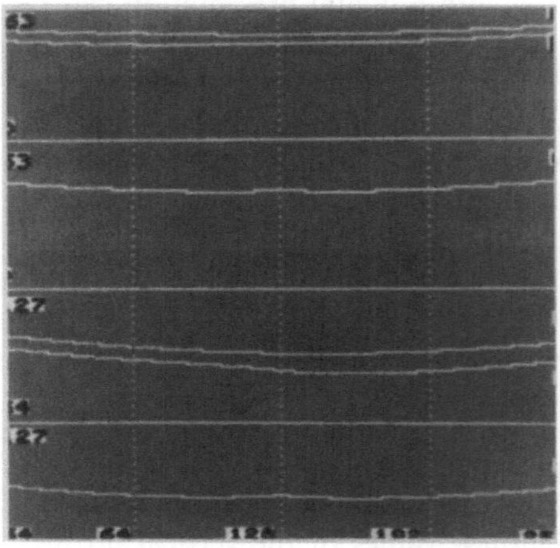

Fig. 6.29

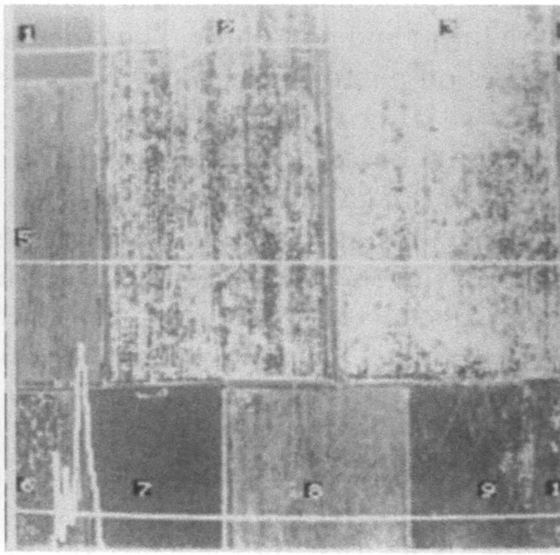

Fig. 6.30

Fig. 6.29. Profiles of the brightnesses of the atmosphere-distorted images. Profiles A, B, C are for a homogeneous surface with albedo A = 0.03 (see Fig. 6.28a); profiles D, E, F with A = 0.12 (Fig. 6.28b). The first and fourth broken curves from the top are for profiles A and D (left bounds of Fig. 6.28a, b). The abscissa axis gives the numbers of image pixels in the profile. The image pixels for the vertical profiles are numbered downward; for the horizontal profiles from left to right. The ordinate axis gives the brightnesses of image pixels

Fig. 6.30. An original image which models the fields of various crops: *2, 3, 4* wheat; *5, 6, 9* barley; *1, 7* beets; *8* grass; *10* shelter belts. The histogram of the relative spatial frequency distribution is displayed at the bottom part of the image

spond to the real situation. Figure 6.31 shows horizontal profiles of brightness G, H, I of the initial model image which give an idea of its structure: averages and fluctuations of the brightnesses of each of the model fields and shelter belts. Atmospheric distortions were calculated from formula (6.84). A value of albedo A in the interval $[A_{min}, A_{max}]$ was taken respectively for the brightness I of each pixel of the initial model image. This was made using a simple formula:

$$A = B_1 I + B_2 \tag{6.85}$$

where $B_1 = (A_{max} - A_{min})/(I_{max} - I_{min})$.

$$B_2 = A_{min} - [(A_{max} - A_{min})/(I_{max} - I_{min})]\, I_{min} \;.$$

I_{min} and I_{max} are, respectively, minimum and maximum brightness values for pixels of the initial model image of crops. The outgoing radiation intensities for each pixel were calculated from formula (6.84) with using a set of approximation coefficients corresponding to the albedo interval values, to which the albedo A of the processed pixel, belongs. The obtained distorted values of intensity were multiplied by a scale coefficient K estimated earlier. Figure 6.32 illustrates a distorted model image of crops for which, in calculations with (6.85), the following values have been taken: $A_{min} = 0.03$, $A_{max} = 0.12$. For each pixel of an atmosphere-distorted image (Figs. 6.28, 6.32) a square equation relative to A was solved. A value of the brightness of this pixel divided

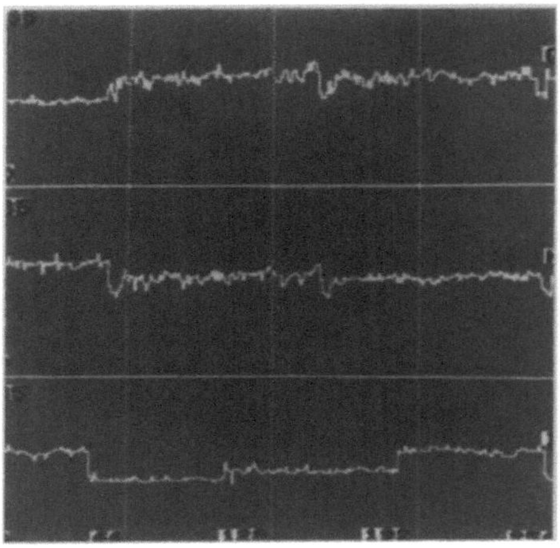

Fig. 6.31

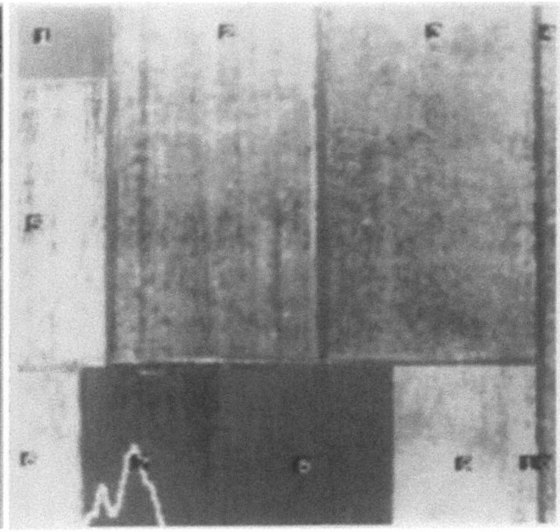

Fig. 6.32

Fig. 6.31. Brightness profiles H, G, I of the original image of crops (see Fig. 6.30)

Fig. 6.32. The atmosphere-distorted original image of crops (see Fig. 6.30) with the histogram shown at the bottom of the image

by the scale coefficient K was taken as a value of I $(A, \zeta, \eta^+, \varphi)$ in formula (6.84). The root of the square equation is found in the following way. The versions of the coefficients of an approximating polynomial are successively examinated. The root is considered true if it lies in the interval $[\alpha - \alpha/10; \beta + \beta/10]$ where α and β are, respectively, the beginning and the end of the interval for which the square equation was solved [77]. The brightness I for the corrected image was estimated with formula (6.85).

The correction of model images of a homogeneous surface (Fig. 6.28) resulted in images with homogeneous brightness. The corrected model image of crops almost coincides with the initial model image and therefore is not considered here.

The effect of atmospheric distortions can be assessed from results of a comparative analysis of the accuracy of crop classification on distorted and corrected images. Based on model images (Figs. 6.30, 6.32) a non-supervised classification was made by the brightness using the modes quantizing technique [165].

The modes of the histogram of the corrected image (Fig. 6.30) correspond to the classes: 1 (crops, maximum modes); 2 (beets, middle-sized mode); 3 (grazing, minimum mode). Only two modes remain on the histogram of the distorted model image (Fig. 6.32). The first mode corresponds to a class involving classes 2 and 3, the second mode to class 1. Comparison of the histograms shows that crops with lower average brightnesses are characterized by merging of their respective classes. This is consistent with the fact that the effect of the atmosphere on the emissions from low-albedo crops is more substantial. The maps of classification are shown in Fig. 6.33a, b. They illustrate the spatial distribution of classes.

Fig. 6.33 a, b. Maps of non-supervized classification of crops shown in **a** Figs. 6.30 and **b** 6.32, from brightness, using the modes quantization technique. *Class 1* (crops) covers a larger area; *class 2* (beets) covers a smaller area; *class 3* (grass) covers the least area

The accuracy matrices P_c and P_a were calculated where the index "c" marks the corrected image and index "a" that distorted by the atmosphere. The matrix element P_{ij} (i, j = 1, 2, 3) is the number of pixels of an image from class i which reached class j:

$$P_c = \begin{pmatrix} 46230 & 18 & 708 \\ 173 & 6040 & 56 \\ 752 & 390 & 5396 \end{pmatrix} \qquad P_a = \begin{pmatrix} 46141 & 80 & 637 \\ 996 & 6368 & 5523 \\ 0 & 0 & 0 \end{pmatrix}$$

With the use of these matrices the accuracy of classification by classes $\alpha_i = P_{ii} / \sum_j P_{ij}$ was calculated and an average accuracy of classification $\alpha = \sum_i P_{ii} / \sum_j P_{ij}$:

$$\alpha_{c1} = 98.45\%, \quad \alpha_{c2} = 96.35\%, \quad \alpha_{c3} = 82.53\%, \quad \alpha_c = 96.49\%$$

$$\alpha_{a1} = 98.47\%, \quad \alpha_{a2} = 49.41\%, \quad \alpha_{a3} = 0\%, \quad \alpha_a = 87.89\%$$

The results obtained show that atmospheric distortion affects the accuracy of classification of crops greatly. Model images of crops exemplify the effect of developed techniques for radiative correction and of cluster analysis on results of thematic processing of images. For instance, whereas the accuracy for class 1 is almost constant, the accuracy for class 2 almost doubles, and for class 3 it increases to 83%. The average accuracy of classification increases by 8.6%. To put into practice the results obtained, the following problems must be solved: the presentation of image pixels in absolute units; the assessment of the actual state of the atmosphere at the moment of survey; the assessment of real sensitivity of the radiation correction scheme, to variations of the state of the atmosphere, with the possible reduction of the sets of approximation coefficients, used in the retrieval of surface albedo. To solve the first of these problems, a detailed analysis is needed of the calibration characteristics of scanners. The second problem can be solved from surface measurements of atmospheric turbidity during the passage of high-altitude aircraft carrying the scanning instruments. The solution of the third problem is connected with further calculations using models of radiation transfer in the atmosphere.

6.11.3 The Morphological and Structural Analysis of Images

A special feature of the procedure of atmospheric correction of the images of agricultural fields is its dependence on the contrasts of these fields and on their geometry. The atmosphere blurs the boundaries of the fields and reduces their contrasts [101]. Numerical modeling of respective effects has shown that, with a given turbidity, the effect of the atmosphere on the classification of surface objects is more substantial in the case of an inhomogeneous surface than in the case of a homogeneous one. Modeling has also shown a reduced separability of classes for an inhomogeneous surface due to the broadening of the function of probability distribution for each class.

The need for a higher accuracy of classification in the combined solution of the problems of radiometric calibration and atmospheric correction based on respective models has been emphasized in [381]. Here the errors have been studied that occur in the process of radiation transfer from the Sun to the receiver, via the track of receiving, recording and processing of data. The effect of atmospheric haze on the accuracy of crop classification ignoring the function of the effect of "adjacent" pixels of image has been demonstrated in [16]. The efficiency of eliminating the atmospheric haze brightness based on the developed techniques has been shown. Another stage of atmospheric correction is a reducing of the boundary blurring of the fields, which further raises the quality of the classification results. Setting the problem of atmospheric correction of video information has been considered in [77]. In this section the problem of enhancing the contrasts of the fields' boundaries with the use of mathematical filtration apparatus will be discussed.

The defocusing effect in scanner images is largely determined by the recording system as well as by atmospheric scattering, towards the scanner, of radiation reflected from the Earth's surface (ES) surrounding the recorded element. The technique of scanner survey used now forecasts a successive detecting of radiation reflected from the elements of ES bands. The width of the band is determined by the diameter of the ES element selected by the solid angle on the observation system.[3]

Radiation in a certain spectral range, reflected from the band is presented as a digital matrix line for an image recorded by the recorder's channel for this particular spectral range. The adjacent line (or column) elements are usually

[3] This section was written with A.S. Barykin and V.P. Popov.

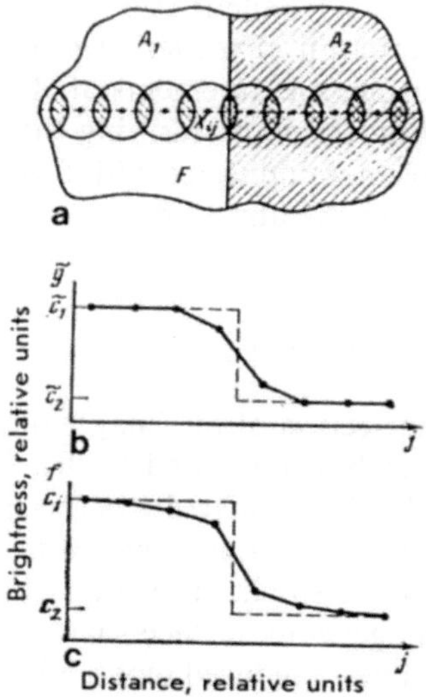

Fig. 6.34 a – c. A scheme of reflection of a surface band into the i-th line of an image f: **a** a fragment F containing a straight interface between two homogeneous fields A_1 and A_2; **b** the brightness profile of the line with the image defocused due to overlapping instant FOVs X_{ij} and X_{ij+1}; **c** the brightness profile of the line with an additional defocusing due to the atmospheric effect

overlapping. The area of overlapping does not exceed 50% of the area of the ES element. Overlapping results in agricultural objects located at the interface between two fields being presented with distorted spectral brightness. Physically, a sharp boundary between two crops with different reflective characteristics will be presented on the image by one or two elements. Figure 6.34a illustrates a fragment F typical of agricultural scenes, containing a straight boundary between two fields. Figure 6.34b shows the effect of smoothing determined by overlapping of adjacent elements of ES. The image in Fig. 6.34 is one-dimensional.

Analysis of space-derived scanner video-images shows that the number of near-boundary elements with spectral brightnesses differing from those of the internal elements of the fields is more than two, equal to four and more (Fig. 6.34c). This circumstance points to the presence of an additional smoothing factor. It is largely determined by the side-effect of radiation reflected from the background. Since for high-resolution images the scanner's viewing angle when surveying from space is small, (e.g., it varies within ±3°) we shall consider the defocusing of the image to be spatially invariant.

The effect of the background light on the recorded image f(i, j) can be written as:

$$f(i,j) = \bar{g}(x_i, y_j) + \iint\limits_{\substack{D \\ (\zeta,\eta)\notin X_{i,j}}} h(\zeta,\eta)\, g(x_i-\zeta, y_i-\eta)\, d\zeta\, d\eta \ . \tag{6.86}$$

Here $\bar{g}(x_i, y_j)$ is the value of brightness of an image pixel which would be recorded by scanner in the absence of the background light; $g(\zeta, \eta)$ is the near-surface distribution of reflected radiance; X_{ij} the instant FOV of the system onto the ES. The center of X_{ij} on the ES has coordinates (x_i, y_j). The ES fragment is taken to be flat, horizontal, with Cartesian coordinates; $h(\zeta,\eta)$ is the function of the defocusing effect of the atmosphere; D is the h carrier.

It has been shown in [101] that the dependence of h on (ζ,η) is close to bell-shaped, since h is largely determined by aerosol scattering with a strongly stretched spindle-shaped phase function.

Equation (6.86) assumes the Earth's surface described by the function $g(\zeta,\eta)$ to be orthotropic. This assumption is based on small sizes of solid angle with the base D and the apex in the place of scanner's location. It follows from the form of the function h and the form of X_{ij}, which can be either rectangular or elliptic, that the smoothing effect of defocusing on the step-wise gradient of brightness transforms it into the convexo-concave shape (Fig. 6.34c). At the boundary there appears a bending point. The convexity of the function f near the boundary of the region with high reflectivity and, respectively, the concavity in the near-boundary region of the adjacent field will be used in the construction of the operator correcting this distortion.

Let the function $g(\zeta,\eta)$ describe the brightness of radiation reflected from ES in a given spectral range in the direction to the scanner. The atmosphere does not exhibit its defocusing effect on $g(\zeta,\eta)$, and the scheme of scanning $(\zeta,\eta)\in X$, X is the ES area represented on the image f, $g(\zeta,\eta)$ means the near-surface distribution of spectral brightness over X.

For the recorded image $f(i,j)$, $i=\overline{1,N_i}$, $j=\overline{1,N_j}$, presented in the form:

$$f(i,j)=\Pi_{X_{i,j}}g(\zeta,\eta)+\iint\limits_{\substack{D\\(\zeta,\eta)\notin X_{i,j}}} h(\zeta,\eta)g(x_i-\zeta,y_j-\eta)\,d\zeta d\eta+\nu(i,j)\ ,$$

it is necessary to construct the operator S to process the image f to improve its quality (i.e., to improve the separability of objects). Here g,h,D,X_{ij} have been estimated, the projecting operator $\Pi_{X_{ij}}$ can be written as:

$$\Pi_{X_{i,j}}g(\zeta,\eta)=\frac{1}{\bar{\mu}(x_{ij})}\iint\limits_{X_{i,j}} g(\zeta,\eta)\,d\zeta d\eta\ ,$$

where $\bar{\mu}(X_{ij})$ is the area of the multitude X_{ij}; $\nu(i,j)$ the natural noise that breaks the homogeneity of the field; N_i, N_j the number of lines and columns in f. Comparison of the quality of images f and Sf is carried out using the following criterion. We shall call F the fragments of image f into which the ES areas are projected containing the straight boundary of two homogeneous fields with different properties, so that on F the brightnesses $g(\zeta,\eta)$ can be presented as:

$$g(\zeta,\eta)=\begin{cases}\tilde{C}_1 & (\zeta,\eta)\in A_1\\ \tilde{C}_2 & (\zeta,\eta)\in A_2\end{cases}$$

where A_1, A_2 realize Boolean division of fragments F, i.e., $A_1\cap A_2\neq 0$, $A_1\cup A_2=F, A=\{A_i\}_1^2$. The criterion of improving the quality of Sf, compared to f, is the existence of such F when an inequality is satisfied:

$$||Sfx_F-\Pi_A Sfx_F||_c<||fx_F-\Pi_A fx_F||_c\ , \qquad (6.87)$$

where $X_F(i,j)$ is the indicator function F determined from the expression:

$$X_F(i,j) = \begin{cases} 1, & (i,j) \in F \\ 0, & (i,j) \notin F \end{cases}$$

$\|\cdot\|_c$ is understood as maximum $\|\cdot\|$ on the image. Use of such norms is determined by the final aim of image processing – to increase the accuracy of crop recognition from spectral brightness. Really, on fragments $\{A_i\}_1^2$ the fulfillment of inequality (6.87) leads to a narrowing of the interval of indication variations. This is equivalent to a decrease of area diameters in the region of spectral indications, corresponding to these crops where the distance between the centers of clusters, remains unchanged. An increased compactness of clusters improves the separability of the objects.

Theoretically, the filtration of atmospheric interference is based on solution of the boundary problem for the equation of radiation transfer in an atmospheric layer with non-orthotropic reflection of ES. However, often it is impossible to apply the solution of the boundary problem for lack of information on the physical properties of the atmosphere in the region of survey. Therefore, to correct images for defocusing, useful algorithms may be those that, instead of physical information, use the topological properties of real agricultural scenes, the quasi-homogeneous distribution of their brightness as well as sharp physical boundaries between the adjacent fields with crops of different reflectivities.

The known techniques for enhancing the sharpness of the boundaries are aimed at improving the psychophysical perception of images [15]. For this purpose, the systems of electronic scanning use high-frequency filters. The technique of non-sharp masking is widely used in which a defocused image f_L is extracted from original image f; f and f_L being weighted. However, this technique lengthens the front, causing local brightness maxima and minima at the boundary [15].

To enhance the sharpness of boundaries, a discrete filtration is used through convolution of the original image f, with a mask H, $L \times L$ in size. The sum of the elements of the mask is equal to unity. The sharpness of boundaries is enhanced also be statistical differentiation [15]. Each pixel of image f is divided by MSD calculated from the surroundings of this pixel. An improved image, in contrast to the original, has large values near the boundary elements, where the image varies rapidly, and smaller values on other sites. Emphasizing the useful boundaries is followed by increased noise.

This discussion suggests the conclusion that all the techniques do not satisfy the criterion of increasing the quality (6.87) of the images of agricultural objects to be classified by spectral indications in the automated regime. Therefore, it is of interest to consider the properties of the smoothing morphological operator S.

Let a single-bound value K be taken on the fragment F, $K \subset F$ being such that the values $\{A_i\}_1^2$ can be presented by combined values obtained from K by means of Euclidean motions. The measure μ of the variable K is determined by the number of elements of F contained in K. At the point x, $x \in F$ is the result of the effect of operator S [15] on the image f determined from:

$$Sf(x) = \Pi_{\bar{g}K(x)} f(\zeta) = \frac{1}{\mu(K)} \sum_{\zeta \in \bar{g}K(x)} f(\zeta) \ ,$$

where $\bar{g}$ is the element from the variations of motions G_K satisfying the relationship:

$$\min_{\substack{g \in G_K \\ gK(x) \subset F}} \| \{f(z) - \Pi_{\bar{g}K(x)} f(\zeta)\} X_{\bar{g}K(x)}(z) \|_{R_N}$$

$$= \| \{f(z) - \Pi_{\bar{g}K(x)} f(\zeta)\} X_{\bar{g}K(x)}(Z) \|_{R_N} \ , \tag{6.88}$$

G_K is the variable such that $\forall g, g \in G_K, x \in gK(x)$.

The operator S is characterized by the following features. On homogeneous areas A_i the operator S smoothes the noise, which follows from the inequality:

$$E\{Sf(x_1) - Sf(x_2\}^2 \leqslant \frac{1}{\mu(K)} E\{f(x_1) - f(x_2)\}^2 \ ,$$

where $x_1, x_2 \in A_i$, E is the symbol of mathematical expectation of a random quantity. An inequality becomes an equality if random quantities $Sf(x_1)$ and $Sf(x_2)$ are independent. Another feature of the operator S is its ability to sense the closeness of the A_1 and A_2 interface. This makes it possible, along with the noise filtering to preserve the sharp brightness gradients between adjacent homogeneous regions A_1 and A_2, i.e., if the point $x \in A_1$ and motions g exist, so that $\bar{g}K(x) \setminus A_1 \neq \emptyset$, the choice of $\bar{g}K(x)$ is more probable, so that the equality $\bar{g}K(x) \setminus A_1 = \emptyset$ is satisfied. The same is valid for $x \in A_2$.

To correct a defocused image f, the property of the operator S is used which makes it possible to enhance the sharpness of the defocused brightness gradient at the interface between two homogeneous areas A_1 and A_2. In the absence of noise, the iterative application of S to f enables one to retrieve the $c_1 - c_2$ gradient.

Really, as has been mentioned, the defocused step-wise gradient in the region A_1 at $c_1 > c_2$ is convex (Fig. 6.34c). It follows from this that, for any $x \in A_1, x \in F$ the variable $\bar{g}K(x)$ will be located on the left of the variable $gK(x)$, $g \in G_K$, since expression (6.88) in this case, will be of less value. Then for $Sf(x)$ we have inequality $f(x) < Sf(x) \leqslant c_1$, if x is such that $f(x) < c_1$. $S^2 f(x)$ is also a convex function, but less non-linear. Inequality $f < Sf < S^2 f \leqslant c_1$ holds for it. With subsequent iterations we obtain similar inequalities. The same will be true at $x \in A_2$, but with an opposite sign, $f > Sf > S^2 f > \ldots \geqslant c_2$. These inequalities result in the solution for condition (6.87) for the operator S.

The presence of noise on the image hinders a correction for defocusing of elements x, multitudes A_1 and A_2 far from the boundary. This is because a weak non-linearity of the function f may not manifest against the noise background, and as a result, the variable $\bar{g}K(x)$ can be located arbitrarily.

Nevertheless, as shown by numerical experiments in the region A_1 (A_2) where elements x are such that a motion g exists for which the condition $gK(x) \cap A_2 \neq \emptyset$ ($gK(x) \cap A_1 \neq \emptyset$) is fulfilled, the effect of noise is negligible if the signal/noise ratio is $(c_1 - c_2)/\sigma > 5/2$ [15]. The number of efficient iterations of processing is also determined by the level of noise. It should be noted that the value of K and carrier D of function h should be consistent. With increasing μ (K) the defocused brightness is corrected more efficiently, however, in this case the geometry of the boundary can be distorted. The larger the nonlinearity and the fewer noise fluctuations, the more accurate will be the retrieved image. This fact is interesting because, it is with large distortions caused by a strongly turbit atmosphere, that the use of the operator S is more efficient. With a negligible distortion its use is not so necessary, it is of neutral character; in this case other methods are known to enhance the sharpness of boundaries [15].

To illustrate the effectiveness of image processing with the operator S in solving this problem, the following numerical experiment was accomplished. A Landsat scanner image of ES was processed, the diameter of the instant FOV of the scanner being about 80 m. Figures 6.35 and 6.36 show fragments F_1 and F_2 containing an agricultural scene and the coastline of a water basin. Figures 6.35 a and 6.36 a illustrate initial images of the fragments F_1 and F_2. The graphs of sections, corresponding to the white dashed line, show the effect

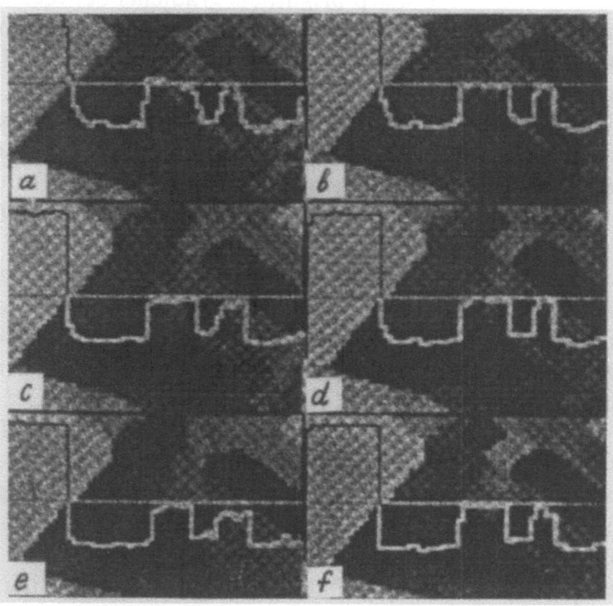

Fig. 6.35 a–f. A fragment F_1 containing an image of crops. **a** Original image; **b** Sf at μ (K) = 2×2; **c** Sf at μ (K) = 3×3; **d** S^2f at μ(K) = 2×2; **e** Sf at μ(K) = 5×5; **f** S^5f at μ(K) = 2×2. The graphs of cross-sections correspond to the *dotted white lines*

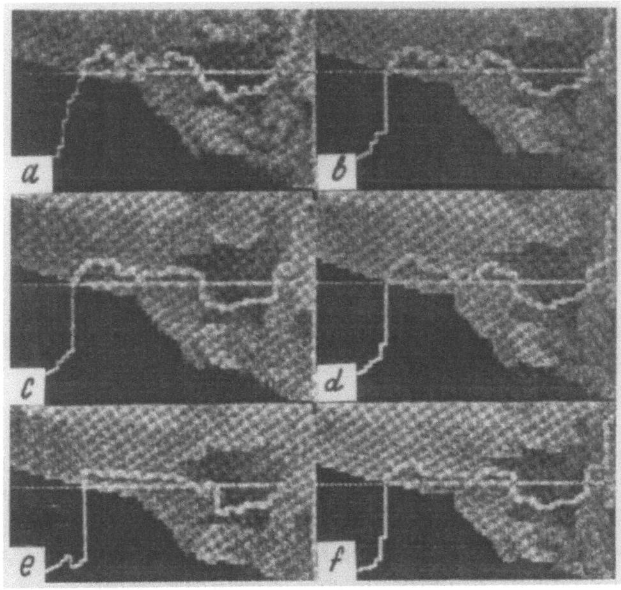

Fig. 6.36 a–f. A fragment F_2 containing an image of a water basin coastline. **a** Original image; **b, d, f** Sf, S^2f, S^5f at $\mu(K) = 2\times2$; **c** Sf at $\mu(K) = 3\times3$; **e** Sf at $\mu(K) = 5\times5$. The graphs of cross-sections correspond to the *dotted white lines*

of defocusing. Maximum gradients of brightness appeared to be smoothed, with the number of transition points more than 5. The recognition of image pixels on the fragments F_1 and F_2 can lead to attributing the transition points to some classes with an intermediate brightness. The processing of F_1 and F_2 by the operator S with the variable K of the form of a 2×2-element square, is shown in Figs. 6.35 b, 6.36 b; these figures illustrate similar sections which show that processing increases the sharpness of brightness gradients between homogeneous regions. The square K was chosen to simplify realization of the operator S, with other conditions formulated in the determination of S being satisfied. The results of iterative processing of the fragments F_1 and F_2 with the number of iterations $m = 1, 2, 5$ are shown in Fig. 6.35 b, d, f and Fig. 6.36 b, d, f, respectively. The graphs of sections illustrate an increase of the sharpness of brightness gradients when the number of iterations grows. To demonstrate the effect of variable K on the result of processing, images Sf obtained with using K 3×3 in size and 5×5 elements for the fragments F_1 and F_2 are shown in Fig. 6.35 c, e and Fig. 6.36 c, e.

The operator S with the variable K with 5×5 elements distorts the image, in the regions containing field corners more than in the case of a 3×3 square. It is seen from Fig. 6.35 and Fig. 6.36 that the size of the variable K should be small to decrease the distortions of the field structure at the corners.

A decrease of the dynamic range of brightness on homogeneous areas was observed in processing of several arbitrarily chosen fields. By means of the

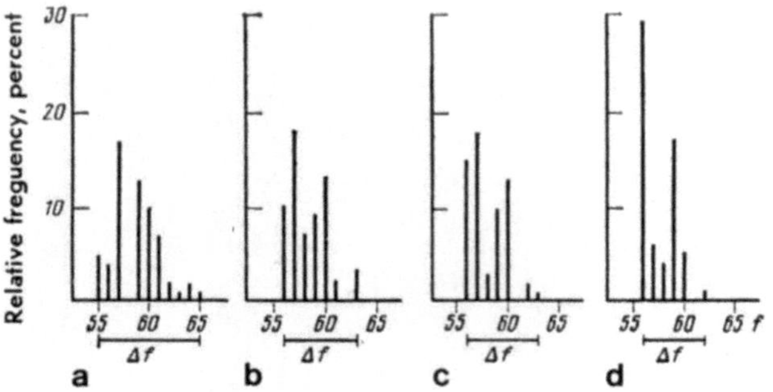

Fig. 6.37 a–d. Histograms of the intra-field brightnesses from images **a** f, **b** Sf, **c** S^2f, **d** S^5f at $\mu(K) = 2\times2$ pixels

basic mathematical provision of the display system, contours were constructed inside these fields, within which the brightness histograms were found. Figure 6.37 shows the histograms of a most typical field after its processing with the operator S using K 2×2 size elements. Figure 6.37 a–d shows the histograms of an initial image f and of the results of its processing Sf, S^2f, S^5f. The range of variations in brightness Δf for Sf, S^2f, S^5f becomes narrower, respectively, by 27%, 27% and 36%, with respect to the f range.

Use of the operator S makes it possible to raise the contrast of defocused gradients of brightness between homogeneous areas with a straight interface, which often takes place for extended boundaries of the fields under crops. With increasing number of iterations m the contrast of boundaries increases; however, with the growing number of iterations, the effectiveness of subsequent iterations decreases. In practice, it is enough to confine oneself to the number $m = 5$. An increase of the measure $\mu(K)$ leads to growing contrasts of direct extended boundaries but distorts the geometry of the field boundaries at the corners. Therefore, the variables K should be used with a small measure $\mu(K) = 2\times2$.

It is necessary to further improve the operator S in order to decrease the distortions at the corners, for instance, by introducing a numerical detector of areas containing the corners and by excluding such areas from processing with the operator. Further studies in this field will be aimed at understanding the alternative possibilities of different approaches to retrieving of the extended boundaries of agricultural fields.

6.11.4 The Technology of Atmospheric Correction of Video Information

As has been said in Section 6.10.2, the problem of recognition of natural objects on aircraft and satellite images proceeds from the existence of a priori alphabet of classes ($k = 1, 2, \ldots, K$; K is the total number of classes), an inventory of indications for prescribed multispectral brightnesses $I = \{I_1, I_2, \ldots, I_M\}$;

(M is the number of channels) and from certain decision rules (classifiers) permitting a separation of classes. Two stages characterize the procedure of recognition [195]: decreasing the dimensionality of multispectral data and optimizing the criteria of successful recognition. The reliability of recognition should serve a criterion of the quality of combined procedures of atmospheric correction and recognition.

When the dimensionality of the data decreases, the intra-class distance minimizes (the compactness of clusters grows), which eventually, leads to increased recognizability of classes. The criterion of compactness K of classes in an initial M-dimensional space can be written as the following measure:

$$D = \sum_{i=1}^{K} = \sum_{\vec{I} \in \Omega_i} R_{jl}^2(i) \ , \qquad (6.89)$$

where:

$$R_{jl}^2(i) = \sum_{p=1}^{M} (\vec{I}_{jp} - \vec{I}_{lp})^2 \qquad (6.90)$$

is the square of Euclidean distance of the j-th spectral measurement $\vec{I}_j$, from the i-th class Ω_i to the l-th spectral measurement $\vec{I}_l$, the closest to an average vector of class Ω_i. As a measure of grouping the $\vec{I}$ vector components, the following quantity was used:

$$\varrho_{ij} = \sum_{p=1}^{M} (|r_{ip}| - |r_{jp}|)^2 \qquad (6.91)$$

where $|r_{ip}|$, $|r_{jp}|$ are modules of the coefficients of correlation between the i-th and β-th components.

The efficiency of these procedures of minimization and grouping is not known a priori. If we proceed from Bayes' strategy of recognition, the function of distribution of random vector $\vec{I}$ is

$$f(\vec{I}) = \sum_{i=1}^{K} P(\Omega_i) f(\vec{I}|\Omega_i) \ , \qquad (6.92)$$

where $P(\Omega_i)$ is the a priori probability of classes Ω_i,

$$f(\vec{I}|\Omega_i) = (2\pi)^{-M/2} |\hat{S}_i|^{1/2} \exp\left[-\tfrac{1}{2}(\vec{I} - \vec{I}_i)^T \hat{S}_i^{-1} (\vec{I} - \vec{I}_i)\right] \qquad (6.93)$$

is the density of normal distribution, T the sign of transposition, $\vec{I}_i$ average vectors, $\hat{S}_i$ the covariance matrix of classes.

Average probability of correct recognition can be estimated as the lower limit [37]:

$$P \geqslant \hat{p} = 1 - 2 \sum_{\substack{i>j \\ i,j=1,2,\ldots,K}}^{K} \sum P(\Omega_i) P(\Omega_j) \exp\left[-B(\Omega_i, \Omega_j)\right] \ , \qquad (6.94)$$

where:

$$B(\Omega_i, \Omega_j) = \frac{1}{4}\left\{2 \ln \left|\frac{\hat{S}_i + \hat{S}_j}{2}\right| - \ln |\hat{S}_i| - \ln |\hat{S}_j|\right\}$$

$$+ \frac{1}{4}\{(\vec{I}_i - \vec{I}_j)^T (\hat{S}_i + \hat{S}_j)^{-1} (\vec{I}_i - \vec{I}_j)\} . \qquad (6.95)$$

An introduction of the matrix $\hat{A}$ $(m \times M)$ of linear transformation of initial space to a smaller space $(m < M)$ under condition (6.95) $(\tilde{\vec{I}}_i = \hat{A}\vec{I}_i, \tilde{\hat{S}}_i = \hat{A}\hat{S}_i\hat{A}^T$, $i = 1, 2, \ldots, K)$ leads to a combination of these two stages of recognition. In this case $\hat{P}_A \geqslant \hat{P}(\tilde{B})$; in (6.94) B is substituted by $\tilde{B}$. Then the probability of correct recognition can be maximized:

$$\frac{\partial \hat{P}_A}{\partial A} = 2 \sum_{i>j}^{K} \sum P(\Omega_i) P(\Omega_j) \exp[-\tilde{B}(\Omega_i, \Omega_j)] \frac{\partial \tilde{B}(\Omega_i, \Omega_j)}{\partial A} , \qquad (6.96)$$

where:

$$\frac{\partial \tilde{B}(\Omega_i, \Omega_j)}{\partial A} = (\tilde{\hat{S}}_i + \tilde{\hat{S}}_j)^{-1} \hat{A}(\hat{S}_i + \hat{S}_j) - \frac{1}{2}(\tilde{\hat{S}}_i^{-1} \hat{A}\hat{S}_i - \tilde{\hat{S}}_j^{-1} \hat{A}\hat{S}_j)$$

$$+ \frac{1}{2}(\tilde{\hat{S}}_i + \tilde{\hat{S}}_j)^{-1} (\tilde{\vec{I}}_i - \tilde{\vec{I}}_j)(\vec{I}_i - \vec{I}_j)^T$$

$$- \frac{1}{2}(\tilde{\hat{S}}_i + \tilde{\hat{S}}_j)^{-1} (\tilde{\vec{I}}_i - \tilde{\vec{I}}_j)(\tilde{\vec{I}}_i - \tilde{\vec{I}}_j)^T (\tilde{\hat{S}}_i + \tilde{\hat{S}}_j)^{-1} \hat{A}(\hat{S}_i + \hat{S}_j) . \qquad (6.97)$$

Coming back to the problem of atmospheric correction of multispectral data, the following should be emphasized. If we proceed from the vectors $\vec{I}^\infty$ and $\vec{I}^0$ at the upper (∞) and lower (0) atmospheric boundaries and from the presentation

$$\vec{I}^\infty = \hat{T}\vec{I}^0 + \hat{D}_0 , \quad \hat{S}_i^\infty = \hat{T}\hat{S}_i^0\hat{T}^T , \qquad (6.98)$$

where $\hat{T}$ is the operator, characterizing atmospheric transparency in different channels (mean constituents of the brightness of radiation reflected by the Earth surface); $\hat{D}_0$ the operator characterizing the atmospheric haze brightness, we may obtain the following linear relationship of variations in spectral brightnesses at a level (∞) and (0) [101]:

$$\delta\vec{I}^\infty = \hat{A}_I \delta\vec{I}^0 + A_\tau \vec{\delta}_\tau + A_\gamma \vec{\delta}_\gamma + \vec{e} . \qquad (6.99)$$

Here $\hat{A}_I$, $\hat{A}_\tau$, $\hat{A}_\gamma$ are dimensionality matrices M; $\vec{e}$ is the vector of the dimensionality measurement error M; (τ, γ) are the optical thickness and phase function of the atmosphere. An important conclusion can be drawn here: average probability of correct recognition expressed by (6.94) with prescribed $\vec{I}_i^\infty = \hat{T}\vec{I}_i^0$, $\hat{S}_i^\infty = E[(\delta\vec{I}^\infty)(\delta\vec{I}^\infty)^T]$ is similar for levels (∞) and (0). Here E is a symbol of mathematical expectation (it is assumed that $E(\vec{\delta}_\tau) = 0$, $E(\vec{\delta}_\gamma) = 0$, and $E(\vec{e}) = 0$). Thus, if this criterion of the quality of recognition is used, the conclusion can be that linear transformations of brightness in atmospheric correction do not improve the introduced measure of quality. However, this example gives us a guaranteed reliability of recognition before and after correction, and the problem of atmospheric correction of video information should be approached proceeding from the chosen criteria of quality.

The earlier related studies have usually proceeded from a representation of atmospheric interference in the multiplicative form (6.98). As we see, this form of atmospheric correction of the aircraft and satellite data has no meaning, because it does not raise the quality of recognition. Besides, the dependence of the vectors and matrix operators from (6.98) on viewing angles should be borne in mind. The right-hand part of (6.98) must be enhanced with a term called "background haze" $\hat{D}_s$ – part of radiation getting into the FOV of the instrument from local environs (often called "side light"). Explicit expressions for operators $\hat{T}$, $\hat{D}_0$, $\hat{D}_S$ have been given in Chapter 5.

There is a need for technological development of these methods of thematic interpretation of remote sensing data, with atmospheric correction as part of this thematic interpretation. An experience and respective difficulties in solving these problems have just been discussed.

Some problems of thematic interpretation of digital air- and space-derived video information, as applied to the current technical means of interactive analysis, have been discussed. The procedure of atmospheric correction of video information is considered as part of the problem of recognition of natural objects from their multispectral images. Next is the problem of introducing the quantitative criteria of the quality of initial information and of the results of thematic interpretation. The prospects for the development of these studies are connected with improving satellite-borne instruments (obtaining high spatial and spectral resolutions), using measures to combine remote sensing data and reference sub-satellite measurements, and with the transition from techniques for assessing the parameters of the state of models of the forecasts of the environmental components. The enumerated complex of scientific problems is being developed together with an improving of technical means of processing and interpretation of video information. This combination of scientific and technical achievements will promote the development of applied methods for space-derived information.

6.12 An Assessment of the State of Natural Objects from Remote Sensing Data

The efficiency of thematic interpretation of the materials of air- and space-based survey is determined by two principal conditions, the quality of information needed in analysis of a particular situation and by available techniques for analysis to reach an objective. The first condition is connected with working out the criteria of quality as applied to various problems for different conditions of survey, different radiation receivers, spectral intervals, etc. The second condition requires the development of both theoretical and empirical methods for the interpretation of measurement data. The first condition puts certain constraints on techniques of interpretation which, in turn, make specific demands on the quality of information, not always reached at the present level of development of remote sounding techniques. Methodological recommendations on the use of the middle-resolution scanner information (resolving elements $280 \times 280\,m^2$) must be substantiated considering smoothing due to this high resolution and the insufficiently reliable calibration of the sensors.

Smoothing shows itself in the fact that a large number of resolving elements appear on the boundaries of agricultural fields, decreasing the reliability of their recognition. Lack of reliable calibration of information determines the difficulty of the transition from recorded signals to energy quantities used in simulation models.

The problem of assessing the state of natural objects on these images is closely connected with constructing models of multifactor regression and interactive classification of soil-vegetation cover.

6.12.1 Models of Multifactor Regression

The problem of establishing relationships between remote sensing data and parameters of natural formations is foremost in the interpretation of the remote sensing results. The measured brightnesses are functions of a certain state of the objects under study. The state of the soil-vegetation system is characterized, for instance, by the following parameters: height, thickness, projective cover, humidity, phytomass volume, etc. [159]. An assessment of the state of crops consists in the determination of the enumerated and other parameters from recorded multispectral brightnesses. In the first approximation, the solution of this problem foresees the use of regression schemes of analysis, which assume the form of the equations of correlation of measured quantities and parameters characterizing the biophytometric state of vegetation cover to be prescribed. The problem is reduced to finding the coefficients of correlation equations based on analysis of the data of combined aircraft and surface measurements.

At this stage of studies, we used a technique of linear step by step regression [151]. The principal idea of the technique consists of the determination of expansion of the vector-response (multispectral brightnesses y_j) from several vector-factors (parameters of the state x_i) in the form of subsequent series of linear expansions, with only one factor being introduced to or removed from expansion at each step. An introduction or removal of the factor in the regression scheme depends on the amount of contribution made by this factor in the equation concerned. If the contribution does not exceed the chosen threshold level, the factor is considered insignificant and is removed from the equation. Then the factors not included into the expansion are analyzed from the viewpoint of the possibility of their inclusion.

Usually, in regression problems the threshold levels in assessing the significance of this or that factor are determined in this technique with the use of F-criterion after Fischer [171]. Here the vector factors and vector responses are considered as random quantities, for which the ratios of mean squares, determined by regression to mean squares due to residual variations, are random quantities with eigen distributions, average dispersions and momenta known. However, the use of this criterion is correct only with the normal distribution of the vector response.

For this purpose, we use a statistical criterion based on the chaos principle. This criterion, consistent with the general principles of constructing statistical criteria of significance, has some different features. It can be used independently of any suppositions about the probability properties of the quantities which constitute the basic sample. Moreover, this criterion can be used when the elements of the sample are neither determined nor random quantities. The applied criterion of the model's efficiency is independent of the volume of this sample, i.e., it can be used both for large and small samples.

As an example of the use of regression models, we shall consider their application to establish correlations between the brightnesses measured by the 7-channel spectrometer (400, 450, 550, 600, 650, 840, and 1015 nm) and biophytometric parameters, found by investigations of agricultural fields.

The fall season survey resulted in the following correlation matrix of the first 6-channel measurement data with the number of stems (the upper halls of the matrix) and the weight of wet phytomass (the lower half of the matrix) of winter wheat:

X_1	X_2	X_3	X_4	X_5	X_6	Y
1.00	−0.13	−0.03	0.17	−0.10	0.14	−0.17
−0.13	1.00	−0.04	0.11	−0.16	0.46	−0.07
−0.03	−0.04	1.00	0.32	−0.20	−0.29	−0.39
0.17	0.11	0.32	1.00	−0.87	−0.27	−0.90
−0.10	−0.16	−0.20	−0.87	1.00	0.37	0.86
0.14	0.46	−0.29	−0.27	0.37	1.00	0.36
−0.14	−0.11	−0.44	−0.85	0.81	0.34	1.00

In the first case (six channels – the number of stems) channel 4 (30% error) and channel 5 (still lower accuracy) were significant. In the second case (six channels – the weight of wet phytomass) only channel 4, with an error of about 42% was significant. In the same period, the matrix of correlation of the weight of dry phytomass (the upper half) and the area of leaves (the lower half) turned out to be:

X_1	X_2	X_3	X_4	X_5	X_6	Y
1.00	−0.13	−0.03	0.17	−0.10	0.14	−0.17
−0.13	1.00	−0.04	0.11	−0.16	0.46	−0.14
−0.03	−0.04	1.00	0.32	−0.20	−0.29	−0.45
0.17	0.11	0.32	1.00	−0.87	−0.27	−0.87
−0.10	−0.16	−0.20	−0.87	1.00	0.37	0.80
0.14	0.46	−0.29	−0.27	0.37	1.00	0.28
−0.12	−0.13	−0.40	−0.86	0.75	0.39	1.00

Here, channels 3 and 4 are statistically significant (error about 10%) and channel 5 with approximately the same accuracy. It is seen from these tables that the phytometric factors considered are connected with results of remote sounding in almost the same way: there is no prevailing effect of the data from any channel (parameter) on the response function. This can be explained by the weak correlation of channels, (except for channels 4 and 5, which are closely connected and, at the same time, the most significant in this case).

In the stem/stalk-growing phase for winter wheat in order to relate the 7-channel measurement data with the volume weight (100 kg/ha), the following correlation matrix was obtained:

X_1	X_2	X_3	X_4	X_5	X_6	X_7	Y
1.00	0.91	0.78	0.85	0.98	0.08	−0.78	−0.80
	1.00	0.85	0.88	0.92	0.14	−0.65	−0.62
		1.00	0.82	0.79	0.14	−0.44	−0.47
			1.00	0.86	−0.12	−0.78	−0.67
				1.00	0.10	−0.78	−0.79
					1.00	0.48	0.33
						1.00	0.92
							1.00

In this case channel 3 became significant (relative error 15%) as well as channels 4 (20%) and 7 (5%). Here, in contrast to the phase of wintering, the following correlation of channels and their relationship with the volume weight of phytomass is observed. Channel 1 was the most significant in correlation equations for 7-channel measurements with the volume weight (relative error about 10%) and the number of plants per unit area (about 7%); the respective coefficients of correlation are given in the upper and lower parts of the following matrix:

X_1	X_2	X_3	X_4	X_5	X_6	X_7	Y
1.00	0.98	0.98	0.98	0.91	0.98	0.65	−0.89
0.98	1.00	0.99	0.99	0.87	0.99	0.64	−0.86
0.98	0.99	1.00	0.99	0.87	0.99	0.66	−0.86
0.98	0.98	0.88	1.00	0.86	0.99	0.62	−0.85
0.98	0.98	0.98	0.99	1.00	0.99	0.64	−0.86
0.98	0.99	0.99	0.99	0.88	1.00	0.67	−0.87
0.65	0.64	0.66	0.62	0.68	0.67	1.00	−0.48
−0.90	−0.89	−0.88	−0.87	−0.79	−0.87	−0.50	1.00

The last two cases of analysis of spectrometric data and biometric parameters of the state of vegetation cover, testify to the promising use of regression models planned for models of crop yield forecasts.

Another sphere of application of regression models is a search for an optimal combination of channels to assess the parameters. Figure 6.38 exemplifies the results of application of one-factor models, where the factors were brightnesses for MKF-6 channels 3 and 4 [191] and vegetation index $\tilde{V}$ [390] and Euclidean distance $\tilde{E}$ [412]:

$$\tilde{V} = \frac{I_4 - I_3}{I_4 + I_3} \; ; \qquad E = \sqrt{I_4^2 + I_3^2}$$

where I_k – is the brightness for channel k.

Table 6.15 lists the coefficient of correlation and the errors in estimating the regression coefficients considering the factor for the same initial data as in Fig. 6.38.

The calculations with one-factor models, where vegetation index and Euclidean distance were taken as a factor (Fig. 6.38a, c), show that the results for Euclidean distance are much better. This can be explained by the non-linear character of the relationships studied. Thus, optical combinations of data for the MKF-6 channels are not confined to the consideration of only the sums or ratios (vegetation indices) which serves the basis for practical use of the information from Landsat multispectral scanners [278, 280, 411]. The studies have shown that non-linear combinations of the type of Euclidean distance in most cases give more stable correlations of recorded brightnesses with the biophotometric parameters of the state of vegetation cover.

Missuse of the hypothesis of the statistical independence (orthogonality) of the parameters to be estimated, can be a possible source of large errors, in assessing the parameters of the state by the classical regression technique of least squares.

In this connection, one of the possible ways to improve the estimates has been used. For multiple linear regression models:

$$y = x\beta + \varepsilon \; .$$

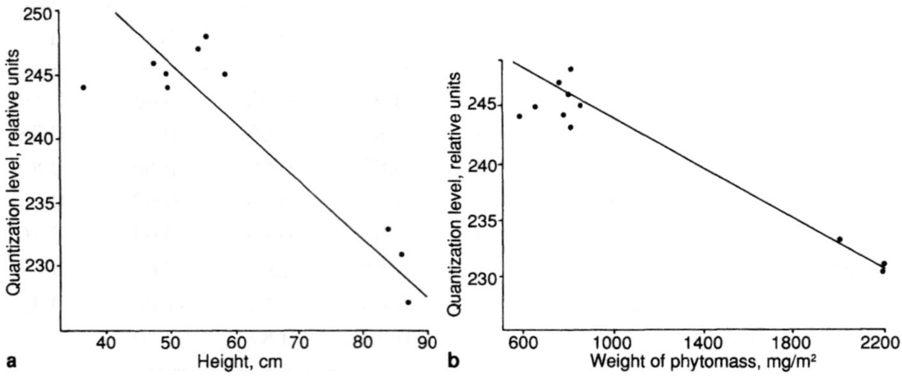

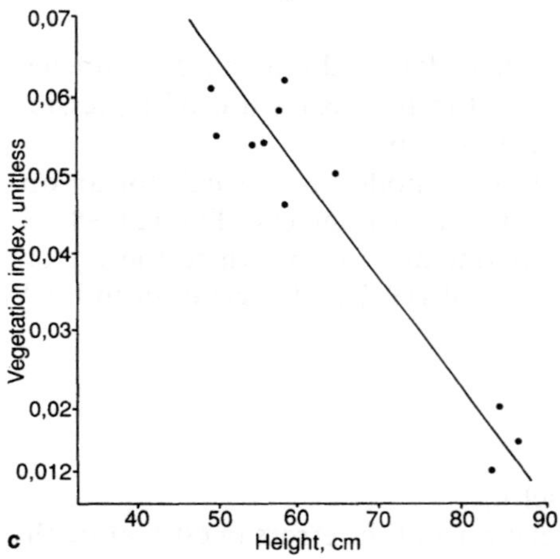

Fig. 6.38 a – c. The dependence of the Euclidean distance between brightnesses for MKF-6 channels 3 and 4 **a** on the average height of vegetation cover; **b** wet phytomass weight and **c** vegetation index

Table 6.15. Correlation coefficients r and MSD of significant factors E, $\tilde{V}$ for the brightnesses from the data of MKF-6 channels 3 and 4 and their combinations with phytometric parameters in a one-factor model

Parameters	E		$\vec{V}$		3 Channel	4 Channel
	r	σ	r	σ		
Wet phytomass	0.97	9%	0.53	–	27%	10%
Dry phytomass	0.92	14%	0.60	44%	36%	11%
LAI	0.89	16%	0.50	–	33%	17%
Thickness	0.31	–	0.39	–	–	–
Height	0.89	17%	0.58	46%	37%	14%
Projective cover	0.91	16%	0.27	–	20%	23%

Note: – means the factor was not included in the models; LAI, leaf area index: relative area of the surface of phytoelements [411]

Here, y – is the vector-predictor of dimensionally i (the response function); x – the non-random matrix of the dimensionality "i×j", matrix of phytobiometric parameters of the state; i – the number of observations; j – the number of parameters; β – the vector of estimated coefficients of equation of dimensionality j; ε – the vector of errors of dimensionality i the mathematical expectation and covariance matrix of which are, respectively, $M\varepsilon = 0$ and $E(\varepsilon\varepsilon^T) = \sigma I_i$ where I_i – is the unit matrix of dimensionality i×i, σ – the dispersion of errors, T – the symbol of transposition. In this case the β can be estimated by minimizing the sum of squares of the differences of the true response function and respective function obtained from the equation:

$$D(\hat{\beta}) = \varepsilon^T\varepsilon = (y-\hat{y})^T(y-\hat{y}) \ ,$$

which gives a non-shifted estimate $\hat{\beta}$ of vector β:

$$\hat{\beta} = (x^T x)^{-1} x^T y \ .$$

However, if a shifted estimate

$$\hat{\beta}^* = (x^T x + kI)^{-1} x^T y \ ,$$

is used to estimate $\hat{\beta}$, this estimate will be more reliable than a non-shifted estimate of $\hat{\beta}$, which is particularly important with non-orthogonal factors-parameters. Studies of the graphs of dependence $\hat{\beta}_l^*(k)$, $l = 1, \bar{j}$ (ridge-traces) make it possible to reveal the extent of relationships between factor-parameters and, choosing a certain k, to improve respective estimates.

Figure 6.39 shows results of related calculations for MKF-6 channel 4 in the period of the winter wheat ear-production phase, where $(x_1 - x_6)$ are the six parameters shown in Table 6.15. Data of Fig. 6.39 suggest that the coefficients of regression equations with small k are unstable: with slight variations in the initial information, the $\hat{\beta}$ coefficients obtained will strongly differ from the initial ones. Many correlations of the parameters estimated lead to larger errors due to different signs of the regression coefficients. Analysis of the results in

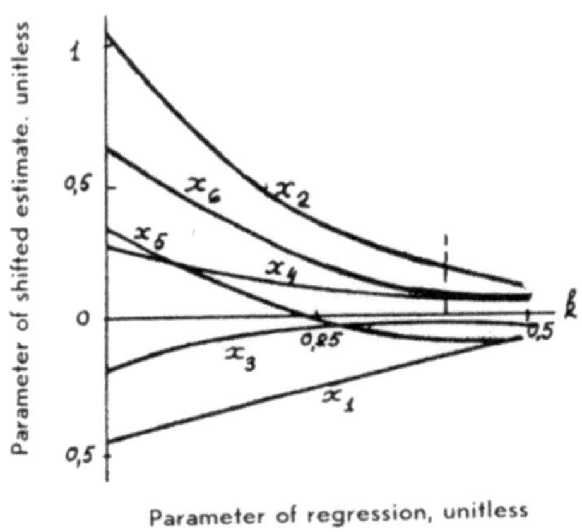

Fig. 6.39. The curves of the shifted estimate of ridge-regression as a function of the biophytometric parameters of vegetation cover. The *arrow* marks the optimal value of the ridge-regression parameter

Fig. 6.39, from the viewpoint of how much the estimates of coefficients β_l^* depend on k – ridge-traces, has shown that the variability of estimates becomes stabilized at $k \geqslant 0.3$. In this case, the correlation of factor-parameters decreases, but their connection with the response function increases.

Further studies make it possible to obtain regressive estimates of the parameters of the state of vegetation cover from materials of air- and space-based survey and to develop a regressive forecast of the crop yield.

6.12.2 Interactive Classification of Soils and Vegetation from Mid-Resolution Images

The solution for this problem includes two principal stages: (1) consideration of natural aspects of the dynamic properties of soil and vegetation [172]; (2) classification by the dialogue display complex based on results in Section 6.10.3.

As an example, we consider results of interactive analysis of satellite images and sub-satellite investigations of agricultural fields during the 1980 Kherson experiment.

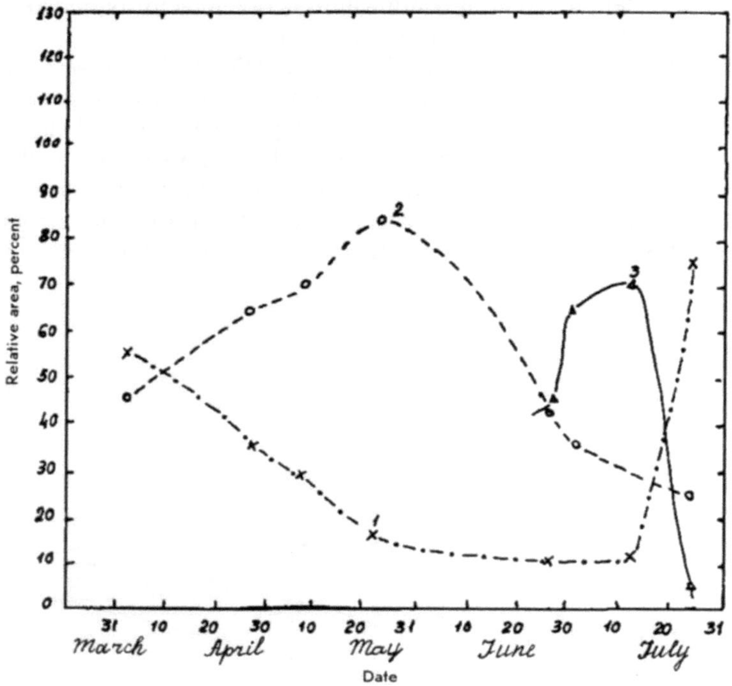

Fig. 6.40. Dynamics of the percentage of: *1* black fallow ground; *2* green vegetation and *3* yellow vegetation for the Kherson region in 1980

Figure 6.40 illustrates data on changes during the vegetation period in percentage of areas in the Kherson region covered with ploughed soil, with green and brown (ripe) vegetation. These results have been obtained using the structure of areas under crops and the phenological phases of principal crop developments in the period considered.

Winter crops and perennial grass are first (compared with other crops) to finish the cycle of development. During the ear-production phase of winter wheat, early spring crops are still producing stem growth, and late spring crops are sprouting – the beginning of stem growth. Due to differences in the development phase these groups of crops differ markedly in the vegetation mass. For instance, on 28 May 1980, from observations at the Kherson meteorological station, the height of winter wheat in the ear-production phase was 70–80 cm, whereas the height of early spring crops (stem growth) did not exceed 25–50 cm, and that of late spring crops (sprouting) about 15 cm (Fig. 6.41).

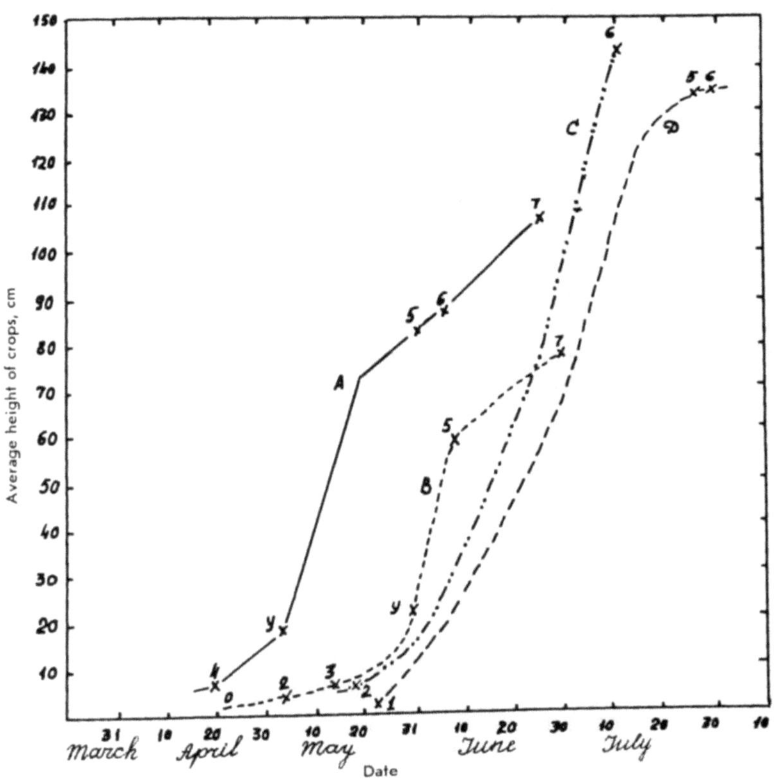

Fig. 6.41. The height and phases of development of some crops in the Kherson region in 1980: *0* soil; *1* shoots, *2* third leaf; *3* bushing; *4* stem growth; *Y* appearance of the lower node of straw; *5* earproduction; *6* flowering; *7* milk ripeness; *A* winter wheat; *B* spring barley; *C* sunflower; *D* maize

Table 6.16. Difference (in days) in the beginning of the development phases of basic crops (from average multi-year data)

Pairs of crops	Phases of development				
	Sprouting	Stem growth	Ear production	Milk ripeness	Wax ripeness
Winter wheat – spring barley	–	19	5	5	4
Spring barley – maize	27	–	40	–	44
Sunflower – maize	15	–	–	–	13

The agricultural fields under these crops can be presumed to differ in structure due to different vegetation mass and different vegetation cover.

The differences between crops at the time of beginning of their vegetation phase for the region studied, are shown in Table 6.16. Notice that winter wheat differs much from late spring crops with reference to the starting time of the same phases of development. The differences between wheat and spring barley in the initial period are more substantial than during the ripening. For instance, the difference in the starting time of their stem-growth phases constituted 19 days, and in the phases ripening only 4 days. The difference in their heights by the time of ripening decreases, too. The late ripening crops (sunflower, maize) differed greatly from early spring crops, in the contrast, in the second half of the vegetation period. This difference between sunflower and maize is substantial only in the ripening period.

Thus, using the phenological calendar, the best periods to recognize crops can be determined. This information about phenophases serves as the basis for the learning process needed for recognition.

With the development of vegetation the curves, similar to those in Fig. 6.41, describe a "tasselled-cap" figure [289] which has been obtained in analysis of Landsat data.

The subsequent interactive analysis of mid-resolution imagery consists of the determination of threshold levels corresponding to different gradations of brightness on digitized images, using histograms of relative frequences of occurrence of objects over a given territory (here the Kherson region). Additional information on the objects studied makes it possible to determine the nature of these "objects". By intuition, the objects on the 28 May 1980 histogram can be assumed to be water surface, ploughed soil and areas with different vegetation covers (Fig. 6.42). In accordance with Figs. 6.40 and 6.41, an interactive analysis has made it possible to recognize the following groups of agricultural crops: winter crops, early spring and late spring crops.

By the threshold levels of the histogram in Fig. 6.42, in accordance with the remarks in Section 6.10.3 regarding selecting of the resolving boundaries, the analyzed image was quantized into five different classes. Comparison of actual data on the structure of soils under crops, obtained by the complex experi-

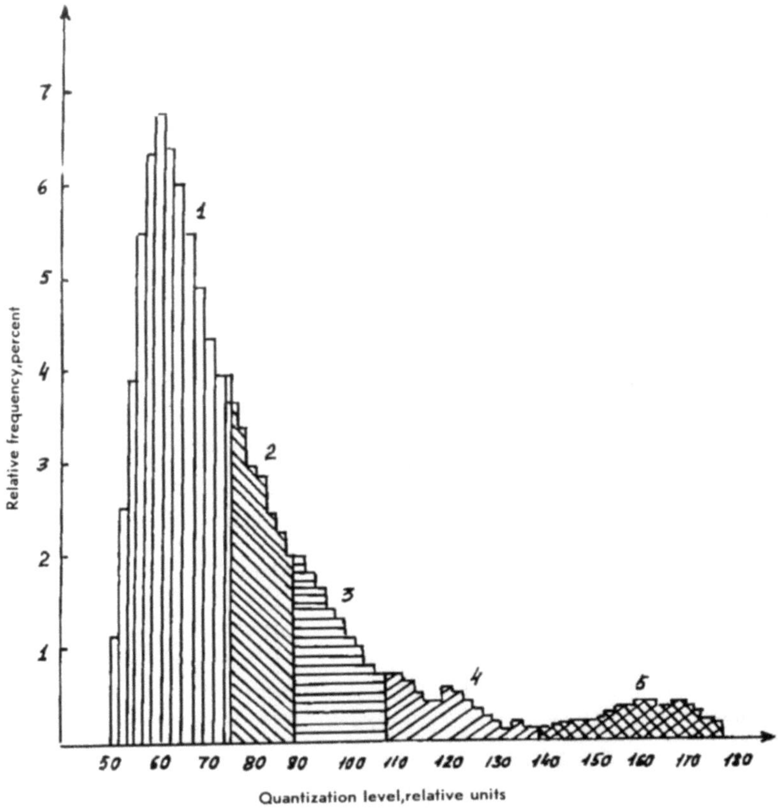

Fig. 6.42. The histogram of an image of the Kherson region as of 28 May 1980: *1* winter crops; *2* early spring crops; *3* late spring crops; *4* soil; *5* water basins

ment, with results of this classification of a mid-resolution image has shown their consistency (Table 6.17).

This suggests that, despite the difficulties in practical use of satellite mid-resolution information: it is possible to classify groups of agricultural fields using the interactive regime approximately with equal success. An increase of the spatial resolution of the applied instruments should raise the accuracy of recognition.

Apart from results of self-learning from satellite images using a priori information (phenological calendar, statistical data on the development of plants, etc.), the data of surface subsatellite measurements supplemented with satellite scanner survey data, are an important source of information [21]. It has been shown in [22, 107] that color coordinates are a useful characteristic to assess the parameters of soil vegetation cover from aircraft spectrometric measurements. The color coordinates are calculated from the formula [45]:

Table 6.17. The structure of agricultural fields of the Kherson area as of 28 May 1980

Land	Area (%)	Area (%)
	From interactive analysis	From field studies
Water basins	8	8
Ploughed soil	12	15
Late spring crops	16	20
Early spring crops	20	11
Winter crops	44	46

Table 6.18. The limits to quantizing densities of the fragments of a mid-resolution image for learning and analyzed samples

Types of objects	Learning	Analysis
Water	Above 202	Above 235
Coniferous forests	184 – 202	222 – 235
Soils	163 – 184	204 – 222
Grass cover	153 – 163	178 – 204
Light colored objects (sands, clouds, etc.)	Below 153	Below 178

$$\begin{Bmatrix} X \\ Y \\ Z \end{Bmatrix} = C \sum_{i=1}^{n} S(\lambda)\varrho(\lambda) \begin{Bmatrix} \bar{x}(\lambda) \\ \bar{y}(\lambda) \\ \bar{z}(\lambda) \end{Bmatrix} \Delta\lambda \ ,$$

where:

$$C = \frac{100}{\sum_{t=1}^{n} S(\lambda)\bar{y}(\lambda_i)\Delta\lambda} \ ,$$

$S(\lambda_i)$ is the relative spectral distribution of incident radiation in the spectral interval $\Delta\lambda = \lambda_{i+1} - \lambda_i$; $\varrho(\lambda_i)$ the SBC; $\bar{x}(\lambda_i)$, $\bar{y}(\lambda_i)$ and $\bar{z}(\lambda_i)$ are summation functions; n the number of spectral intervals. The concept of color coordinates has successfully been used in [22, 107] to estimate the humus content in soils, the chlorophyll content in the leaves of plants, the areas under dead and damaged crops, etc. All these results of sub-satellite referencing are needed to obtain a learning sample in image classification procedure. We shall consider here, an example of a combined satellite-aircraft experiment.

The interactive analysis of a digitized mid-resolution image consists of the color displaying of image, tracing on the fragment with a mouse the route of

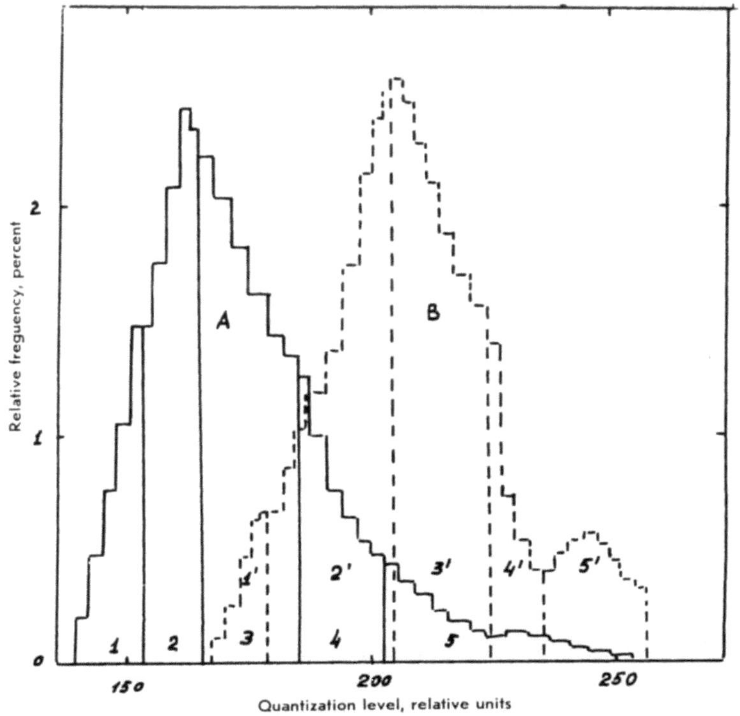

Fig. 6.43. Smoothing of the histogram of the analyzed fragment of the Ukraine territory. *A* With learning; *B* with analysis of the current state. Figures on the graphs correspond to the types of objects given in Table 6.18

aircraft spectrometry; calculating the histograms of relative frequencies, and constructing a learning sample with the use of the color coordinates of subsatellite "points".

Results of interactive classification of a fragment of the Ukrainian territory on 11 August 1981 are given in Table 6.18, where the following classes are selected: water, forest, grass, soil and light colored objects (sands and late ripening crops with a large vegetation mass).

The respective histograms for the learning data base and for additional expeditional studies are shown in Fig. 6.43. The constant shifting in the Table 6.18 data relative to the histograms in Fig. 6.43 is explained by location of the learning and analyzed samples in different natural zones: the learning from the aircraft spectrometric data was accomplished in the forest-steppe zone, and the results of learning were used in the steppe zone where the surface expedition was working.

Table 6.19. Results of comparison of the color coordinates of some soils in the Kherson region

Points	From satellite data	From aircraft data
1	12.6	15.0
2	13.5	13.8
3	19.2	15.9
4	14.5	15.0
5	15.9	17.1
6	15.3	14.5
7	17.0	14.5
8	13.9	18.8
9	15.3	18.0
10	21.2	15.0
11	14.7	15.4
12	15.5	13.3
13	14.3	13.8
14	12.7	14.7

Differences in the brightnesses of ploughed soils were identified with variations in their humus content [22, 107], through of course, a complex characteristic of the state of soil is needed. If the learning data base on color coordinates contains a description of other parameters of the state of soils, for instance, their erosion, the content of moisture, iron compounds, salts and other components, variations in these parameters can be traced [198]. Table 6.19 compares the color coordinates obtained from the mid-resolution imagery interactive analysis, with the use of the sub-satellite reference sample, aircraft observations and laboratory soil samples taken in this region. The learning sample was based on the color coordinates determined by the SBC for the fragment A (Fig. 6.43) and by summation functions from laboratory analyses of the samples. Individual post-learning elements of the analyzed image were presented as the color coordinates of the learning fragment.

An extrapolation of the learning results on another area of image (a fragment in Fig. 6.43), for which the color coordinates were available for the preceding period of variations (the right-hand column, Table 6.19) made it possible to express the elements of this fragment through the color coordinates as of the data of survey (the left-hand column, Table 6.19). Table 6.19 data refer not to individual agricultural fields but to the sets of ploughed fields in an approximately similar state. These data demonstrate a satisfactory agreement between the results of comparison, for instance, in 9 of 14 cases relative errors do not exceed 10%, with maximum errors of the data compared no worse than 6%.

A comparatively good coincidence of results in Table 6.19, may be explained by a slight variability of data in the applied learning sample of color coordinates and by sufficiently reliable recognition of soils and vegetation on images. Both these circumstances point to objective possibilities of combined solution of the problems of interactive analysis and sub-satellite referencing. Therefore, the usefulness of these techniques of analysis of satellite and aircraft data, to assess the state of the soil and vegetation cover, can be considered proved.

6.13 The Structure and Principal Elements of the Automated System of Processing Digital Aerospace Video Information

The atmospheric radiative correction of satellite images (spectra), the retrieval of the albedo of natural formations and the determination of the optical parameters of the atmosphere from the data of aircraft and satellite measurements, are part of the general problem of automated processing, interpretation and analysis of space-derived video information. To solve this most important problem of remote sensing of the environment, several operations must be realized successively to input, accumulate, process, archive and distribute the information to assess the state of the environment and its resources [343–345]. The information is needed to monitor ecological processes and economic activities.

Aerospace information processing, in the interests of ecology and economic activity, is divided into two principal interrelated stages: interdisciplinary processing and disciplinary thematic processing.

The interdisciplinary processing includes the following operations (1) radiometric (photometric) correction of video information; (2) correction of geometric distortions of images, their geographical attributing from ballistic data, and transformation of satellite images into certain cartographic projections; (3) annotation and registration of satellite data after their interdisciplinary processing.

Consideration of atmospheric effect on the aerospace video information is an important problem of its own, interesting both for interdisciplinary processing and for thematic (disciplinary) interpretation of satellite data.

Problem-oriented transformations of aerospace video information after its interdisciplinary processing, underlying characteristic contours and boundaries, selecting genetically homogeneous areas, stretching the dynamic range, equalizing of images are also an important aspect of automated processing in the disciplinary (thematic) interpretation of satellite data [401–403].

Information about spectral brightnesses of natural formations does not solve the problem of identification and classification of objects. Therefore, one of the principal concerns of disciplinary thematic processing is the solution of these problems, with due regard to the users' needs. Other principal objectives of the disciplinary thematic interpretation of satellite data consists of the complex assessment of the states of complicated ecological-economic objects: constructing the models to control these objects; and to make decisions. In this case, the selection of the most informative indications, construction and assessment of the quality of the functional models of classification and identification of objects make it possible to optimize the initial video information (using real spectral, spatial and temporal distributions of the radiation fields measured).

At the output of the automated system of the satellite data processing the video terminals, displays, etc., must provide the users and the data bank with thematic maps (of a given scale, format and projection) in terms of parameters characterizing the state of the object.

As a rule, the automated systems of processing are constructed as "universal" designed to solve one or several disciplinary thematic problems but not those rigidly attributed to a definite class of digital videodata. At the same time, data for particular aspects are usually specific and have features of their own, which requires problem-oriented automated systems of processing for various spheres of application.

Problem-oriented, applied-automated processing systems consist of numerous techniques and means of data processing, they analyze various input information, and actively use the operator's possibilities. This results in various functional units and sub-systems, designed not only to solve the image processing problems themselves, but also to process other types of initial data: accumulation, archiving and presentation of results; organization of interaction with the users and with other systems. Various components aimed at this objective determine the transformation of the present automated processing systems into the integrated systems. Both external and internal integration is one of the basic trends of the development of the present automated processing systems which cannot be considered individually, without means of accumulation and transformation of videodata, processing systems, and analysis of other types of data. External integration leads to an appearance in the automated processing systems of the means of communication, sub-systems of accumulation and archiving of both input and intermediate data and of the final results. Internal integration concerns various components of the automated processing systems – means and techniques for processing and the data processed [6].

An integration of the processing means implies a combination in one system of various universal or specialized arrangements permitting parallel operations with the data. For instance, with the use of independent specialized processors, each performing its own operation: filtration, calculation of the object's area, calculation of the perimeter, segmentation, etc. An integration of the processing techniques implies the presence, in the system, of functional subsystems and processing algorithms to solve various problems and to reach an objective in different ways (possibly, parallel). In this case, an analysis of imagery becomes a kind of multiple experimentation with initial data. Such an experiment is characterized by a combined use of image processing techniques, of mathematical statistics, multidimensional regressive analysis, artificial intelligence, analysis of the scenes, etc., all available data, including intermediate results, being involved in the processing. This necessitates an organization of an efficient data exchange in the system, of well-developed archiving and easy access to the data (including video-data bases). An important perspective problem in the integration of processing techniques is a search for standardized successions of the processing algorithms for various situations. These make it possible to specify successively the results obtained, reaching reliability not because of more initial data but due to more complete information. In its turn, this determines stricter requirements in displaying the output information, their diverse possibilities and flexibility, including a parallel presentation of results through the "window" systems on video terminals.

An integration of the processing techniques is closely connected with an integration of the data themselves, i.e., with a combined analysis of all available information (archived videodata, data on test areas, qualitative or quantitative characteristics – indications). One of the trends observed here, is a development of techniques of combined analysis of various classes of data, for instance, a presentation of the scene and its description. Another trend of data integration is a development of universal structures (presentations) of videodata, one for various types of processed images – graphic, semitonic, spectrozonal, color, etc. With these structures, the processing techniques become constant with respect to videodata type, which clearly aids the possibility of their combined use.

Along with specialized image processors, the structure of the present automated processing systems involves moderate, small and personal computers oriented with the programmed videodata processing. Special processors are fast-operating compared to matrix or network systems due to series processing of the pixels of one image (matrix) and of several images (network).

An important direction, of the development of the program provision of automated processing systems, is the development of efficient systems of archiving of and searching for videodata bases and banks. At present these systems often exist independently of the systems of analysis and processing, i.e., they are not integrated with these. This gap can be bridged by creating common languages of processing and manipulating of the videodata bases.

It is important to consider carefully the program provision for interaction between automated processing systems and users. Apart from interaction languages, command languages were developed based on the choice of an alternative (table of solutions) suggested by the system and techniques to describe the state and the system's potential were developed.

Most of the changes in the various components of the automated processing systems are connected with the users' desire to achieve maximum productivity from these systems and their efficient interaction with the users. As applied to automated processing systems, the problem of videodata presentation is in reality the problem of studies of the interior data structure, with which the apparatus and program modules of the system interact. The productivity of algorithms and the number of problems potentially solved by the system depend on this structure. A well-chosen single interior presentation of videodata makes it possible to compress both the transmitted information and that archived in the system, to avoid their cumbersome transformation (for instance, from spectral presentation to syntactic). It is, therefore, desirable that present multifunctional automated processing systems have common memory or common type of structured data which is actually an intrasystem interface.

Emphasis is now placed on the presentation and processing of images based on pyramidal-recursive structures [6]. This is explained by adjusting such structures to construction of apparatus (in particular, with the use of integral technology) and by constructive principles of the development of the program provision. The potential of simultaneous in series computation in both spatial and temporal structures, will ensure their maximum efficiency among the computational structures known. The development and studies, of new ways of presentation of videodata on computers, are now one of the main directions of fundamental studies of automated image processing, which is a source of new concepts and methods for processing of the digital aerospace video information.

Chapter 7 Models of Interaction of the Shortwave and Longwave Components of the Radiation Budget with the Atmosphere and the Earth Surface

Chapter 2 dealt with the general theory of interaction between radiation and natural media represented as a random cloud of scatterers, a random continuum medium, and a rough surface. Extreme transitions have been shown from the general theory of interaction to the radiation transfer theory as a partial case of the general theory with certain supposition concerning the radiation wavelength, the size of scattering centers and the coherence functions. Proceeding from the theory of shortwave radiation transfer, Chapters 3 to 5 were dedicated to the theory of the transfer function of the atmosphere-object-background system, and Chapter 6 discussed the atmospheric correction of digital air- and space-derived video information.

This chapter will deal with the problem of radiation transfer, mindful of (Chapter 8) a consideration of one more important problem of the hydrometerological interpretation of the remote sensing data – the problem of retrieval of the components of the Earth radiation budget characterizing the relationships between the short-wave radiation (wavelengths 0.3 – 2.5 μm) and longwave radiation (wavelengths 2 – 100 μm). Therefore, in this Chapter we shall analyze the boundary problems of the short- and longwave radiation transfer from the point of view of application of the results obtained in Chapter 6.

7.1 The Boundary Problems of Shortwave Radiation Transfer

In solving the boundary problems of shortwave radiation transfer, the radiation field is usually presented as three groups of photons: photons repeatedly scattered in the atmosphere and not reaching the Earth's surface; photons singly reflected from the surface; photons repeatedly re-reflected by the atmosphere and by the surface. The third group of photons is usually small compared to the two others.

Another possibility of describing the outgoing shortwave radiation field concerns in presentation of radiation as a mean constituent characterizing the system's brightness for a horizontally-homogeneous terrestrial surface, and as a fluctuating constituent due to variations in brightness in the case of an inhomogeneous surface. Here, too, the atmospheric haze brightness may be specially selected in order to reduce the original image to some "reference" one (without atmospheric interference). However, more often the atmospheric haze brightness constituent is combined with the mean constituent of reflected radiation, this sum being called the mean brightness of the system (see Chaps. 3–5). We shall adhere to the last point of view, since a representation of radiation field as a sum of the mean and fluctuating constituents is practically convenient for atmospheric correction (see Chaps. 4 and 5). Obtaining a "reference" image and selecting photons scattered in the atmosphere and not reaching the Earth's surface are, to some extent, artificial procedures whose practical realization has its specific difficulties. At the same time, in the atmospheric correction of air- and space-derived information the recorded brightnesses can easily be separated into the mean and fluctuating constituents.

Note also that, the boundary problem for the mean constituent of the intensity of radiation scattered in the atmosphere, is solved by selecting a source of direct solar radiation in the right-hand part of the transfer equation. In the respective boundary problem for the fluctuating constituent this source is absent in the right-hand part of the equation, since fluctuations of the radiation field appear due to horizontal inhomogeneities of the surface with a coefficient of reflection $\bar{R}+R'$ at the incident radiation intensity $I+I'$, with subsequent neglect of the mixed correlation moment $R'I'$.

As a result, the boundary problems of the shortwave radiation transfer in the system "surface-atmosphere", ignoring polarization effects for plane-parallel geometry, are of the following form:

for the mean constituent of radiation intensity within the spectral interval $\Delta\lambda$:

$$\eta \frac{\partial \bar{I}(z,\eta,\varphi)}{\partial z} + \sigma_t(z) I(z,\eta,\varphi) = \sum_{p'=1}^{P'} \sigma_s^{p'}(z) \sum_{k=0}^{\infty} b_k^{p'}(z) \frac{2k+1}{4\pi}$$
$$\times \left\{ \int_0^{2\pi} d\varphi' \int_{-1}^{1} d\eta' \bar{I}(z,\eta',\varphi') P_k(\vec{\Omega}\vec{\Omega}') + S_0 P_k \left[\eta \cos\theta_0 \right.\right.$$
$$\left.\left. + \sqrt{1-\eta^2}\sqrt{1-\cos^2\theta_0}\cos(\varphi-\varphi_\odot)\right] \exp\left[-\int_z^H \sigma_t(z') dz' \right] \right\} , \tag{7.1}$$

$$\bar{I}(H,\eta^-,\varphi) = 0 , \tag{7.2}$$

$$\bar{I}(0,\eta^+,\varphi) = D_0 \bar{R}(\eta^+,\varphi,\cos\theta_0,\varphi_\odot) + \int_0^{2\pi} d\varphi' \int_0^1 \bar{I}(0,\eta',\varphi')$$
$$\times \bar{R}(\eta^+,\varphi,\eta^-,\varphi')\eta^- d\eta^- , \tag{7.3}$$

for linear deviations of radiation intensity from averages within the same spectral interval:

$$\eta \frac{\partial I'(z,\eta,\varphi,x,y)}{\partial z} + \sqrt{1-\eta^2} \left[\cos\varphi \frac{\partial I'(z,\eta,\varphi,x,y)}{\partial x} + \sin\varphi \frac{\partial I'(z,\eta,\varphi,x,y)}{\partial y} \right]$$
$$+ \sigma_t(z) I'(z,\eta,\varphi,x,y) = \sum_{p'=1}^{P'} \sigma_s^{p'}(z) \sum_{k=0}^{\infty} b_k^{p'}(z) \frac{2k+1}{4\pi}$$
$$\times \int_0^{2\pi} d\varphi' \int_{-1}^{1} d\eta' I'(z,\eta',\varphi',x,y) P_k(\vec{\Omega}\vec{\Omega}') , \tag{7.4}$$

$$I'(H,\eta^-,\varphi,x,y) = 0 \ , \tag{7.5}$$

$$I'(0,\eta^+,\varphi,x,y) = \int_0^{2\pi} d\varphi' \int_0^1 \bar{I}(0,\eta^-,\varphi') R'(\eta^+,\varphi,\eta^-,\varphi',x,y)\eta^- d\eta^-$$
$$+ \int_0^{2\pi} d\varphi' \int_0^1 I'(0,\eta^-,\varphi',x,y) \bar{R}(\eta^+,\varphi,\eta^-,\varphi')\eta^- d\eta^- \ . \tag{7.6}$$

The boundary problems (7.1–7.3) and (7.4–7.6) can be solved for instance, using an advanced method of spherical harmonics (Sect. 3.8) with the use of prescribed characteristics (sections and phase functions) of the interaction of radiation with gaseous and aerosol substances. However, the solution of problems (7.4–7.6) is difficult because of the presence of the second term in the left-hand part of Eq. (7.4). The respective systems of algebraic equations cannot be solved for each individual azimuthal harmonic, and this, in turn, necessitates the use of high-order matrices in calculations. Finally, the calculation algorithm for the boundary problem (7.4–7.6) turns out to be complicated. A simplified solution for this problem can be obtained using the mathematical apparatus of the theory of linear systems (see Chap. 5).

7.2 Cross-Sections of Interaction and Phase Functions

The transfer equation is valid only for the monochromatic radiation intensity, and real photometric measurements are made for spectral intervals of finite width [13]. The transfer equation for monochromatic radiance could be repeatedly solved and then the obtained solutions could be averaged within the required spectral interval, not very effectively using calculation means because of the necessity to calculate the fields of radiation for a large number of subintervals in a given spectral interval. Also, calculation of the atmospheric radiation fields is difficult to achieve because of numerous spectral lines in a given interval, and macrosections of interaction depend on meteorological parameters of the atmosphere (for instance, on temperature and pressure). However, the equation of radiation transfer for a spectral interval of a finite width, can be of the same form as for the monochromatic radiation intensity, if effective cross-sections of interaction and phase function, averaged within the given interval with a certain weight function are introduced. Hence, a real possibility appears to archive non-monochromatic cross-sections and phase functions on magnetic tape for their multiple use in calculations of radiation fields in the atmosphere.

In boundary problems (7.1–7.3) and (7.4–7.6) these effective cross-sections of interaction and phase functions are being used. For calculations of monochromatic microsections of scattering and absorption and phase functions of the particles of atmospheric aerosol and cloud formations, the Mie theory was used, whose calculation formulas for different ratios of particle's radius and incident radiation wavelength have been given in Chapter 1. The results described are valid for polydisperse ensembles of aerosol and cloud particles characterized by a complex refraction index, particle-size distribution and by a certain concentration of paticles within each of their radii ranges. The sections and indicators of Rayleigh scattering in the atmosphere and microsections of absorption for each of p′-substances of the atmosphere, in the visible and in the near IR spectra have already been discussed. Thus, here we shall not reconsider these but only refer to expressions (1.38–1.40) to calculate the microsections of scattering (absorption) and coefficients of expansion of the phase functions of atmospheric aerosol and cloud formations for a spectral interval of the width $(\lambda_1, \lambda_{1+1})$.

The absorption microsections for each of the p′-substances of the atmosphere averaged over the frequency interval (ν, ν_{1+1}), will be:

$$k^{p'}_{l,1+1} = \frac{\sum\limits_{m} \int\limits_{\nu_l}^{\nu_{l+1}} \chi(\nu)\, k^{p'}_{m}(\nu, T, p)\, d\nu}{\int\limits_{\nu_l}^{\nu_{l+1}} \chi(\nu)\, d\nu}, \tag{7.7}$$

where the coefficient of absorption $k^{p'}_{m}$ in each of m-lines is characterized by a certain contour of the line, its half-width, intensity and position of its center on the frequency axis [85]. The transformation of the solar radition spectrum and longwave radition in the atmosphere is considered in (1.38–1.40) and (7.7), through the respective weighted functions $F(\lambda)$ and $\chi(\nu)$.

On the assumption of the Lorenzian contour of the absorption line:

$$k^{p'}_{n}(\nu, T, p) = \frac{S_n(T)}{\pi} \frac{\alpha_n(p, T)}{(\nu - \nu_{on})^2 + \alpha_n^2(p, T)}, \tag{7.8}$$

where (T, p) is temperature and pressure, respectively; α_n, S_n, ν_{on} are half-width, intensity and position of the center of the respective absorption line:

$$\alpha_n = \alpha_{on} \left(\frac{P}{P_0}\right) \left(\frac{T_0}{T}\right)^{n_{p'}}, \tag{7.9}$$

$$S_n = S_{on} \left(\frac{T_0}{T}\right)^{j_{p'}} \exp\left[E''_n \left(\frac{1}{T_0} - \frac{1}{T}\right)\right], \tag{7.10}$$

$n_{p'}, j_{p'}$ are empirical coefficients [85], E''_n the energy of the lower energy level. The $\alpha_{on}, S_{on}, \nu_{on}, E''_n$ values have been taken from magnetic tape, where they have been tabulated for normal conditions (T_0, p_0) from $\lambda = 0.8\ \mu m$ towards longer wavelengths [330].

For the principal atmospheric gases (water vapor, oxygen, ozone, carbon dioxide) the microsections of absorption average over individual intervals can be written in the form:

$$k^{p'}_{l,l+1} = \left(\frac{T_0}{T}\right)^{j_{p'}} \left[f_1(T) + \left(\frac{p}{p_0}\right) \left(\frac{T_0}{T}\right)^{n_{p'}} f_2(T) + \left(\frac{p}{p_0}\right)^3 \left(\frac{T_0}{T}\right)^{3n'_p} f_3(T) \right] = \left[\frac{cm^2}{g}\right] , \tag{7.11}$$

where for the functions $f_1(T), f_2(T), f_3(T)$ rather cumbersome expressions have been obtained, approximated then by polynomials [76], orthogonal for discrete sets of real temperature values:

$$f_j(T_k) = \sum_{n=0}^{N} a_n^j \tilde{\varphi}_n(t_k) , \quad j = 1,2,3 ,$$

$$\tilde{\tilde{\varphi}}_n(t) = \sum_{i=0}^{n} (-1)^{i+n} \times \frac{(i+n)(i+n-1)\ldots(2i+n-1)(j+t)(j+t-1)\ldots(j+t+i+1)}{(i!)^2(2j)(2j-1)\ldots(2j+i-1)} , \tag{7.12}$$

$$\sum_{i=-j}^{j} \tilde{\tilde{\varphi}}_n^2(i) = \frac{(2j+n+1)!\,(2j-n)!}{(2n+1)\,[(2j)!]^2} , \quad t = \frac{2T_k - a - b}{b-a} ,$$

$2j$ is the number of intervals into which the temperature range with the lower a and upper b boundaries is divided; T_k is the temperature in the k-th point of the range.

The approximation coefficients for each of the p′-substances of the atmosphere for the spectral interval 0.8 to 2 μm obtained by data processing described above, have already been given in [83]. Table 7.1 lists these coefficients for spectral intervals 0.8 – 7 μm and 2 – 100 μm.

In the tables given the last three symbols denote the sign and the order of the respective numbers. Only the first two coefficients $a_n^j (j = 1, 2)$ for the first two functions $f_1(T), f_2(T)$ have been used.

The coefficients $a_n^{j=3}$ appeared to be negligible compared to $a_n^{j=1}$ and $a_n^{j=2}$ for the given spectral intervals.

The microsections of radiation absorption by atmospheric gases are retrieved with formulas (7.11) and (7.12). Thus, Table 7.1 data resulting from difficult to apply characteristics of magnetic tape [75] give a thorough idea of the absorbing properties of the principal optically-active gaseous-substances of the atmosphere and, at the same time, are very effective in practical use.

The macroscopic sections of scattering and absorption are calculated by multiplying these microscopic sections by the densities of the respective atmospheric substances. Here, note that the substitution of the usually applied functions of transmission of atmospheric gases by their macrosections of absorption leads to different results in radiation field calculations, not exceeding 2%. The use of macrosections is a physically more substantiated procedure than the use of transmission functions applied only in the absence of scatter-

Table 7.1. Coefficients of approximation a_n^j (j = 1, 2) of functions $f_1(T)$ and $f_2(T)$ for spectral intervals in the short- and longwave spectral intervals

Order of priority	Water vapor		Carbon dioxide		Oxygen	
	$a_n^{j=1}$	$a_n^{j=2}$	$a_n^{j=1}$	$a_n^{j=2}$	$a_n^{j=1}$	$a_n^{j=2}$
			$\Delta\lambda = (0.8-7)\ \mu m$			
0	1.1054+00	9.1338−03	1.0213−02	−1.6522−06	2.1118−04	−3.4201−08
1	7.8934−01	−8.9095−04	6.4614−03	1.0562−05	5.6623−05	−2.0415−08
2	5.1355−01	−1.9867−04	4.7520−04	6.6599−06	1.2682−07	−5.8611−09
3	2.9837−02	−2.2311−04	8.4419−06	1.4490−06	1.8104−08	−8.7224−10
4	7.6504−04	2.5676−05	−3.7122−06	8.0404−08	7.3328−10	1.0797−11
5	−2.9960−04	9.8650−07	2.6061−07	−1.2636−08	−1.2001−10	1.0127−11
			$\Delta\lambda = (2-100)\ \mu m$			
0	3.4945+00	−1.1356−02	8.2845−02	−1.6516−05	1.0851−05	−8.6432−10
1	2.2249+00	−4.6958−03	9.9220−02	−1.9354−05	2.7612−06	2.2228−10
2	2.9866−01	7.8730−05	3.3435−02	−7.2237−07	1.6465−09	−1.0809−12
3	2.9337−02	1.1609−05	4.8786−03	−9.9298−07	−6.1203−10	3.9091−14
4	4.9056−02	−4.3239−06	−1.2009−04	−2.8963−08	−5.0115−11	3.3331−15
5	1.0432−04	4.0867−06	−5.1605−04	1.0415−09	−1.4533−12	2.4684−12

Table 7.2. The stratification of the lower atmospheric layers for the summer mid-latitudes

z, km	T, K	p, mbar	ϱa, g/m^3	ϱH_2O, g/m^3	$\varrho oz \cdot 10^{-5}$ g/m^3
0	294	1013	1191	14	6
1	290	902	1080	9.3	6
2	285	802	975.7	5.9	6
3	279	710	884.6	3.3	6.2
4	273	628	799.8	1.9	6.4
5	267	554	722.1	1.0	6.6
6	261	487	648.7	0.61	6.9
7	255	426	583.0	0.37	7.5
8	248	372	522.5	0.21	7.9
9	242	324	433.3	0.12	8.6
10	235	281	415.9	0.064	9.0
11	229	243	369.3	0.022	11
12	222	209	326.9	0.0060	12
13	216	179	288.2	0.0018	15
14	216	153	246.4	0.0010	18
15	216	130	210.4	76×10^{-5}	19
16	216	111	179.7	64×10^{-5}	21
17	216	95	153.5	56×10^{-5}	24
18	216	81.2	130.5	50×10^{-5}	28
19	217	69.5	111.0	49×10^{-5}	32
20	218	59.5	94.53	45×10^{-5}	34

ing, the effective macrosections being calculated with better averaging of absorption of individual lines with the weight function [expression (7.7)].

Table 7.2 shows the vertical stratification of aerosols, water vapor, ozone, and Table 7.3 that of aerosols. Table 7.4 shows the vertical change of total sections σ_t for the data from Tables 7.2 and 7.3. The vertical distribution of σ_t, together with the distribution of the quantities:

$$C_n = \left(\sum_{p'=1}^{P'} \sigma_s^{p'} b_n^{p'} \right) \Big/ \sigma_t$$

totally characterize the scattering and absorbing properties of a real stratified atmosphere. The estimates of $C_n (20 \geqslant n \geqslant 0)$ for the two spectral intervals and two atmospheric heights are given in Table 7.5.

The Table 7.5 data make it possible to judge the extent of stretching of real phase function in the atmosphere (in C_1 value) and estimate the relationship between scattering and absorption (in C_0 value). Table 7.5 data show that the real phase function differs from the Rayleigh function, since the C_n values with $n \approx 20$ differ little from C_0, i.e., these values with large indices in the expansion of phase functions in Legendre polynomials contribute to general in-

Table 7.3. The stratification of the lower layers of atmospheric aerosols

Δz, km	ϱ_0[a]	ϱ_{1+2+3}	$\beta_{d=1}$	$\beta_{d=2}$	$\beta_{d=3}$
0 – 1	2×10^4	150	3.9	3.4	4.7
1 – 2	3×10^3	80	3.7	3.2	4.8
2 – 3	1×10^3	27	3.5	3.0	4.8
3 – 4	5×10^2	20	3.8	2.9	4.9
4 – 5	5×10^2	15	3.2	2.8	4.9
5 – 6	3×10^2	3.1	3.0	2.8	5.0
6 – 7	2.5×10^2	2.1	2.9	2.7	5.0
7 – 8	2.5×10^2	1.2	2.8	2.7	5.0
8 – 9	2.5×10^2	1.0	2.6	2.7	5.1
9 – 10	2.3×10^2	0.6	2.5	2.7	5.2
10 – 11	2.0×10^2	0.6	2.4	2.7	5.2
11 – 12	1.2×10^2	0.18	2.2	2.8	5.2
12 – 13	70	0.15	2.1	2.8	5.3
13 – 14	20	0.15	1.9	2.8	5.3
14 – 15	10	0.20	1.8	2.8	5.4
15 – 16	8	0.27	1.6	2.8	5.4
16 – 17	3	0.3	1.5	2.8	5.5
17 – 18	1	0.35	1.5	2.9	5.6
18 – 19	0.8	0.45	1.4	2.9	5.6
19 – 20	0.7	0.42	1.4	2.9	5.6
20 – 21	0.7	0.40	1.5	3.0	5.6

[a] Here, ϱ_0 is the concentration of Aitken particles (cm^{-3}); ϱ_{1+2+3} the concentration of large and giant particles (cm^{-3}); $\beta_{d=1,2,3}$ the Yunge indices, respectively, for the ranges of aerosol particles radii $\Delta r = 0.1 - 1.0$; $1.0 - 2.5$, and $2.5 - 6.0\ \mu m$.

Table 7.4. Total sections of interaction (km^{-1}) for the illustrated atmospheric models

Δz, km	$\Delta\lambda = 1.4 - 1.5\ \mu m$	$\Delta\lambda = 0.9 - 1.0\ \mu m$	$\Delta\lambda = 0.5 - 0.55\ \mu m$
0 – 1	1.4425 + 01	1.1458 + 00	1.8109 – 01
1 – 2	1.0761 + 01	8.3164 – 01	8.4339 – 02
2 – 3	7.3189 + 00	5.5039 – 01	3.6821 – 02
3 – 4	4.7623 + 00	3.6124 – 01	2.7307 – 02
4 – 5	2.9549 + 00	2.2884 – 01	2.3268 – 02
5 – 6	1.8202 + 00	1.4261 – 01	1.6901 – 02
6 – 7	1.1145 + 00	8.8668 – 02	1.3115 – 02
7 – 8	6.7055 – 01	5.3806 – 02	9.7839 – 03
8 – 9	3.9998 – 01	3.3271 – 02	8.4591 – 03
9 – 10	2.3468 – 01	2.0888 – 02	7.8076 – 03
10 – 11	1.2841 – 01	1.0807 – 02	4.8746 – 03
11 – 12	6.7641 – 02	5.6593 – 03	3.6354 – 03
12 – 13	3.5277 – 02	3.0895 – 03	2.9981 – 03
13 – 14	1.8633 – 02	1.7816 – 03	2.5140 – 03
14 – 15	1.0196 – 02	1.1942 – 03	2.2588 – 03
15 – 16	1.3637 – 02	8.3227 – 03	9.2216 – 03
16 – 17	1.2152 – 02	8.8292 – 03	9.6666 – 03
17 – 18	1.2412 – 02	1.0133 – 02	1.0851 – 02
18 – 19	1.4593 – 02	1.2610 – 02	1.3159 – 02
19 – 20	1.3357 – 02	1.1745 – 02	1.2239 – 02
20 – 21	3.6879 – 03	2.7928 – 03	3.4957 – 03

Table 7.5. The estimates of coefficients C_n for illustrated atmospheric models for spectral intervals 1.0 – 1.1 and 0.62 – 0.65 μm

No.	1.0 – 1.1 μm		0.62 – 0.65 μm	
	Δz = 0 – 1 km	Δz = 19 – 20 km	Δz = 0 – 1 km	Δz = 19 – 20 km
0	3.556 – 01	5.781 – 01	9.062 – 01	8.574 – 01
1	2.842 – 01	5.145 – 01	5.574 – 01	6.680 – 01
2	2.565 – 01	4.786 – 01	4.778 – 01	5.961 – 01
3	2.301 – 01	4.466 – 01	3.517 – 01	4.741 – 01
4	2.156 – 01	4.270 – 01	3.283 – 01	5.630 – 01
5	2.037 – 01	4.081 – 01	2.862 – 01	4.127 – 01
6	1.945 – 01	3.923 – 01	2.811 – 01	3.846 – 01
7	1.856 – 01	3.754 – 01	2.592 – 01	3.846 – 01
8	1.776 – 01	3.595 – 01	2.540 – 01	3.779 – 01
9	1.697 – 01	3.433 – 01	2.412 – 01	3.625 – 01
10	1.623 – 01	3.275 – 01	2.341 – 01	3.530 – 01
11	1.549 – 01	3.120 – 01	2.260 – 01	3.419 – 01
12	1.478 – 01	2.968 – 01	2.189 – 01	3.316 – 01
13	1.409 – 01	2.820 – 01	2.128 – 01	3.230 – 01
14	1.342 – 01	2.676 – 01	2.060 – 01	3.125 – 01
15	1.277 – 01	2.537 – 01	1.997 – 01	3.033 – 01
16	1.213 – 01	2.400 – 01	1.941 – 01	2.944 – 01
17	1.151 – 01	2.267 – 01	1.875 – 01	2.844 – 01
18	1.090 – 01	2.137 – 01	1.826 – 01	2.764 – 01
19	1.030 – 01	2.010 – 01	1.764 – 01	2.670 – 01
20	9.710 – 01	1.884 – 01	1.716 – 01	2.594 – 01

teraction of radiation with the atmosphere, and in the Rayleigh phase function only the first two terms ($n \leqslant 2$) are not zero.

Table 7.6 shows the spectral distributions of the atmospheric optical thickness $\tau = \sigma_t H$ based on data in Tables 7.2 and 7.3 about the size distribution of particles in its different layers. The visible spectral intervals in Table 7.6 approximately correspond to the color lines with the boundaries in the region of the node points of ozone absorption; in the near IR they are taken every 0.1 μm to obtain the macrosections of absorption in the form (7.11).

The energy characteristics of the spectrum of solar radiation together with the scattering properties of the atmosphere as a mixture of gases and the absorbing properties of ozone in the Chappui bands are given in Table 7.7. Table 7.7 data enable estimation of the contribution of the characteristics of the Rayleigh scattering and absorption of solar radiation by ozone, calculated from analytical formulas without approximation, to the total optical thickness of the atmosphere (Table 7.6).

Table 7.6. The spectral change of atmospheric optical thickness τ for the summer mid-latitudes

No.	$\Delta\lambda$, μm	τ
1	1.90 – 2.00	9.2030 + 01
2	1.80 – 1.90	2.2095 + 02
3	1.70 – 1.80	2.8875 + 00
4	1.60 – 1.70	3.2766 – 01
5	1.50 – 1.60	4.7682 – 01
6	1.40 – 1.50	4.4809 + 01
7	1.30 – 1.40	1.1665 + 02
8	1.20 – 1.30	4.6188 – 01
9	1.10 – 1.20	9.5405 + 00
10	1.00 – 1.10	5.4832 – 01
11	0.90 – 1.00	3.5408 + 00
12	0.80 – 0.90	4.4085 – 01
13	0.76 – 0.80	3.4340 – 01
14	0.7036 – 0.7600	3.5403 – 01
15	0.6500 – 0.7036	3.7517 – 01
16	0.6220 – 0.6500	3.9777 – 01
17	0.5872 – 0.6220	5.0173 – 01
18	0.5754 – 0.5872	4.3871 – 01
19	0.5512 – 0.5754	4.7370 – 01
20	0.5060 – 0.5512	4.9774 – 01
21	0.4874 – 0.5060	5.4513 – 01
22	0.4551 – 0.4874	5.9864 – 01
23	0.4200 – 0.4551	6.9168 – 01
24	0.3900 – 0.4200	8.1688 – 01
25	0.3500 – 0.3900	1.0464 + 00

7.3 The Solution of the Boundary Problem for Shortwave Radiation

One of the possible techniques for solving the boundary problem (7.1 – 7.3) is an expansion of radiance in series in Legendre polynomials (the spherical harmonics technique):

$$\bar{I}(z,\eta,\varphi) \approx \sum_{n=0}^{2N+1} \frac{2n+1}{4\pi} \Psi_{n,0}(z) P_n(\eta) + 2 \sum_{n=1}^{2N+1} \sum_{m=1}^{\min(n,M)} \frac{2n+1}{4\pi} \times \frac{(n-m)!}{(n+m)!} P_n^m(\eta) \left[X_{n,m}(z) \cos m\varphi - Y_{n,m}(z) \sin m\varphi\right] , \quad (7.13)$$

where $X_{n,m}(z) = \mathrm{Re}\,\Psi_{n,m}(z)$; $Y_{n,m}(z) = \mathrm{Im}\,\Psi_{n,m}(z)$, N, M are the boundary indices of the considered number of spherical harmonics, the abnormal order of the number of terms of the series is determined by the Marshak form of presentation of the boundary conditions [27], discussed next.

As a result of this expansion, the problem is reduced to the solution of the system of conventional differential equations with respect to the vectors of

Table 7.7. Estimates of the spectral solar constant of the air scattering macrosections under normal conditions and of the ozone absorption microsections for the spectral interval given in Table 7.6

No.	$\int_{\lambda_l}^{\lambda_{l+1}} F(\lambda)\,d\lambda, \frac{\text{cal}}{\text{cm}^2\text{min}}$	$\sigma_s^a(T_0, p_0)$, km^{-1}	$k_{l,l+1}^{oz}$, cm^{2}g^{-1}
1	1.686 − 02	7.616 − 05	0
2	2.002 − 02	9.411 − 05	0
3	2.393 − 02	1.176 − 04	0
4	2.880 − 02	1.490 − 04	0
5	3.489 − 02	1.916 − 04	0
6	4.263 − 02	2.507 − 04	0
7	5.259 − 02	3.343 − 04	0
8	6.499 − 02	4.558 − 04	0
9	7.932 − 02	6.376 − 04	0
10	9.537 − 02	9.204 − 04	0
11	1.156 − 01	1.382 − 03	0
12	1.443 − 01	2.171 − 03	0
13	6.788 − 02	3.020 − 03	0
14	1.072 − 01	3.925 − 03	8.115
15	1.154 − 01	5.384 − 03	19.672
16	6.528 − 02	6.885 − 03	39.089
17	8.558 − 02	8.471 − 03	56.708
18	3.014 − 02	9.902 − 02	56.813
19	6.322 − 02	1.128 − 02	51.157
20	1.229 − 01	1.467 − 02	29.571
21	5.288 − 02	1.884 − 02	99.556
22	9.906 − 02	2.346 − 02	4.014
23	1.001 − 01	3.166 − 02	0.270
24	7.165 − 02	4.314 − 02	0
25	7.015 − 02	7.420 − 02	0

moments of the angular distributions of radiation intensity [$n = 0, 1, \ldots, 2N+1; m = 0, 1, \ldots, \min(n, M)$]:

$$(n+1-m)\frac{d\begin{Bmatrix} X_{n+1,m}(z) \\ Y_{n+1,m}(z) \end{Bmatrix}}{dz} + (n+m)\frac{d\begin{Bmatrix} X_{n-1,m}(z) \\ Y_{n-1,m}(z) \end{Bmatrix}}{dz}$$

$$+\sigma_t(z)[1-C_n(z)](2n+1)\begin{Bmatrix} X_{n,m}(z) \\ Y_{n,m}(z) \end{Bmatrix}$$

$$= S_0 \exp\left[-\frac{1}{\zeta}\int_z^H \sigma_t(z')\,dz'\right]$$

$$\times\,\sigma_t(z)C_n(z)(2n+1)P_n^m(\cos\theta_0)(-1)^{n-m}\begin{Bmatrix} \cos m\varphi_\odot \\ -\sin m\varphi_\odot \end{Bmatrix}. \qquad (7.14)$$

The classical form of the solution of the problem obtained consists of the following six subsequent stages:

1. General solution of the respective homogeneous systems for each i-zone of the atmosphere (within each zone the physical parameters $\sigma_t^{i-1,i}$ and $C_n^{i=1,i}$ characterizing the scattering and absorbing properties of the atmosphere are vertically constant):

$$\psi_n^i(z) = \sum_{j=1}^{N+1} G_n(x_j^i)\,[A_j^i e^{x_j^i \sigma_t^{i-1,i} z} + (-1)^n A_{-j}^i e^{-x_j^i \delta_t^{i-1,i} z}] \;. \tag{7.15}$$

2. Polynomials $G_n(x)$ meet the following recurrent relationships:

$$(n+1)G_{n+1}(x) + nG_{n-1}(x) + \frac{1-C_n}{x}(2n+1)G_n(x) = 0 \;,$$
$$G_0(x) = 1, G_i(x) = \frac{C_0 - 1}{x}, G_n(-x) = (-1)^n G_n(x) \;, \quad (n = 1,2,\ldots,2N+1) \;, \tag{7.16}$$

where $\{x_j\}$ is the positive root of the function $G_{2N+2}(x)$; the form of the polynomials $G_n(x)$ is independent of the order of P_{2N+1} approximations.

3. Three matrices are constructed:

$$(\hat{R}^{i-1,i})_{n,k} = (-1)^{(k-1)(n-1)} G_{n-1}\left[x_E^i\left(\frac{k+1}{2}\right)\right] \;, \tag{7.17}$$

$$(\hat{\Pi}^{i-1,i})_{n,k} = \frac{(-1)^{(n-1)(k-1)}(1 - C_{k-1}^{i-1,i})G_{k-1}\left[x_E\left(\frac{n+1}{2}\right)\right](2k-1)}{h\left[x_E\left(\frac{n+1}{2}\right)\right]} \;, \tag{7.18}$$

$\hat{Q}^{i-1,i}(x^i\tau)$ with diagonal elements $e^{x_1^i\tau}$, $e^{-x_1 i_\tau}, \ldots, e^{x_{N+1}^i\tau}$, $e^{-x_{N+1}^i\tau}\ldots$; $n,k = 1,2,\ldots,2N+2$; $E(d)$ is the whole part of the number d, $h(x)$ is found from the condition of orthogonality:

$$h(x) = 2(N+1)x^2 G_{2N+1}(x)\frac{dG_{2N+2}(x)}{dx} \;. \tag{7.19}$$

4. The vectors are introduced:

$$\vec{\psi}^i(z) = \left\{\psi_0^i(z), \psi_1^i(z), \ldots, \psi_{2N+1}^i(z)\right\} \tag{7.20}$$

$$\vec{\varphi}^i(z) = \left\{A_1^i e^{x_1^i \sigma_t^{i-1,i} z} \;, \quad A_{-1}^i e^{-x_1^i \sigma_t^{i-1,i} z} \;, \ldots, A_{-(N+1)}^i e^{-x_{N+1}^i \sigma_t^{i-1,i} z}\right\} \tag{7.21}$$

so, that the vector matrix of the solution becomes:

$$\vec{\psi}^{i}(z) = \hat{R}^{i-1,i}\vec{\varphi}^{i}(z) \ . \tag{7.22}$$

5. It is proved that:

$$(\hat{\Pi})^{i-1,i} = (\hat{R}^{i-1,i})^{-1} \ . \tag{7.23}$$

In addition $\vec{\psi}^{i}(z_i) = \hat{p}^{i-1,i}\vec{\psi}^{i}(z_{i-1})$, (7.24)

where z_i corresponds to the upper boundary of the i-th zone, z_{i-1} to its lower boundary,

$$\hat{p}^{i-1,i} = \hat{R}^{i-1,i}\hat{Q}^{i-1,i}[x^i(z_i - z_{i-1})\sigma_t^{i-1,i}](\hat{R}^{i-1,i})^{-1} \ . \tag{7.25}$$

6. Using the properties of continuity of the vectors of moments of the angular distributions at the interior and exterior boundaries of the respective atmospheric layers, we obtain:

$$\begin{aligned}\vec{\psi}^{I'+1}(H) &= \hat{p}^{I',I'+1}\vec{\psi}^{I'+1}(z_{I'}) = \hat{p}^{I',I'+1}\vec{\psi}^{I'}(z_{I'}) \\ &= \hat{p}^{I',I'+1}\hat{p}^{I'-1,I'}\vec{\psi}^{I'-1}(z_{I'-1}) = \ldots = \hat{p}\vec{\psi}^{1}(0) \ ,\end{aligned} \tag{7.26}$$

where:

$$\hat{p} = \hat{p}^{I',I'+1}\hat{p}^{I'-1,I'}\hat{p}^{I'-2,I'-1}\hat{p}^{0,1} \tag{7.27}$$

is the matrix of transmission for an atmospheric thickness H (I′ is total number of layers with constant properties).

The computer realization of this technique for calculations of the angular and spectral distributions of radiation in the atmosphere has shown that:

- due to high orders of the matrix of transmission, for individual atmospheric layers, in the region of substantial scattering, the arithmetic arrangement can be overloaded even in case of a 20-layer atmosphere;
- if the system to be solved is higher than that of the 30th order, then the functions $G_{2N+2}(x)$ have two and more close roots; in this case it is impossible to perform computations without a double control of accuracy.

Both these substantial limitations take place in a real stratified atmosphere, both in the short- and longwave spectral regions.

Of these stages of solution, we shall take only expressions (7.24–7.27), and the matrix $\hat{p}^{i-1,i}$ will be found from other considerations. This makes it possible to reduce computing time, to realize an advanced technique of spherical harmonics, compared to the classical technique. The computer time can additionally be cut down by reducing the calculating order of approximation.

7.4 Possibilities of Reducing Computer Time

A finite-difference representation of expression (7.14) could formally by used for the zero azimuthal harmonic $\psi_n = \psi_{n,0}$ (by substituting, within individual i-zones of the atmosphere, the integral expression on the left-hand side with the square formula for a trapezium, and on the right-hand side with that for rectangles), to obtain the following systems of three-diagonal algebraic equations for the lower z_{i-1} and upper z_i boundaries of the respective i-zone:

$$\psi_{n+1}(z_{i-1}) - (1 - C_n^1)\Delta_{i-1}^i \sigma_t^i \frac{2n+1}{2(n+1)} \psi_n(z_{i-1}) + \frac{n}{n+1} \psi_{n-1}(z_{i-1})$$

$$= \psi_{n+1}(z_i) + (1 - C_n^i)\Delta_{i-1}^i \sigma_t^i \frac{2n+1}{2(n+1)} \psi_n(z_i) + \frac{n}{n+1} \psi_{n-1}(z_i) + Q_n^{(i)} \ , \tag{7.28}$$

where:

$$Q_n^{(i)} = S_0 \cos\theta_0 C_n^i (-1)^n P_n(\cos\theta_0) \frac{2n+1}{n+1} (g_{i+1} - g_i) \ ,$$

$$\Delta_{i-1}^i = z_i - z_{i-1} \ ,$$

$$g_i = \exp\left\{\left(\sum_{j=i}^{I'+1} \Delta_{j-1}^j \sigma_t^j\right) \Big/ \cos\theta_0\right\} \ , \qquad i = 1, 2, \ldots, I'+1 \ ,$$

$$g_{I'+2} = 1 \ ,$$

$$C_n = \frac{1}{\sigma_t(z)} \sum_{p'=1}^{P'} \sigma_s^{p'}(z) b_n^{p'}(z) < 1 \ .$$

Here, we shall use the balance technique or the integral-interpolation method [173] expressing the law of energy preservation on the chosen grid of the finite-difference scheme. This method makes it possible to obtain schemes where coefficients in all grid-nodes are calculated with the same formulas as average coefficients of differential equation (7.14) round a grid node. The balance technique results in the schemes of a special iterative procedure in the class of non-monotonic coefficients.

Special problems of approximation, convolution and stability of the scheme (7.28) need further studies. However, this scheme does have second-order accuracy, and the balance method ensures its convolution to the initial scheme (7.14). Writing (7.28) in the form:

$$\tilde{A}_n \psi_{n-1} - \tilde{B}_n \psi_n + \tilde{C}_n \psi_{n+1} = -\tilde{F}_n \ , \qquad n = 1, 2, \ldots, 2N+1 \ , \tag{7.28a}$$

where $\tilde{A}_1 = \tilde{C}_{2N+1} = 0$ (other coefficients are not zero), we obtain, for sufficiently small optical thicknesses $\tau_i = \Delta_{i-1}^i \sigma_t^i \lesssim 1$, difficulties are faced with respect to meeting the condition of theoretical stability of calculations based on the successive technique:

$$|\tilde{A}_n| + |\tilde{C}_n| \leq |\tilde{B}_n| \ , \qquad n = 1,2,\ldots,2N+1 \ ,$$
$$\tilde{A}_n = n, \tilde{C}_n = n+1, \tilde{B}_n = (1-C_n)\tau_i \frac{2n+1}{2} \ . \qquad (7.28\,b)$$

From the vector matrix writing of expression (7.28) as:

$$\hat{T}_i^{(1)} \vec{\psi}(z_i) = \hat{T}_i^{(2)} \vec{\psi}(z_{i-1}) + \vec{Q}^{(i)}$$

or:

$$\vec{\psi}(z_i) = \hat{p}^{i-1,i} \vec{\psi}(z_{i-1}) + [\hat{T}_i^{(1)}]^{-1} \vec{Q}^{(i)} \qquad (7.29)$$

there follows its identity to expression (7.24) and possibility to determine all the terms of the matrix $\hat{p}^{i-1,i} = \hat{T}_i^{(2)} [\hat{T}_i^{(1)}]^{-1}$ without any matrix inversions, in contrast to (7.25), by prescribing a system of vectors $\vec{\psi}(z_{i-1})$ in the form $\{1,0,0,\ldots,0\}$; $\{0,1,0,\ldots,0\}$; $\ldots$; $\{0,0,0,\ldots,1\}$. One can see that:

$$\psi_n = \beta_{n+1} \psi_{n+1} + \tilde{Z}_{n+1} \ ,$$

where:

$$\beta_{n+1} = (\tilde{B}_n - \tilde{A}_n \beta_n)^{-1} \tilde{C}_n \ ,$$
$$\tilde{Z}_{n+1} = (\tilde{B}_n - \tilde{A}_n \beta_n)^{-1} (\tilde{F}_n + \tilde{A}_n \tilde{Z}_n) \ .$$

The conventional scheme of solving the problem in the case of the successive technique is as follows:

1. Starting from arbitrary $\beta_1, \tilde{Z}$, we determine using recurrent formulas:

$$\beta_1, \beta_2, \beta_3, \ldots, \beta_n, \ldots, \beta_{2N+1}, \beta_{2N+2} = 0 \ ,$$
$$\tilde{Z}_1, \tilde{Z}_2, \tilde{Z}_3, \ldots, \tilde{Z}_n, \ldots, \tilde{Z}_{2N+1}, \tilde{Z}_{2N+2} \ .$$

2. Successive calculation of ψ_n towards decreasing index n, proceeding from calculated value of $\tilde{Z}_{2N+2}$:

$$\psi_{2N+1} = \tilde{Z}_{2N+2}, \psi_{2N}, \psi_{2N-1}, \ldots, \psi_n, \ldots, \psi_2, \psi_1 \ .$$

Condition (7.28 b) ensures the stability of calculations, providing $(\tilde{B}_n - \tilde{A}_n \beta_n)$ at all $n = 1,2,\ldots,2N+1$, if $|\beta_l| \leqslant 1$ $(l = 2,3,\ldots,n)$, but $|\beta_{l+1}| < 1$ [173]. However, calculations have shown that in practice the stability of calculations is preserved with "weak satisfaction" of the latter condition and condition (7.28 b).

This has been confirmed as the results obtained at $|\beta_l| \to 1$ with agree well with control calculations by other authors based on different techniques, with the same initial data. With this search for the limit of real stability the prescription of optical thicknesses from compromise considerations turns out to be decisive: its decrease ensures a more accurate retrieval of the laws of radiation-atmosphere interaction; its increase is needed for the stability of calculations. In experiments with the basic model given in Section 7.2, it has been determined that this limit for dividing the atmospheric layers is $\tau = 0.1$, close to total optical thickness of a Rayleigh atmosphere in the visible, only 7–8 values of β_n being alternatively, above and below unity, for $n > 8$ $|\beta_n| \to 1$.

Thus, the successive technique to solve the systems of three-diagonal algebraic equations using specially chosen vectors $\vec{\psi}(z_{i-1})$ makes it possible to determine all the elements of the matrix $\hat{p}^{i-1,i}$ for individual atmospheric layers selected based on these considerations. A similar procedure of solving the three-diagonal systems can be used to retrieve vectors $\vec{Q}_n^{(i)}$ by reconstructing the calculation scheme in "n" at a given z_i, in contrast to the reconstruction in "i" at different z_i and z_{i-1}, when it was taken that:

$$Q_0 = Q_1 = \psi_1(z_{i-1}) = 0 \ ; \quad \psi_0(z_{i-1}) = 1 \ .$$

With the use of these procedures we obtain the following relationship between even and odd components of the vectors of moments of the angular distribution of radiance:

$$\vec{\psi}_{even}(H) = \hat{p}_1 \vec{\psi}_{even}(0) + \hat{p}_2 \vec{\psi}_{odd}(0) + \vec{Q}_{even} \tag{7.30}$$

$$\vec{\psi}_{odd}(H) = \hat{p}_3 \vec{\psi}_{even}(0) + \hat{p}_4 \vec{\psi}_{odd}(0) + \vec{Q}_{odd} \tag{7.31}$$

which, together with the boundary conditions presented in the Marshak form [27] with Lambertian reflection ($\bar{R} = \bar{A}/\pi$):

$$\vec{\psi}_{odd}(H) = \hat{p}_5 \vec{\psi}_{even}(H) \ , \tag{7.32}$$

$$\vec{\psi}_{odd}(0) = D_0 \frac{4\bar{A}}{1+\bar{A}} \vec{R}_0 + \left(-\hat{p}_5 + \frac{4\bar{A}}{1+\bar{A}} \hat{p}_6 \right) \vec{\psi}_{even}(0) \tag{7.33}$$

enables one to find a solution of the problem considered in the form:

$$\vec{\psi}_{even}(0) = -\frac{4\bar{A}}{1+\bar{A}} D_0 \hat{p}_7^{-1} (\hat{p}_4 - \hat{p}_5 \hat{p}_2) \vec{R}_0 - \hat{p}_7^{-1} (\vec{Q}_{odd} - \hat{p}_5 \vec{Q}_{even}) \ , \tag{7.34}$$

where:

$$(\hat{p}_1)_{n,k} = (\hat{p})_{2n-1,2k-1}; \ (\hat{p}_2)_{n,k} = (\hat{p})_{2n-1,2k} \ , \tag{7.35}$$

$$(\hat{p}_3)_{n,k} = (\hat{p})_{2n,2k-1}; \ (\hat{p}_4)_{n,k} = (\hat{p})_{2n,2k} \ , \tag{7.36}$$

$$(\hat{p}_5)_{n,k} = (4n-3) \varkappa_{2n-2,2k-1} \ , \tag{7.37}$$

$$(\hat{p}_6)_{n,k} = (4n-3) \varkappa_{0,2k-1} \varkappa_{2n-2,1} \ , \tag{7.38}$$

$$(\vec{R}_0)_n = \varkappa_{0,2k-1} \qquad (n; k = 1, 2, \ldots, N+1) \ , \tag{7.39}$$

$$\varkappa_{n,l} = \int_0^1 P_n(\mu) P_l(\mu) d\mu = \begin{cases} \dfrac{1}{2n+1}, n = l \\ 0, n \neq l, (n-l) - \text{even} \\ \dfrac{(-1)^{1/2(n+l-1)} n! \, l!}{2^{n+l+1} (l-n)(n+l+1) [(n/2)! \, ((l-1)/2)!]^2} \ , \quad \begin{matrix} n - \text{even} \\ l - \text{odd} \end{matrix} \end{cases} \tag{7.40}$$

$$[\vec{\psi}_{even}(0)]_n = \psi_{2n-2}(0),\ [\vec{\psi}_{odd}(0)]_n = \psi_{2n-1}(0)\ , \tag{7.41}$$

$$[\vec{\psi}_{even}(H)]_n = \psi_{2n-2}(H),\ [\vec{\psi}_{odd}(H)]_n = \psi_{2n-1}(H)\ , \tag{7.42}$$

$$(\vec{Q}_{even})_n = Q_{2n-2},\ (\vec{Q}_{odd})_n = Q_{2n-1}\ , \tag{7.43}$$

$$\hat{p}_7 = \hat{p}_3 - \hat{p}_5\hat{p}_1 - (\hat{p}_4 - \hat{p}_5\hat{p}_2)\left(\hat{p}_5 - \frac{4\bar{A}}{1+\bar{A}}\hat{p}_6\right)\ . \tag{7.44}$$

Use of solutions of the system of 3-diagonal equations in the calculation scheme, for the third time, makes it possible to estimate $\vec{\psi}(H)$ from (7.30), (7.31) from $\vec{\psi}(0)$ calculated from (7.33–7.34), considering (7.35–7.44).

The calculation scheme described above enabled reduction of the computer time onehundredfold, compared to the classical method mentioned.

The initial data for some calculation results are given in Tables 7.8 and 7.9. We have proceeded from the fact that [322] present results of model calculations of radiation transfer in the atmosphere using different techniques for idealized initial data: the phase function characterizes the model of haze L after Deirmendjian [43]; the total optical thickness is unity; the current optical thickness connected with the height of measurement level varies discretely from zero to unity; three versions of values of the probability of quantum sur-

Table 7.8. The Elterman-1968 model phase functions used in calculations

θ°	Haze model L	Aerosol phase functions		Rayleigh phase functions	Atmospheric phase functions	
		λ = 0.55 μm	λ = 0.85 μm		λ = 0.55 μm	λ = 0.85 μm
0	300.332	269.80	150.50	1.500	194.25	138.58
10	19.179	11.840	11.960	1.477	8.922	11.121
20	8.362	5.295	5.361	1.412	4.202	5.045
30	3.572	3.028	3.066	1.312	2.545	2.926
40	1.620	1.846	1.520	1.190	1.661	1.494
50	0.798	1.162	1.177	1.060	1.133	1.168
60	0.430	0.753	0.763	0.937	0.805	0.777
70	0.252	0.504	0.511	0.838	0.598	0.537
80	0.162	0.352	0.356	0.773	0.470	0.389
90	0.114	0.257	0.260	0.750	0.395	0.299
100	0.0889	0.199	0.202	0.773	0.361	0.248
110	0.0768	0.167	0.169	0.838	0.356	0.223
120	0.0739	0.152	0.157	0.937	0.373	0.219
130	0.0791	0.152	0.155	1.060	0.408	0.227
140	0.0919	0.168	0.171	1.190	0.456	0.252
150	0.110	0.205	0.209	1.312	0.517	0.297
160	0.128	0.258	0.263	1.412	0.583	0.355
170	0.114	0.357	0.363	1.477	0.674	0.452
180	0.126	0.536	0.544	1.500	0.807	0.621

Table 7.9. Vertical stratifications of macrosections of the gaseous and aerosol substances of the atmosphere

z, km	λ = 0.55 μm			λ = 0.85 μm		
	Aerosol scatter.	Rayleigh scatter.	Ozone absorp.	Aerosol scatter.	Rayleigh scatter.	Water vapor absorp.
0	1.58 − 1	1.16 − 2	3.28 − 4	1.24 − 1	2.06 − 3	9.76 − 3
2	3.00 − 2	9.55 − 3	2.70 − 4	2.35 − 2	1.69 − 3	5.95 − 3
4	6.66 − 3	7.77 − 3	2.08 − 4	6.20 − 3	1.38 − 3	2.21 − 3
6	3.54 − 3	6.26 − 3	1.99 − 4	2.28 − 3	1.11 − 3	8.89 − 4
8	3.39 − 3	4.50 − 3	2.10 − 4	2.65 − 3	8.85 − 4	2.53 − 4
10	3.17 − 3	3.92 − 3	3.22 − 4	2.48 − 3	6.97 − 4	1.02 − 4
12	3.12 − 3	2.96 − 3	5.71 − 4	2.45 − 3	5.03 − 4	5.64 − 5
14	2.82 − 3	2.16 − 3	8.80 − 4	2.21 − 3	3.84 − 4	1.12 − 5
16	2.52 − 3	1.58 − 3	9.48 − 4	1.97 − 3	2.80 − 4	6.19 − 6
18	2.41 − 3	1.15 − 3	1.12 − 3	1.89 − 3	2.05 − 4	3.32 − 6
20	1.49 − 3	8.43 − 4	1.51 − 3	1.16 − 3	1.49 − 4	1.05 − 6
25	4.15 − 4	3.80 − 4	1.66 − 3	3.25 − 4	6.75 − 5	7.13 − 7
30	1.25 − 4	1.75 − 4	8.31 − 4	9.98 − 5	3.30 − 5	4.52 − 7
35	3.29 − 5	8.03 − 5	3.97 − 4	2.57 − 5	1.43 − 5	1.92 − 7
40	8.66 − 6	3.79 − 5	1.71 − 4	6.75 − 6	6.73 − 6	8.09 − 8
45	2.28 − 6	1.87 − 5	5.30 − 5	1.78 − 6	3.21 − 6	6.16 − 8
50	6.02 − 7	9.74 − 6	1.71 − 5	4.71 − 7	1.73 − 6	4.80 − 8

vival W_0 and solar zenith angle cosine $\mu_\odot$ are used in calculations: (1; −1), (0.9; −1), and (0.9; −0.5), the latter version being computed for three values of azimuth angles (0°, 90°, and 180°).

The results of calculations given in [322] should it seems, be further developed through the respective calculations for coordinated experimental data on the stratified spectral change of macrosections of scattering, absorption and phase function. Having obtained spectral distributions of sections and phase functions, having estimated the respective calculation accuracies, and having made correct averaging over the spectrum of these distributions, one can obtain sections and phase functions in a two-group approximation, i.e., to simplify, as much as possible, the procedure of reduction of satellite radiation budget measurements to the surface level.

Here to illustrate the results of calculations, we use the values of sections for λ = 0.55 μm and 0.85 μm for the Elterman-1968 model and the values of phase functions consistent with them, after the Mie theory. Comparison of results obtained with these technique, with some results from [322] is shown in Table 7.10 which reveals a reasonable agreement of the advanced method of spherical harmonics (more efficient in computer time terms) and the classical method.

The efficiency of calculations of short-wave and long-wave radiation in the atmosphere can be further increased by decreasing the calculation order of ap-

Table 7.10. Comparison of the results of calculations of the angular distributions of the outgoing radiation intensity (upper lines in each of the five versions) with the data from [322]: classical method of spherical harmonics, method of multiple scattering, and Monte-Carlo simulations (the second, third, and fourth lines in each version, respectively)

μ^+		0.4	0.6	0.8	1.0
$W_0 = 0.9$, $\mu_0 = -0.5$	$\varphi = 0°$	1.760−1	6.819−2	2.670−2	9.348−2
		1.752−1	6.811−2	2.661−2	9.344−2
		1.745−1	6.784−2	2.650−2	9.340−2
		1.815−1	6.970−2	2.645−2	9.645−2
	$\varphi = 90°$	3.961−2	2.258−2	1.393−2	9.352−3
		3.954−2	2.249−2	1.385−2	9.344−3
		3.905−2	2.228−2	1.377−2	9.340−3
		3.700−2	2.170−2	1.360−2	8.750−3
	$\varphi = 180°$	2.990−2	1.964−2	1.391−2	9.349−3
		2.986−2	1.951−2	1.389−2	9.344−3
		2.953−2	1.938−2	1.384−2	9.340−3
		2.750−2	1.810−2	1.310−2	8.750−3
$W_0 = 1$; $\mu_\odot = -1$		5.422−2	3.123−2	2.165−2	1.769−2
		5.414−2	3.117−2	2.156−2	1.761−2
		5.372−2	3.126−2	2.172−2	1.782−2
		5.240−2	3.070−2	2.135−2	1.680−2
$W_0 = 0.9$; $\mu_\odot = -1$		3.685−2	2.215−2	1.617−2	1.375−2
		3.683−2	2.210−2	1.610−2	1.374−2
		3.687−2	2.218−2	1.620−2	1.390−2
		3.780−2	2.240−2	1.600−2	1.345−2

proximation through selecting single scattering in the scheme of calculations. In particular, Figure 7.1 shows some results of calculations: calculations were made using the total scheme in the P_{71}-approximation (A′), in the approximation of single scattering with the exact phase function (B′), using total scheme but in the P_3-approximation (C′), and in the approximation of single scattering with the three term phase function (D′). In this case the equation should be satisfied:

$$A' = B' + C' - D' \; .$$

The respective diagnosis of errors between the exact solution (A′) and each of approximate solutions (B′, C′, D′ and their combination written above), together with possible errors due to neglecting the azimuthal dependence and the vertical inhomogeneity of the atmosphere is shown in Fig. 7.2. Such an analysis of the calculation scheme has made it possible to decrease the calculation order of approximation with the selection of single scattering for a more complicated case of 0.55 μm wavelength (left-hand parts of the graphs in

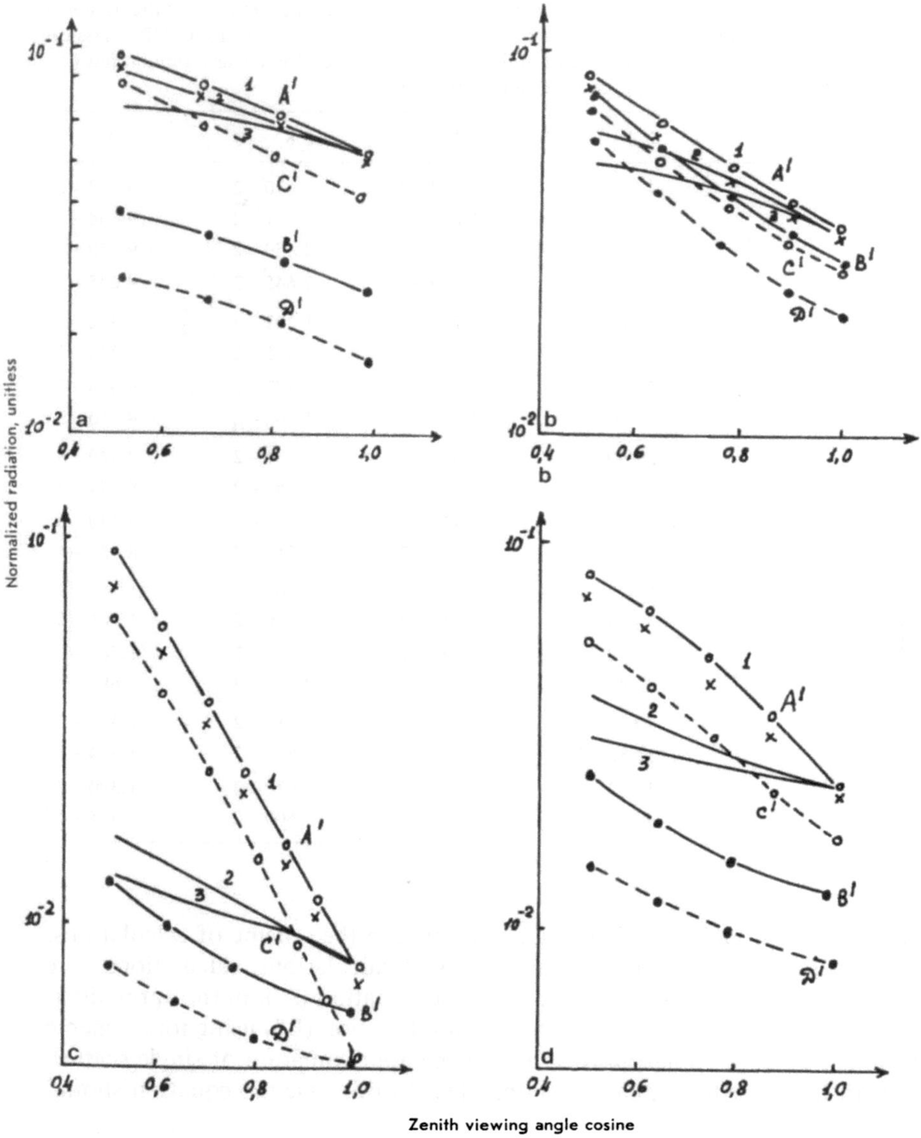

Fig. 7.1 a–d. Comparison of the angular distributions of the outgoing radiation intensity at the upper atmospheric boundary. *1* $\varphi = 0°$; *2* $\varphi = 90°$; *3* $\varphi = 180°$; *4* one-layer atmosphere. Curves *A*, *B* correspond to $\cos\theta_0 = 1$; *C*, *D* $\cos\theta_0 = 0.5$; **a**, **c** for $\lambda = 0.55$ μm; **b**, **d** for $\lambda = 0.85$ μm

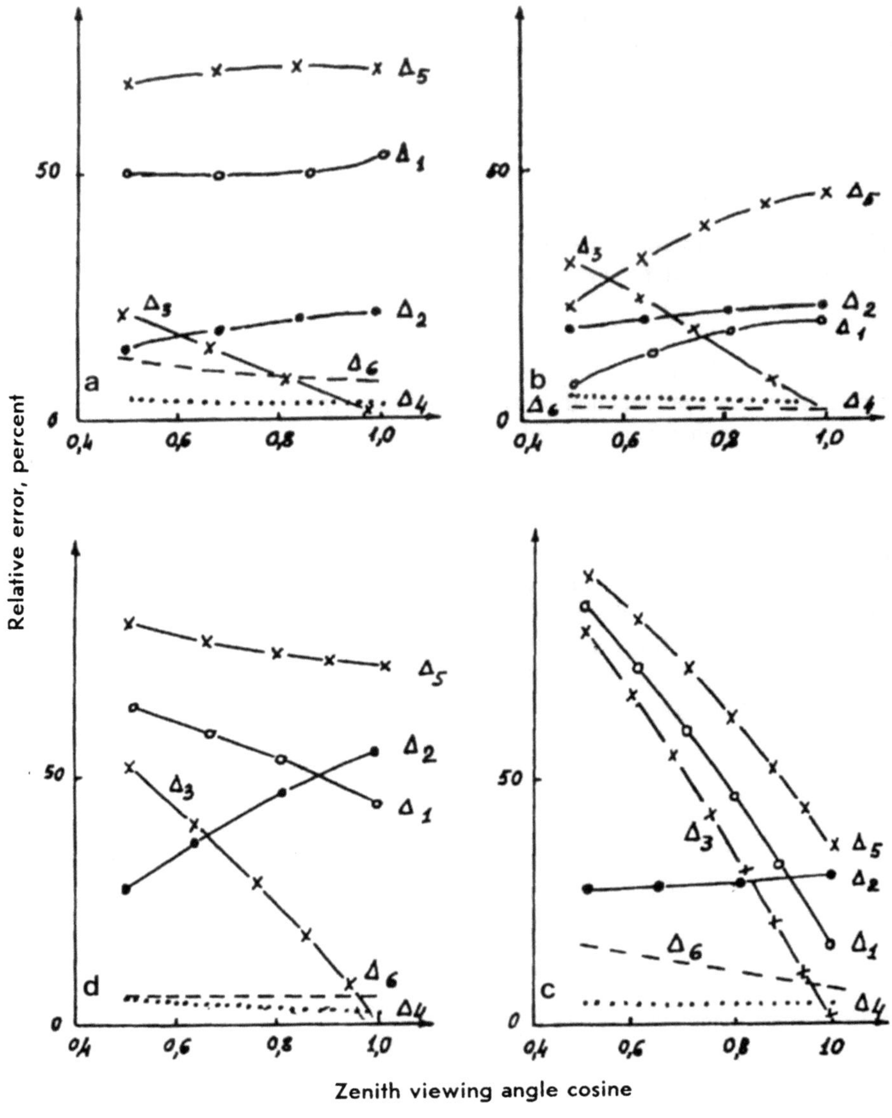

Fig. 7.2 a–d. Relative errors of calculations shown in Fig. 7.1: Δ_1 – between A′ and B′; Δ_2 – between A′ and C′; Δ_3 – between curves 1 and 3 for A′; Δ_4 – between curves 1 and 4 for A′; Δ_5 – between A′ and D′; Δ_6 – between A′ and B′+C′−D′

Figs. 7.1, 7.2) at $\cos\theta_0 = 1$ to the P_7-approximation, at $\cos\theta_0 = 0.5$ to the P_{21}-approximation. Here, the error between calculations after the complete scheme in the P_{71}-approximation and approximations with selected single scattering, did not exceed 43%. The computer time expenditure was cut down by between 10 (small solar zenith angles) to 20 times (large solar zenith angles).

Thus, the use of the results of calculations makes it possible to reduce the order of approximation with prescribed errors between results of numerical and analytical solutions. This enables further reduction of the computer time needed for the advanced technique of spherical harmonics.

7.5 Consideration of Non-Orthotropicity and Horizontal Inhomogeneity of the Earth-Surface

Most substantial progress, in the use of the theory of radiation transfer in the atmosphere, has been reached on simplified assumptions of Lambertian reflection and homogeneous terrestrial surface. The experimentally observed effects of non-orthotropicity and horizontal inhomogeneity of the surface limit greatly the possibilities of using analytical and numerical methods of solving the boundary problems of radiation transfer in the atmosphere. The advanced method of spherical harmonics allows generalization of the results of calculations of the outgoing shortwave radiation field for the two factors, without too much complication of the calculation scheme.

With the non-orthotropic reflection considered, a vector-matrix representation is used for 3-diagonal algebraic equations in the form of (7.29), where $\vec{\psi}$ is the lines of the following matrix:

$$\begin{bmatrix} \psi_{0,0} & \psi_{1,0} & \psi_{2,0} & \dots & \psi_{2N-1,0} & \psi_{2N,0} & \psi_{2N+1,0} \\ 0 & \begin{Bmatrix} X \\ Y \end{Bmatrix}_{1,1} & \begin{Bmatrix} X \\ Y \end{Bmatrix}_{2,1} & \dots & \begin{Bmatrix} X \\ Y \end{Bmatrix}_{2N-1,1} & \begin{Bmatrix} X \\ Y \end{Bmatrix}_{2N,1} & 0 \\ 0 & 0 & \begin{Bmatrix} X \\ Y \end{Bmatrix}_{2,2} & \dots & \begin{Bmatrix} X \\ Y \end{Bmatrix}_{2N-1,2} & 0 & 0 \\ & & & \dots & & & \\ & & \underbrace{0\dots0}_{N} & \begin{Bmatrix} X \\ Y \end{Bmatrix}_{N,N} & \begin{Bmatrix} X \\ Y \end{Bmatrix}_{N+1,N} & \underbrace{0\dots0}_{N} & \end{bmatrix} \begin{matrix} m=0 \\ m=1 \\ m=2 \\ \\ m=N \end{matrix}$$

The vector-matrix representation of the upper boundary condition (7.2), with non-zero azimuthal harmonics considered, is similar to (7.32):

$$\begin{Bmatrix} \vec{X}(H) \\ \vec{Y}(H) \end{Bmatrix}_{2n-1+m,m} = \hat{p}_5^{(m)} \begin{Bmatrix} \vec{X}(H) \\ \vec{Y}(H) \end{Bmatrix}_{2n-2+m,m} , \tag{7.45}$$

where:

$$(\hat{p}_5^{(m)})_{n,k} = \frac{(2n-2)!}{(2n-2+2m)!} (4n-3+2m) \varkappa^{(m)}_{2n-2+m,2k-1+m} , \tag{7.46}$$

$$\varkappa^{(m)}_{n,k} = \int_0^1 P^m_n(\eta) P^m_k(\eta)\, d\eta =$$

$$(m \leqslant n, m \leqslant k)$$

$$= \begin{cases} \dfrac{1}{2n+1} \dfrac{(n+m)!}{(n-m)!}\,; \quad n = k \\ 0,\ n \neq k,\ (n-k)-\text{even} \\ \dfrac{(-1)^{(k-n+1)/2}(n+m)!\,(k+m)!}{(n-k)(n+k+1)2^{n+k-1}\left[\left(\dfrac{n+m}{2}\right)!\left(\dfrac{n-m}{2}\right)!\left(\dfrac{k+m+1}{2}\right)!\left(\dfrac{k-m-1}{2}\right)!\right]^2} \,. \\ (n-k)-\text{odd},\ (m+n)-\text{even} \end{cases} \quad (7.47)$$

A generalization of the lower boundary condition (7.3) when non-orthotropic reflection is ignored:

$$\left(1 - \frac{4\pi\bar{R}}{1+d}\right) \begin{Bmatrix} \vec{X}(0) \\ \vec{Y}(0) \end{Bmatrix}_{2n-1+m,m} = -\hat{P}^{(m)}_5 \begin{Bmatrix} \vec{X}(0) \\ \vec{Y}(0) \end{Bmatrix}_{2n-2+m,m}$$

$$+ \frac{4\pi\bar{R}}{1+d} \hat{P}^{(m)}_6 \begin{Bmatrix} \vec{X}(0) \\ \vec{Y}(0) \end{Bmatrix}_{2n-2+m,m} + 4\pi^2 \frac{\bar{R} D_0}{1+d} \vec{R}^{(m)}_s \begin{Bmatrix} 1 \\ -1 \end{Bmatrix} \quad (7.48)$$

follows from the representation of a two-parameter set of reflection indicators (parameters d, g) in the form by Henye-Greenstein:

$$\frac{d}{1+d}\left\{1 + \frac{1-g^2}{d[1-g^2-2g(\vec{\Omega}\vec{\Omega}')]^{3/2}}\right\} \equiv \frac{d}{1+d}\left\{1 + \frac{1}{d}\sum_{n=0}^{\infty}\left[P_n(\eta^+)P_n(\eta^-)\right.\right.$$

$$\left.\left. + 2\sum_{l=1}^{n} \frac{(n-l)!}{(n+l)!} P^l_n(\eta^+)P^l_n(\eta^-)\cos l(\varphi-\varphi')\right](2n+1)g^n\right\} \,. \quad (7.49)$$

In (7.48) the components of matrix $\hat{p}^{(m)}_6$ and of vector $\vec{R}^{(m)}_s$ are:

$$(\hat{p}^{(m)}_6)_{n,k} = (4n-3+2m)\frac{(2n-2)!}{(2n-2+2m)!} \varkappa^{(m)}_{m,2k-1+m}(2m+1)$$

$$\times \frac{g^m}{(2m)!} S^{(m)}_{2n-1+m,m} \,, \quad (7.50)$$

$$(\vec{R}^{(m)}_s)_k = g^m \varkappa^{(m)}_{m,2k-1+m} P^m_m(\cos\theta_0) \,, \quad (7.51)$$

where:

$$S^{(m)}_{n,k} = -\frac{k-m+1}{2k+1}\varkappa^{(m)}_{n,k+1} - \frac{k+m}{2k+1}\varkappa^{(m)}_{n,k-1} \quad (7.52)$$

from the equation:

$$P_m^n(-\eta) = (-1)^{n-m} P_m^n(\eta) \tag{7.53}$$

and recurrent formula:

$$(2k+1)\eta P_k^m(\eta) = (k-m+1) P_{k+1}^m(\eta) + (k+m) P_{k-1}^m(\eta) \ . \tag{7.54}$$

Writing in the explicit form, the matrix $(\hat{p}_5^{(m=1)})_{n,k}$, consisting of N+1 and the same number of columns:

$$(\hat{p}_5^{(m=1)})_{n,k} = \frac{4n-1}{2n(2n-1)} \begin{bmatrix} \varkappa_{1,2}^{(1)} & \varkappa_{1,4}^{(1)} & \dots & \varkappa_{1,2N+2}^{(1)} \\ \varkappa_{3,2}^{(1)} & \varkappa_{3,4}^{(1)} & \dots & \varkappa_{3,2N+2}^{(1)} \\ \vdots & \vdots & & \vdots \\ \varkappa_{2N+1,2}^{(1)} & \varkappa_{2N+1,4}^{(1)} & \dots & \varkappa_{2N+1,2N+2}^{(1)} \end{bmatrix} \begin{matrix} n=1 \\ n=2 \\ \vdots \\ n=N+1 \end{matrix}$$

$$k=1 \qquad k=2 \qquad\qquad k=N+1$$

we obtain recurrent formulas relating to the elements of matrix $(\hat{p}_5^{(m=1)})_{n,k}$ at $m>0$. We have:

$$\varkappa_{n,k+2}^{(m>0)} = \varkappa_{n,k}^{(m>0)} \frac{(k+m+2)(n-k)(n+k+1)}{(n-k-2)(n+k+3)(k-m+1)} \ , \tag{7.55}$$

that is, knowledge of the value $\varkappa_{1,2}^{(1)} = 0.75$ is enough to find the first lines of matrix $(\hat{p}_5^{(m=1)})_{1,k}$.

Calculating the ratio $\varkappa_{n+2,k}^{(m>0)}/\varkappa_{n,k}^{(m>0)}$, we obtain:

$$\varkappa_{n+2,k}^{(m>0)} = -\varkappa_{n,k}^{(m>0)} \frac{(n-k)(n+k+1)(n+m+2)}{(n-k+2)(n+k+3)(n-m+1)} \ , \tag{7.56}$$

which makes it possible to determine all the element of matrix $(\hat{p}_5^{(m=1)})_{n,k}$.

Finally, a recurrent transformation of the elements of this matrix in the transition from these m to others has the form:

$$(\hat{p}^{(m+1)})_{n,k} = (\hat{p}_5^{(m)})_{n,k} \frac{(4n+2m-1)(n+k+m-1)(2k+2m+1)}{2(4n+2m-3)(n+m)(n+k+m)} \ , \tag{7.57}$$

$(m = 1, 2, \dots, N-1;\ n, k = m, m+1, \dots, N+1-m)$.

The elements of matrix $(\hat{p}_6^{(m)})_{n,k}$, according to (7.50) and (7.52), are transformed into the form:

$$(\hat{p}_6^{(m)})_{n,k} = (\hat{p}_5^{(m)})_{n,k} \left[-\frac{g^m}{(2m)!} \varkappa_{2n-1+m,m-1}^{(m)} \delta_{2n-1+m,m+1} \right] \ , \tag{7.58}$$

where $\delta_{n,k}$ is the Kronecker symbol. Performing operations with (7.51) similar to (7.55–7.57), we obtain:

$$(\vec{R}_s^{(m=0)})_k = \varkappa_{0,2k-1} \ , \tag{7.59}$$

$$(\vec{R}_s^{(m+1)})_k = (\vec{R}_s^{(m)})_k\, g \frac{P_{m+1}^{m+1}(\cos\theta_0)}{P_m^m(\cos\theta_0)} \frac{(2m+1)(zk+2m+1)(m+k)}{m+k+1} , \quad (7.60)$$

$(m = 1, 2, \ldots, N-1;\ k = m, m+1, \ldots, N+1-m)$.

Thus, the scheme of calculation of the transformation in the atmosphere of radiative images of natural objects with anisotropic scattering is a generalization of the calculation scheme for radiation transfer for the zero harmonic of expansion of intensity in Legendre polynomials. Here we do not dwell upon details of the calculation scheme to solve the boundary problem (7.1 – 7.3) with non-isotropic reflection, discussed in [76].

It appears that the solution for the boundary problem (7.1 – 7.3) can also be generalized, for a horizontally-inhomogeneous terrestrial surface. For this purpose we rewrite the boundary problem (7.4 – 7.6) for the fluctuating constituent of the Fourier transformation of radiance in the form:

$$\eta \frac{\partial I'(z,\vec{p},\vec{S})}{\partial z} + [\sigma_t(z) - i(\vec{p},\vec{S})]\, I'(z,\vec{p},\vec{S})$$
$$= \sum_{p'=1}^{P'} \sigma_s^{p'}(z) \int_{\Omega} I'(z,\vec{p},\vec{S})\gamma_{p'}(z,\vec{S},\vec{S}')\, d\vec{S}' , \quad (7.61)$$

$$I'(H,\vec{p},\vec{S}) = 0 , \quad \vec{S} \in \Omega \quad (7.62)$$

$$I'(0,\vec{p},\vec{S}) = \int_{\Omega_-} \bar{R}(\vec{S},\vec{S}')\, I'(0,\vec{p},\vec{S})\, d\vec{S}'$$
$$+ \int_{\Omega_-} R'(\vec{S},\vec{S}')\, \bar{I}(0,\vec{p},\vec{S})\, d\vec{S}' , \quad \vec{S} \in \Omega_- \quad (7.63)$$

where:

$$\vec{p} = \{p_x, p_y\},\ \vec{S} = \{\eta, \sqrt{1-\eta^2}\cos\varphi, \sqrt{1-\eta^2}\sin\varphi\} .$$

All computations for $\bar{I}(z,\eta,\varphi)$ mentioned in Sections 7.3 – 7.4 hold for real and imaginary parts of $I'(z,\vec{p},\vec{S})$. In particular, Eq. (7.61) of the boundary problem (7.61 – 7.63) is written as:

$$\vec{\psi}(H) = \hat{p}\,\vec{\psi}(0) , \quad (7.64)$$

where vector $\vec{\Psi}$ (a line of matrix) becomes "longer" with the dimensionality $2(2N+1)N$:

$$\vec{\psi} = \{\vec{\psi}^{Re}, \vec{\tilde{\psi}}^{Im}\}$$
$$= \{\psi_{0,0}^{Re}; \psi_{1,0}^{Re}; \ldots, \psi_{2N+1,0}^{Re}; \ldots; \psi_{m,m}^{Re}; \ldots;$$
$$\psi_{2N+1-m,m}^{Re}; \ldots; \psi_{N,N}^{Re}; \psi_{N+1,N}^{Re};$$
$$\tilde{\psi}_{0,0}^{Im}; \tilde{\psi}_{1,0}^{Im}; \ldots; \tilde{\psi}_{2N+1,0}^{Im}; \ldots; \tilde{\psi}_{m,m}^{Im}; \ldots;$$
$$\tilde{\psi}_{2N+1-m,m}^{Im}; \ldots; \tilde{\psi}_{N,N}^{Im}; \tilde{\psi}_{N+1,N}^{Im}\} ,$$

matrices $\hat{p}_1, \hat{p}_2, \hat{p}_3, \hat{p}_4$ become cellular:

$$\hat{p} = \begin{pmatrix} \hat{p}_1 & \hat{p}_2 \\ \hat{p}_3 & \hat{p}_4 \end{pmatrix} .$$

The boundary conditions (7.62) and (7.63) in this case become:

$$\vec{\psi}^{Re}_{even}(H) = -\hat{p}_{51}\vec{\psi}^{Re}_{odd}; \; \vec{\psi}^{Im}_{even}(H) = -\hat{p}_{51}\vec{\psi}^{Im}_{odd}(H) , \tag{7.65}$$

$$\vec{\psi}^{Re}_{even}(0) = \hat{p}_{61}\vec{\psi}^{Re}_{odd}(0) + \bar{\vec{R}}_0, \; \vec{\psi}^{Im}_{even}(0) = \hat{p}_{61}\vec{\psi}^{Im}_{odd} + \bar{\vec{R}}_0 \tag{7.66}$$

where:

$$\hat{p}_{51} = \begin{pmatrix} \hat{p}_5 & 0 \\ 0 & \hat{p}_5 \end{pmatrix} , \quad \hat{p}_{61} = \begin{pmatrix} \hat{p}_6 & 0 \\ 0 & \hat{p}_6 \end{pmatrix} ,$$

and the solution for the boundary problem (7.61 – 7.63):

$$\vec{\psi}^{Re}_{even}(0) = -\hat{p}_7^{-1}(\hat{p}_4 - \hat{p}_{51}\hat{p}_2)\bar{\vec{R}}_0 , \tag{7.67}$$

$$\vec{\psi}^{Im}_{even}(0) = -\hat{\bar{p}}_7^{-1}(\hat{\bar{p}}_4 - \hat{\bar{p}}_{51}\hat{\bar{p}}_2)\bar{\vec{R}}_0 . \tag{7.68}$$

We shall not go into details of this calculation scheme described in [83]. The results of calculations of radiances normalized against solar constant for spectral interval $\Delta\lambda = 0.5 - 55$ µm are shown in Figs. 7.3 – 7.5. Figure 7.3 shows

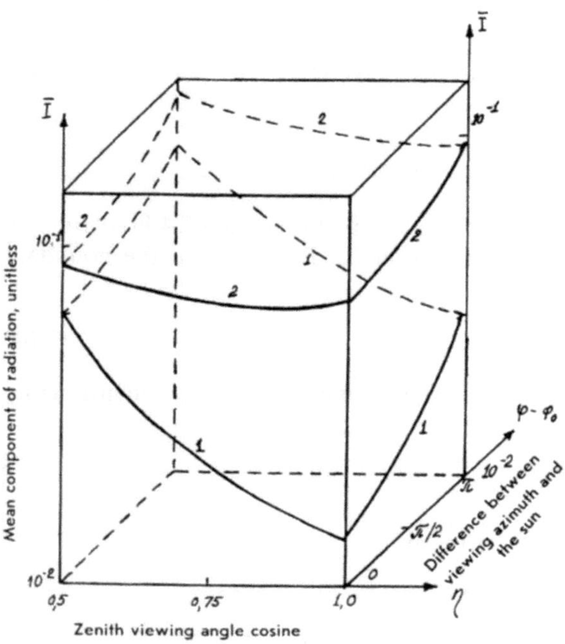

Fig. 7.3. Dependences of the mean constituent of the outgoing radiation intensity on the angular conditions of survey and on variations in average surface albedo from zero *1* to unity *2*

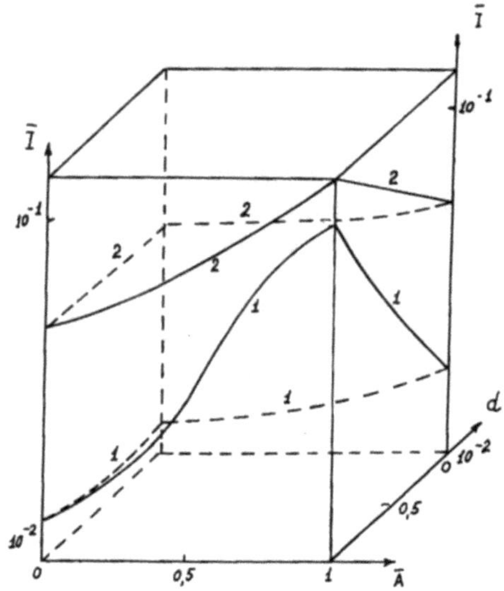

Fig. 7.4. Dependences of the mean constituent of the outgoing radiation intensity on the reflective properties of the Earth surface for *1* transparent and *2* turbid atmosphere

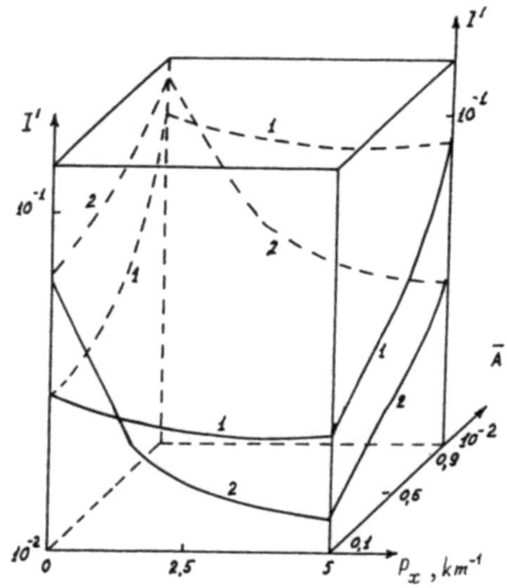

Fig. 7.5. Dependences of the fluctuating constituent of the outgoing radiation intensity on spatial frequency and average surface albedo with varying atmospheric turbidity: *1* transparent atmosphere; *2* turbid atmosphere

variations in the outgoing radiation intensities $\bar{I}$ for a moderately turbid atmosphere depending on the angular conditions of survey (η^+, $\cos\theta_0$, $\varphi - \varphi_\odot$) with the albedo of a homogeneous Lambertian surface varying from zero to unity. The dependence of $\bar{I}$ on η^+ coincides with that on $\bar{I}$ on $\cos\theta_0$ based on the "optical theorem". Figure 7.3 shows the values for real angular coordinates of satellite-borne scanning systems of observations, at which the applied plane-parallel geometry of the problem is valid.

Figure 7.4 estimates the effect of the atmospheric state and of the homogeneous but non-orthotropic terrestrial surface on the outgoing radiation intensity at $\cos\theta_0 = 0.7$. The parameter $d = 0$ corresponds to a Lambertian surface, $d = 1$ to specularly reflecting surface. The state of the atmosphere varies from transparent (curve 1) to turbid (curve 2).

Figure 7.5 illustrates the character of the dependent nature of the outgoing radiation intensity on the horizontal distribution of natural objects over the Earth's surface determined through average albedo $\bar{A}$ and spatial frequency p_x, with varying atmospheric states. The atmosphere is a low-frequency filter smoothing the high-frequency constituent of the spectrum of horizontal inhomogeneities. The respective boundary frequency decreases with growing optical thickness of the atmosphere τ, which is also shown in Fig. 7.5.

7.6 The Longwave Radiation Transfer

The solution for the boundary problems of longwave radiation transfer is, as well be shown later, a natural generalization of the considered boundary problems of shortwave radiation transfer [250].

Let us calculate the intensity of longwave radiation in a plane-stratified atmosphere of a finite thickness H, with the following constraints:

- the non-polarized radiation field is stationary;
- scattering in the medium does not change the frequency characteristics of the radiation field;
- processes of radiant energy scattering in the medium are constant relative to rotation, and the phase function can be expanded in a finite series in Legendre polynomials;
- a lower atmospheric layer with $H \lesssim 50$ km is considered, where conditions of local thermodynamic equilibrium are satisfied, since collisions play the dominating role here in excitation of atoms and molecules;
- the surface, black for longwave radiation, reflects the incident radiation in the shortwave region in a certain way and contributes to the field of outgoing radiation at a certain frequency;
- the atmospheric layer contains P' substances (air, aerosol, water vapor, water droplets, etc.), characterized by partial sections of scattering, absorption and by phase functions; the layer contains $I'+1$ zones within which the macroparameters of processes are vertically constant.

This setting of the problem differs from the commonly assumed concept of atmospheric transmission functions by considering the transfer problem together with radiation scattering (particularly substantial in a cloudy atmosphere). The applied macrosections of absorption for each of the p′-substances of the atmosphere coincide with the mass coefficients of absorption, and the transmission function is an exponent, with the product of macrosection of absorption by the density of the respective atmospheric substance with the "minus" sign taken as its power.

In some problems, especially, in equations of atmospheric dynamics, the radiative effects are considered only in the expression for radiative heat flux divergence per unit atmospheric volume and in the heat balance equation on the surface (see Section 7.7). In both cases, radiation intensity being integrated by azimuth (the flux through unit sphere and unit area, respectively). Therefore, it is worthwhile attempting to write an equation of transfer and condition at the atmospheric boundaries for mean azimuthal intensity $\tilde{I}_\nu(z, \mu)$.

Here we return from radiance within a certain spectral interval $\Delta\nu$ to the intensity of monochromatic radiation, to get a deeper insight into the laws of formation of the outgoing longwave radiation field, connected with the eigen emission of the medium and with solar shortwave radiation as a source of eigen emission.

We discuss a two-group approximation: for the shortwave radiation corresponding to direct and scattered solar energy with conditional wavelength boundaries 0.3 – 7 μm; and for the longwave radiation corresponding to eigen emission of the atmosphere and surface with conditional boundaries from 2 to 100 μm.

The boundary frequency ν_b separates shortwave (SW) and longwave (LW) spectral intervals.

The azimuthally averaged intensities of short- and longwave radiation are, respectively:

$$I_{SW}(z,\eta) = \int_{\nu_b}^{\infty} I_\nu(z,\eta)\,d\nu;\; I_{LW}(z,\eta) = \int_{0}^{\nu_b} I_\nu(z,\eta)\,d\nu;$$

$$I_\nu(z,\eta) = \frac{1}{2\pi}\int_0^{2\pi} I_\nu(z,\eta,\varphi)\,d\varphi\ .$$

Since the azimuthal averaging of radiation intensity is meant here, there is no need to separate intensities $I_\nu(z,\eta,\varphi)$ into average and fluctuating constituents. Hence, there is no need to consider such factors as non-orthotropic reflection and horizontal inhomogeneity of the Earth surface.

If variations in the radiation field in the atmosphere as a function of height and angular conditions of survey, are weakly connected with frequency variations, then, the intensity of monochromatic radiation can be written in the form:

$$I_\nu(z,\eta) = \tilde{I}(z,\eta)\,\chi(z,\nu)\ , \tag{7.69}$$

where $\chi(z,\nu)$ is the weight function of averaged sections; phase functions and albedos equal to the function of solar energy distribution for wavelengths (0.3 – 2) μm; to the Planck function of the absolute blackbody emissions for wavelengths (7 – 100) μm; and to the superposition of these two functions for wavelengths (2 – 7) μm. As a result, averages of sections and coefficients of expansion of phase functions in series in Legendre polynomials, and of albedo will be estimated from the following expressions (the "$\sim$" sign corresponds to longwave radiation):

$$\tilde{\sigma}^{p'}_{s,c,t}(z) = \int_0^{\nu_b} \tilde{\sigma}^{p'}_{\nu(s,c,t)}\chi(z,\nu)\,d\nu \Big/ \int_0^{\nu_b} \chi(z,\nu)\,d\nu \ ;$$

$$\sigma^{p'}_{s,c,t}(z) = \int_{\nu_b}^{\infty} \sigma^{p'}_{\nu}(s,c,t)\chi(z,\nu)\,d\nu \Big/ \int_{\nu_b}^{\infty} \chi(z,\nu)\,d\nu \ ,$$

$$b^{p'}_{n}(z) = \pi\int_0^{\infty}\int_{\nu_b}^{\infty} b^{p'}_{n,\nu}\left(z, 2\pi\frac{\tilde{r}}{\lambda}, \tilde{m}\right)\tilde{r}^2\tilde{K}_p\left(2\pi\frac{\tilde{r}}{\lambda}\right)$$

$$\times \tilde{f}(\tilde{r})\chi(z,\nu)\,d\nu\,d\tilde{r} \Big/ \int_{\nu_b}^{\infty} \sigma^{p'}_{s,\nu}(z)\chi(z,\nu)\,d\nu$$

(the expression for $\tilde{b}^{p'}_{n}(z)$ is similar to that for $b^{p'}_{n}(z)$ with the substituted limits of frequency intergration from zero to ν_b),

$$A = \int_{\nu_b}^{\infty} A_\nu \chi(z=0,\nu)\,d\nu \Big/ \int_{\nu_b}^{\infty} \chi(z=0,\nu)\,d\nu \ .$$

Considering these remarks and assumptions, the equations for the boundary problem will be:

$$\eta\frac{\partial I_\nu(z,\eta)}{\partial z} + \begin{Bmatrix}\sigma_{\nu t}(z)\\ \tilde{\sigma}_{\nu t}(z)\end{Bmatrix} I_\nu(z,\eta)$$

$$= \sum_{p'=1}^{P'} \begin{Bmatrix}\sigma^{p'}_{\nu s}(z)\\ \tilde{\sigma}^{p'}_{\nu s}(z)\end{Bmatrix} \sum_{k=0}^{2k+1} \begin{Bmatrix} b^{p'}_{\nu k}(z)\\ \tilde{b}^{p'}_{\nu k}(z)\end{Bmatrix} \frac{2k+1}{2} P_k(\eta)$$

$$\times \int_{-1}^{1} I_\nu(z,\eta')P_k(\eta')\,d\eta'$$

$$+ \begin{Bmatrix} S_{0\nu}P_k(\vec{\Omega}\vec{\Omega}_\odot)\exp\left[-1/\zeta\int_z^H \sigma_t(z')\,dz'\right] & \nu > \nu_b \\ \sigma_{\nu c}(z)E_\nu[T(z)] & \nu \leqslant \nu_b \ , \end{Bmatrix}$$

$$I_\nu(z=H,\eta^-) = 0, \eta^- > 0 \tag{7.71}$$

$$I_\nu(0,\eta^+) = \begin{cases} \dfrac{A_\nu}{\pi}D_0 + 2A_\nu\displaystyle\int_0^1 I_\nu(0,\eta^-)\eta^-\,d\eta^- \ , & \nu > \nu_b \\ \dfrac{2E_\nu(T_g)}{\sigma_{SB}T_g^4}\left[(1-A)\displaystyle\int_0^1 I_{SW}(0,\eta^-)\eta^-\,d\eta^- \right. & \\ \left. +\displaystyle\int_0^1 I_{LW}(0,\eta^-)\eta^-\,d\eta^- + B(T_g)/(2\pi)\right] \ , & \nu \leqslant \nu_b \ . \end{cases} \tag{7.72}$$

The solution for the boundary problems of each of the two groups of photons has the form:

$$\begin{Bmatrix} I_{SW}(z,\eta) \\ I_{LW}(z,\eta) \end{Bmatrix} \approx \frac{1}{4\pi} \sum_{n=0}^{2N+1} (2n+1) P_n(\eta) \begin{Bmatrix} \psi_n^{SW}(z) \\ \psi_n^{LW}(z) \end{Bmatrix} , \tag{7.73}$$

where:

$$\begin{Bmatrix} \psi_n^{SW}(z) \\ \psi_n^{LW}(z) \end{Bmatrix} = 2\pi \int_{-1}^{1} P_n(\eta) \begin{Bmatrix} I_{SW}(z,\eta) \\ I_{LW}(z,\eta) \end{Bmatrix} d\eta .$$

We do not use notations in the following equations for SW and LW radiation in vectors of the moments of angular distributions of intensity: the "$\sim$" sign corresponds to LW radiation, and its absence over $\psi(z)$ to SW radiation.

Integrating (7.70–7.72) in frequency over $(0, \nu_b)$ and using (7.75), as well as recurrent formula for Legendre polynomials

$$(n+1) P_{n+1}(\eta) + n P_{n-1}(\eta) = (2n+1)\eta P_n(\eta)$$

and the value of tabulated integral $\varkappa_{n,m}$ (7.40), we reduce the integral-differential equation (7.72) and boundary conditions (7.71–7.72) to the following linear system of equations:

$$(n+1)\frac{d\tilde{\psi}_{n+1}(z)}{dz} + n\frac{d\tilde{\psi}_{n-1}(z)}{dz} + \tilde{\sigma}_t(z)[1-\tilde{C}_n(z)]$$
$$\times (2n+1)\tilde{\psi}_n(z) = 4\delta_{on}\tilde{\sigma}_c(z)\sigma_{SB}T^4 , \quad (n = 0,1,\ldots,2N+1) , \tag{7.74}$$

$$\frac{d\tilde{\psi}_{2N+2}}{dz} = 0 , \tag{7.75}$$

$$\tilde{\psi}_{2k+1}^{l'+1}(H) = \sum_{n=0}^{N} (4n+1)\varkappa_{2n,2k+1}\tilde{\psi}_{2n}^{l'+1}(H) , \tag{7.76}$$

$$\tilde{\psi}_{2k+1}^{1}(0) = 2\varkappa_{0,2k+1}[B - \psi_1^1(0)] + \sum_{n=0}^{N}(4n+1)$$
$$\times (2\varkappa_{2n,1}\varkappa_{0,2k+1} - \varkappa_{2n,2k+1})\tilde{\psi}_{2n}^{1}(0) , \quad (k = 0,1,\ldots,N) . \tag{7.77}$$

By definition,

$$\tilde{C}_n(z) = \frac{\tilde{B}_n(z)}{\tilde{\sigma}_t(z)} , \quad \tilde{B}_n(z) = \sum_{p'=1}^{P'} \tilde{b}_n^{p'}(z)\tilde{\sigma}_s^{p'}(z) , \tag{7.78}$$

$$\tilde{\sigma}_{t,s,c}^{p'}(z) = \frac{\int_0^{\nu_b} \sigma_{\nu(t,s,c)}^{p'}(z)\chi(z,\nu)d\nu}{\int_0^{\nu_b}\chi(0,\nu)d\nu} . \tag{7.79}$$

Condition (7.75) is needed to coordinate the number of unknown functions $\tilde{\psi}_n(z)$ with the order of system (7.74). Note that (7.77) contains surface albedo in a non-explicit form, since in derivation of this system relationship (7.33) was used:

$$\psi^1_{2k+1}(0) = -\sum_{n=0}^{N}(4n+1)\left(\varkappa_{2n,2k+1} - \frac{4A}{1+A}\varkappa_{0,2k+1}\varkappa_{2n,1}\right)$$
$$\times \psi^1_{2n}(0) + D_0\frac{4A}{1+A}\varkappa_{0,2k+1} \ , \qquad (7.80)$$

which represents the spectral form of writing the boundary condition at the surface for SW radiation.

A system of equations:

$$\vec{\psi}^{\,I'+1}_{odd}(H) = \hat{p}\,\vec{\psi}^{\,I'+1}_{even}(H) \ , \qquad (7.81)$$

$$\vec{\psi}^{\,1}_{odd}(0) = 2\vec{S}_{odd} + (2\hat{p}_6 - \hat{p}_5)\,\vec{\psi}^{\,1}_{even}(0) \ , \qquad (7.82)$$

in which the elements of matrices $\hat{p}_5$ and $\hat{p}_6$ and vectors $\vec{\psi}^{\,I'+1}_{odd}(H)$, $\vec{\psi}^{\,I'+1}_{even}(H)$, $\vec{\psi}^{\,1}_{odd}(0)$, $\vec{\psi}^{\,1}_{even}(0)$, $\vec{S}_{odd}$ are determined by the relationships:

$$(\hat{p}_5)_{n,k} = (4n-3)\varkappa_{2n-2,2k-1};\ (\hat{p}_6)_{n,k} = (4n-3)\varkappa_{2n-2,1}\varkappa_{0,2k-1} \ ; \qquad (7.83)$$

$$[\vec{\psi}^{\,I'+1}_{odd}(H)]_k = \psi^{I'+1}_{2k-1}(H);\ [\vec{\psi}^{\,I'+1}_{even}(H)]_k = \psi^{I'+1}_{2k-2}(H) \ ;$$
$$[\vec{\psi}^{\,1}_{odd}(0)]_k = \psi^1_{2k-1}(0);\ [\vec{\psi}^{\,1}_{even}(0)]_k = \psi^1_{2k-2}(2) \ ;$$
$$(\vec{S})_k = [B - \psi^1_1(0)]\varkappa_{0,2k-1} \ , \qquad (7.84)$$

is a compact vector-matrix writing of boundary conditions (7.76) and (7.77). It is seen that vector:

$$(\vec{M}^i)_n = 4\sigma_{SB}(T^4)^{i-1,i}\delta_{1n} \ ; \qquad n = 1,2,\ldots,2N+2 \qquad (7.85)$$

is a partial solution for system (7.74) in the i-th zone with constant physical parameters. The general solution for a homogeneous system (see Sect. 7.4) is totally determined by the matrix of transmission $\hat{p}^{i-1,i}$. Using (7.85) and the technique for solution described in Section 7.4, we relate vectors $\vec{\psi}^{\,i}(z)$ at the upper z_i and lower z_{i-1} boundaries of the zone considered:

$$\vec{\psi}^{\,i}(z_i) = \hat{p}^{i-1,i}\vec{\psi}^{\,i}(z_{i-1}) + (\hat{E} - \hat{p}^{i-1,i})\vec{M}^i \ , \qquad (7.86)$$

where $\hat{E}$ is a unit matrix.

Equation (7.86), together with the condition of continuity of the vector of moments $\vec{\psi}(z)$ at the boundaries of the zones, makes it possible to derive a relationship between the vectors of moments at the upper $\vec{\psi}^{\,I'+1}(H)$ and lower $\vec{\psi}^{\,1}(0)$ boundaries of the atmosphere:

$$\vec{\psi}^{\,I'+1}(H) = \hat{p}\,\vec{\psi}^{\,1}(0) + \vec{M}_{tot} \ , \qquad (7.87)$$

where:

$$\vec{M}_{tot} = \sum_{i=I'+1}^{1}\hat{K}^i(\hat{E} - \hat{p}^{i-1,i})\vec{M}^i \ , \qquad (7.88)$$

$$\hat{K}^i = \begin{cases} \hat{E} \ , & \text{if} \quad i = I'+1 \\ \hat{p}^{I',I'+1} \ , & \text{if} \quad i = I' \\ \hat{p}^{I',I'+1}\hat{p}^{I'-1,I'}\ldots\hat{p}^{i,i+1} \ , & \text{if} \quad 1 \leqslant i < I' \ . \end{cases} \qquad (7.89)$$

Construct four new matrices and two new vectors of an order of N + 1 from the elements of matrix $\hat{\mathrm{p}}$ and vector $\vec{M}_{\mathrm{tot}}$:

$$(\hat{\mathrm{p}}_1)_{k,n} = (\hat{\mathrm{p}})_{2k-1,2n-1};\ (\hat{\mathrm{p}}_2)_{k,n} = (\hat{\mathrm{p}})_{2k-1,2n}\ ; \tag{7.90}$$

$$(\hat{\mathrm{p}}_3)_{k,n} = (\hat{\mathrm{p}})_{2k,2n-1};\ (\hat{\mathrm{p}}_4)_{k,n} = (\hat{\mathrm{p}})_{2k,2n}\ ; \tag{7.91}$$

$$(\vec{M}^{\mathrm{odd}}_{\mathrm{tot}})_k = (\vec{M}_{\mathrm{tot}})_{2k-1},\ (\vec{M}^{\mathrm{even}}_{\mathrm{tot}})_k = (\vec{M}_{\mathrm{tot}})_{2k}\ . \tag{7.92}$$

In notations of (7.84), (7.88 – 7.92), Eq. (7.87) can be written as the following system:

$$\vec{\psi}^{\,l'+1}_{\mathrm{even}}(H) = \hat{\mathrm{p}}_1\,\vec{\psi}^{\,1}_{\mathrm{even}}(0) + \hat{\mathrm{p}}_2\,\vec{\psi}^{\,1}_{\mathrm{odd}}(0) + \vec{M}^{\mathrm{odd}}_{\mathrm{tot}}\ , \tag{7.93}$$

$$\vec{\psi}^{\,l'+1}_{\mathrm{odd}}(H) = \hat{\mathrm{p}}_3\,\vec{\psi}_{\mathrm{even}}(0) + \hat{\mathrm{p}}_4\,\vec{\psi}_{\mathrm{odd}}(0) + \vec{M}^{\mathrm{even}}_{\mathrm{tot}}\ . \tag{7.94}$$

Substitute (7.93) and (7.94) into (7.81). Excluding from the obtained equation vector $\vec{\psi}^{\,1}_{\mathrm{odd}}(0)$ with the help of (7.82), and solving it with respect to $\vec{\psi}^{\,1}_{\mathrm{even}}(0)$, we find:

$$\begin{aligned}\vec{\psi}^{\,1}_{\mathrm{even}}(0) = {} & [(\hat{\mathrm{p}}_5\hat{\mathrm{p}}_1 - \hat{\mathrm{p}}_3) + (\hat{\mathrm{p}}_5\hat{\mathrm{p}}_2 - \hat{\mathrm{p}}_4)(2\hat{\mathrm{p}}_6 - \hat{\mathrm{p}}_5)]^{-1} \\ & \times [\vec{M}^{\mathrm{even}}_{\mathrm{tot}} - \hat{\mathrm{p}}_5\vec{M}^{\mathrm{odd}}_{\mathrm{tot}} - 2(\hat{\mathrm{p}}_5\hat{\mathrm{p}}_2 - \hat{\mathrm{p}}_4)\vec{S}_{\mathrm{odd}}]\ .\end{aligned} \tag{7.95}$$

With $\vec{\psi}^{\,1}_{\mathrm{even}}(0)$ known, using (7.92) we calculate $\vec{\psi}^{\,1}_{\mathrm{odd}}(0)$ and, hence, $\vec{\psi}^{\,1}(0)$ at the surface, and using (7.86), we calculate vectors $\vec{\psi}(z_i)$ at the remaining boundaries of the zones. With the coefficients of expansion known, it is easy to calculate the angular distribution (7.82). All the conclusions in Section 7.4 about the applicability of solutions of 3-diagonal systems of algebraic equations being valid in this case.

7.7 Radiation Fluxes and Heat Flux Divergences in Radiation-Balance Models

Studies of the Earth's surface radiation budget (ERB) SW and LW components (its albedo and temperature) are a key problem in the application of information in the interests of hydrometeorology. Both these components, characterizing the relationship between scattered solar radiation and eigen thermal emission of the surface-atmosphere system, determine the sources and sinks of radiative energy which govern general circulation of the atmosphere and of the ocean, which is particularly important in long-range weather and climate forecasts.

The role of satellite measurements of the ERB grows in connection with the detection in the World ocean of individual zones of longterm weather anomalies and climate changes, called "energy-active zones". A mathematical apparatus of conjugated equations in thermohydrodynamics developed by Marchuk [119, 120] makes it possible to describe climate changes through observed anomalies of processes in these zones. It is possible to use conventional observation data (first, observations from ships) to monitor the development of processes in energy-active zones, and satellite data to get a deeper insight into the processes that govern long-term weather anomalies and climate changes [124]. Use of both types of observations to study climate changes is a component of the programme "Sections" [125], the principal objective of which is to study the role of the ocean in short-period climate changes.

Existing methods for retrieving the longwave component of radiation budget are based on empirical connections between outgoing longwave radiation and efficient surface emission. Retrieving the surface albedo is a more difficult problem. However, in this case regression models are usually used. Though the experience of satellite data interpretation points to real possibilities of retrieving the ERB, the respective empirical techniques do not provide sufficient accuracy of retrieval. Results of model calculations of radiation transfer using these techniques are, as a rule, subsidiary.

Fluxes in the atmosphere and at the surface are determined by integrating radiant intensities, used in Sections 7.1–7.7, in angular coordinates and the wavelengths. In particular, the flux through unit sphere connected with radiative heat flux divergence in the atmosphere is assessed using the expression:

$$F_A = \int_0^{2\pi} d\varphi \int_{-1}^{1} d\eta \int_0^{\infty} \sigma_{\nu c}(z) I_\nu(z,\eta,\varphi)\, d\nu \tag{7.96}$$

and the flux across unit area connected with the heat balance at the surface is:

$$F_s = \int_0^{2\pi} d\varphi \int_0^{1} d\eta\, \eta \int_0^{\infty} I_\nu(z,\eta,\varphi)\, d\nu \ . \tag{7.97}$$

Then, on the basis of the 2-group approximation for SW and LW radiation we obtain:

$$\begin{Bmatrix} F_A^{SW}(z) \\ F_A^{LW}(z) \end{Bmatrix} = 2\pi \begin{Bmatrix} \sigma_c^{SW}(z) \\ \overline{\sigma_c^{LW}(z)} \end{Bmatrix} \int_{-1}^{1} \begin{Bmatrix} I_{SW}(z,\eta) \\ I_{LW}(z,\eta) \end{Bmatrix} d\eta = \begin{Bmatrix} \sigma_c^{SW}(z)\,\psi_0^{SW}(z) \\ \overline{\sigma_c^{LW}(z)\,\psi_0^{LW}(z)} \end{Bmatrix} , \tag{7.98}$$

where in sections σ_c^{SW}, σ_c^{LW} the effects of absorption by each of the p′-substances of the atmosphere are considered.

The radiative heat flux divergence in the right-hand part of the equation for heat flux divergence in AGCMs is:

$$\begin{aligned} Q_{rad}(z) &= \frac{1}{\varrho(z) C_v} \left\{ \int_{-1}^{1} d\eta \int_0^{\infty} I_\nu(z,\eta)\, \sigma_{\nu c}(z)\, d\nu - 4\pi \int_0^{\infty} \sigma_{\nu c}(z) E_\nu[T(z)]\, d\nu \right\} \\ &= \frac{1}{\varrho(z) C_v} [\sigma_c^{SW}(z)\,\psi_0^{SW}(z) + \sigma_c^{LW}(z)\,\psi_0^{LW}(z) - 4\sigma_{SB} T^4(z)] \ . \end{aligned} \tag{7.99}$$

As is seen, in calculations of radiative heat-flux divergences, it is important to know the real macrosections of absorption and vectors of the moments of the angular distributions with zero index.

Since the balance of non-radiative fluxes in the lower boundary condition (7.72) is determined by the expression:

$$B(T_g) = -(E^T + L^{ev} E^q + \Pi) \tag{7.100}$$

and, at the same time, is equal to:

$$B(T_g) = F^{\uparrow}(z=0) - F^{\downarrow}(z=0) - S^-(z=0) + AS^-(z=0) \ , \tag{7.101}$$

it can be expressed through the components of the vectors of the moments of the angular distributions with index 1:

$$B(T_g) = \psi_1^{LW}(z=0) + \psi_1^{SW}(z=0) \ . \tag{7.102}$$

Here:

$$AS^-(z=0) = S^+(z=0) \ ,$$

$$S^-(z) = \int_0^1 I_{SW}(z,\eta)\eta\, d\eta \ ,$$

$$F^{\downarrow}(z) = \int_0^1 I_{LW}(z,\eta)\eta\, d\eta \ .$$

$$F^{\uparrow}(z) = \int_0^1 I_{LW}(z,-\eta)\eta\, d\eta$$

$$A = \int_0^1 I_{SW}(z,-\eta)\eta\, d\eta \Big/ \int_0^1 I_{SW}(z,\eta)\eta\, d\eta \ .$$

This makes it possible, using the solutions mentioned for boundary problems for SW radiation, to retrieve the surface albedo, and using solutions for SW and LW radiation, to retrieve surface balance $B(T_g)$ from space-measured $B[T(H)]$.

Table 7.11 shows coefficients of approximation for the effect of the cloudless atmosphere on radiation budget

$$\begin{aligned} -B_\infty/B_{max} &= \psi_1^{SW}(z=H) + \psi_1^{LW}(z=H) = -(1-A_\infty)S_0 + F_\infty \\ &= 2\pi \int_{-1}^{1} [I^{SW}(z,\mu) + I^{LW}(z,\mu)\mu\, d\mu] \approx 2\pi[\tilde{a}_1 + \hat{\tilde{T}}_1(\tilde{B}_g) \\ &+ a_3\hat{\tilde{T}}_1(\cos\theta_0) + \tilde{a}_4\hat{\tilde{T}}_2(\tilde{B}_g) + \tilde{a}_5\hat{\tilde{T}}_1(\tilde{B}_g)\hat{\tilde{T}}_1(\cos\theta_0) + \tilde{a}_6\hat{\tilde{T}}_2(\cos\theta_0)] \ , \end{aligned} \tag{7.103}$$

where $\hat{\tilde{T}}_{1,2}$ are Chebyshev's polynomials, orthogonal on the length $[-1,1]$: $\tilde{B}_g = B_g/B_{max}$; B_{max} is a maximum of radiation budget (assumed to be 300 W/m^2).

Table 7.11. Estimates of coefficients $\tilde{a}_n$ $(n = 1, \ldots, 6)$ and respective estimates of the difference Δ_0 between retrieved $B_g^{retr.}$ and climatic B_g^{cl}, divided by $B_g{}^{av}$ averages (fourth and fifth symbols after comma characterize, respectively, the sign and the power with base 10)

	Mid-latitudes summer	Mid-latitudes winter	Mid-latitudes spring/summer	The tropics
$\tilde{a}_1$	3.912 − 3	3.995 − 2	8.121 − 3	2.048 − 3
$\tilde{a}_2$	2.203 − 3	4.116 − 3	2.225 − 3	1.986 − 3
$\tilde{a}_3$	− 1.007 − 2	− 1.130 − 3	− 1.124 − 2	− 4.110 − 3
$\tilde{a}_4$	− 9.966 − 3	− 1.075 − 2	− 1.189 − 2	− 8.998 − 3
$\tilde{a}_5$	4.014 − 3	1.026 − 3	5.255 − 3	3.685 − 3
$\tilde{a}_6$	− 5.156 − 3	− 4.802 − 3	− 5.314 − 3	− 4.447 − 3
Δ_0	0.17	0.28	0.32	0.05

Along with the model of the atmosphere in Section 7.2, typical of the summer mid-latitudes, coefficients $\tilde{a}_n$ $(n = 1, \ldots, 6)$ were calculated for three other models from [83]. With $\tilde{a}_n$ known, the B_g values are retrieved from measured B_∞ and solar zenith angles known on the date of survey.

The bottom line in Table 7.11 contains maximum deviations between results of B_g retrieval with formula (7.103) for monthly mean satellite data [106] and climatic estimates made from heat balance atlases [24]. It can be seen that for the tropics, in contrast to mid-latitudes, the applied model of the cloudless atmosphere need no specification. In mid-latitudes, it is necessary to introduce empirical corrections for the probability of different types of cloud formations. The respective results will be considered later. Here we note that techniques for solving the problems of the transfer theory are the most important sphere of application of both current and climatic satellite information. Basic features of satellite data are considered in the next chapter.

Chapter 8 Observations of the Earth Radiation Budget from Space

8.1 The Status of the Problem of Retrieving the Heat and Radiation Budgets of the Earth's Surface

In this chapter we consider the western experience in the application of spaceborne hydrometeorological systems to assess the components of radiation budget at the Earth's surface level.

Based on Meteosat scanning radiometer (channel 0.4–1.1 μm) data on the brightness of the surface-atmosphere system, L_s, on ground observations of global radiation E_G in individual locations, and on theoretical calculations, the following parameterization to retrieve α, has been tested in [358]: $\pi L_s = C + A\alpha + B\alpha^2$ (α is the albedo of a Lambertian homogeneous surface). Here $A = E_G^2 / E_s (E_s$ is the extra-atmospheric insolation). Coefficients $B = -\alpha_s A (\alpha_s$ is the spherical albedo of the system) and $C = \alpha_a E_s$ (α_a is the atmospheric albedo) are calculated (in this case solar zenith angles and viewing angles are considered to be equal). Analysis of the 2 July 1979 measurements in the Sahel region in Upper Volta (Burkino Faso) revealed reliably recovered albedos and showed that retrieval accuracy is largely determined by errors in global radiation observations.

Since a daily mean value of global radiation was used, this confined retrieval to clear-sky conditions. However, with available data on the observed day time change, the albedo can be retrieved where clouds whose optical thickness does not exceed 0.75 are present. Each error (in brackets) in spherical albedo (100%), atmospheric albedo (10%), global radiation (3%) and measured brightness (3–4%) is equivalent to a 5% error in albedo retrieval. The principal advantage of this technique that it allows estimation of spatially averaged albedo on the averaging scale of a grid step used in atmospheric general circulation models (AGCMs).

The albedo of the clear-sky surface-atmosphere system (CSSA) is an important climatic characteristic because, in particular, the influence of clouds on the shortwave radiation budget of the Earth is determined by the difference between the cloudy- and clear-sky albedos. In this connection, a comparison has been undertaken [222] of the CSSA values obtained from satellite observations and from model computations for 1°-sided squares for realistically distributed albedos of the major surface types, varying optical properties of the atmosphere, and a wide range of solar zenith angles.

Observed CSSA values have been obtained from the GOES-2 satellite in November 1978 and largely cover a region 45°S–45°N; 30°W–125°W characterized by widely changing surface types (forests, bushes, grass cover, wet soil, deserts, marshes, ocean). The technique for retrieving total albedo of the system (wavelength interval 0.2–4.5 μm) has been discussed in detail, based on satellite data, on the angular distribution of shortwave radition in a narrow spectral interval of 0.55–0.75 μm. The CSSA calculations were made with the shortwave radiation range divided into four bands: 0.2–0.5; 0.5–0.7; 0.7–0.85; and 0.85–4.0 μm. Comparison of observed and calculated CSSA values over the ocean has shown that the difference between absolute values does not exceed ±1%, i.e., it is within the allowed errors of measurements and calculations. The CSSAs over the ocean vary substantially as a function of latitude. About 50% of CSSA variability are explained by variations in the ocean surface albedo due to varying height of the sun. Albedos over the land agree within ±2%, with surface albedos varying widely, from 13% in the Amazon basin to 24% over the mountains of the western North America. These differences do not exceed the uncertainties of observations and calculations, too. An important factor of CSSA variability over the land is the spectral dependence of vegetation albedo. Specific geography of CSSA variations can be characterized based on three categories of natural conditions: (1) broad-leaved and coniferous ever-green forests as well as cultivated tropical lands (low albedo); (2) mountains, grass cover, and bushes (high albedo); (3) decidious forests and high-grass cover (intermediate albedo values).

Comparison of synchronous data on outgoing shortwave radiation (OSR) obtained from GOES and of results of ground-based pyranometric observations of global radiation (GR) in the region of the Great Plains (USA) revealed the possibility of retrieval of the hourly sums of GR from OSR data with an error of about 10–15% [292]. With available 4–5 near-noon OSR values, it is possible to estimate the diurnal sums of GR with an error of less than 15%. In both cases, retrieved GR estimates refer to 40 km^2-side squares in the center of the cells of a 1°×1° grid. The technique for retrieving the GR hourly sums F is based on the combined use of the clearly sky OSR values B_0 for 1°-side squares calculated using an approximation algorithm and directly measured spatially averaged intensity B_m. The final semi-empirical algorithm for retrieving the GR hourly sums is as follows:

$$F = F_0(r_0/r)^2 \cos Z_0 [a + b \cos Z_0 + c \cos^2 Z_0 + d(B_m^2 - B_0^2)] ,$$

where F_0 is the hourly sum of extra-atmospheric insolation (4921 kJ/m^2); $(r_0/r)^2$ the coefficient which determines variations in the extra-atmospheric insolation caused by varying distance between the Earth and the sun; Z_0 the solar zenith distance. The dimensionless coefficients of regression are: $a = 0.4147$; $b = 0.7165$; $c = -0.3909$; $d = -1.630$ kJ/m^2. The last term in this formula describes the effect of cloudiness. The given value of d is applicable to the land but it must be changed in the case of water surfaces (in this case the last term should be 10–15% less). Attempts to use more complicated (and physically more informative) retrieval algorithms have proved abortive, since they ensure only a slight decrease in retrieval uncertainty. Based on the use of the formula given above, the GR distribution over the United States, Mexico and South America has been mapped. The GR monthly means were estimated with an accuracy better than 5%, which makes it possible to analyze the practical application potential of solar energy in the regions without actinometric observations and for studies in the field of radiative climatology. Comparison of the retrieved GR field with the U.S. actinometric network data revealed some underestimation of the values averaged over the period 1982–1983, especially in the central and south-western United States. The regions of minimum GR dispersion are located in the regions of maximum GR, like deserts, where high GR values are persistent.

A statistical (regression) model has been developed [229] to retrieve GR from meteorological satellites data based on the assumption that GR is determined by total cloud amount. First, the cloud over index $n^t(i,j)$ is determined at a time t, for each pixel (i, j) of a $5 \times 5\ km^2$ cloud cover image. This value depends on the prescribed characteristic surface albedo $\varrho(i,j)$, surface albedo $g^t(i,j)$ retrieved from satellite data, and average cloud albedo ϱ_c: $n^t(i,j) = [\varrho^t(i,j) - \varrho(i,j)]/[\varrho_c - \varrho(i,j)]$. There is line dependence of the cloud cover index (it characterizes the part of resolving element covered with clouds) on the ratio $K(i,j) = G(i,j)/G_0(i,j)$ of GR to insolation outside the atmosphere. This enables the obtaining of relationship $K^t(i,j) = a(i,j)n^t(i,j) + b(i,j)$ in which coefficients $a = K_b(i,j) - K_c(i,j)$ and $b = K_c(i,j)$ can be found using ground actinometric observations (indices b and c determine the correspondence of the quantities to the conditions of continuous cloudiness and of clear skies). Results of testing the technique illustrate its reliability. In most cases, the coefficient of correlation of retrieved and directly measured GR was above 0.8.

As far as interpretation errors and spatial resolution of solar radiation network are concerned, Burt [225] has reported that at present the USA have 35 actinometric stations, i.e., one station covers about $10^5\ km^2$. In this connection, he disussed the problem of interpolation of the data from such a thin network. Analysis has been made of various solutions for this problem. It has been shown that, in the worst case, the errors in the extrapolation of the GR diurnal sums from the data of available network are 21%, but for seasonal averages, they decrease to 5%. This network makes possible the adequate assessment of the spatial variability only on characteristic scales of more than 2000 km. Even resolution below 1500 km is impossible in practice. The GR variability has been estimated on spatial scales below the limit of the available network.

Buriez et al. [224] analyzed the effect of various parameters which characterize the observational conditions, on the accuracy of GR retrieval from the NOAA AVHRR data for channels 1 (0.5–0.7 μm) and 2 (0.7–1.1 μm), and from the Meteosat scanner data in the visible spectrum. With this aim, theoretical studies have been carried out of the effect of various parameters (sun height, the content of optically active gaseous and aerosol components of the atmosphere, surface albedo, types of clouds) on the ratio between measured normalized OSR and atmospheric transmission. Then analysis was performed on the sensitivity of the retrieved/directly measured GR difference with the use of NEPHOS data for Western Europe and the East Atlantic (February 1982–January 1983): temperature, pressure, humidity and cloudiness from the AVHRR data and from data of hourly pyranometric observations at 77 stations in Great Britain, FRG and France. The results obtained show that the use of the AVHRR channel 1 data provides the highest accuracy of GR retrievals. Most important is consideration of variability of such input parameters as sun height, on which, radiation absorbed by the atmosphere, depends substantially. Water vapor is the dominating factor of molecular absorption. The principal reason for retrieval uncertainties is connected with the lack of reliable data on the volume and optical properties of aerosols. There is no doubt that independent measurements of aerosol optical thickness are quite urgent. Calculations have shown that the effect of the type of cloud over on relationships between normalized OSR (system's albedo) and transmission is slight (for instance, variations in the height of cloud top may affect only where there is strongly varying aerosol content and properties in the above-cloud atmosphere). The effect of surface albedo can be considered by retrieving its values using the minimum-albedo technique, but this is only possible with accurate coordinate referencing of the satellite-derived OSR data (the fact that the albedo from channel 1 is generally small, is lucky). Comparison of the hourly values of retrieved and directly measured GR showed poor agreement, largely determined by errors in geographical referencing and by spatial and temporal variability of cloud cover.

However, the errors are substantially reduced by diurnal averaging and, especially by monthly averaging, though considerable differences remain due to an reliable information about the atmosphere (water vapor and aerosol content, etc.). Therefore, systematic errors in the estimation of atmospheric transparency varied between -0.018 in March and $+0.014$ in September. Time averaging does not decrease the errors that determine the errors of GR retrieval surpassing the permissible level of 10 W/m^2.

A technique has been suggested [359] for retrieving the albedos of surface areas, about 150×200 km in size, from the Meteosat scanning radiometer data on the OSR in clear-sky conditions. The technique is based on the use of data on the ratio of OSRs for adjacent areas, considered to be identical to the ratio of albedos.

Availability of only one reference point for ground albedo observations, makes it possible to retrieve absolute albedos from in-pair OSR ratios. The condition for successful realization of the technique is surface isotropicity and the absence of substantial horizontal inhomogeneity of the optical properties of the atmosphere, which determines the transformation of surface brightness by atmospheric thickness. The surface albedo mapping has been exemplified for the region of West Africa for different days, between 18 February and 2 June 1979. Comparison with data of ground pyranometric observations has shown that maximum differences do not exceed 15%. Retrieved albedos varied between 0.18 and 0.6. The albedo of the Sahara turned out to be higher than those usually prescribed in climate numerical modeling. The albedo of the

Sahel decreased greatly in the transition from dry to wet season. It has been shown [359] that further studies are needed to substantiate the technique for the transition from retrieved albedos in a limited wavelength interval (0.4 – 1.1 μm) to total albedo.

Comparison of diurnal sums of GR directly measured at 20 stations in Texas (USA) and retrieved from satellite data for the period 1981 – 1983 has been made in [246] to estimate the errors of retrieval using the Tarpley technique and to analyze the ways in which the difference between GR values in two locations depends on the distance between them. Differences between directly measured and retrieved GRs vary from $\pm 1\ \mathrm{mJ/m^{-2}}$ day in northern Texas (about two thirds of the territory of this stage) to $\pm 2\ \mathrm{nJ/m^2\ day^{-1}}$ in the southern part of the state. The MSD is $2.5\ \mathrm{mJ/m^{-2}\ day^{-1}}$, which constitutes 12% with respect to global mean GR for the territory of Texas. Drawing the dependence curves for normalized (with respect to mean GR) MSD in GRs revealed a rapid increase, when the distance grows to about 200 km, with subsequent rapid slowing down. In accordance with the results of preceding studies, the coefficient of autocorrelation of GRs normalized with respect to possible ones, decreases exponentially with the growing distance, but much faster than it had been observed earlier. The coefficient of autocorrelation decreases more slowly if 5-day averages of GR are used.

Based on the use of GR data from seven actinometric stations in Hungary for 21 months (November 1981 – July 1983) and data of cloud amount estimated by processing space-derived cloud images, empirical relationships have been obtained [352] which express the dependence of relative GR daily mean (compared to extra-atmospheric insolation) on cloud amount x (in parts of unit) in the form: $G = a - bx - cx^2$. The values of coefficients a, b, c vary depending on the chosen region (in the Great Plains and the river Dunay). There is a linear relationship between GR and daily mean value of relative brightness Y inferred from satellite data: $G = c - dY$. Obtained empirical formulas make it possible examine in detail the GR fields constructed using ground data only. The GR values have been mapped [352] for Hungary. With some corrections, bearing in mind the dependence of GR on visibility L (atmospheric transparency conditions): $G = -8.3 + 1.9 \lg(9.25\ L)$, the errors in GR retrievals do not exceed 3 – 6%.

Using observation data from six Hungarian actinometric stations and Meteosat cloud amount data for the period November 1981 – July 1983, it has been demonstrated [368] that it is possible to approximate (by a second-order polynomial), the dependence of relative GR, G (calculated with reference to extra-atmospheric insolation) on cloud amount x.

For the region of Budapest and for Hungary, generally, $G = a + bx + cx^2$. For the Hungarian region the coefficients are: $a = 3.63 \cdot 10^{-7} N + 4.45 \exp(-0.0357 N + 56.0)$; $b = -0.13 - 0.85 \cos\{(N - 152) \times 360/365\}$; $c = -0.39 + 0.1 \cdot \cos\{(N - 152) 360/365\}$, where N is the number of days, starting from 1 January. The coefficient of correlation with directly observed values is 0.949 here.

Raschke [366] reviewed retrieval possibilities for the components of surface radiation budget from satellite measurements, bearing in mind that, permissible errors in the determination of global and monthly mean values, vary within $2-20\ W/m^2$. Now, reliable techniques have been developed to retrieve GR, which need, however, further verification by comparison with ground observations data (especially with snow or ice cover). Use of digitized images makes it possible to trace surface albedo changes, but the problem of transition from directly measured spectral albedos to total albedo remains still unsolved. The problem of retrieving the atmospheric thermal emission and effective emission is only now being solved. In all cases, ground-based control observations on specially chosen key sites are very important.

Kriebel et al. [317] analyzed the technique for retrieving surface albedos for 20-km-side squares from satellite-derived clear-sky surface-atmosphere system brightness, with a-priori knowledge of phase functions for typical surfaces which form the given site. The technique is based on comparison of calculated brightnesses of the system with measured ones to choose the value of albedo which ensures an agreement between measured and calculated quantities. Atmospheric transparency and moisture content estimated from ground network data are used as input parameters for the calculation model. Maps of land use and vegetation cover and Landsat data enable assessment of the share of the various surfaces which constitute the study site. The technique provides for an error in albedo retrieval better than ± 0.01. As stated in [317], further studies of the phase functions for various natural formations are needed in view of their importance.

An additional review has been made in [357] of possibilities for retrieving total surface albedo from space-derived spectral brightness of the surface-atmosphere system. This retrieval needs successive (stage-by-stage) solution of the problems of atmospheric correction, of the transition from brightness to radiation flux, and from spectral quantities to integral ones. It has been shown that though the principal ways of solving the problem of atmospheric correction are clear, there are many unsolved problems in practical correction. Models of reflection indicators for typical natural formations need substantiation.

Based on NOAA-6 AVHRR data (channels $0.55-0.68$; $0.73-1.1$; $10.5-11.5\ \mu m$), brightness temperatures (BT) and surface albedo of the Sinay desert have been estimated [351] for sand desert and semidesert (arid steppe). The intensity of outgoing shortwave radiation over semidesert turned out to be about 25% lower than over desert, where the albedo (from Landsat data) constituted 0.42. With both surfaces considered Lambertian, the difference in albedos was also about 25%, but if semidesert is modeled as a surface with vertical columns of vegetation, the albedo of semidesert will be below 42%. In the morning hours (about 07:00 local time), the semidesert BT was 2.5 K above the BT of desert, which points to a more intensive turbulent heat exchange between the semidesert surface and the atmosphere, controlled by surface albedo (in this case the effect of evapotranspiration is negligible).

Within the AgRISTARS programme aimed at developing techniques to assess the state of and to forecast the crop yield, based on satellite remote sensing data, an operative technique has been substantiated [291] for retrieving the hourly and diurnal sums of GR incident onto a horizontal surface R for $1 \times 1°$ squares from the OSR measurement data B with the use of 2-channel scanning radiometers carried by GOES geostationary satellites. Starting from March 1983, the following specified empirical algorithm of retrieval has been used:

$$R = R_0(z) + d(B_m^2 - B_0^2) \ ,$$

where z is the solar zenith angle; R_0 the clear-sky GR; B_m an average OSR for a 40 km-side square; B_0 the minimum clear-sky OSR. This relationship is substantiated by comparing data of 7200 synchronous groundbased (pyranometric) and satellite observations. Use of the specified algorithm has favoured a decrease of MSD of retrieved hourly (daily) sums of GR from 19 to 16% (from 12 to < 10%) and almost completely excluded systematic errors (overestimation of small and underestimation of large values of GR observed earlier). The discussed algorithm of retrieval cannot be applied near the land-water interface and cannot distinguish between clouds and snow cover, which makes it useless in winter conditions.

Beginning from October 1981, the US National Weather Service makes forecasts of GR sums for periods 24–60 hours, using the multiple linear regression technique. Various meteorological parameters serve as predictors. Testing the reliability of these forecasts for the Atlantic region for a one year period gave errors from 25% (24 h before) to 30% (60 h before), whereas the error in the inertial forecast (from data for the previous day) reached 47%.

It follows from [369] that, with satellite-derived cloud cover imagery as a source of information to estimate GR, relative brightness of clouds determined from relationship $B_1 = [(F - \min_0)/(\max_0 - \min_0)] \cdot 10$, is the most useful parameter. Here F is the reflected solar radiation; min $(\max)_0$ are solar radiation fluxes for cloudless and cloudy areas, with very high albedos (in the same region). Processing Meteosat visible imagery enables one to calculate the quantity $B_2 = [(f - f_{min})/(f_{max} - f_{min})] \cdot 10$, where f is the space-derived brightness. Based on the data for Hungary, the technique for estimating B_1 from measured B_2 has been illustrated. Retrieving the GR daily mean relative (calculated with reference to extra-atmospheric insolation) value G is possible, due to its linear relationship with B_1. From averaged ground and satellite data for the period November 1981 – May 1982, $G = 62.9 - 5.68\,B_1$ with a correlation coefficient of 0.943.

Darnell et al. [241] proposed a technique for estimating the atmospheric thermal emission from satellite remote sensing data using an approximate scheme of parameterization of the thermal emission transfer in the atmosphere. This technique proceeds from the fact that 99% of thermal emission is from below 700 hPa level. The principal factors determining this emission are vertical profiles of temperature and water vapor, and the amount and base height of clouds. Except for the latter parameter, the remaining can be found from remote sensing data. Since these data contain information also about the top height of clouds, the height of the cloud base can be calculated from prescribed climatological quantities of cloud over thickness (assumed to

Table 8.1. The sensitivity of atmospheric thermal emission to the errors in the input parameters

Parameter	Probable error in parameter	Error in thermal emission (%)
Surface and tropospheric temperature	±2.5 K	±3.0
Total water vapor content in the atmosphere	±30%	±4.6
Cloud amount	±20%	±3.1
Cloud base height:		
650 hPa	±100 hPa	±1.29
800 hPa	±100 hPa	±1.35

be 50 hPa). Table 8.1 characterizes the sensitivity of atmospheric thermal emission to the errors in the input parameters.

Comparison of retrieved values of thermal emission (for 2.5°-side squares) and data of ground sub-satellite pyranometric observations for the period 15 April – 15 May 1979 (41 orbits) revealed MSD about 20 W/m^2 (6.5%) due to errors in remote sensing and in the assessment of local meteorological parameters.

A technique for retrieving GR and cloud cover characteristics from Meteosat OSR measurements has been suggested by Möser and Raschke [337]. The technique is based on the use of results of statistical processing of numerical modeling data of the OSR field in clear sky and continuous cloud conditions. Comparison with data from 19 pyranometric stations in the FRG has shown that maximum MSD of retrieved values relative to those measured reaches 90 W/m^2 at noon (about 11% GR at clear skies). With 1-month averaging, the difference decreases to 30 W/m^2.

From GOES satellite data averaged over 50-km side squares in the visible spectrum (0.55 – 0.75 μm) for the summer of 1977, an analysis has been made [354] of the solar zenith angle dependence of the clear-sky surface-atmosphere system albedo for 33 sites in the Great Plains (USA). Absolute calibration of satellite scanning radiometer has been made through comparison with sub-satellite aircraft measurements. Analysis of results has shown that the albedo increases with growing solar zenith angle and that the 33 sites can be classified into 3 categories of surfaces corresponding to different longitudinal intervals. For each of the categories, the following parameterization of albedo A, as a function of zenith angle θ holds: $A = a + b\theta + c\theta^2$, but coefficients of regression a, b, c are different.

The surface albedo retrieval (with atmospheric correction) suggested that with solar zenith angle varying within 10°–80°, variations in albedo may reach 25% (results of observations refer mainly to agricultural fields).

Comparison was made of surface radiation budget observations at 4 Canadian actinometric stations and results of retrieving the surface-atmosphere system radiation budget, based on the GOES-E August – September 1981 data averaged over 50-km-side squares in the regions of the stations. From this, the following regressive relationship was established between the diurnal sums of surface radiation budget GNET, system radiation budget PNET, and long-wave outgoing radiation LWD [355]: GNET = 0.3636 (PNET) + 0.247 (LWD) + 23.407. A similar relationship holds for the hourly values of the radiation budget. The estimates of the coefficient of correlation between the diurnal and hourly sums of the system's radiation budget gave a maximum of 0.96, provided the hourly sums refer to 14:00, which corresponds, approximately, to the time of passage of the available polar-orbit satellites.

Pinker and Corio [356] consider the results of comparison of groundbased measurements of the surface radiation budget and satellites observations of the surface-atmosphere system radiation budget for the Great Plains (USA), starting from summer 1977. The surface radiation budget was estimated from the GR network observations, characteristic values of surface albedo, and from effective emission calculated using the Monteith formula. The NOAA scanning radiometer data (channels 0.55 – 0.75 μm and 10.5 – 12.5 μm) averaged over 50-km side squares, served as the source of information about the system's radiation budget. Upon excluding unreliable data, three regression equations were obtained:

1. $R_0 = 101.814 + 1.195\, R_s$ (R_0 is the surface radiation budget; R_s the system's radiation budget; the coefficient of correlation is 0.69);
2. $R_0 = 0.484 + 0.374\, R_s + 0.601 \cdot 10^{-4} (R_s)^2$ (the coefficient of correlation is 0.69);
3. $R_0 = 0.462 \cdot 10^2 + 0.285\, R_s + 0.862\, F_s$ (F_s is the OLR; the coefficient of multiple correlation 0.76).

Comparison of radiation budgets of the surface and the system has shown that the averaged albedo of the system is not always the best indicator of OLR (as a rule, there is a negative correlation between these two quantities). So, for instance, in the regions of New Mexico, Colorado and Wyoming (USA), situations have been observed when the albedos were practically equal but the OLRs varied substantially, being higher over deserts than over mountains. There is no doubt that GR is not an important factor of the surface radiation budget variability, as is sometimes assumed, when regression relationships between surface radiation budget and GR are used.

Halpern [274] has analyzed a technique for retrieving GR from OSR measured in the visible spectrum, with the GOES scanning radiometers, by using the OSR numerical modeling results and by selecting a GR value, which would correspond to the situation when calculated and measured OSR estimates, agree best. Numerical modeling was made using a version of the spherical harmonics technique developed by Dave and Braslaw (1975). The reliability of GR calculations has been illustrated by comparison with ground observations in Sacramento (California, USA) for three days, 7 – 9 April 1978 (in this period, the observation conditions varied from clear skies to continuous cloudiness). Since the GR calculations were made only for extreme cases of clear skies and overcast skies, estimates for the case of broken cloudiness were taken as weighted means, with the cloud amount taken into account.

In comparison with satellite data, an averaging was made over 2×2-pixel squares. Differences in the hourly mean retrieved and observed GR values did not exceed 3%, as a rule. Practical use of this technique is hindered by the following circumstances: (1) lack of reliable schemes of transition from the scanning radiometer's readings to the OSR values; (2) need for data on surface albedo; (3) the technique for GR calculations is one-dimensional, which excludes estimates for conditions of broken cloudiness.

By approximating the results of numerical modeling of the vertical profiles of the clear-sky shortwave radiation fluxes, a relationship has been obtained [233] between the albedo of the surface-atmosphere system α and the surface albedo R_g: $\alpha = a + bR_g$, where $a = 0.0587$, $b = 0.730$. The MSD of this approximation is 0.028. The equation $\alpha = R_g$ takes place at $R_g = 0.22$. Comparison of calculated albedos of the system, with clear-sky zonal-mean monthly mean albedos of the system from satellite data, gave MSD = 0.5. With the introduced zenith angle dependence of coefficients a and b the MSD of approximation decreases to 0.017 (at zenith angle 0°); 0.018 (60°), and 0.021 (85°).

Powell et al. [363] proposed a simple technique for retrieving global radiation K using empirical approximation: $K = K_0 P - Q(OPQ)^2 - R_m$, where K_0 is the possible global radiation; OPQ the amount of opaque clouds assessed visually (using a circle radius 64 km, with a grid) from satellite cloud imagery; P, Q, R are empirical coefficients calculated for each month; m the atmospheric mass. The number of GOES images processed every day averaged 5. Data processing for 8 months (1977–1978) for 20 locations in Arizona (USA) has shown that mean-square, mean-absolute and systematic errors in retrieving the GR diurnal sums constitute, respectively, 13.2, 8.6, and 1.1%.

Möser and Raschke [338] developed a technique for retrieving the GR field by processing Meteosat digitized images. The technique is based on the relationship between satellite-derived OSR and GR obtained by model calculations. The parameters of this relationship are "adjusted" by comparison with data of pyranometric observations at several stations. Comparison of retrieved monthly mean diurnal sums of GR with the data of the West European actinometric network for July 1979 (April 1982) gave MSD = 0.25 (0.28) kWhr/m^2, which is approximately equivalent to 5%–6% with respect to the clear-sky diurnal sum. The spatial resolution is 25×25 km^2. Based on satellite data, the distribution of GR sums has been mapped for West Europe and North Africa, for different time intervals. Estimation of retrieval errors has shown that, to map the weekly and monthly sums, it is enough to use four to five images for each day every 3 h, and to obtain the diurnal sums, six images are needed (with a 2-h interval).

For three models of the atmosphere (tropical, standard, and winter midlatitude), calculations have been made [332] of the spectral distribution of effective emission of the smooth water surface in the wavelength interval 4–100 μm using the matrix-operator technique. Calculations have shown that about 72% of effective emission fall onto the 8–13 μm transparency window.

Hence, the retrieval of effective emission from space-derived OLR is only possible from data for transparency windows. The empirical multiple regression technique has been developed based on the NOAA-7 AVHRR data for spectral intervals 10.3 – 11.3 and 11.5 – 12.5 μm. The processing of results of calculations of effective emission and OLR for 77 cases covering the various clear-sky meteorological conditions (for instance, relative humidity varied between 20% – 90%) has shown that for the data base considered, the coefficient of multiple correlation reaches 0.9, and the MSD of approximation is below 10 W/m^2.

Taking the instrumental noise into account does not affect the error estimates in practice, and with averaging over 4 pixels (this is equivalent to the 2×2-km^2 resolution), it is completely preserved. An example of processing the 23 August 1981 realistic data for the water basin north of the Azores islands (it covers 512×512 pixels, averaging is made over 16 pixels) has demonstrated that, in cloudless regions, the retrieved effective emission field is quite realistic. From Meteosat-2 analogue images in the visible spectrum for the period November 1981 – July 1983 in Hungary, Paál [353] obtained empirical regression relationships which express the dependence of diurnal mean net radiation G, determined with respect to extra-atmospheric insolation (results of observations at six actinomeric stations were used), on two characteristics of cloud amount: (1) cloud amount, X (10-point scale), assessed by manual processing of images; (2) relative brightness, Y, which is a ratio between the difference in the brightness of the cloudy and cloudless areas of image and a similar difference for two fixed sites.

In the first case: $G = A + BX + CX^2$, where coefficients A, B, C depend on N – the number of days starting from 1 January; in the second case: $G = D + EY$, where D and E are also functions of N. Estimation of empirical coefficients from data for individual regions and for the whole of Hungary has led to greatly different values. In some cases, there is a considerable annual change in these coefficients. The retrieval accuracy for GR can be raised by introducing a correction for meteorological visibility L (km), which characterizes changes in atmospheric transparency: $G = -8.3 + 1.9 \lg(9.24L)$. Testing the reliability of regression relationships obtained from independent data gave satisfactory results.

A study has been undertaken in [367] (1) to assess reliability of three techniques for retrieving GR at the surface level from OSR measured with scanning radiometers carried by geostationary meteorological satellites; (2) to analyze the possibility for using these techniques to retrieve the mesoscale spatial distribution of GR in the regions of Greater Vancouver (Canada) and of the Fraser river estuary.

Two of the techniques considered (by Hay and Hanson 1978, technique HH; by Tarpley 1979, T) are empirical-statistical and based on the use of regression relationships obtained from ground-based and space-based observations. Technique HH is based on empirical relationship between atmospheric transparency T_r (estimated with due regard to cloudiness) and measured (normalized) OSR (SR): $T_s = a - b$ SR, where a, b are coefficients of regression.

In this case GR can be calculated as $I_s = I_0 \cos\theta\, T_r$ (I_0 is the extra-atmospheric direct solar radiation, θ the solar zenith angle). In technique T based on distinguishing between three types of observation conditions (clear sky – cloud amount $n < 0.4$; $0.4 \leqslant n < 1.0$; $n = 1.0$), for clear skies the OSR (minimum OSR) is parameterized as a function of solar zenith angle and the azimuth between the direction of the sun and viewing direction. The OSR in conditions of cloudy skies is determined by retrieved (using the two-threshold technique) cloud amount. The final result is presented in the form of three regression relationships for I_s as a function of n and measured OSR (including normalized minimum OSR). The third technique proposed by Gautier, Diak and Masse (GDM) implies an approximate parameterization of the dependence of GR on the main factors which characterize the optical properties of the atmosphere surface albedo and cloud amount. Distinguishing between clear and cloudy sky conditions using the threshold technique is very important for successive realization of the GDM technique. Parameters of all regression relationships have been obtained from observations at 12 actinometric stations in the region mentioned, $45 \times 70\ km^2$ in size (the distance between the stations varies from 4 to 74 km). The GOES-2 (135 °W) data for shortwave channel 0.55 – 0.75 μm averaged over 5×5-pixel sites, served as a source of information on the OSR. A very important stage of initial data processing was their precise geographical referencing with errors not more than ±1.25 km in longitude and ±1.71 km in latitude. The spatial resolution of space-derived OSR data in longitude (latitude) constituted 0.82 (1.67) km, which determined the size of pixel i.e., $1.37\ km^2$. Analysis of the results obtained (an estimation of errors in retrieved GR values with the use of independent data of ground and satellite observations) has shown that the GDM technique provides the best results in retrieving the GR hourly sums.

Use of the HH technique revealed a systematic error, which can be avoided with newly found values of regression coefficients. This situation shows a limited applicability of the empirical-regression approach manifested through spatial and temporal variability of regression coefficients. The T technique can be applied in clear sky conditions but it is less reliable in conditions of cloudiness. An examination of regression coefficients provided a decrease in the errors of retrieval. On the whole, the GDM technique may be considered preferable with partial or continuous cloudiness, whereas the T technique is more reliable in clear skies. Table 8.2 data illustrate the errors in retrievals using the GDM technique (observations for 9 days). Naturally, the errors decrease greatly when the period of averaging grows.

An increase in errors with clouds present is interesting. A consideration of mesoscale GR fields retrieved from the data of 9 days has led to the conclusion that these data are reliable and more informative than the results of fragmentary ground based actinometric observations. Maximum possible spatial resolution of retrieved GR fields depends on special features of the retrieval technique and of observational data. Since for the GDM technique the MSD of retrieved hourly (daily) sums constitutes ±17.4% (±8.6%), resolution better than 15 (12) km cannot be reached. Use of a longer observation series

Table 8.2. Errors of retrieval of the hourly (in brackets) and diurnal sums of global radiation in conditions of clear skies, partial and continuous cloudiness, with the use of the Gautier-Diak-Messe technique. The hourly (daily) sums are expressed in $kJ/m^2 h^{-1}$ ($MJ m^{-2} day^{-1}$)

Observation conditions	MSD	Correlation coefficient	Averages	
			Observed	Calculated
Clear skies	4.3 (7.1)	0.999 (0.993)	19.5 (1.693.7)	20.0 (1738.4)
Partial cloudiness	11.95 (37.0)	0.751 (0.919)	8.9 (714.6)	8.7 (693.0)
Continuous cloudiness	26.4 (41.5)	0.743 (0.836)	4.9 (399.4)	5.4 (446.3)
All cases	8.6 (17.4)	0.995 (0.982)	12.3 (1027.1)	12.6 (1057.1)

(19 days) with more cloudy days caused a decrease of resolution. The HH and T techniques provide smaller resolutions. To increase the spatial resolution, the technique for retrieving must be improved. As far as the GDM technique is concerned, the prospects for its improving are connected with three possibilities: (1) more reliable identification of cloud conditions; (2) specified consideration of the shortwave radiation absorption by clouds; (3) account of aerosol.

Based on the use of results of calculations of downward longwave radiation (DLWR) with clear skies for 180 climatological models of the atmosphere, a parameterization has been proposed [271] for the DLWR depending on the parameters determined from meteorological satellites. Since numerical modeling has shown that about 99% DLWR are formed by a tropospheric layer located below 700 hPa, the following parameters have been chosen as the principal ones which determine the DLWR: the moisture content U_1 of the surface – 700 hPa layer and mean temperature T of this layer estimated from retrieved T_0 values (surface temperature); T_1 (mean temperature of the surface – 850 hPa layer); and T_2 (850–700 hPa). In this case the DLWR:

$$F(W/m^2) = (A_0 + A_1 V + A_2 V^2 + A_3 V^3) T^{3,7} ,$$

where $V = \ln U_1$, and coefficients of regression $A_0 = 1.91 \times 10^{-7}$; $A_1 = 1.59 \times 10^{-8}$; $A_2 = 2.13 \times 10^{-10}$ and $A_3 = 9.63 \times 10^{-10}$ with the coefficient of correlation 0.99 and dispersion about 1.1%. The obtained relationship was tested from independent satellite remote sensing data for one month, averaged over a 5°-sided square (near Ann Arbor, Michigan, USA), by comparing it with the results of precise calculations of the DLWR. The coefficient of correlation was 0.94, dispersion 12 W/m^2 (about 4% with respect to the DLWR). A decrease of the accuracy of parameterization is explained here by the effect of near-surface inversions and super-adiabatic vertical temperature lapse rates in the daytime. In the absence of such anomalies of temperature stratification, the reliability of the parameterization under discussion is always sufficiently high. Maschoff et al. [324] specified the technique for processing the data of Nimbus-7 wide-angle sensors (WAS) which measure the fluxes of total (wavelength region 0.2 to 50 nm) radiation and OSR (0.2–3.8; 0.695–2.8 μm).

Specifications resulting from laboratory studies of similar sensors and from analysis of realistic data, forecast examination of the inhomogeneity of the Earth's radiation field within the WAS viewing angle, selectivity of the sensors and variations in the temperature of the sensors, contribution of direct solar radiation, radiative heat exchange of receiving surfaces and hemispherical filters with outer space.

Preliminary results of the development of the technique for determining the atmosphere-absorbed shortwave radiation (AASR) from combined data of ground based and space based observations have been discussed by Justus [290]. An application of this technique is illustrated by processing the data from GOES-E satellite and eleven ground stations in the USA for August–December 1980. Ground observations include information about GR hourly values, total cloud amount and opaque cloud amount. The data of satellite measurements of the brightness of the surface-atmosphere system were averaged over 5×5-element sites, in the centers of which there were ground based observational stations. The required AASR value is determined from the relationship: $A = (1 - r_p) - T(1 - r_g)$, where $r_p = F_r/F_i$ (F_r is the OSR; F_i the extra-atmospheric insolation); $T = F_t/F_i$ (F_t the global radiation); r_g the surface albedo. If the observational data are divided for conditions of clear and cloudy skies, the T and AASR values can be found separately for these two observational conditions. The transition from measured spectral intensity to the total OSR flux is based on the use of numerical modeling results for different atmospheric models.

As shown in [264], the technique for retrieving GR from the surface-atmosphere system's brightness measured in the visible with GOES scanning radiometers applied so far, is based on averaging of retrieved values over 8×7-element sites (spatial resolution about 1.7 km). Since the processing of large data bases from geostationary satellites is very cumbersome and laborious, possibilities have been analyzed for its simplification by larger spatial averaging. Comparison of the results of retrieving obtained on various scales of averaging (up to 64×63 pixels) with data of ground-based pyranometric observations has shown that in retrieving the GR hourly means an averaging up to 64×64 pixels may be done, which corresponds to a grid step of 100 km.

Summing up the review of the western literature on the retrieval of the radiative characteristics of the terrestrial surface, there is a need for the respective information base of satellite observations to specify the models of conversion of the data to surface level and the transformation functions of these data, based on theoretical and empirical studies.

8.2 Empirical Analysis of the Earth's Radiation Budget Components

The previous sections have dealt with techniques for obtaining processing and analysis of air- and space-derived information obtained with narrow angle radiation sensors. Problems of atmospheric and oceanic circulation need results of analysis of radiation fluxes either recorded by wide angle sensors or recalculated from results of measurements of the angular distribution of radiance [123]. Here, an analysis of relationships and redistributions of solar shortwave radiation fluxes and longwave thermal emission is of particular importance. These components of the surface-atmosphere system radiation budget determine possible anomalies of temperature, cloudiness and other characteristics of the general circulation of the atmosphere and ocean.

Principal equations to calculate radiation fluxes in the atmosphere and at the surface have already been considered. This section will be aimed at retrieving the surface radiation budget components from satellite measurements and at statistical analysis of initial and retrieved fields of radiation budget.

A special feature of the analysis of radiation fields here is the global scale of these fields represented through flux values. This differentiates the results of this section from those preceding, since angular coordinates are not included here into the quantities to be analyzed. They were, however, important when obtained, so they were excluded from averaging. Therefore, the outgoing radiation field anisotropy, which is rather substantial in the problems of processing and interpreting "instant" measurements, has no value here.

Use of satellite information in climate studies largely depends on the availability of the data of observations from ships in the zones where the ocean exerts maximum effect on short-term climate changes – energy-active zones of the ocean (EAZO). The EAZO concept verified by numerical modeling results has been developed in the programme "Sections" [125], aimed at studies of the role of the ocean in short-term climate changes [106].

Satellite-derived data on radiation budget can promote the solution of the problem considered, since they contain information about the sources and sinks of radiation energy which governs the general circulation of the atmosphere and the ocean. However, possible manifestations of the processes in EAZOs resulting from statistical analysis of satellite data on the radiation budget, are still poorly understood. The results of this chapter are aimed at solution of this problem.

8.3 The Variability of the ERB Components from Satellite Measurements

Analysis of the available ERB satellite data has shown that distributions of radiation budget dispersion – an information measure of both the annual and interannual variability – are of much greater interest than distributions of average fields [106]. The dispersion of all the satellite data, represented as monthly mean, is written for individual radiation budget components as:

$$\sigma_R = \frac{1}{12} \sum_{m=1}^{12} \left[\frac{1}{J} \sum_{i=1}^{J} (R_{i,m} - \tilde{\bar{R}}_m)^2 \right]^{1/2} ,$$

where $\tilde{\bar{R}}_m = \frac{1}{J} \sum_{i=1}^{J} R_{i,m}$ is monthly means over J years, in contrast to $\tilde{\bar{R}} = \frac{1}{12} \sum_{m=1}^{12} \frac{1}{J} \sum_{i=1}^{J} R_{i,m}$ – averages over the entire set of measurements. The MSD σ_R, along with the normalized MSD $\tilde{\sigma}_R = \frac{1}{R} \sigma_R$, is a measure of interannual variability of individual radiation budget components.

In analysis of the intraannual variability, the time series of monthly means are used most often. In this case the amplitudes of monthly means are usually calculated in Fourier series:

$$S_{\tilde{p}} = \frac{1}{k} \sum_{t=0}^{k-1} \tilde{\bar{R}}_t \exp\left(-i \frac{2\pi}{k} \tilde{p} t \right) ,$$

where $k = 12$; $\tilde{p} = 1, 2, \ldots, 6$; $\tilde{p} = 1$ characterizes the amplitude of the annual change, $\tilde{p} = 2$ the amplitude of the semi-annual change, etc. Nevertheless, of particular interest is a sampled MSD

$$\sigma^2 = \frac{1}{k} \sum_{t=0}^{k-1} \tilde{\bar{R}}_t^2$$

and a periodogram (the percentage of the sampled MSD falling on a frequency interval with an index $\tilde{p}$):

$$h_{\tilde{p}} = \frac{k \Delta t}{2\pi} |S_{\tilde{p}}|^2$$

where $\pi/\Delta t$ is the Nyquist frequency, the step Δt is determined from the theorem of information counting. With the $h_{\tilde{p}}$ periodogramme known, one can, by smoothing, estimate spectral density, and by summing up the Fourier harmonics – the spectral function.

If one considers the monthly means of realization of a stationary random process, the periodogram is expressed in terms of a sampled covariance function m_τ of the Fourier time series:

$$h_{\tilde{p}} = \frac{\Delta t}{2\pi} \left(m_o + 2 \sum_{\tau=1}^{k-1} m_\tau \cos \omega_{\tilde{p}} \tau \right) = \frac{\Delta t}{2\pi} \sum_{\tau=-(k-1)}^{k-1} m_\tau e^{-i\omega_{\tilde{p}}\tau} ;$$

$$\omega_{\tilde{p}} = \frac{2\pi}{k} \tilde{p} \; ; \quad \tilde{p} = 0, 1, \ldots, n \text{ [entier } (\tilde{p}/2)] ,$$

$$m_\tau = \begin{cases} \dfrac{1}{k} \sum\limits_{j=0}^{k-1-|\tau|} \tilde{\bar{R}}_j \tilde{\bar{R}}_{j+|\tau|} & \tau = 0, \pm 1, \ldots, \pm (k-1) \; ; \\ 0 & |\tau| \geqslant k \; . \end{cases}$$

The dependence of the periodogram on the sampled covariance function, is rather important in the theory of estimation of spectral densities, as applied to analysis of two-dimensional stationary fields of radiation budget, will be considered below (see Sect. 8.6).

Analysis of the most adequate, at present, monthly mean satellite measurements [125] covering the period 1964–1987, has shown the existence of vast areas of substantial variability of individual components of the ERB. These are areas near the east coast of the Asian continent, the zone of the Indian monsoon, a region west of South America, the Sahara desert, and some areas of the Atlantic Ocean. However, it is very difficult to select from these data the EAZOs of the northern Atlantic Ocean, due to their inadequacy as compared to the total area of the region analyzed and, probably, because of substantial inhomogeneity of the information used in analysis: data of the first meteorological satellites, data of some US operational satellites, and Nimbus-satellite data.

The interannual variation of the ERB components was analyzed using results of NOAA satellite measurements between June 1974 and March 1978 in the form of monthly mean maps [123]. This information has already been used by several authors to analyze the interannual variability of radiation budget, to study the cloud-radiation feedback mechanism, the monsoon circulation, and to solve some other problems. Results from all these studies have been systematized [106]. The MSD of the 45-month data base with respect to monthly means within a $2.5° \times 2.5°$ grid step were considered as a measure of variability.

Sun synchronous orbits have made it possible to take measurements in the same territory of the globe twice a day: 09:00 and 21:00 local time. Further processing of satellite data consisted of presenting the results of scanner measurements as monthly mean, outgoing short- and longwave radiation fluxes and in the subsequent recording of information in a certain format for the prescribed cartographic projection. Eventually, for analysis of the interannual variability the data were used on a monthly mean radiation budget, its short- and longwave components and incoming solar radiation within a step of the $2.5° \times 2.5°$ global grid.

The results of calculations of the MSD of radiation budget and its components with respect to their monthly means, are shown in Figs. 8.1–8.3.

Examination of the Fig. 8.1 data reveals the following areas of substantial interannual variability of the ERB MSD: the Sahara desert (35 W/m^2), the Kuroshio current (28 W/m^2), the Gulf Stream region (20 W/m^2), the Peru upwelling basin, near the Horn Point and in the Guinea Bay (16 W/m^2), the region east of Newfoundland, the tropical Atlantic Ocean near South America, the deserts of Middle Asia and South Africa, individual areas of the Pacific ocean (15 W/m^2), the Norwegian Sea (13 W/m^2), etc.

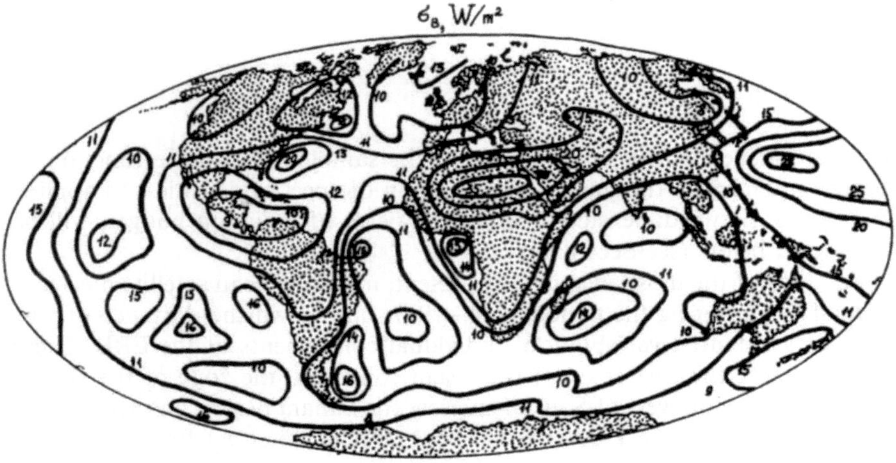

Fig. 8.1. The interannual mean-square deviations of radiation budget

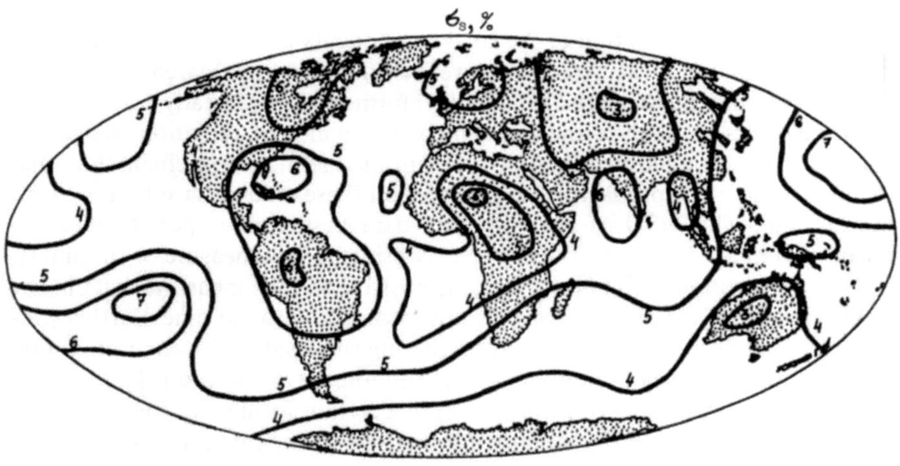

Fig. 8.2. The interannual mean-square deviations of the shortwave component

The MSD in the remaining parts of the globe average 9 – 12 W/m^2. Note, that the enumerated potential energy-active zones, interesting from the viewpoint of possible manifestations of climatic signals, are not always zones of substantial variability of heat exchange between surface and the atmosphere. They can be connected with anomalies of cloudiness, with large values of the energy and mass exchange in the atmosphere, with dynamic atmospheric effects, and other factors.

The albedo component of the ERB (Fig. 8.2) has other features: cloudiness can be considered the principal reason of its variability. It is seen from Fig. 8.2

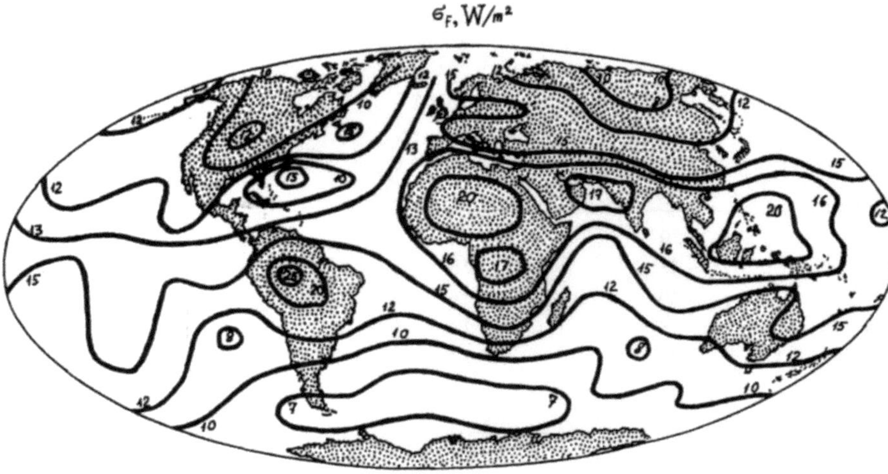

Fig. 8.3. The interannual mean-square deviations of the longwave component

that this component varies most in the near-equatorial regions of the Pacific, Indian and, to some extent, Atlantic Oceans. There are two regions of maximum variability centered at 10°N, 120°W, and 20°N, 160°E, where it reaches 7%; the Sahara desert, Central Brazil, Indian monsoon, Northern Europa and Northern Canada, where the variability constitutes 6%, and other regions. Minimum variabilities about 3% correspond to Central regions of Asia and Australia. A relative zonal change of this parameter in the Southern Hemisphere, should be noted at latitudes above 50°S.

The interannual variability isolines for the ERB longwave component (Fig. 8.3) correspond, to some extent, to the isolines of the ERB variability (Fig. 8.1): the variability of longwave component is close to maximum (15 – 20 W/m^2) in the Sahara desert, in the regions of the eastern coastline of the Asia continent and North America. However, there are substantial differences in the isolines in Figs. 8.1 and 8.3: for instance, maximum variability of the ERB longwave component is observed in the continental regions of Brazil (up to 25 W/m^2), that does not show in the data in Fig. 8.1; the region of the Indian monsoon is characterized by strongly varying longwave component, and in the Southern Hemisphere for latitudes higher than 50°S the pattern of isolines in Fig. 8.3 is, perhaps, close to the data in Fig. 8.2. Note, that the MSD of the longwave component equal to 7 W/m^2 (minimum variability) correspond to minimum variations in radiative (efficient) temperature of the system (RTS) about 1.5 – 2 K, and the respective MSD 25 W/m^2 (maximum variability) to RTS variations constituting 6 – 7 K (Table 8.3). In this case, practically, all the interannual variability of the surface-atmosphere system can be explained by the contribution of the thermal regime variability to the ERB variability. The latter can be connected, however, not with the system's temperature variations but, for instance, with the substantial effect of various cloud

Table 8.3. The ERB longwave component ($F = \sigma_{SB}T^4$) as a function of the surface-atmosphere system temperature (t^0_{s-a}) and absolute temperature (T), variations $\Delta T = T\bar{T}$ and ΔF and their MSD $\sigma_T = \sqrt{\overline{T^2} - \bar{T}^2}$ and $\sigma_F = \sqrt{\overline{F^2} - \bar{F}^2}$ in the realistic range of changes of t^0_{s-a} from $-30\,°C$ to $-10\,°C$

T, K	t^0_{s-a}, C	F, W/m^2
243	−30	199
248	−25	216
253	−20	234
258	−15	253
263	−10	273

ΔT, K	σ_T, K	ΔF, W/m^2	σ_F, W/m^2
1	0.8	3.5 − 4	2.9 − 3.7
2	1.6	7 − 8	5.5 − 6.9
3	2.4	10 − 12	9.0 − 10.5
5	4.1	17 − 20	14.2 − 16.0
7	5.7	24 − 28	20.5 − 23.1
10	8.2	35 − 39	29.2 − 34.4

formations which lead to the prevailing variability of the shortwave component over the longwave one.

Thus, it follows from the Fig. 8.3 data that the satellite data under discussion have made it possible to distinguish between the regions of interannual variability with the amplitude of RTS contrasts 4 − 5 K. In this connection, the prospects can be evaluated for available satellite systems assessing climatic signals in the atmosphere, since the errors in the "instant" measurements of such systems reach 1.5 − 2 K, and by presenting the measured fields monthly means, these errors can be reduced.

8.4 The Retrieval of the Radiation and Heat Budgets of the Earth's Surface

The available techniques for retrieval of the surface radiation budget are of empirical character, as a rule [106]. In this connection, mention must be made of numerous regressive relationships, which relate the radiation budget to its components at the lower and upper atmospheric levels. Several effective techniques have now been developed to solve the boundary problems of radiation transfer in the Earth-atmosphere system, and of techniques for approximation of the angular and spectral distributions of radiation using orthogonal polynomials. The posing of the problem of reducing monthly means to surface level differs somewhat from that of atmospheric correction of "instant" satellite data. In the first case, for most of the global regions (except for vast deserts, for instance) it is necessary to know the statistical characteristics of cloud fields, but information about atmospheric aerosols, although necessary in the problem of atmospheric correction, plays a secondary role here. The angular anisotropy of the outgoing emission field is not very important in the use of monthly mean satellite data. It is vital, however, for primary processing of data for individual "orbits" of satellite information, and in calculations of the fields of shortwave $\tilde{I}(z, \mu)$ and longwave $\tilde{I}(z, \mu)$ radiation (z is the vertical coordinate, μ the cosine of zenith angle).

The Nimbus-6 HIRS and SCAM microwave data enable retrieval of the following parameters of summertime cloudiness [125]: the top height of the upper and lower clouds; the cloud layer thickness; surface emissivity; and water content of clouds. The results of retrieval corresponded, basically, to the data on cloudiness obtained from near-ground and radar measurements.

Information on the characteristics of cloudiness and cloud cover is needed, in at least three scientific studies: in numerical models of weather forecast (clouds as a prognostic factor); in remote sensing of the environment (clouds as an "interfering" factor); in climatic studies (clouds as a factor of atmospheric energetics). Each of these three groups of studies requires information with different levels of spatial and temporal resolution. In view of different requirements of cloud in information and difficulties in adjusting these requirements to the three study groups, the following recommendations have been made: data on radiation (and not only on cloudiness) should be archived directly from satellite observations, without their averaging; archives are needed with the maximum possible, spatial and temporal resolution; detailed comparative estimates of cloudiness are needed from ground based and satellite measurements.

The most comprehensive archive of data on cloud amount has been obtained from a 3-D nephanalysis of clouds based on the US aircraft measurements [291]. Comparison of these data with monthly mean albedos of the surface-atmosphere system from NOAA scanning radiometer within a $2.5° \times 2.5°$ grid step has shown potential possibilities for using these satellite data to assess the cloud amount over the oceans. The results are very important from the viewpoint of using satellite data in climate studies.

Along with the enumerated data archives on cloudiness, other data have been used: both from ground network [290], and from satellite measurements [338]. However, information from geostationary satellites is most promising. In particular, at four local times the Meteosat visible and IR data were recorded on magnetic tapes [338] upon primary processing for cloud cover analysis. A specialized system of processing and analysis in operation for one year has provided the users with the data needed in models of atmospheric dynamics and in climate studies. The distributed data have been analyzed within individual segments 32×32 IR resolving elements, with constructing 2-D (visible and IR channels) histograms and subsequent grouping of elements into clusters. Further analysis, with results are distributed among users, consists of interactive control of the quality of information, its presentation as false-color images and in localization of the zones of cloudiness based on cluster-analysis.

From the representation of the solution for boundary problems of radiation transfer as:

$$\begin{Bmatrix} I(z,\mu) \\ \tilde{I}(z,\mu) \end{Bmatrix} = \frac{1}{4\pi} \sum_{n=0}^{2N+1} (2n+1) \begin{Bmatrix} \psi_n(z) \\ \tilde{\psi}_n(z) \end{Bmatrix} P_n(\mu) \ ,$$

where $(\psi_n, \tilde{\psi}_n)$ are vectors of the moments of angular distributions of intensity, P_n Legendre polynomial, the possibility follows to retrieve the surface radiation budget B_g from the first moments of distributions [123]:

$$B_g = \psi_1(z=0) + \tilde{\psi}_1(z=0) \ .$$

Values $\psi_1, \tilde{\psi}_1$ are tabulated beforehand for different atmospheric models taking into account radiative characteristics of the climate [150]. The problem of the choice of a particular model of retrieval is not great for a monthly means' analysis, due to double smoothing of the functions used in analysis: from individual realizations during one month measurements and from individual models, since statistical information is available on the state of the atmosphere but not except parameters.

For illustration purposes, one can draw a parallel with smoothing from NOAA data for 1974–1978 [106], for the radiation budget longwave component, whose detailed analysis has been given above.

These data were twice averaged during each day of measurements (as has been mentioned, the time of crossing the equator for these data is 09:30 and 21:30), and over 99 models of the transition mission function (in the transition from intensities to radiation fluxes). It appears that despite large errors of "instant" measurements and substantial variations in the state of the atmosphere during survey sessions, the linear regression of intensity measurements and calculated radiation fluxes explains 98% of total variability [106], i.e., the systematic error in the representation of the ERB longwave component as mean monthly maps constitutes only 2%, though the measurement error and the uncertainties connected with inadequate knowledge of the actual state of the atmosphere and the transmission function are much higher.

Another systematic error of initial data for analysis of the ERB interannual variability is the disappearance of sensitivity in the shortwave region during 45 months of measurements: it constitutes about 1% per year (see Winston J.S. et al. 1978). Note, that, in contrast to the longwave component, results of measurements of the albedo component of the radiation budget for the spectral interval 0.5–0.7 μm can be transformed into total radiation flux for the region 0.2–4.0 μm.

In this connection, it should be mentioned that, based on analysis of available and planned measurement instruments within the programme of the US 3-satellite ERB experiment (ERBE) and other experiments, the following problems have been considered [106]: (1) combined normalization and calibration of data from different satellites; (2) simultaneous measurements of downward and upward radiation fluxes from several satellites and with different instruments. It has been shown that a successful realization of such combined experiments provides much more useful information for climatic studies than a considerable increase of absolute accuracy of measurements.

Based on analysis of measurement errors, the values of mathematical expectations of the error field relative to the spatial resolution of the measuring instruments, have been obtained upon filtration. The results make it possible to assess variations in the measurement errors with varying spatial and angular resolution of the instruments, which is needed in retrievals of the Earth's radiation budget [358].

In conclusion of the discussion of the errors in measurements and retrievals of the ERB components, Table 8.4 [124] gives a classification of cloud formations from Meteosat data, as well as the estimates of rms errors (two right-hand columns) for the shortwave and longwave radiation (SWR, LWR). Knowledge of the cloud top height, its temperature and cloud amount are necessary but insufficient conditions for possible retrieval of the surface radiation budget.

Data of the two right-hand colums of Table 8.4, characterizing the maximum effect of the uncertainty of the vertical extent of clouds, make it possible to assess the role of results of analysis of the surface radiation budget variability with due regard to retrieval errors.

Note, that a maximum error of 10.3% for the budget (Table 8.4) still preserves the importance of the retrieved values of the surface radiation budget.

Let us see what can be taken from the results of the ERB retrieval to analyze the ocean heat fluxes. It should be borne in mind that we speak about retrieving an "instant" picture of the atmosphere, ocean heat exchange from averaged fields of satellite measurements, presented, for instance, as monthly mean maps of radiation budget. Thus, we are interested only in the ocean surface radiation budget which is made equal to the sum of heat fluxes (turbulent heat, evaporation and fluxes to the ocean), neglecting possible non-additive manifestations of these processes during the period over which the satellite data are averaged. Since sensible and latent heat fluxes can be assessed from ground (ship) measurements, an addition of these measurements to retrieval results makes it possible to estimate heat fluxes to the soil (ocean). Thus, the balance of heat fluxes at the interface is:

$$B = -(E^{T} + L^{ev} E^{q} + \Pi) \ ,$$

where the eddy heat flux in the atmosphere:

$$E^{T} = -C_{p} \varrho v^{v} \frac{\partial T}{\partial z} \ ,$$

latent heat flux (evaporation):

$$E^{q} = -\tilde{k} \varrho v^{v} \frac{\partial q}{\partial z}$$

Table 8.4. The classification of cloud formations from meteosat data

Type of clouds	IR radiation thresholds for different models (W/m^2)						MSD of regression (%)	
	Top height (km)	Albedo threshold (%)	Mid.-lat. summer	Mid-lat. spring, fall	Mid-lat. winter	The tropics	Shortwave radiation	Longwave radiation
Low	<2	>50	>64	>55	>47	>67	2.1	1.5
Middle	2–6	>50	38–64	35–55	37–47	40–67	3.8	3.0
Deep convective and frontal	>6	>50	<38	<35	<31	<40	6.1	4.2
Cirrus	>7	$\leqslant 50$	<38	<35	<31	<50	2.5	1.9

($\tilde{k}$ is the coefficient that takes into account the soil moisture content; for the ocean $\tilde{k} = 1$; $\partial q/\partial z$ is the humidity gradient),

$$\Pi = \begin{cases} \lambda_g \left(\dfrac{\partial T}{\partial z}\right)_g & \text{for the soil} \\ -\alpha^v C_p^{H_2O} \varrho_0 v_0^v \left(\dfrac{\partial T}{\partial z}\right)\Bigg|_0 & \text{for the ocean} \end{cases}$$

It is seen that when the fluxes of turbulent and latent heat in the atmosphere are known, the flux Π, which is difficult to measure directly, can be estimated from retrieved $B(T_g)$.

Note also, that we consider the retrieval of radiation budget $B(T_g)$ but not the retrieval of explicit temperature T_g. However, it is possible, along with fluxes E^T and $L^{ev}E_q$, to estimate T_g from on-ship measurements. These measurements, together with satellite measurements of radiation budget and results of surface budget retrieval, can be used either to estimate the temperature of the lower boundary of the upper mixed turbulent layer of the ocean or, on the contrary, to estimate the coefficient of vertical eddy mixing v_0^v, if the temperature of the lower boundary is known (for instance, from acoustic sounding data). In this case, the possibility appears for analysis of the temperature fields below the ocean-atmosphere interface by using satellite measurement data, results of the surface radiation budget retrieval and ship measurements of heat fluxes in the atmosphere, temperature and probably, acoustic sounding data.

Thus, data of satellite measurements in the optical range represented as radiation fluxes, can be very important for analysis of the oceanic thermal regime with the use of accompanying measurements from ships.

8.5 The Interannual Variability of Radiation Budget at the Upper and Lower Boundaries of the Atmosphere

Taking analysis of the initial fields of radiation budget and retrieved fields as an example, and based on the results and considerations mentioned, we consider closely the behavior of the isolines of averages and MSD of these fields. This analysis of the anomalies of the radiation budget components makes it possible to trace the formation of the radiative regime of an EAZO. We analyze the averages and MSDs as measures of the interannual variability for a territory confined to (10°S–70°N, 0°–80° W) with a 2.5° step, both in longitude and in latitude. The choice of this territory with the prescribed grid step is determined by the use of the Fourier 2-D 32×32-element transformation in statistical analysis of the spatial impact of the energy zones of the Northern Atlantic Ocean, shown in Figs. 8.1 and 8.3.

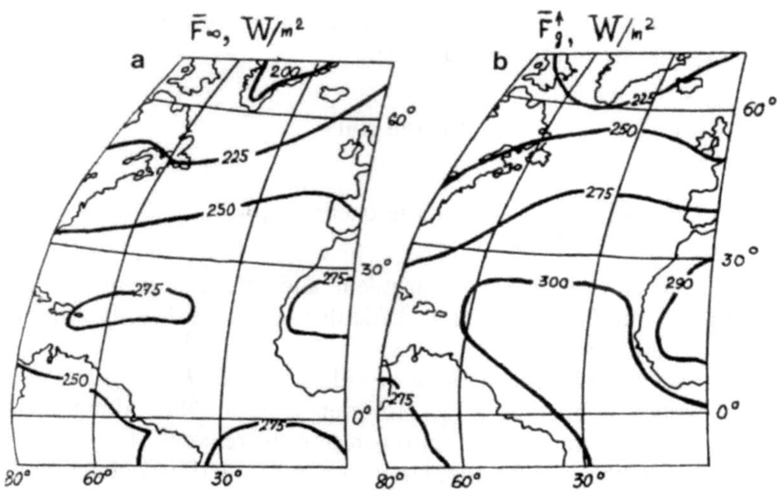

Fig. 8.4a, b. The longwave component averages (see text for details)

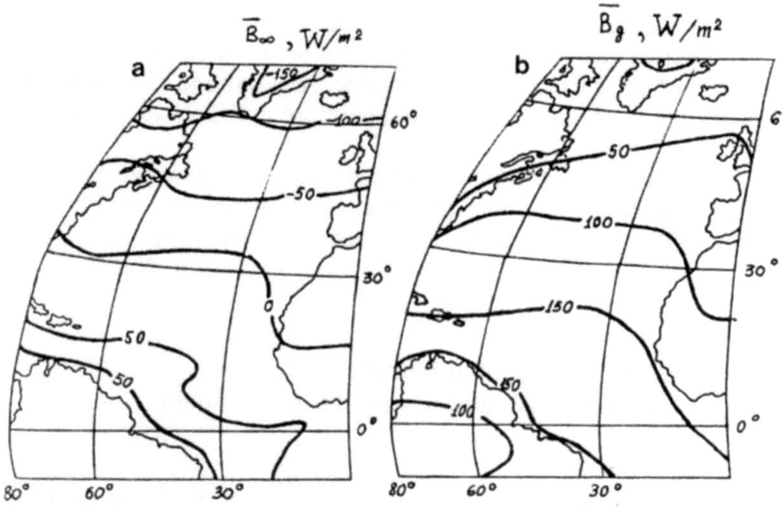

Fig. 8.5a, b. The radiation budget averages (see text for discussion)

Mean values of the initial data and those reduced to the surface level, on albedo, outgoing longwave radiation, radiation budget and system-absorbed solar radiation are given in Figs. 8.4 – 8.6 (a and b, respectively). The respective MSDs for these ERB components are given in Figs. 8.7 – 8.9.

Figure 8.6a characterizes the land-sea albedo contrasts, and Fig. 8.6b shows a sufficiently homogeneous sea surface albedo north and south of 30°N. Figure 8.4a illustrates a certain zonality of the system's own emission for

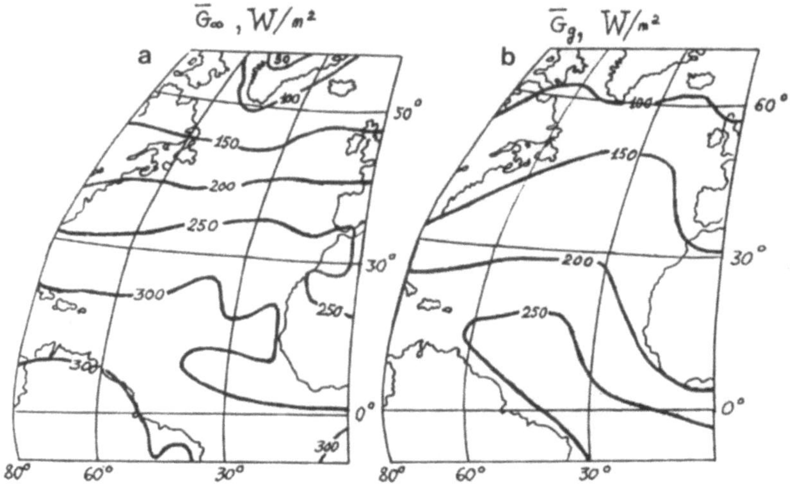

Fig. 8.6 a, b. The absorbed solar radiation averages (see text)

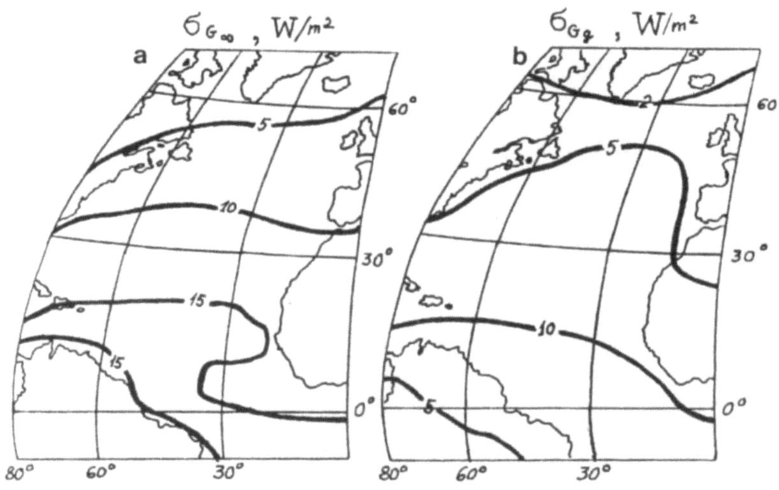

Fig. 8.7 a, b. Mean-square deviations of the absorbed solar radiation (see text)

latitudes higher than 30°N that is broken at lower latitudes. The pattern of isolines in Fig. 8.4 b corresponds, basically, to that in Fig. 8.4 a. The isolines in Fig. 8.5 a, b have a slightly different character, they point to a broken zonality of the ERB change at latitudes below 40°N, connected with the effect of land-ocean contrasts. For latitudes above 30°N, in contrast to satellite-derived negative ERB values (Fig. 8.5 a), the surface radiation budget (Fig. 8.5 b) is positive everywhere, which is connected with the contribution of the atmospheric downward longwave radiation.

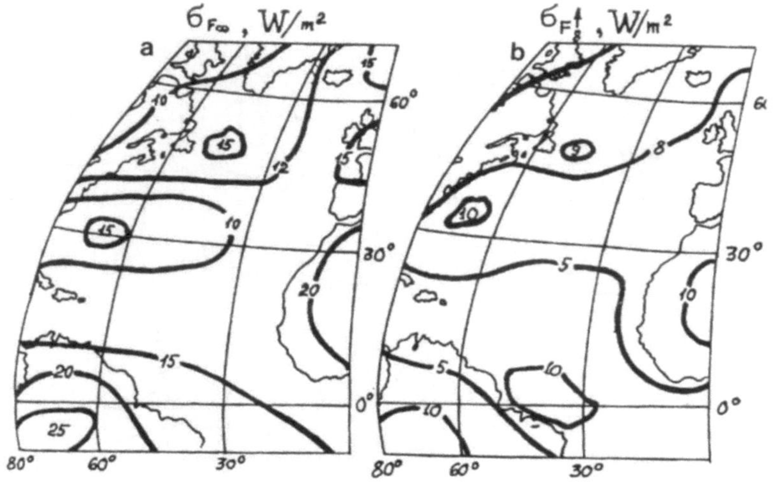

Fig. 8.8 a, b. Mean-square deviations of the longwave component (see text)

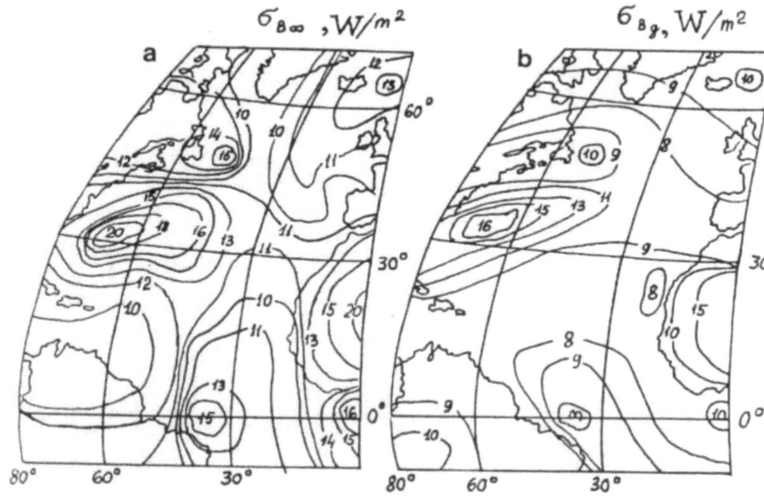

Fig. 8.9 a, b. Mean-square deviations of radiation budget (see text)

The data of Figs. 8.4–8.6 for the ERB component averages do not suggest anything definite about the sources and sinks of radiative energy: these data characterize some climatic variables and are quite independent of the anomalies of the sources of heating. The isolines of MSD of these components are quite different (Figs. 8.7–8.9). The isolines in Fig. 8.7 a are characterized by the appearance of 3–4 "centers of atmospheric activity": the zone of the Gulf Stream, the region of the Saharan dust outbreaks to the Atlantic Ocean, the tropical zone of the ocean, some land regions. Following a change in the path of the curves in Fig. 8.7 b, compared to Fig. 8.7 a a conclusion can be draw about 3 – 4 different types of cloud formations, associated with these "centers of activity" mentioned.

Specific features of the isolines' variability, compared to the data on absorbed solar radiation in Fig. 8.7, are shown by the ERB longwave component (Fig. 8.8). Along with strong variability in the Saharan regions and the northern part of South America, regions showing substantial OLR variability can be traced in the Sargasso Sea, a place where the Gulf Stream and cold Labrador current converge, near the equatorial Atlantic, and also in the Norwegian Sea. Reducing these data to surface level does not, change the location of the zones of substantial variability of radiation much, but shifts them slightly (it is particularly noticeable with the Sargasso Sea EAZO).

Still more specific characteristics of EAZOs are shown by the ERB MSD values (Fig. 8.9). Here, the isolines are particularly concentrated in the region of the Gulf Stream, near Newfoundland, and in two locations of the equatorial Atlantic. Figure 8.9 a shows typical regions of maximum radiation budget variability in the areas considered: the Sahara desert, the western part of the Sargasso Sea, an area east of Newfoundland, two regions of the near-equatorial Atlantic near South America and Africa, as well as a less variable region east of Iceland.

A systematic error in the data of Fig. 8.7 a caused by the sensitivity trend of space-borne receivers in the shortwave region, constitutes 1% per year, with an accuracy of transition from spectral measurements of the thermal radiation to the total fluxes under analysis of about 2%. At the same time, for the enumerated EAZOs of the Atlantic Ocean, mean values of the radiation budget are, respectively, -10, -50, 50, 30, and -120 W/m^2 (Fig. 8.5 a). The conclusion can be drawn, therefore, about overestimation of variations caused by natural variability (Fig. 8.9 a); of the uncertainties determined by measurement errors and by the representation of satellite information as monthly mean maps. Note, that for the data in Fig. 8.9 b these variations are within measurement errors, and do not exceed them.

The data of Fig. 8.9 b are interesting because of the possibility of representing the atmosphere, as an almost additive clutter, that weakly deforms the zones of strong variability of the radiation budget, and slightly shifts the variability maxima. Comparison of the data of Figs. 8.7 – 8.9 suggests that the pattern of annual variability, characteristic of Fig. 8.9, is mainly caused by the longwave component and, to a lesser extent, by the absorbed solar radiation (Fig. 8.8). The EAZOs of the North Atlantic shown in Fig. 8.9 are, firstly, connected with a substantial variability of the ocean-atmosphere heat exchange.

The comparison of the data of Figs. 8.5 b, 8.9 b with results of the RB direct on-ship measurements for 1972 – 1980, represented as monthly means for the region 30°–20°N, 20°–40°W [106], revealed results that coincide well: relative errors did not exceed 10 – 12% for both averages and MSDs. Despite the fact that individual components of the surface radiation budget differed in this case by a factor of 1.5 – 2 (the albedo being overestimated, the longwave radiation, underestimated), there is a clear usefulness of combined studies of the radiation budget and its components.

Comparison of Figs. 8.4b–8.6b with the data of the Earth's Surface Heat Budget Atlas [24] confirmed the trend mentioned previously towards an overestimation of retrieved surface albedos and under estimation of surface emission, with a clear coincidence of radiation budget values. The reason for systematic errors in radiation budget components have to be studied further.

Thus, the concept of the oceanic energy active zones, initially substantiated by numerical modeling results, is now experimentally verified, based on analysis of satellite data on the radiation budget and its components.

8.6 Cloudiness as a Factor of Interannual Variability of the Radiation Budget

The great stability of global climate shows that, on the global scale, heat losses due to longwave radiation emitted by the surface-atmosphere system, are in approximate equilibrium with the solar shortwave radiation absorbed by the system. Both these ERB components determine the sources and the sinks of radiative energy which govern the general circulation of the atmosphere and the ocean, but in turn, depend on meteorological conditions.

Analysis of the effect of clouds on radiation budget is very important for understanding the feedback mechanism responsible for the formation of SST anomalies. An increase of cloud amount reduces the amount of solar energy received by the system. At the same time, heat losses due to longwave outgoing radiation of the system decrease. The difference between these quantities characterizes the regulating role of clouds in the formation of the ERB and of climate changes.

Observations from satellites permit continuous monitoring of the radiation budget together with continuous observations of cloud cover, which has been mentioned in Section 8.3. First observations from satellites have made it possible to obtain extensive information on clouds. However, a variety of cloud shapes and their variability hinder the use of single-valued criteria for classifying the types and shapes of cloud formations in analysis of satellite data.

The role of satellite measurements of the radiation budget and of cloudiness grows in connection with the detection in the World ocean of individual zones of the formation of longterm weather anomalies and climate changes – EAZOs. The mathematical apparatus of conjugated equations of thermo-hydrodynamics [119] permits a description of climate change through observed anomalies of the processes taking place in these zones. Anomalies of heat fluxes in the ocean, are closely connected with cloud anomalies, which necessitate the solution of the problems of the effect of EAZOs, cloudiness and the radiation budget of the system.

In analysis of the dependence of the system's radiation budget on cloudiness, a parameterization of the budget's variations is usually based on the difference between partial derivatives of each component [91]. In this case, the combined effect of growing albedo and decreasing outgoing thermal emission with increasing cloudiness, is analyzed. In most related studies it has been shown that the short-wave component dominates the longwave one. In some studies, however, the mutual compensation of these effects has been mentioned [91]. This refers to average values of the radiation budget and cloud amount and not to their intra-annual and interannual variability. Nevertheless analysis of the radiative effect of EAZO, reveals substantial MSDs of the radiation budget as a measure of the interannual variability [120]. Average values do not appear to aid solution of the problem (see Section 8.4).

Possibilities of comparing the data of satellite measurements with direct shipboard measurements of radiation appear the satellite data is reduced to surface level. The expediency of using the solutions of direct problems of radiation transfer in the atmosphere has been mentioned. Here it is of primary importance is the need to consider the effect of cloudiness from satellite-derived monthly means. The difficulty grows due to a lack of current information about cloudiness, though attempts have been made to identify clouds based on these data. Two kinds of difficulty appear: how to classify the types and the shapes of cloud formations in analysis of satellite-derived monthly means; what are the optical characteristics of cloud formations used in reducing satellite data to the surface level?

Analysis of averages and MSDs of initial and transformed satellite data on the radiation budget revealed the existence of zones of substantial interannual variability, closely connected with EAZOs. The characteristic feature of isolines of the radiation budget MSD is the potential to represent the atmosphere, when analyzing the interannual variability, as an almost additive interference which weakly deforms the zones of strong variability and slightly shifts these zones (Fig. 8.9). This suggests that, despite the considerable transformation of radiation by clouds, the formation of the zones of substantial variability of the radiation budget (EAZOs) is determined by the change of thermocline over the ocean in the autumn-to-spring period, which is masked by definite shapes of clouds. Assuming that the atmosphere behaves as a low-frequency filter [125], simple estimates can be obtained of the sensitivity of the retrieval scheme to variations in the atmospheric state based on analysis of functional dependences of initial and retrieved data.

Figure 8.10 shows retrieved averages of the surface radiation budget that depend on satellite data on the radiation budget [125] for the territory (10 °S – 70 °N, 0 °– 80 °W). To analyze the sensitivity of the radiation budget data to variations in the state of the atmosphere, the lines of regression between average values of satellite ($\bar{B}_\infty$) and retrieved ($\bar{B}_g$) data for the cloudless summertime mid-latitude atmosphere over the ocean are shown also in Fig. 8.10 as are models of desert aerosols for overestimated cloud amounts (continuous cloudiness of the shapes most typical for the given territory).

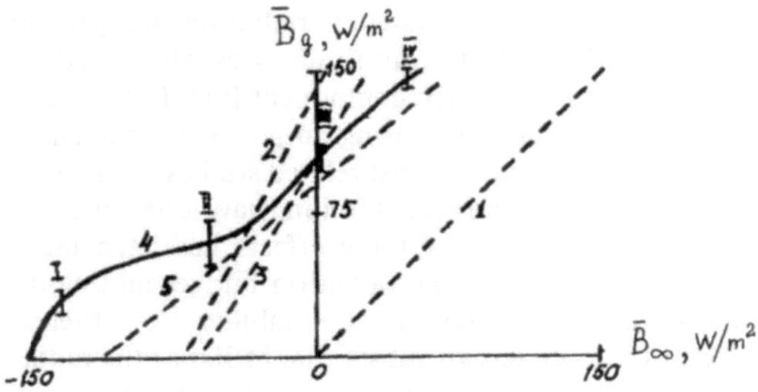

Fig. 8.10. Interrelationships between averages of the radiation balance at the atmospheric top and base: *1* not atmosphere; *2, 3* cloudless atmosphere (*2* sea surface, *3* desert); *4* real cloudiness; *5* continuous cloudiness. Individual sections of curve *4* correspond to the following regions with the limits of variability for this curve shown; *I* regions of Greenland; *II* latitudinal belt 40–55 °N; *III* the Sahara desert; *IV* the tropical Northern Atlantic

The data in Fig. 8.10 suggest the following three conclusions. First, the dependence of $\bar{B}_g$ on $\bar{B}_\infty$ (curve 4) differs greatly from cloudless conditions (2, 3) and less from continuous cloudiness conditions (5) towards increasing $\bar{B}_g$, with prescribed $\bar{B}_\infty$. This fact reflects the presence of broken cloudiness in realistic observation conditions. Second, there are four characteristic regions for which the correlation between $\bar{B}_g$ and $\bar{B}_\infty$ is substantially different: these are an area near Greenland, mid-latitudes, the Sahara desert, and tropical latitudes of the Atlantic Ocean. Here the $\bar{B}_g$ variability with prescribed $\bar{B}_\infty$ values is a maximum at mid-latitudes, which can be explained by maximum variability of the types and the shapes of clouds for the given region, compared to the three others. Third, with curve 4 in Fig. 8.10 smoothed down, the atmosphere can be considered almost an additional interference, shifting $\bar{B}_g$ towards overestimated values with prescribed $\bar{B}_\infty$ compared to continuous cloudiness. Thus, the data in Fig. 8.10 show the prevailing of the shortwave component over the longwave one, with an increased cloud amount (the transition from curve 4 to curve 5), according to the studies of other authors mentioned [91]. In contrast to their model studies, we estimated this effect here based on analysis of satellite data.

Figures 8.11–8.13 show the distribution of the MSD of retrieved radiation budget (Fig. 8.11), its longwave component (Fig. 8.12), and absorbed shortwave radiation (Fig. 8.13) as a function of the respective characteristics based on satellite data. The data in Figs. 8.12–8.13 refer to the same territory as the data in Fig. 8.10. In Fig. 8.11 the points are grouped into four classes since they belong to different areas of MSD: the Sahara desert, high and tropical latitudes, and an area of western boundary currents in the Gulf Stream region.

The data in Fig. 8.12 are quite different uniformly distributed along the MSD of retrieved values 9 W/m², with the MSD of initial data from 10 to 20 W/m². This relatively weak sensitivity of the MSD of the retrieved long-

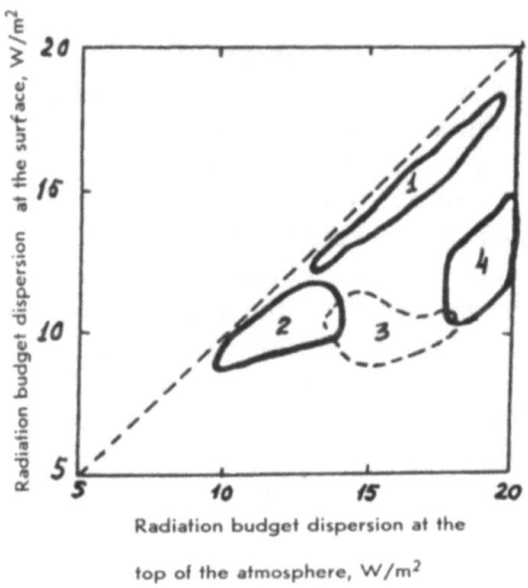

Fig. 8.11. A combined distribution of the radiation balance dispersions at the atmospheric top and base over territories 10°S–70°N, 0°–80°W: *1* the Sahara desert; *2* high latitudes; *3* tropical latitudes; *4* western boundary currents

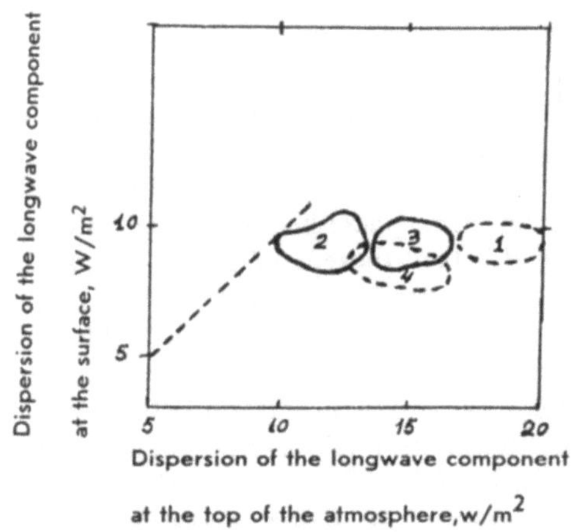

Fig. 8.12. The same distributions as in Fig. 8.11, for the longwave component of the radiation balance

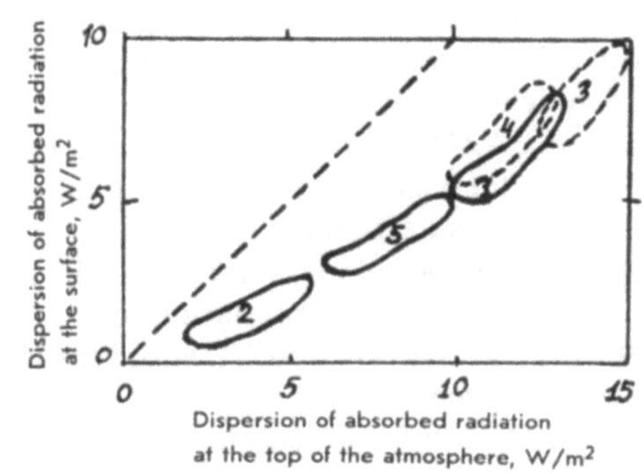

Fig. 8.13. The same distributions as in Fig. 8.11, for absorbed solar radiation (*1–4* correspond to Fig. 8.11, *5* for the territory 30°–55°N)

wave component could be explained by the strong smoothing effect of the atmosphere in the scheme of retrieval, where the isolines of retrieved MSDs did not correspond mostly, to the isolines of initial data [106]. The latter favors a similar sensitivity of the scheme of MSD retrievals to variations in both components of the radiation budget.

Figure 8.13 data demonstrate a systematic shift of retrieved MSDs towards their smaller values, compared to initial MSDs, with weakly differing areas of land and ocean in tropical and near tropical latitudes. The data in Figs. 8.12 and 8.13 confirm the strong sensitivity of the surface-atmosphere system to variations in the shortwave component, with varying cloudiness, compared to the longwave component, where these effects are not mutually compensated. Possible occurrences of EAZOs during climate changes are connected with the solution of the problem of detecting the climatic signal, against the noise background, caused by natural variability of the ocean-atmosphere system.

8.7 Statistical Characteristics of Anomalies of the 2-D Fields of the Radiation Budget of the North Atlantic

Along with the determination of the zones of strong interannual variability of the radiation budget, an interesting analysis concerns the spatial distribution of the effects of these zones which are the regions of strong variability of the ocean-atmosphere heat exchange. When examining the characteristic scales of the effects of these EAZOs, a 2-D analysis of the fields of the radiation budget and its components can be rather useful. This analysis is based on the 2-D Fourier transformation of the fields represented as standard anomalies, empirical procedures of the 2-D filtration of anomalies, and on the anomaly fields expansion in natural orthogonal components. Satellite date are useful for the solution of the problem in question due to approximately synchronous measurements over large territories. These make possible the formation of high-resolution fields allowing the laws connected with small spatial wavelengths to be studied.

If, using the 2-D statistical analysis of satellite information, the characteristic spatial wave structure of the radiation budget field and its components could be determined, it would not only useful for the development of theoretical concepts of wave formations and characteristics scales of the impact of EAZOs. In addition, it would promote the use of shipboard measurements in these zones to construct a learning sample, allowing all the fields of satellite measurements, within certain spheres of impact, to be represented in terms of ship board measurements. The test-sample learning serves as the basis for the thematic interpretation of satellite data. The point concerns the "radiative effect" of EAZOs, while the meteorological effect is also connected with the effects of the dynamics of the atmosphere and the ocean, of the phase transformation of water, etc. For the North Atlantic region in question, the total meteorological impact can be shown, for instance, in asynchronous correlations of cloudiness and temperature anomalies in the European region of the USSR [125] and a 6–8 months' time lag.

The 2-D analysis of the anomaly fields of the radiation budget and its components is based on the supposed homogeneity of these fields with spatial filtering of their trend constituent, determined by natural variability of the latitudinal-longitudinal change in the radiation budget. With this assumption, the spectral density, characterizing the distributions of 2-D spatial harmonics of the anomalies of the respective fields, and the correlation function for different harmonics are related by Fourier transformation.

The 2-D spectral density is calculated using the spatial smoothing of the 2-D periodogram [106]:

$$H_{pq} = \frac{N\Delta\theta' k\Delta\varphi'}{4\pi^2} |C_{pq}|^2 ,$$

where $\theta' = 0, 1, \ldots, N-1$; $\varphi' = 0, 1, \ldots, k-1$. In our case, in analysis of radiation budget anomalies for a territory (0°–80°W, 10°S–70°N), the latitudinal, $\Delta\theta'$, and longitudinal, $\Delta\varphi'$, grid step is 2.5°, $N = k = 32$. The amplitudes of the 2-D Fourier expansion of the anomaly fields, $Y_{\theta'\varphi'}$, are:

$$C_{pq} = \frac{1}{Nk} \sum_{\theta'=0}^{N-1} \sum_{\varphi'=0}^{k-1} Y_{\theta'\varphi'} e^{-i(\Omega_p \theta' + \omega_q \varphi')}$$

where $\Omega_p = \frac{2\pi}{N} p$, $\omega_q = \frac{2\pi}{k} q$ are spatial frequencies in latitude and in longitude, respectively, $Y_{\theta',\varphi'}$ – are the anomalies of the 2-D fields of the radiation budget, in which a constituent is filtered with the use of normals (mean values of the radiation budget $\bar{R}_{\theta',\varphi'}$) and standards (radiation budget dispersions $\sigma_R(\theta', \varphi')$ where $R_{\theta'\varphi'}$ are initial fields):

$$Y_{\theta',\varphi'} = (R_{\theta',\varphi'} - \tilde{R}_{\theta',\varphi'})/\sigma_R(\theta', \varphi') .$$

The main goal of the spectral and correlation analysis of homogeneous random fields is an estimation of averages, spectral densities and correlations functions using discrete realizations in the finite node system. Smoothing a periodogram is equal to an estimation of the spectral density [106]:

$$\bar{H}_{pq} = \sum_{j=-2}^{r} \sum_{l=-S}^{S} \alpha_{j,l} H_{p+j,q+l} ,$$

where $\alpha_{j,l}$ is the 2-D filter (the spectral window of smoothing). To ensure non-shifted estimates and minimum MSDs, a regression second-order filter with a window of five-point width can be used. The smoothed field MSD $\overline{\sigma_{pq}^2}$ is less than that of the initial field, σ_{pq}^2:

$$\overline{\sigma_{pq}^2} = \sigma_{pq}^2 \sum_{j=-5}^{5} \sum_{l=-2}^{2} \alpha_{j,l}^2 < \sigma_{pq}^2 .$$

A sampled covariance function of the 2-D radiation budget fields anomalies is calculated with the formula:

$$\tilde{K}_{\nu'\tau'} = \begin{cases} \dfrac{1}{Nk} \sum\limits_{\theta'=0}^{N-1-|\nu'|} \sum\limits_{\varphi'=0}^{k-1-|\tau'|} Y_{\theta',\varphi'} Y_{\theta'+|\tau'|,\varphi'+|\nu'|} ; \\ \qquad \nu'\tau' \geqslant 0, |\nu'| \leqslant N-1, |\tau'| \leqslant k-1 \\ \dfrac{1}{Nk} \sum\limits_{\theta'=0}^{N-1-|\nu'|} \sum\limits_{\varphi'=0}^{k-1-|\tau'|} Y_{\theta',\varphi'+|\tau'|} Y_{\theta'+|\nu'|,\varphi'} ; \\ \qquad \nu'\tau' < 0, |\nu'| \leqslant N-1, |\tau'| \leqslant k-1 \\ 0 ; \qquad |\nu'| \geqslant N \quad \text{or} \quad |\tau'| \geqslant k \end{cases}$$

can be estimated by the 2-D Fourier transformation of $\bar{H}_{pq}$. The MSD-normalized covariance function is usually called the correlation function:

$$K_{\nu\tau} = \tilde{K}_{\nu\tau} / \overline{\sigma^2} \ .$$

If, instead of infinite limits in Fourier transformation in the spectral density – correlation function transition, we introduce finite frequency limits and calculate the fraction of MSD for this limited frequency interval, we obtain the spectral function instead of spectral density, which can be expressed in the 1-D case as [106]

$$J(\omega_p) = \frac{4\pi}{k\Delta t} \sum_{\varphi=1}^{p} h_\varphi / \sigma^2 \ .$$

This characteristic is easily used due to its linear frequency dependence [106]. It uses the criteria obtained in estimating the distribution function for random values: in particular, the confidence intervals obtained from estimated deviation of the empirical functions from the theoretical ones (testing the hypothesis of belonging in time series to a white-noise sample). The related results from analysis of normalized spectral functions as functions of time (a) – 45 months and latitude (b) – 32 locations, are given in Fig. 8.14. This Figure shows that calculations of the spectral function do not exceed a 95% white-noise confidence limit in temporal variations, and in more than two thirds of realizations they are beyond this limit in the latitudinal change of the radiation budget, for a totality of 45-month changes (b). Therefore, the variability of the 2-D analysis of the radiation budget fields in time-latitude and time-longitude coordinates is rather questionable, though, in this case, there are possibilities significant factors may be used for the 2-D analysis. At the same time, statistical significance of results is suggested, from the 2-D analysis in latitude-longitude coordinates.

Figure 8.15 shows spectral densities and correlation functions of the radiation budget calculated using the 45-month database (a) for average fields (b) average Januaries (c), for initial data (solid lines) and for data reduced to the surface level (dotted line), with the technique described in Chapter 6.

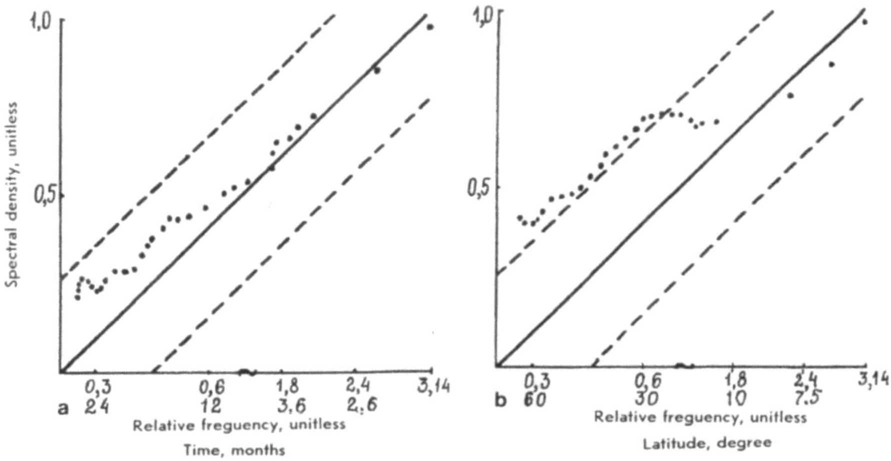

Fig. 8.14a, b. Normalized spectral functions in coordinates of: **a** time; **b** latitude of location

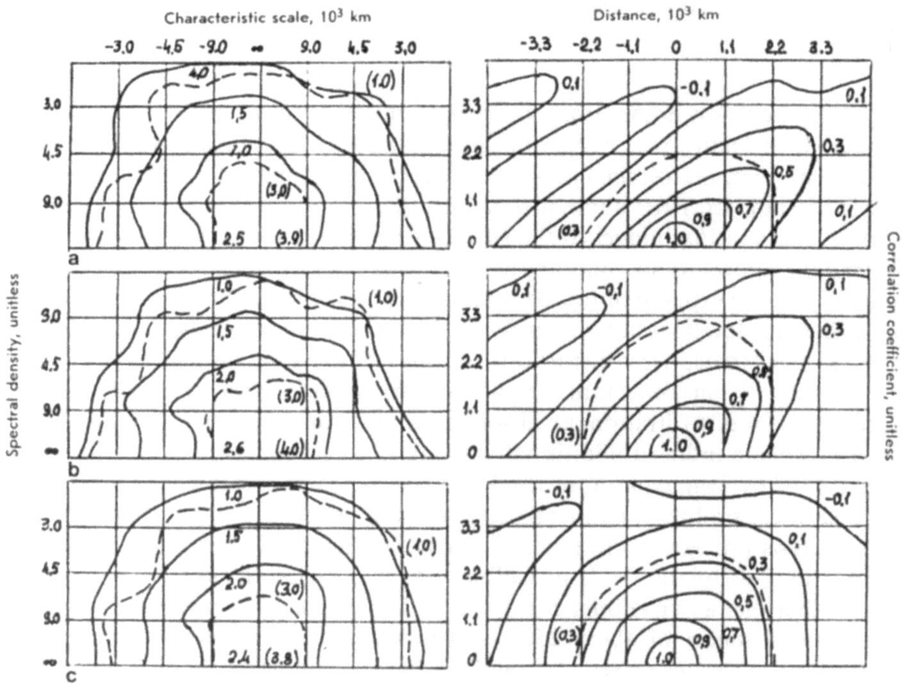

Fig. 8.15a–c. The 2-D distributions of spectral densities $\tilde{S}$ as a function of characteristic scales $\tilde{T}$ of wave formations and correlation functions $\tilde{\varkappa}$ depending on distance from the coordinates origin: **a** for the whole set of monthly means, **b** for July averages, **c** for January averages. *Solid lines* for the surface-atmosphere system, *dashed lines* for the surface

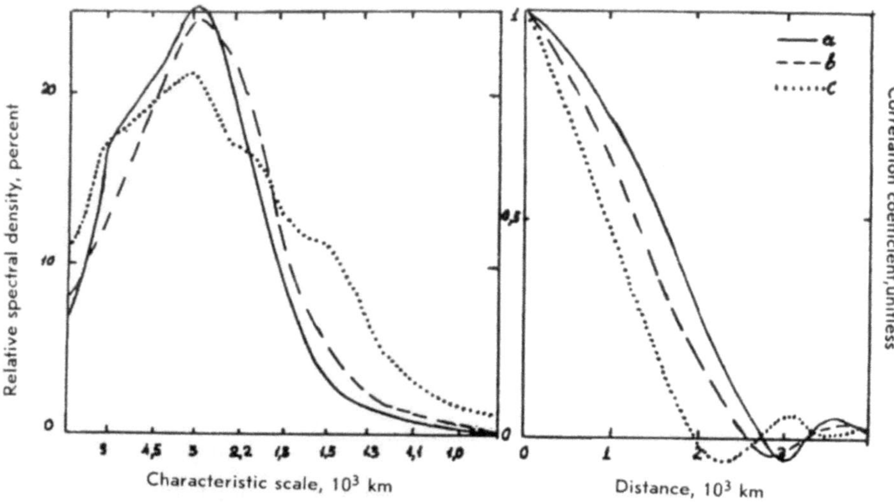

Fig. 8.16. Spectral densities and correlation functions reduced to the first quarter for the surface-atmosphere system: *a* with the latitudinal trend filtered with the help of the norms and standards for the period 1974–1978; *b* data for the first 2-D harmonic without the filtered trend; *c* corresponds to a but for the surface

The conclusion drawn from analysis of initial spectral density curves suggests that the major part of spatial variability (2-D MSD) falls on the first four 2-D harmonics, i.e., on the horizontal scale of more than 2000 km: about 85% MSD of standard ERB anomalies fall on these four harmonics. Upon reduction of the radition budget to the surface level, the number of significant 2-D harmonics grows to six, and about 80% MSD fall on them. It happens, apparently, at the expense of (a), additional smoothing impact of the atmosphere as a filter for low spatial frequencies [106], though not, as in the case of filtration of the atmospheric distortions, during instant satellite measurements.

Along with an appearance of a "fine structure" in the respective curves after atmospheric filtration, an absolute value of low-frequency anomalies due to atmosphere-induced additional systematic errors grows. All the curves in Fig. 8.15 show a maximum in the low-frequency area (scale about 9000 km), caused, most probably, by incomplete filtering of the latitudinal trend, i.e., by the analysis procedure itself (this is most clearly seen in Fig. 8.16).

An analysis of the fields of isolines of the correlation function (Fig. 8.16) shows that the radiative impact of EAZOs almost completely vanishes at characteristic distances of about 2500 km (the normalized correlation coefficient at these distances decreases to about 0.3). For the atmosphere-filtered data (the dotted line in Fig. 8.15), this impact disappears at a distance of about 500 km. This fact can be explained probably by a greater radius of correlation of cloud formations in a given region, as compared to the region of the thermal effect of EAZOs.

Note also a characteristic anisotropy of the effect of these zones at the atmospheric top level, most typical of monthly means, and a decrease of this anisotropy after atmospheric filtration (Fig. 8.15). Comparison of the isocorrelants in Fig. 8.15 with the isolines of interannual MSD in Fig. 8.9 shows that the observed breaking of anisotropy appears to be connected not only with the substantial effect of the Gulf Stream EAZO (Fig. 8.9 markedly shows a maximum family of isolines from initial and filtered data), but also with a particular effect of the near-equatorial Atlantic EAZO. (Here the cluster of isolines in Fig. 8.9b becomes thinner, compared to Fig. 8.9a, but the anisotropy of the isocorrelants in Fig. 8.15 also decreases.)

The plots in Fig. 8.15 represented for the first and second quarters (the respective curves for the first quarter are symmetrically reflected on the third quarter, and for the second quarter – on the fourth one), based on the known symmetry of periodogrammes [105], can be reduced to the first quarter. In this case spectral densities and correlation functions can be represented in a more familiar way (Fig. 8.16).

Figure 8.16 shows the data in Fig. 8.15 reduced to the first quarter (a), with spectral densities and correlation functions obtained using the first 2-D harmonic, without the latitudinal trend being filtered, by extracting a field-averaged value (b). Here the curves are calculated from filtered data reduced to surface level through "climatic" norms (c), i.e., the norms characterizing the whole 45-month database.

The conclusion can be drawn from comparison of the curves of Fig. 8.16a and b that the low-frequency maximum for characteristic scales 9000 km mentioned is caused by dividing the radiation budget into measured and random components. This is a partially artificial procedure, since no "pure" undisturbed latitudinal trend exists in nature. Also, Fig. 8.16 shows that with an increased number of significant 2-D harmonics, this false low-frequency maximum weakens, since the weight of the remaining wave harmonics of the radition budget anomalies grows. For instance, wave formations with a characteristic scale of about 2000 km, and possibly, about 1500 km, start appearing. As for the correlation function, it is seen in Fig. 8.16 and as mentioned previously when the data is reduced to the surface level, the radius of correlation of the EAZO effect is slightly diminished: it lies within 2000–2500 km and approximately corresponds to the distance noted between the Atlantic EAZOs.

We have seen that in the expansion of the fields of 2-D anomalies in dual Fourier series, four harmonics are significant for initial satellite data, and six harmonics upon their reduction to the surface level. The reverse situation is observed when using another mathematical approach, often used in meteorology and climatology. This is the expansion of correlation functions in Figs. 8.15 and 8.16, using the system of natural orthogonal components, i.e., eigen vectors of correlation matrices. This approach provides an optimal expansion of the respective fields and allows representation of the 2-D fields of the radiation budget anomalies as a finite set of eigen functions.

Table 8.5 shows eigen vectors φ_i, eigen numbers λ_i, and the distribution of radiation budget anomalies MSDs d_i, whose distributions are given in

Table 8.5. The values of eigen vectors φ_i, eigen numbers λ_i, and MSDs of 45-month averages of the radiation budget anomalies d_i for characteristic horizontal scales, at the atmospheric top (a) and base (b)

$r \cdot 10^3$ km	φ_1 a	φ_1 b	φ_2 a	φ_2 b	φ_3 a	φ_3 b
1	0.130	0.142	0.435	0.502	−0.420	−0.332
1.5	0.212	0.288	0.396	0.424	−0.095	−0.009
2	0.289	0.395	0.201	0.338	0.126	0.102
2.5	0.441	0.464	0.102	0.260	0.442	0.426
3	0.409	0.290	−0.001	0.195	−0.009	0.055
λ_i	36.64	27.30	14.46	8.36	3.23	2.68
d_i, %	69.38	77.78	80.05	86.53	91.28	94.22
1	−0.335	−0.274	−0.311	−0.222	0.313	0.203
1.5	0.154	0.105	0.102	−0.007	0.070	−0.008
2	0.428	0.324	−0.114	0.246	−0.160	−0.119
2.5	0.504	0.403	−0.023	0.105	0.098	0.084
3	−0.020	0.009	0.049	−0.116	0.108	0.189
λ_i	1.91	1.49	1.34	1.24	1.15	1.02
d_i, %	95.42	99.40	97.48	100.0	100.0	100.0

Figs. 8.15 and 8.16. A system of ortho-normalized functions $\varphi_\alpha(r_k)$ of radiation budget anomalies at a multitude of points $r_k (k = 1, 2, \ldots, n)$ being introduced, which meet the requirement:

$$\sum_{k=1}^{n} \varphi_\alpha(r_k)\varphi_\beta(r_k) = \delta_{\alpha\beta} = \begin{cases} 1 & \text{at} \quad \alpha = \beta \\ 0 & \text{at} \quad \alpha \neq \beta \,. \end{cases}$$

Since here in the expansions of anomalies the deterministic constituent has been filtered out, the effect of all 32×32 points in latitude and longitude is equally probable, which makes it possible at any point r_k to expand in series the field of radiation budget anomalies Y_k either at the upper or lower boundary of the atmosphere:

$$Y_k = \sum_{\alpha=1}^{n} a_\alpha \varphi_{\alpha k} \,,$$

where $Y_k = Y_R(r_k)$, $\varphi_{\alpha k} = \varphi_\alpha(r_k)$.

The coefficients of expansion (principal components of the field of radiation budget) are:

$$a_\beta = \sum_{k=1}^{n} Y_k \varphi_{\beta k}$$

and in addition,

$$\sum_{j=1}^{n} K_{kj}\varphi_{\beta j} = \lambda_\beta \varphi_{\beta k} \;; \quad k = 1, 2, \ldots, n \;; \quad \beta = 1, 2, \ldots, n \,,$$

i.e., the functions $\varphi_\beta(r_k)$ represent eigen vectors of the covariance matrix K_{kj}, and eigen numbers λ_β of this matrix are equal to the MSD of the expansion coefficients: $\lambda_\beta = \overline{\alpha_\beta^2}$. The MSD distribution in Table 8.5 is determined by relationships:

$$d_1 = \frac{\overline{Y_1^2}}{\sum_{k=1}^{n} \overline{Y_k^2}}, \ldots, d_i = d_{i-1} + \frac{\overline{Y_i^2}}{\sum_{k=1}^{n} \overline{Y_k^2}} .$$

Table 8.5 data show that $\varphi_1(r_k)$ is positive everywhere and constitutes about 70% of the MSD of the radiation budget anomalies. The first four eigen functions largely explain the variability of the fields of anomalies reduced to the surface level. For the initial fields of anomalies, about six functions are needed, the eigen values becoming close to unity.

On the whole, the transition from the right-hand fields of Fig. 8.15 and 8.16 to the data of Table 8.5 makes it possible to perform and easier spatial analysis using a limited set of functions φ_i, listed in Table 8.5. Using these functions, one can interpret "instant" fields of satellite radiation data anomalies using synchronous shipboard measurements of meteorological parameters in EAZOs, though obtaining functions φ_i is based on the statistics of the spatial distributions of anomalies.

Thus, along with the description of the spatial impact of the energy-active zones, the empirical analysis mentioned allows representation of the field of radiation budget anomalies in terms of natural orthogonal components (empirical orthogonal functions) and relates the data of continuous satellite and ship measurements in EAZOs. The performed statistical analysis determines not only the location of the zones of substantial interannual variability of the radiation budget but also the characteristic scale of impact of these zones on the entire North Atlantic territory under analysis.

8.8 Anomalies of the Radiation Budget and the EAZO Problem

Analysis of the MSD of satellite-derived monthly averages of the radiation budget and its components for 45 months of continuous observations between 1974 and 1978 revealed the zones of substantial interannual variability, closely connected with EAZOs. An additional analysis of measurement accuracy and the representation of satellite data in the form of monthly mean maps of total outgoing radiation fluxes have shown a statistical significance of the results obtained, which demonstrate the violation of zonality of MSD of the radiation budget and its longwave component in the regions of EAZOs. Statistically significant also are the MSDs of the data on radiation budget components reduced to the surface level, using results of the solution of the boundary problems of radiation transfer in the atmosphere (see above).

With the North Atlantic as an example, we shall consider in more detail the causes of characteristic patterns of MSDs which serve as a measure of interannual variability of the radiation budget and its components. Their mean values are characterized by a zonal change (breaking of the zonal change is mainly caused by the distribution of continents and oceans). However, the interannual variability of MSDs of the radiation budget and its longwave components has a clear non-zonal character (Figs. 8.8, 8.9): clusters of isolines are observed in the region of the Sargasso Sea (here the amplitude of the radiation budget is about 20 W/m^2); in the region of Newfoundland and near the equator at 40°W (15 W/m^2); and in the Norwegian Sea (13 W/m^2); with annual mean values of the radiation budget for the rest of the territory in question (10°S–70°N, 0°–80°W) about 10–12 W/m^2 (except for Africa, where the amplitudes also reach 20 W/m^2).

Figure 8.17 shows the annual course of mean values of the radiation budget and its components for the central parts of the EAZOs mentioned; those for oceanic basins along 30 °W and at a distance of 1000–4000 km from EAZOs are shown in Fig. 8.18, and for regions near the eastern coastlines of Greenland, North and South America in Fig. 8.19. Comparison of the data in Figs. 8.17–8.19 has shown that the annual change of the radiation budget and its longwave component varies insignificantly in the three EAZOs enumerated, apart from those in the tropics where these variations are more substantial. The annual course of mean values is highly variable for the shortwave ERB component, which may be determined by cloud cover variations in these EAZOs.

The annual changes of the interannual MSDs of the radiation budget and its components (Figs. 8.20–8.22) are quite different. Here, the month to month distribution of the radiation budget and its longwave component for all the three EAZOs, except for the tropical zone, is mainly inhomogeneous (Fig. 8.20). A characteristic bimodal distribution of the annual change of MSD with two maxima in April and in October is also shown for oceans near the Newfoundland EAZO and east of the Gulf Stream (Fig. 8.21), in the absence of the regularities mentioned above for the Norwegian Sea and in the equatorial region. The annual march of the radiation budget and its longwave component for the land adjacent to an EAZO (Fig. 8.22) differs greatly from that presented in Figs. 8.20 and 8.21.

A specific feature of the bimodal distributions (Fig. 8.20) consists that for the data presented in Fig. 8.20a, the contribution of individual components to the interannual variability for Aprils and Octobers is approximately equal, while for the same months in Fig. 8.20b the contribution of the anomalous year 1976 prevails. It is this that has led to the exceeding of the longwave component MSD, compared with those of the radiation budget in Fig. 8.20b. It follows from Fig. 8.20c that the amplitude of the interannual variability for the whole 45-month data base on the radiation budget, equal to 20 W/m^2 (for the longwave component 15 W/m^2) is mainly determined by two maxima in April and in October and by minima in July and in December. These changes of the ERB longwave component can lead to temperature variations of the surface-

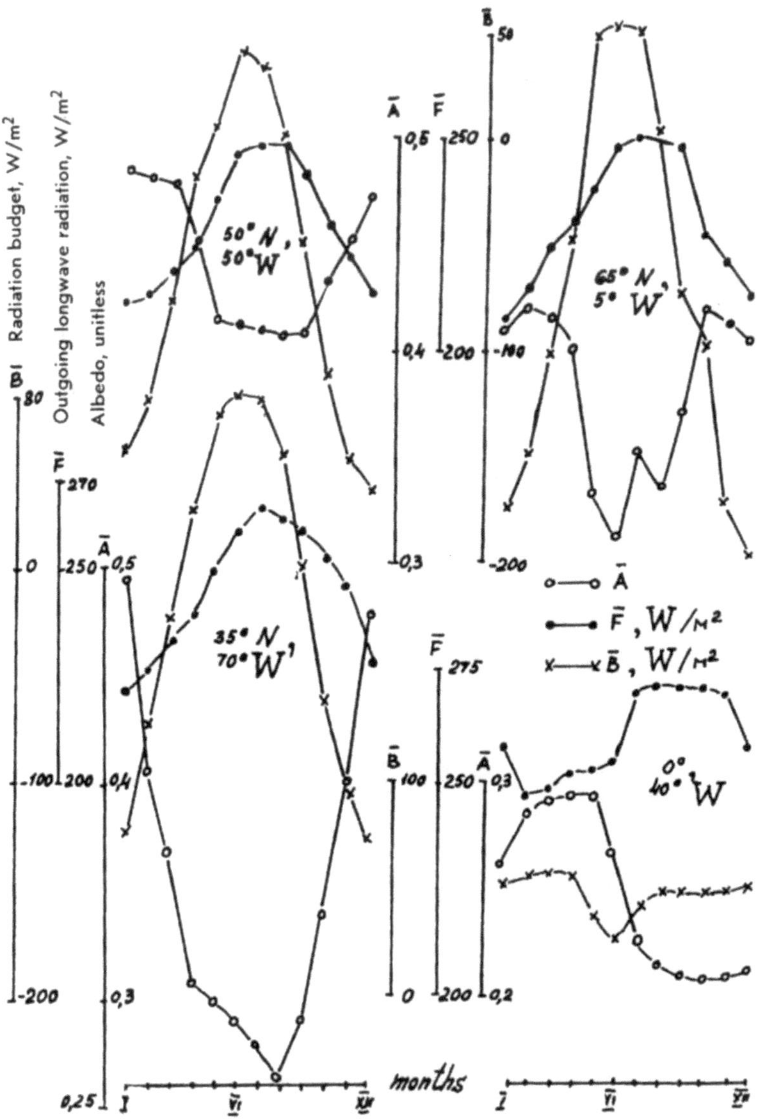

Fig. 8.17. The annual change of average values of the radiation budget and its components for the four EAZOs of the Northern Atlantic

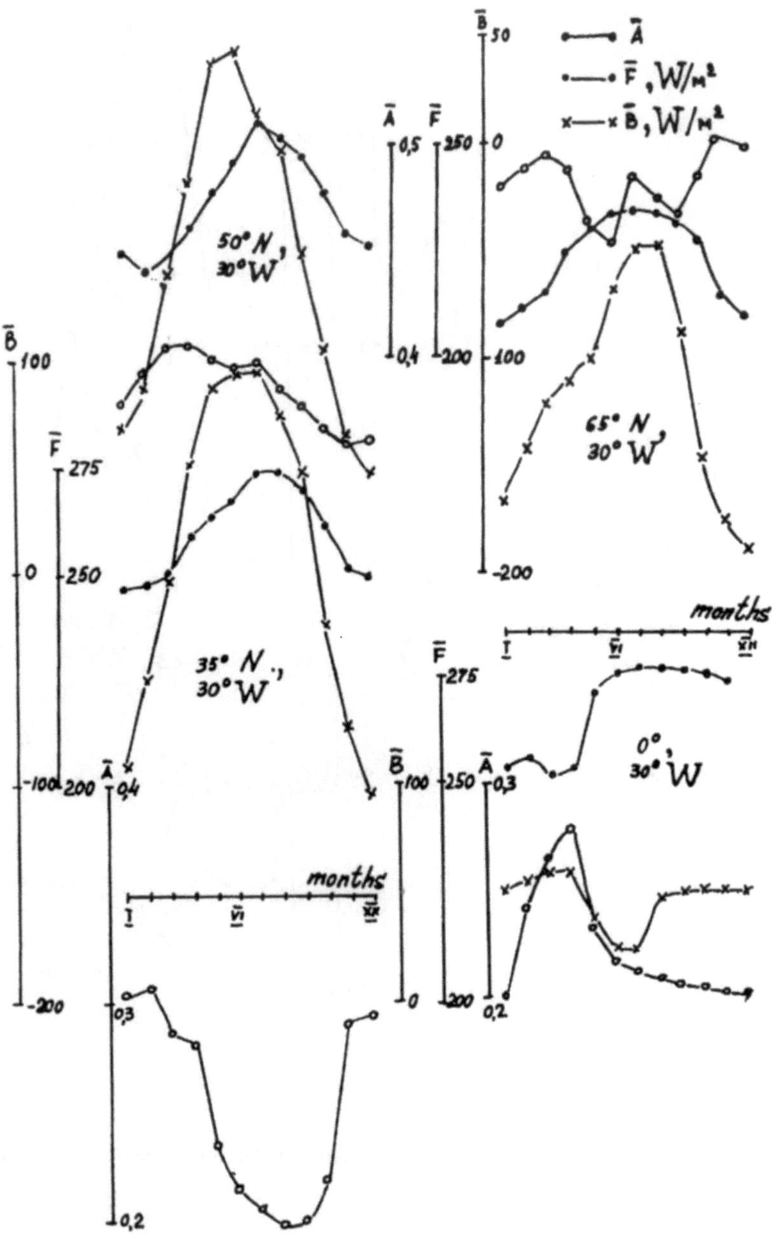

Fig. 8.18. The longitudinal section along 30 °W of the characteristics shown in Fig. 8.17

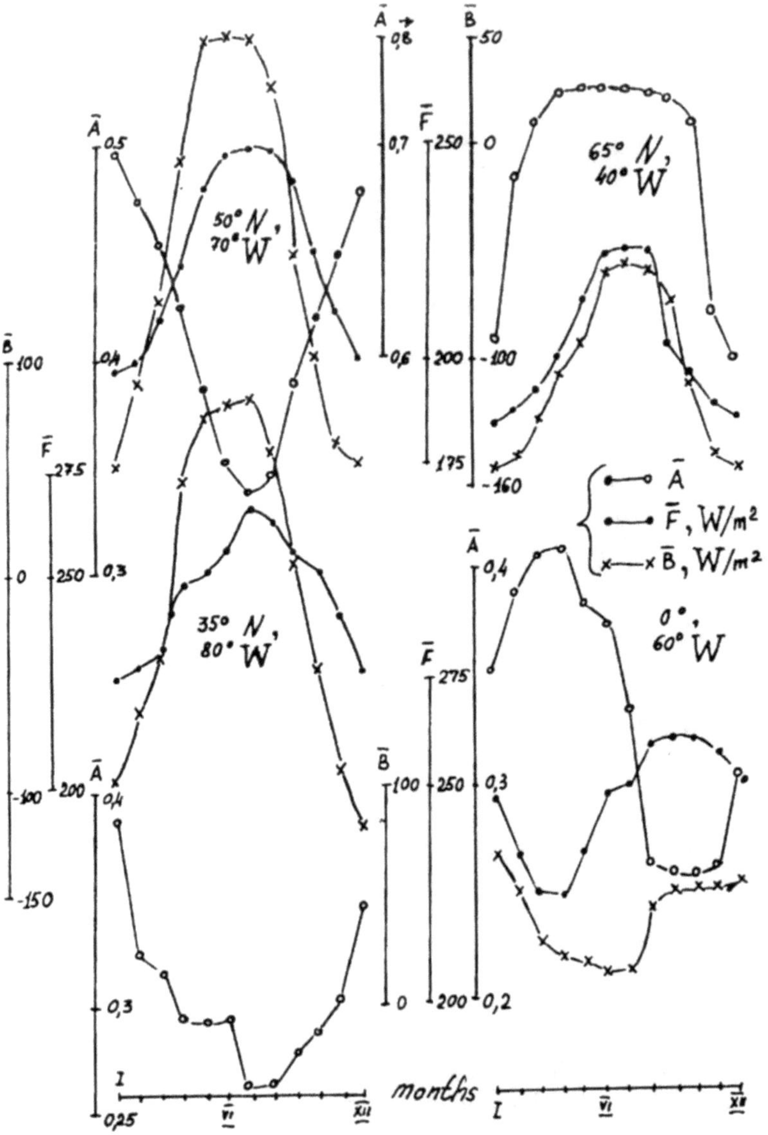

Fig. 8.19. The annual change of the characteristics shown in Fig. 8.17 for the land territories adjacent to EAZOs

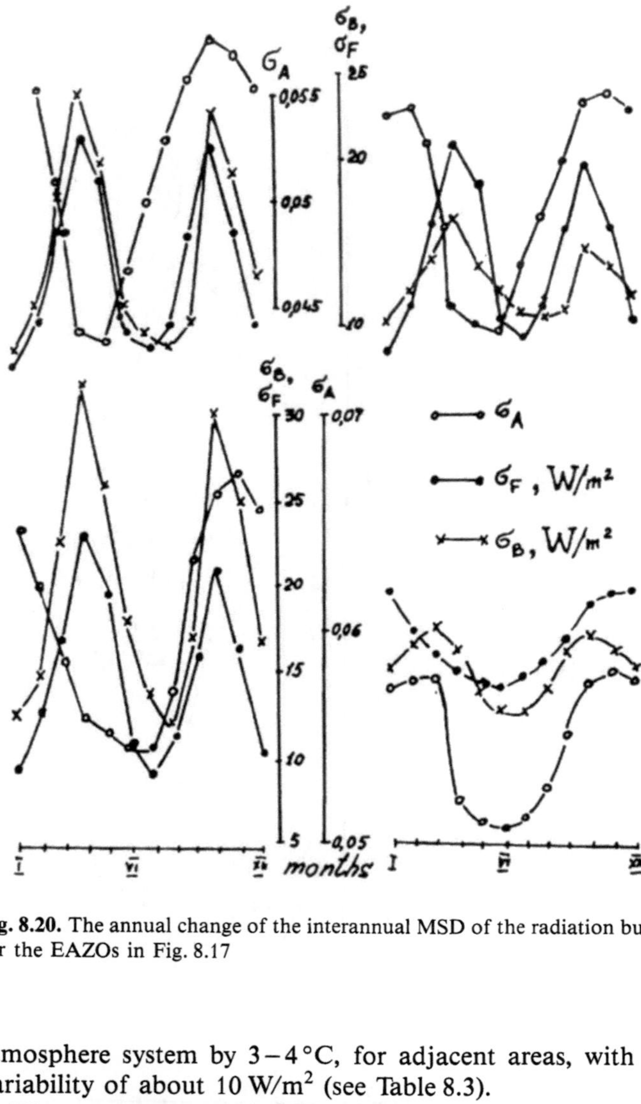

Fig. 8.20. The annual change of the interannual MSD of the radiation budget and its components for the EAZOs in Fig. 8.17

atmosphere system by 3 – 4 °C, for adjacent areas, with a month to month variability of about 10 W/m^2 (see Table 8.3).

The temperature variations of the system at given scales of averaging are hard to explain by specific cyclone tracks or by anomalies of heating sources outside the territory under study, which affect it through advection. Also, the influence of the polar front shifting or other dynamic effects are hard to believe. The ERB variability presented in Fig. 8.20a, b is explained perhaps by the impact of the ocean in the zone of the Gulf Stream, the cold Labrador current near Newfoundland and the North Atlantic Drift in the Norwegian Sea. The absence of a marked correlation between the variabilities of the short- and longwave ERB components favors the oceanic origin of this variability. Comparison of the data in Figs. 8.20 and 8.22 for the continental territories adja-

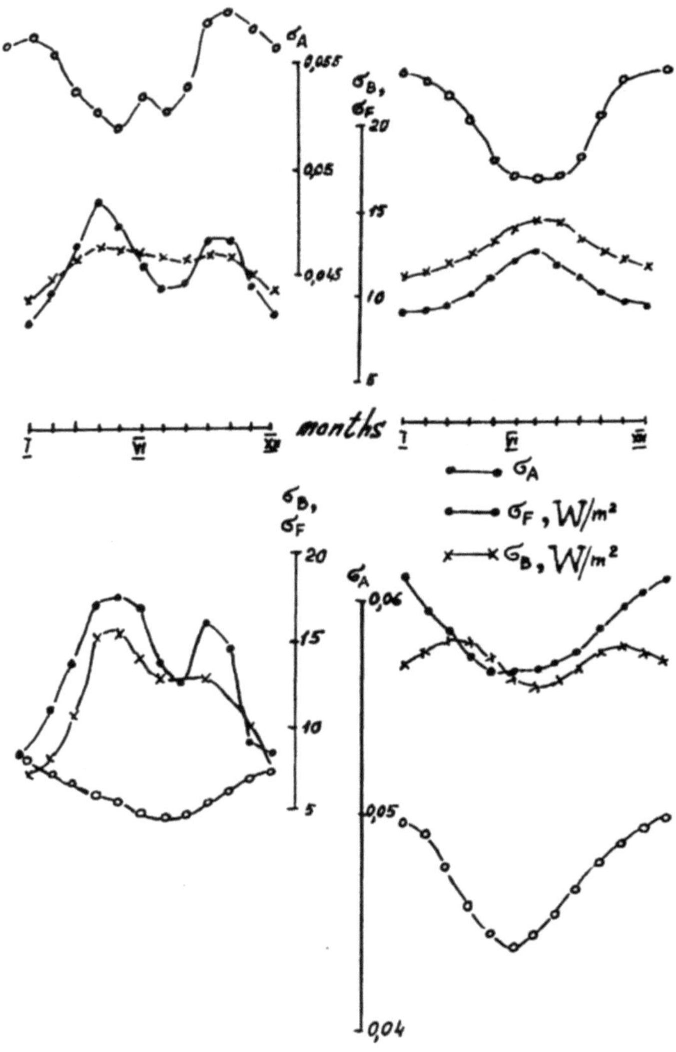

Fig. 8.21. The annual change of the interannual MSD of the radiation budget and its components for the territories in Fig. 8.18

cent to EAZOs confirms this fact: here the interannual variability may be of different origin, compared to the oceans, since for the data in Fig. 8.22, the bimodal distribution typical of Fig. 8.20, vanishes. Assuming that the data variability in Figs. 8.20d and 8.21d is also of oceanic origin, the respective maxima in the winter Northern Hemisphere can be explained by the warm water from the Southern Hemisphere crossing the equator.

These assumptions must further be verified both experimentally and theoretically from independent observations and modeling. The analyzed ERB

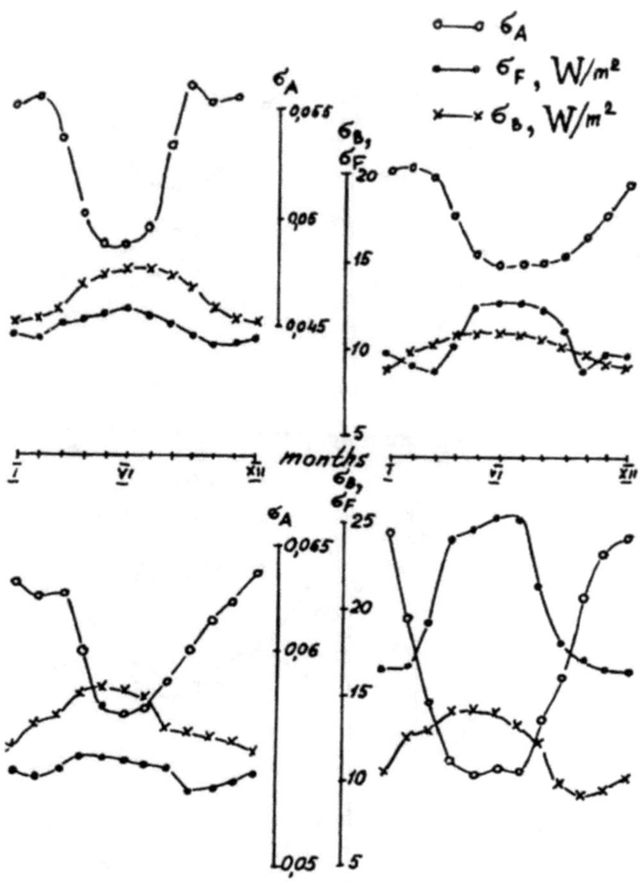

Fig. 8.22. The annual change of the interannual MSD of the radiation budget and its components for the territories in Fig. 8.19

values are, no doubt, more sensitive to variations in the ocean-atmosphere system taking place during the transition periods (spring, fall) than to the dynamic effects of the heating sources anomalies, where changes are more characteristic of winter and summertime conditions. This circumstance suggests an important role of the effect of seasonal variations of the oceanic thermocline in the region considered of the North Atlantic during the transition seasons. These variations, possibly not typical of other scales of averaging, are shown by monthly mean averaging of the data of ERB observations.

The conclusions on the basis of empirical analysis of satellite data must be verified. These conclusions are preliminary because of the comparatively short time period of satellite-derived data analyzed and they only hold for prescribed scales of monthly mean averaging. Background data for the longwave ERB component was obtained by averaging for every point within the 2.5°×2.5°

grid step data of measurements at 09.00 and 21.00 local time, and for the shortwave component only measurements at 09.00. Additional errors in the data analyzed appear in the transition from radiance to fluxes and in the subsequent retrieval of total hemisphere short- and longwave radiation fluxes from data of observations in narrow spectral intervals.

Despite these problems of obtaining and interpreting satellite data, there is general confidence in the usefulness of these data for analysis of probable developments of EAZOs in the outgoing radiation fluxes recorded from satellites. At the same time, analysis of Figs. 8.17–8.22 suggests the prevailing effect of the EAZO in the Gulf Stream region for the whole of the Northern Atlantic under study.

8.9 Teleconnections Between the ERB Anomalies in Tropical and Mid-Latitudes

The problem of applying the time series of satellite measurements to climatic studies is closely connected with estimating the climatic noise level and subsequent analysis of the statistical significance of the anomalies of the ERB and its components in different regions of the globe. The problem is reduced to an identification of the geographical regions where the observed interannual variability is much higher than the climatic noise level. The detection of teleconnections between the ERB anomalies contributes to solving this problem of identification of various signals of short-term climate changes.

In this connection, we explain the results of analysis of the interannual variability of the radiation budget together with analysis of the noise component in the data of NOAA and TIROS-N satellites. The sampled variability of measurement data necessitates the transition from theoretical statistics techniques [106] (at an infinite number of realizations with respective asymptotic transitions) to realistic estimations from limited samples. The climatic noise results from limited samples.

As with the natural variability of the measured fields, it may be estimated from the autocorrelation function.

In practice, lower daily variabilities and longer correlation times are observed at a large spatial averaging. As a result, the combination of data for certain territories at grid nodes permits enlargement of the volume of sampling and obtaining of anomaly distributions, which will be smoothed down as if an analysis were made with long observation series. This methodological approach, together with analysis of instrumental errors, sensitivity trends of sensors and cloud-caused variations, broadens the prospects for use of satellite-derived information to detect a climatic signal against noise.

The mathematical expectance of square difference between current data and monthly means is [107]:

$$\mathrm{Var}(\bar{X}_T) = \frac{S_D^2}{n}\left[1 + 2\sum_{\tilde{\tau}=1}^{n}\left(1 - \frac{\tilde{\tau}}{n}\right)R_x(\tilde{\tau})\right] , \tag{8.1}$$

where $X_k (k = 1, 2, \ldots, n)$ is a set of n-observations for a given month (not only the radiation budget components but also other parameters), $\bar{X}_T = \frac{1}{n}\sum_{k=1}^{n} X_k$ are monthly means, $\tilde{\tau}$ is determined by the number of independent samplings; the measure of variability – non-shifted sampled MSD, is

$$S_D^2 = \frac{1}{n-1}\sum_{k=1}^{n}(X_k - \bar{X}_T)^2 , \tag{8.2}$$

and the autocorrelation function is:

$$R_x(\tilde{\tau}) = \frac{1}{n-\tilde{\tau}}\sum_{k=1}^{n-\tilde{\tau}}(X_k - \bar{X}_T)(X_{k+\tilde{\tau}} - \bar{X}_T)/S_D^2 . \tag{8.3}$$

It is shown in Fig. 8.23 for the variability of the outgoing longwave radiation averaged over 2.5°-sided squares for the Central Pacific for 1974–1981 [125].

The S_D values serve as a measure of observational data variability, provided they are day to day independent, and $\mathrm{Var}(\bar{X}_T)$ is also a measure of climatic noise. If observations do not correlate, then the expected variability of monthly means depends only on the number of observations used to obtain the monthly means, i.e., the expected variability of monthly means is determined by day to day variations. The climatic noise level can be estimated from MSDs and autocorrelation functions.

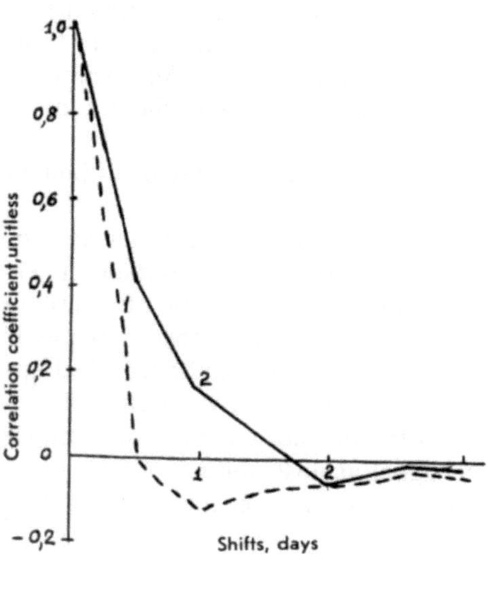

Fig. 8.23. Autocorrelation functions of the outgoing longwave radiation for winter (1; $S_D = 27\ \mathrm{W/m^2}$) and summer (2; $S_D = 25\ \mathrm{W/m^2}$) conditions

Figure 8.23 shows that, despite the fact that daily mean variability in summer and in winter for the central tropical Pacific is approximately equal, a substantial difference of autocorrelation functions for these seasons has led to substantially different values of the signal/noise ratio, which is 2.7 for summer and 4.0 for winter. Along with noise-loaded data characterized by the Var $(\bar{X}_T)$ parameter, the MSD for current monthly means were calculated:

$$S_{0,1}^2 = \frac{1}{n-1} \sum_{k=1}^{n} (X_k^{(l)} - \bar{X}^{(l)})^2 \; , \tag{8.4}$$

where:

$$\bar{X}^{(l)} = \frac{1}{J} \sum_{i=1}^{J} X_i^{(l)} \; , \quad l = 1, 2, \ldots, 12 \; ;$$

J is equal to the number of monthly means used in observations.

In Section 8.7 in analysis of the spectral density curves for different seasons, a greater variability of mean July values of this characteristic was revealed, compared to mean January ones. This fact could have been interpreted from the viewpoint of more substantial climatic occurrences of radiation budget anomalies in summer. However, the considerably noise loaded summertime data do not allow the drawing of a unanimous conclusion about the increased significance of climatic signals in analysis of summertime anomalies. Further studies in this direction are needed.

Note, that for the Gauss process, 95% of measurement data variability fall within $\pm 2\,\mathrm{Var}(\bar{X}_T)^{1/2}$, i.e., in this case the confidence interval is estimated with the use of well known significance criteria, considering varying MSDs. The expected variability (in contrast to those found) is determined by the character of variations from one calculation to another. If observations are not correlated, then Var $(\bar{X}_T)$ will only depend on the number of measurements n used to find $\bar{X}_T$ averages. An effective number of independent measurements is determined by $\tilde{\tau}$ and governs the summation law in the expression for Var $(\bar{X}_T)$. Analysis of the data [125] has shown that more than 25% of anomalies lie outside the boundaries, determined by MSDs, expected for the given region.

A 95% confidence interval characterizes the limits within which true values of anomalies lie (with 0.95 probability) from calculated sampled MSD, considering the noise loading of data.

Analysis of data [125] is based on consideration of the character of their internal correlation, from data of individual intramonthly measurements. An analysis performed in Sections 8.3–8.8 was based on monthly means which could not incorporate the intramonthly variability. We have only 45 time realizations for each of the points at which the annual change can be filtered out. However, a methodical approach suggested in [125], can be used, by enhancing the number of points at the expense of an enlarged area of analysis. On the other hand, the time correlation (the expected covariance) of monthly means can be estimated from 45 realizations, with the annual course filtered out. The expression for the expected correlation of monthly means is as follows [125]:

$$R_{\bar{X}_T \bar{Y}_T}(0) = \frac{\mathrm{Cov}(\bar{X}_T, \bar{Y}_T)}{[\mathrm{Var}(\bar{X}_T)]^{1/2}[\mathrm{Var}(\bar{Y}_T)]^{1/2}}$$

$$= \frac{R_{XY}(0) + \sum_{\tilde{\tau}=1}^{n-1}\left(1-\frac{\tilde{\tau}}{n}\right)R_{XY}(\tilde{\tau}) + \sum_{\tilde{\tau}=1}^{n-1}\left(1-\frac{\tilde{\tau}}{n}\right)R_{XY}(-\tilde{\tau})}{\left[1+2\sum_{\tilde{\tau}=1}^{n-1}\left(1-\frac{\tilde{\tau}}{n}\right)R_x(\tilde{\tau})\right]^{1/2}\left[1+2\sum_{\tilde{\tau}=1}^{n-1}\left(1-\frac{\tilde{\tau}}{n}\right)R_Y(\tilde{\tau})\right]^{1/2}}, \tag{8.5}$$

where $R_X(\tilde{\tau})$ and $R_Y(\tilde{\tau})$ are the autocorrelation of observations made at different times; $\mathrm{Cov}(\bar{X}_T, \bar{Y}_T)$ the covariance of observations at different times; $R_{XY}(\tilde{\tau})$ and $R_{YX}(\tilde{\tau})$ the cross-correlation of observations with different shifts $\tilde{\tau}$:

$$R_{XY}(\tilde{\tau}) = \frac{1}{n-\tilde{\tau}}\sum_{k=1}^{n-\tilde{\tau}}(X_k - \bar{X}_T)(Y_{k+\tilde{\tau}} - Y_T)/S_{XY}, \tag{8.6}$$

$$R_{YX}(\tilde{\tau}) = \frac{1}{n-\tilde{\tau}}\sum_{k=1}^{n-\tilde{\tau}}(X_{k+\tilde{\tau}} - \bar{X}_T)(Y_k - \bar{Y}_T)/S_{XY}, \tag{8.7}$$

$$S_{XY} = \left[\frac{1}{n-1}\sum_{k=1}^{n}(X_k - \bar{X}_T)^2\frac{1}{n-1}\sum_{k=1}^{n}(Y_k - \bar{Y}_T)^2\right]^{1/2}. \tag{8.8}$$

Here $T = n\Delta t$ is the length of a time series ($n = 45$), Δt the interval between observations (1 month), $R_{XY}(0) = R_{YX}(0)$ the coefficients of mutual correlation of observations made at different times.

Now we consider teleconnections between monthly mean anomalies for the radiation budget and its components at different points. Simultaneous observations (in Sect. 8.4 we connected them with instant measurements) at adjacent grid points are correlated due to the finite size of weather disturbances shown by radiation measurements from satellites. Monthly means have a natural trend to increased correlation for certain directions in which the disturbances move.

The existence of the signal of the short-term climate variability, revealed in the tropics by the characteristics of cloudiness and of the ERB longwave component field, has been proved by many authors [106]. These and other studies form the basis of the TOGA programme. The various teleconnections between the key regions obtained from the longwave component fluctuations in the equatorial central Pacific using correlations with zero-shift from (8.6) and (8.7), are given in Fig. 8.24.

The significance of the mutual correlation coefficients, corresponding to a 95% significance level, was determined from the condition:

$$R_{XY}(0) > 2\,\mathrm{SD}(C_{XY})/\sigma_F, \tag{8.9}$$

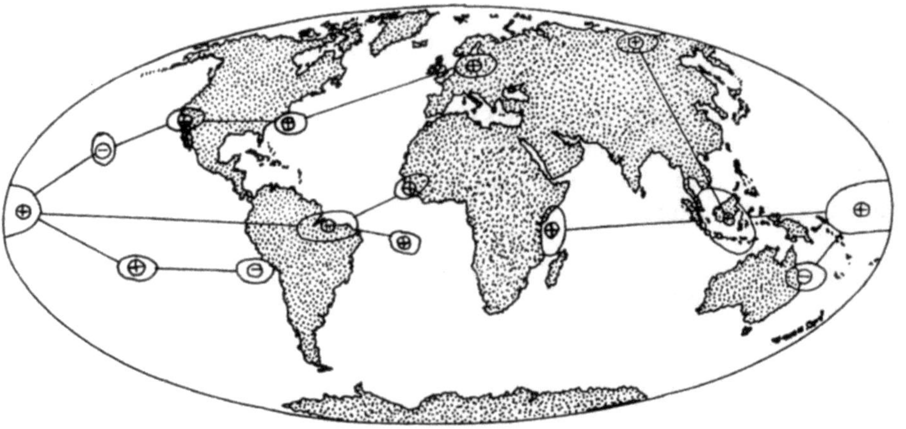

Fig. 8.24. A schematic image of key global regions from the point of view of teleconnections between radiation budget anomalies and initial fluctuations in the central tropical Pacific

where standard deviations of the covariance of two different-time observations, which can be represented in the form of the Markov processes of the first order, are equal to

$$SD(C_{XY}) = \sigma_F \sqrt{\frac{2}{n\{1-\exp[-(\lambda_X+\lambda_Y)]\}}} , \tag{8.10}$$

$C_{XY}(\tilde{\tau})$ is the sampled estimate of the cross-covariance of two discrete processes X and Y:

$$C_{XY}(\tilde{\tau}) = \frac{1}{n-|\tilde{\tau}|} \sum_{k=1}^{n-|\tilde{\tau}|} (X_k - \bar{X}_T)(Y_{k+\tilde{\tau}} - \bar{Y}_T) , \tag{8.11}$$

λ_X and λ_Y are the indices of the exponent of the Markov random processes.

Expression (8.10) follows from the theoretical representation of the covariance of the estimates of $C_{XY}(\tilde{\tau}_1)$ and $C_{XY}(\tilde{\tau}_2)$ as

$$\begin{aligned} \mathrm{Cov}[C_{XY}(\tilde{\tau}_1), C_{XY}(\tilde{\tau}_2)] = \frac{1}{n_{XY}} \sum_{k=-\infty}^{\infty} & [\gamma_{XX}(k)\gamma_{YY}(k+\tilde{\tau}_1+\tilde{\tau}_2) \\ & + \gamma_{XY}(k+\tilde{\tau}_1)\gamma_{YX}(k+\tilde{\tau}_2)] , \end{aligned} \tag{8.12}$$

where $\gamma_{XX}, \gamma_{YY}, \gamma_{XY}, \gamma_{YX}$ – are non-normalized autocorrelation and cross-covariance functions connected with the numerators of (8.6) and (8.7), n_{XY} an effective number of the pairs of processes X and Y at shifts $\tilde{\tau}_1$ and $\tilde{\tau}_2$, respectively. If we assume that the processes do not correlate ($\gamma_{XY} = \gamma_{YX} = 0$), then from (8.12) and from the theory that these processes are autoregressive (Markov) of the first order, there follows expression (8.9). Parameters λ_X and λ_Y are incorporated in the expressions:

$$\gamma_{XX}(\tilde{\tau}) = \sigma_F^2(X)e^{-\lambda_X|\tilde{\tau}|} , \tag{8.13}$$

$$\gamma_{YY}(\tilde{\tau}) = \sigma_F^2(Y)e^{-\lambda_Y|\tilde{\tau}|} , \tag{8.14}$$

where $\sigma_F^2(X)$ and $\sigma_F^2(Y)$ are the MSDs of the longwave component for different times of observations.

The quantitative measure of data teleconnectivity in Fig. 8.24 was determined using the expression:

$$Q(X, \tilde{\tau} = 0) = \frac{1}{N(X, \tilde{\tau} = 0)} \sum_Y H(|R_{\tilde{X}_T\tilde{Y}_T}(0)| - |\hat{R}_{\tilde{X}_T\tilde{Y}_T}(0)|)\,|R_{\tilde{X}_T\tilde{Y}_T}(0)| , \tag{8.15}$$

where:

$$\hat{R}_{\tilde{X}_T\tilde{Y}_T}(0) = 2\,SD(C_{XY}) , \tag{8.16}$$

$H(\alpha)$ is the Heavyside step function, $N(X, \tilde{\tau} = 0)$ the total number of observations, for which $|R_{\tilde{X}_T\tilde{Y}_T}(0)| \geqslant |\hat{R}_{\tilde{X}_T\tilde{Y}_T}(0)|$.

It is seen from Fig. 8.24 that teleconnections between the equatorial Pacific and the following geographical regions (the sign $\pm$corresponds to synphase/counterphase variations are most significant):

	(+)	(−)
The tropics	Equatorial Central Pacific, East Africa	Polynesia, Australia, Peru coastline, northeastern Brazil
Extratropical latitudes	The western North Atlantic, West Europe, Siberia	Hawaii, arid zone of the South Pacific

Teleconnections between these regions of the tropical Pacific and the Gulf Stream EAZO, most significant from the viewpoint of its maximum amplitudes of the radiation budget and its longwave component, are most important to us. Now we calculate the shift correlation with zero lagging of this most important EAZO of the North Atlantic for the radiation budget components, using formula (8.10). Results of the respective calculations are given in Figs. 8.25–8.27. They differ from similar calculations [123–125] for the total number of initial daily measurements from satellites.

The data in Figs. 8.25–8.27 shows that the effect of the Gulf Stream EAZO on the radiation budget and its longwave component in the North Atlantic is at the limit of climatic significance determined from (8.9) and corresponding to 95% confidence level. In accordance with (8.9), the indicated limit is determined by significant coefficients of correlation reaching a maximum close to 0.6, and moves (for the radiation budget) along the Gulf Stream current at a distance of about 1500 km, and drops to 0.3 at a distance of about 2500 km.

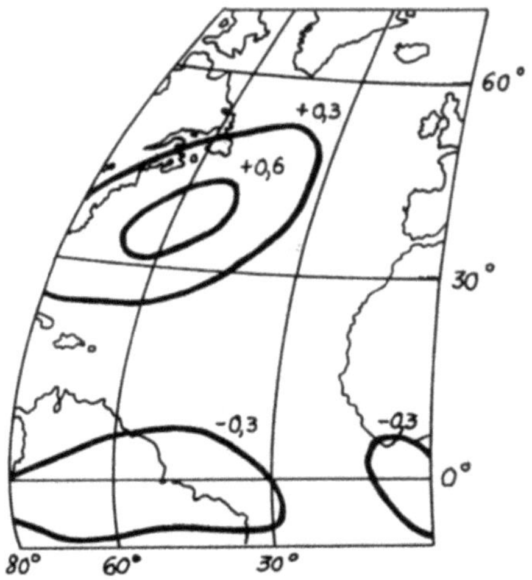

Fig. 8.25

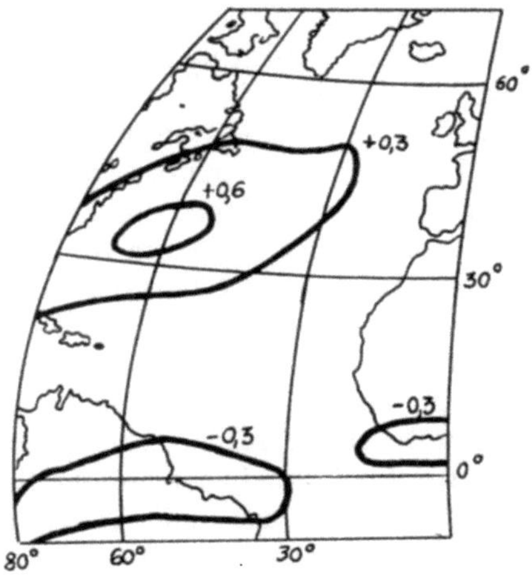

Fig. 8.26

Fig. 8.25. Isolines of the shift correlation (with zero lagging) of the radiation budget of the North Atlantic relative to the Gulf Stream EAZO

Fig. 8.26. Isolines of the shift correlation (with zero lagging) of the radiation budget longwave component for the North Atlantic with respect to the Gulf Stream EAZO

Fig. 8.27. Isolines similar to those in Fig. 8.26 for the shortwave component

These results correspond to results of the 2–D analysis of anomalies (see Fig. 8.15). In this figure similar regularities are observed for the correlation function calculated over the entire territory (10°S–70°N, 0°–80°W), without special selection of the Gulf Stream area.

The result obtained explains the anisotropy of the curves in Fig. 8.15 a, b for correlation functions, based on the prevailing effect of the Gulf Stream EAZO and its special climatic significance for the given territory.

Similar conclusions can be drawn from analysis of the contours of the shift correlation for the longwave component (Fig. 8.26). The respective levels (Fig. 8.27) of significance of the cross-correlation for the shortwave component are less than those for the longwave component and radiation budget. This fact confirms the assumption of particular importance of anomalies in this EAZO for the territory considered.

It has been mentioned that the broken anisotropy of the isolines of correlation functions for the North Atlantic, typical of the summer season, can also be connected with a certain effect of the near-equatorial Atlantic EAZO. However, the effect of this EAZO grows in the winter Northern Hemisphere (it was qualified as the transfer of warm waters across the equator), causing the more isotropic change of isocorrelants in Fig. 8.15. In addition, in the transition season the Gulf Stream EAZO exhibits maximum effect. These remarks show that the final conclusions about the formation of teleconnections between radiation budget anomalies in mid-latitudes in different season need further studies. The relatively small-scale implication of the most significant, from the viewpoint of the remote effect of the tropical Pacific, Northern Atlantic EAZO (the effect of this zone, as follows from Figs. 8.25 – 8.27, is shown only within the Atlantic Ocean) can be explained by smoothed monthly means. For initial fields of radiation budget and its components, the horizontal effect of this EAZO would probably be more substantial.

There is some uncertainty of the calculation procedure considered (this uncertainty is connected with the absence of instant measurements data and with obtaining estimates on the basis of smoothed monthly mean fields) and of respective estimates for 45 monthly mean fields characterized by smoothed monthly observations. Despite this, an assumption can be made that further

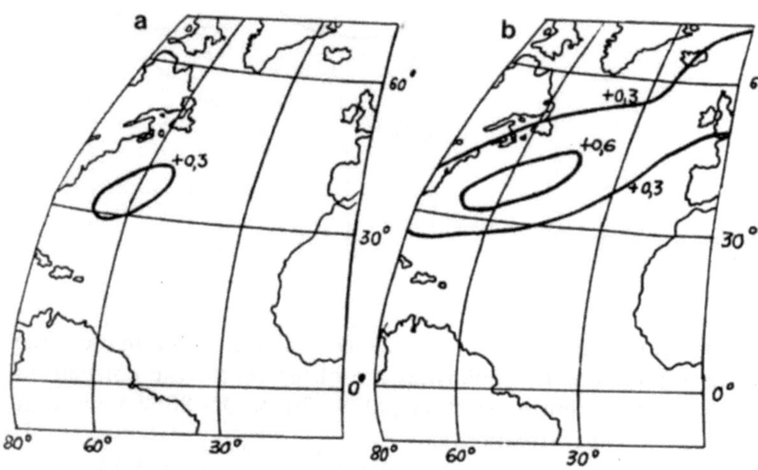

Fig. 8.28 a, b. Isolines of cross-correlation of the longwave component with respect to the Gulf Stream EAZO with: **a** a-5-month shift and **b** +5-month shift

increasing the length of the time series will increase the indicated significance and allow study of correlations between the considered EAZO and other mid-latitude regions. This confirms the conclusion of the existence of cross-correlation about ±5 months for radiation budget anomalies with initial reference level in the Gulf Stream EAZO (Fig. 8.28).

Figure 8.28 shows, in particular, the presence of lagged correlations in mid-latitudes of the North Atlantic relative to the Gulf Stream EAZO, in which the effect of disturbances, as follows from Fig. 8.24, is secondary (the original source of the ERB anomalies is the tropical Pacific manifested in the Southern Oscillation/El Nino events [106, 125]).

The Fig. 8.28 data show a certain increase in the distance of the effect of Gulf Stream EAZO teleconnections with a lagged correlation as compared to zero shift. Further studies in this direction will make it possible to perform a more detailed analysis of correlations between anomalies of the radiation budget and its components in the North Atlantic and respective anomalies of meteorological parameters in continental Europe.

8.10 Prospects for Using Satellite Measurements in Climate Studies

With analysis of time series of satellite-measured radiation budget and its components as an example, potential has been demonstrated of using these measurements in studies of the processes of the atmosphere-ocean interaction. A major problem which needs further investigation, is to understand the mechanism of the feedback between cloudiness and the components of the radiation budget of the ocean-atmosphere system [125]. Examples of the effect of cloudiness of the fields of radiation budget given can be an initial approximation to the solution of such a problem. Specified data are needed on relationships between microphysical and radiative characteristics of cloudiness as well as specific calculations of the radiation transfer in cloudy conditions. Cumbersome calculations of the radiation fields in the presence of clouds are hardly expedient without subsequent approximation of radiation fields by orthogonal polynomials.

There is a real possibility of using the data of the optical range from polar orbit and geostationary satellites in climate studies. Part of the archived satellite information on radiation budget has been used by some authors in experimental studies of the atmospheric and oceanic circulation and in the solution of some climatic problems. However, the practical use of satellite information in climate studies is still in its initial stage, because of short series of observations, low accuracy of measurements, etc. A necessary condition for further studies in this direction is study of the diurnal change of the radiation budget components by combining the data of geostationary and polar orbit satellites and of the satellites in shifting orbits. In this connection, completion of a special satellite experiment on radiation budget (ERBE) may be useful. Results of the transition from the data of scanning radiometers to radiation fluxes represented in certain map projections have to be specified. The information content of the data of different types of space-borne instruments must be assessed.

From the climatic point of view, it is important to assess the phytobiometric resources of vegetation and ocean. The climatic effect of increased carbon dioxide has not been considered here, but an estimation of the biometric parameters of the vegetation state, directly concerns this problem.

Interactive procedures of classification are useful not only in the assessment of the soil state and vegetation cover from air- and space-derived data but also in the classification of cloud formations from the view point of assessing the effect of cloudiness in studies of the atmospheric general circulation and of climate. Another important sphere of application of interactive procedures of analysis of satellite information is a "learning" from measurements over test areas, for instance, EAZOs, for statistical analysis of the global fields of radiation budget and its components.

Of particular importance, is the development of climate models, which would include estimates of the signals of short-term climate changes and estimates of the noise constituents of the observational data. Results of analysis of the interannual variability of radiation budget and its components can serve as the methodological basis for such studies, in which the understanding of the radiation-cloudiness relationships is a key problem.

As an example, we consider the sensitivity of climate models based on the radiation budget data to variations in cloudiness.

In radiative climatology, the sensitivity of the radiation budget to variations in cloud amount n_c is usually investigated. This parameter is estimated in the following way:

$$\delta = \frac{\partial B}{\partial n_c} = \frac{\partial G}{\partial n_c} - \frac{\partial F}{\partial n_c} \ . \tag{8.17}$$

The partial derivative $\partial G/\partial n_c$ characterizes the albedo (short-wave) effect of clouds; $\partial F/\partial n_c$ their effect on the outgoing longwave radiation of the atmosphere (greenhouse effect). It has been mentioned that, understanding the mechanism for the cloudiness-radiation feedback, is based on assessments of the radiation budget of the surface-atmosphere system with varying cloudiness. In most studies it has been shown that variations in the shortwave component prevail over those in the longwave ($\delta < 0$). Some studies, however mention a mutual compensation of these effects ($\delta \approx 0$).

A characteristic feature of studies on satellite climatology is a representation of planetary albedo as:

$$A = A_c n_c + A_s(1 - n_c) \ ,$$

where A_c, A_s are, respectively, the albedos of clouds and cloud-free territories. The next assumption presupposes independence of A_s and A_c of the cloud amount, and then:

$$\frac{\partial G}{\partial n_c} = -S_0 \frac{\partial A}{\partial n_c} = -S_0(A_c - A_s) \ . \tag{8.18}$$

Since

$$\frac{\partial F}{\partial n_c} = \frac{\partial F}{\partial A}\frac{\partial A}{\partial n_c} = (A_c - A_s)\frac{\partial F}{\partial A} \ , \tag{8.19}$$

the problem of analysis of the sensitivity of the radiation budget to cloud cover variations is, thus, reduced to finding the $A_c - A_s$ difference and to estimating the derivative $\partial F/\partial A$.

Despite the excellent possibility of estimating the $A_c - A_s$ difference from satellite data, no estimates have been obtained so far. This is explained by two reasons: difficulties in classification of cloud formations and in obtaining the surface albedo from satellite data. Here the concept of minimum albedo is usually used [338]; a minimum value of recorded brightness within the

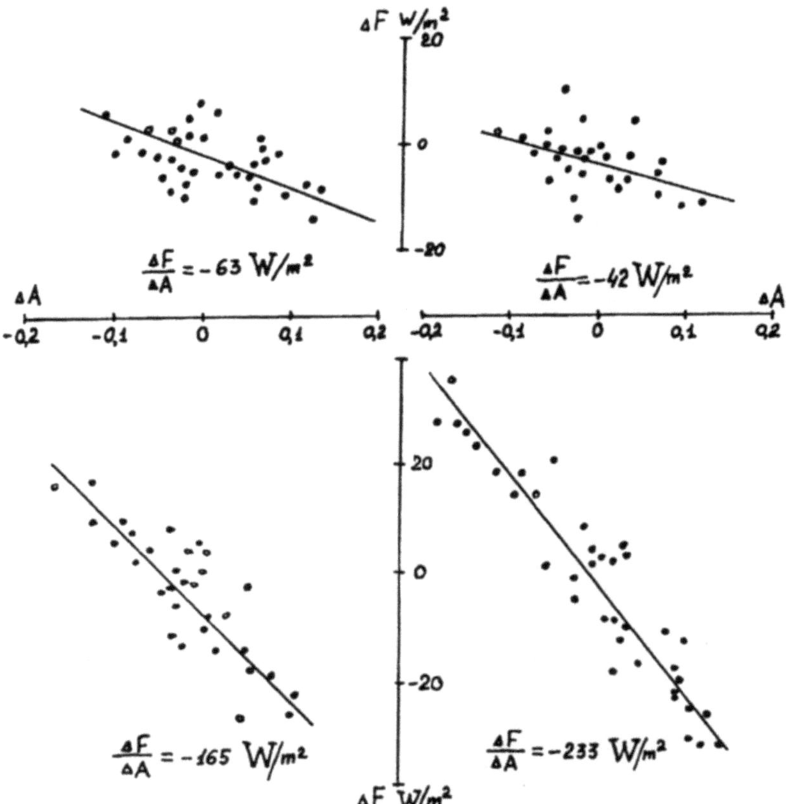

Fig. 8.29. Regressive estimates of the $\partial F/\partial A$ derivatives for the four EAZOs shown in Fig. 8.17

horizontal scales 250 – 500 km is ascribed to A_s; the values for A and n_c are taken from climatological data, and A_c is estimated with the use of binary classification.

The results obtained make it possible to solve the problem in general, without using these suppositions and assumptions. It is possible to identify four classes or cloud formations by respective energy levels which have been used in Sect. 8.4 (Table 8.4) in analysis of the MSD fields of radiation budget. Also, it is possible to obtain A_s for given classes of cloud formations and for atmospheric aerosols using the scheme of surface radiation budget retrieval. An additional analysis of regression relationships between the interannual variability of the longwave component and albedo makes it possible to estimate the derivative $\partial F/\partial A$.

Figure 8.29 illustrates the regression estimates of derivatives for the 45-monthly mean observations and four EAZOs used in the programme "Sections" mentioned above. The data in Fig. 8.29 reveal substantially different effects of clouds in these zones: the slope of the regression line varies from $-42\ W/m^2$ for the Norwegian EAZO to $-233\ W/m^2$ for the tropical test area. These variations, may be explained by substantially different effects of cirrus clouds in the first case and of thick convective clouds in the second case.

Figures 8.30 – 8.32 show the isolines of partial derivatives $\partial F/\partial n_c$, $\partial A/\partial n_c$ and $\partial B/\partial n_c$ calculated for the NOAA data using, respectively, formulas (8.17 – 8.19) and the results discussed in Sect. 8.4. The data in Figs. 8.30 – 8.32 do not allow definite conclusions about the radiation-cloudiness interaction in

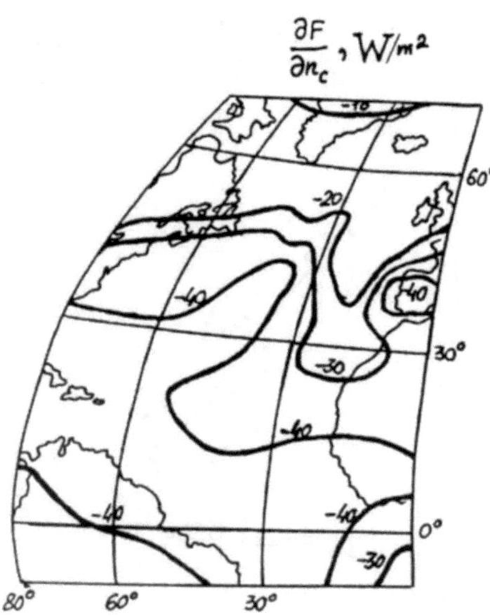

Fig. 8.30. Isolines of the $\partial F/\partial n_c$ derivatives

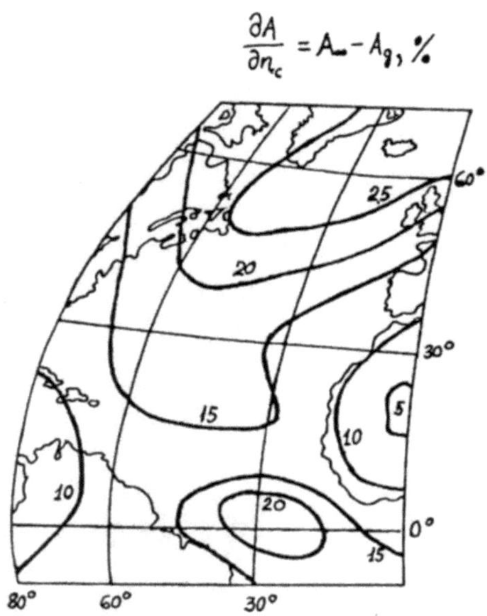

Fig. 8.31. Isolines of the $\partial A/\partial n_c$ derivatives

Fig. 8.32. Isolines of the $\partial B/\partial n_c$ derivatives

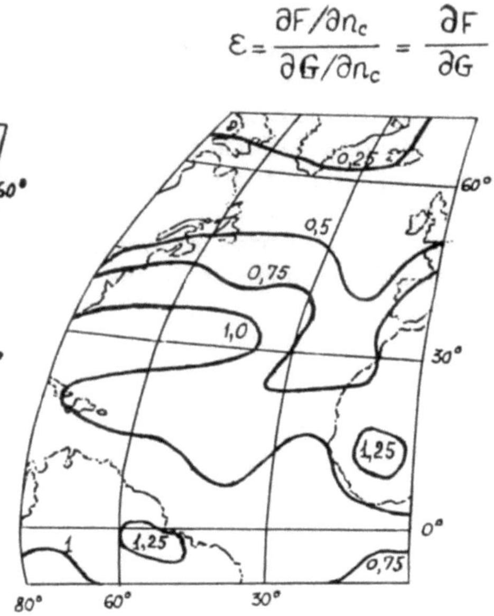

Fig. 8.33. Isolines of the sensitivity of the long- and shortwave components of the radiation budget to variations in cloud amount

EAZOs. Only a cluster of isolines $\partial F/\partial n_c$ can be observed in the region of the Gulf Stream transfer to the Northern Atlantic Drift, a development of the tropical Atlantic and the land-ocean interface in the $\partial A/\partial n_c$ values, and to some extent, a combined effect of these phenomena in $\partial B/\partial n_c$ isolines. Clearly, the mechanism for the EAZO effect on the formation of characteristic cloud systems in the region of the Northern Atlantic needs further study.

However, the estimates of derivatives can be useful in analysis of the following climatological characteristic:

$$\varepsilon = \frac{\partial F/\partial n_c}{\partial G/\partial n_c} = \frac{\partial F}{\partial G},$$

involving combined variations in the longwave and shortwave components of the radiation budget when cloudiness varies.

Figure 8.33 shows the isolines of the estimates of this characteristic. It is seen that $\varepsilon > 1$ only in a narrow area of the ocean basins near South America and the Gulf Stream current with maximum $\varepsilon \approx 1.25$. For the remaining part of the territory $\varepsilon < 1$ and reaches a minimum of $\varepsilon \approx 0.3$ in the region of Greenland. On average, for the whole analyzed territory (10°S–70°N, 0°–80°W) $\varepsilon \approx 0.74$, which confirms the prevailing effect of variations in the shortwave component over the longwave component with varying cloudiness.

Thus: the results of the interannual variability of the radiation budget components; of the surface radiation budget retrieval; of analysis of cloudiness as a factor of the interannual variability; and anomalies of the radiation budget, in the light of the EAZO concept, can be considered an initial stage of the use of satellite observational time series in climate studies. The subsequent development of results of empirical analysis of 2-D fields of anomalies, from the learning samples based on accompanying measurements in EAZOs can, in the future, serve as the basis for hydrometeorological interpretation of satellite and ship board measurements in the interests of studies of short-period climate changes.

Conclusion

Successful development of complex experimental and theoretical studies in the interests of the Earth's resources and environmental studies using remote spectral-polarization optical techniques, has brought forth space-based spectrophotometry to the front line. It is involved in various scientific, technical and economic programmes with the help of satellites, manned spacecraft, and permanent orbital stations. The purpose of the book has been to generalize the available data, to perform a comparative analysis of the existing techniques, and to develop new algorithms to consider the effect of the aerosol atmosphere on space-derived spectrophotometric information. Use of an advanced spherical harmonics technique, similarity ratios and exact solutions of the initial equation of radiation transfer has made it possible to model the spatial angular and spectral distribution of the radiation fields of the atmosphere-object-background system. This is necessary to solve a number of urgent problems in the optics of natural media. These problems first appeared after successful spectrophotometry of the Earth from Soyuz-5, Soyuz-7 manned spacecraft and from the Solyut-1 orbital station, and after complex subsatellite optical experiments.

Emphasis has been laid on the following problems:

1. Theoretical modeling of the effect of the aerosol atmosphere on spectral brightnesses, brightness contrasts, reflective characteristics, color and extent of polarization of natural formations, extended backgrounds and objects when observing the Earth from space in the spectral interval $\lambda = 400-800$ nm, in the interests of studies of the Earth's resources and the environment.
2. Developing techniques and algorithms for atmospheric correction of space-derived optical information.
3. Practical use of the techniques for atmospheric correction of spectrometric information from Meteor and Kosmos satellites in automated computers for image processing.

Comparison of Sobolev's modified classical technique with an advanced spherical harmonics technique has shown that they are identical within a 5% uncertainty. This result is of principal importance for numerical solution of direct and inverse problems of the radiation transfer theory, since it allows restriction to a consideration of the matrices of the optical-physical parameters of natural media not higher than the third order. In addition, the new calculation scheme permits a more detailed study of the interaction of shortwave radiation with a realistic stratified aerosol atmosphere based on the latest achievements in the field of atmospheric optics.

The conclusions about the effect of the vertical optical inhomogeneity of the aerosol atmosphere and non-orthotropicity of reflection from the surface on the field of reflected radiation at different viewing angles, sun illumination and wavelengths are of great importance.

The horizontal diffusion of photons, in the system "atmosphere-object-background" for realistic natural media at high sun elevations and nadir viewing, is weak as shown by results of estimation.

Comparison of the effects of the horizontal inhomogeneity and non-orthotropicity of the surface suggests that, consideration of the surface non-orthotropicity, is very important in solving the problem of atmospheric correction of space-derived information. This important conclusion enables use of a simpler 1-D theory of radiation transfer in modeling the radiation fields in space-based spectrophotometry and in the development of atmospheric correction of satellite images.

Another important conclusion is the need to use realistic transformation functions in atmospheric correction which involve the basic laws of outgoing radiation field formation. The conventional empirical transformation functions are mostly connected with a sufficiently arbitrary presentation of the atmosphere as a multiplicative interference. In contrast the atmospheric correction techniques developed in this book have made it possible to define the real forms of these transformations proceeding from the results of the radiation field modeling in the surface-atmosphere system. One more substantial result consists of demonstrating atmosphere correction techniques in real time, using specialized information processing means, by approximating the results of solution of the transfer theory direct problems by orthogonal polynomials.

One of the principal spheres of application of the results discussed, is the solution of several inverse problems connected with determination of the radiative characteristics of natural formations from the outgoing radiation intensity recorded by the satellite. Knowing the spectral intervals where the distorting effect of the atmosphere is small, and with available estimates of this effect as a function of the outgoing radiation intensity, the spectral albedo of natural formations, their viewing angle, and solar zenith angle, the information content and improve the quality of air- and space-derived images used in studies of the environment and of natural resources can be enhanced.

The recommendations, algorithms and techniques for considering the distorting effect of the atmosphere suggested here can be used in the development and improvement of operational complexes to study the Earth's resources, involving both the related satellite-based instruments and the techniques for interpreting the data of radiation measurements. The applied approach to the radiation transfer can be used in other spectral regions: UV, IR, microwave. For instance, in the thermal spectral interval this can be realized only if the parameters of the boundary problem are changed.

The interest in modeling the radiation fields has recently grown in connection with studies of the impact of human activity and environmental pollution on global climate. It is important that the suggested approach can be used, in calculations of radition fluxes and heat flux divergences, to parameterize them in models of atmospheric general circulation and numerical weather forecasts.

Also, it can be used to determine the characteristics of the roughness zones in the ozean and to analyze the optical data obtained from the ocean water basins, based on satellite measurements.

Not less important is a study of the radiative effects in assessing the possible impact of human activity and environmental pollution on the global-scale heat balance. With the help of techniques developed by the authors one can estimate the combined effect of aerosol, carbon dioxide, cloudiness and other factors on the Earth's radiation budget and climate.

Results of calculations of the radiation fields at different levels in the atmosphere with varied meteorological situations and in different spectral intervals are also needed in analysis of the information obtained from balloons, radio sondes, meteorological rockets, etc.

In the latter case, its interpretation can promote successful solution of inverse problems in the optics of atmospheric aerosol. These calculations can be useful in the interpretation of groundbased astrophysical measurements. It should be emphasized that the results considered here may be used directly to study the light and heat regime of other planets in the solar system (Venus, Mars, etc.) based on the data for atmospheric composition of these planets obtained recently from automatic interplanetary stations and descending modules.

References

1. Avanesov GA (1980) Stud Earth Space 2:66
2. Avaste OA, Atroshenko VS (1960) Izv Akad Nauk SSSR Ser Geogr 3:265
3. Avaste OA, Darbinian RA (1975) Izv Akad Nauk SSSR FAO 2:1030
4. Avaste OA, Mullamaa YAR, Shifrin KS (1964) Studies on atmospheric physics of the Est. SSR Acad Sci Inst Atmos Phys 6:245
5. Alexandrov VV, Gorsky ND (1983) Algorithms and programmes of the structural method for data processing. Nauka Press, Leningrad
6. Alexandrov VV, Gorsky ND (1985) Presentation and processing of images. Recursive method. Nauka Press, Leningrad
7. Ambartsumian VA (1966) Scientific studies, vol 1. Erevan
8. Apresian LA, Kravtsov YA (1983) Radiation transfer theory. Nauka Press, Moscow
9. Kondratyev KY (ed) (1978) Atmospheric aerosol and its effect on radiation transfer. Gidrometeoizdat, Leningrad
10. Babak EV, Beliaev AS, Guitin YA (1981) Opt Spect 2:349
11. Badaev VV, Georguievsky YS, Pirogov SM (1975) Izv Akad Nauk SSSR FAO 11:522
12. Barabanenkov YN (1975) Adv Phys Sci 117:49
13. Barabanenkov YN, Kravtsov YA, Rytov SM, Tatarsky VI (1970) Adv Phys Sci (USSR) 102:3
14. Barret E, Kurtis L (1979) Introduction to cosmic land use. Progress Press, Moscow
15. Barykin AS, Ermolaev AG, Pytiev YP et al. (1980) Method for adaptive noise suppression based on morphological information about images. Identification of papillar patterns on the optical-digital complex. DVGU Press, Vladivostok
16. Barykin AS, Kozoderov VV, Popov VP (1985) Stud Earth Space 4:78
17. Barykin AS, Kozoderov VV, Popov VP (1987) Stud Earth Space 1:77
18. Belov VV, Krekov GM (1981) Cosmic Invest 19:139
19. Belchansky GI, Kondratyev KY, Kozoderov VV, Sazonov NV, Topchiev AG, Fedchenko PP (1984) Stud Earth Space 5:8
20. Biriukov YL, Krylov YV (1974) Izv Akad Nauk SSSR FAO 10:1231
21. Borisoglebsky GI, Kozoderov VV (1982) Meteorol Hydrol 3:94
22. Borisoglebsky GI, Kozoderov VV, Fedchenko PP (1984) Soil Sci 5:140
23. Buznikov AA, Kondratyev KY, Smokty OI (1975) Dokl Akad Nauk SSSR 222:821
24. Budyko MI (1974) Climate changes. Gidrometeoizdat, Leningrad
25. Beliu, Stulinger E (1977) Skylab orbital station. Mashinostroenie Press, Moscow
26. Van de Hulst G (1961) Scattering of light on small particles. IL, Moscow
27. Veinberg A, Vigner E (1960) Physics of nuclear reactors. IL, Moscow
28. Kondratyev KY et al. (1974) Vertical profiles of the parameters of the atmospheric transfer operator. In: Problems in atmospheric physics. LGU Press, Leningrad
29. Vladimirov VS (1961) Mathematical problems of one-velocity theory of particles transport. Trans Steklov Math Inst USSR Acad Sci 61:120
30. Vladimirov VS (1967) Equations of mathematical physics. Nauka Press, Moscow
31. Kondratyev KY et al. (1975) Izv Akad Nauk SSSR FAO 11:348
32. Kondratyev KY et al. (1977) Izv Akad Nauk SSSR FAO 13:471
33. Gavrilova LA (1979) Izv Akad Nauk SSSR FAO 15:110
34. Gakhov FD (1977) Boundary problems. Nanka Press, Moscow

35. Germogenova TA (1973) Numerical methods for solving the problems of multiple light scattering in the atmosphere. In: Summaries Rep 24th IAF Congr, USSR, Baku, 7–13 October 1973. VINITI, Moscow
36. Germogenova TA, Sushkevich TA (1969) Solution of the transfer equation using the average fluxes technique. In: Physical aspects of reactors protection, vol 3. Atomizdat, Moscow
37. Golovko VA (1985) The effects of an adequate account of the atmosphere and errors in spectral measurements on the reliability of identifying the state of natural objects. In: Pakhomov LA, Mishev DN (eds) Remote sounding of the Earth from "Meteor-Priroda" satellite. Gidrometeoizdat, Leningrad
38. Goncharsky AV, Yagola AG, Leonov AS (1971) J Calc Math Math Phys 11:1296
39. Gorelik AL, Scripkin VA (1977) Recognition techniques. High School Press, Moscow
40. Grenander U (1978) Lectures on the theory of images, vol 1. Synthesis of images, vol 2. Analysis of images. Mir Press, Moscow
41. Goody R (1966) Atmospheric radiation. Mir Press, Moscow
42. Gutshabash SD (1963) Izv Akad Nauk SSSR, Geophysics 7:1117
43. Deirmendjian D (1971) Scattering of electromagnetic radiation on spherical polydisperse particles. Mir Press, Moscow
44. Denisov NG, Tatarsky VI (1963) Izv VUZov, Radiophysics 6:488
45. Dzhadd D, Vyshetski G (1978) Color in science and technology. Mir Press, Moscow
46. Pakhomov LA, Mishev DN (1985) Remote sounding of the Earth from "Meteor-Priroda" satellite. The Soviet-Bulgarian Experiment "Bulgaria-1300-II". Gidrometeoizdat, Leningrad
47. Alekseev AS (ed) (1983) Remote sounding: a quantitative approach. Nedra Press, Moscow
48. Dolguinov AZ, Gnedin YN, Silantyev NA (1979) Propagation and polarization of radiation in cosmic medium. Nauka Press, Moscow
49. Dosov VN, Kozoderov VV, Pakhomova LA (1985) A combined analysis of SMP-32 and MSU-M data. In: Pakhomov LA, Mishev DN (eds) Remote sounding of the Earth from "Meteor-Priroda" satellite. Gidrometeoizdat, Leningrad
50. Devison B (1960) Theory of the transfer of neutrons. Atomizdat Press, Moscow
51. Diabin YP, Tantashev MV et al. (1977) Izv Akad Nauk SSSR FAO 13:1205
52. Ermakov SM (1971) Monte-Carlo simulations and related problems. Nauka Press, Moscow
53. Zelmanovich IL, Shifrin KS (1968) Tables on light scattering, Vol 3. Gidrometeoizdat, Leningrad
54. Zolotukhin VG, Usikov DA, Grushin VA (1980) Stud Earth Space 3:58
55. Zuev VE (1970) Propagation of visible and infrared waves in the atmosphere. Sov Radio Press, Moscow
56. Zuev VE (1981) Propagation of laser emission in the atmosphere. Radio Commun Press, Moscow
57. Zuev VE, Kabanov MW (1977) The transport of optical signals in the terrestrial atmosphere (in the interfering conditions). Sov Radio Press, Moscow
58. Zuev VE, Savelyev BA, Fadeev VY, Kozlov VA (1975) Radiative phase functions for water aerosols in the spectral interval 0.34–1.118μm. Meteorol Stud 22, Nauka Press, Moscow
59. Zuev VE, Ivlev LS, Kondratyev KY (1978) Izv Akad Nauk SSSR FAO 9:371
60. Zuev VE, Kabanov MV et al. (1978) Izv Akad Nauk SSSR FAO 14:1268
61. Zuev VE, Belov VV, Krepov GM (1982) Dokl Akad Nauk SSSR 266:1353
62. Ivanov AP (1969) Optics of scattering media. Nauka Tech Press, Minsk
63. Ivanov AI, Karimova LM, Korovchenko VA (1980) Izv Akad Nauk SSSR FAO 16:869
64. Ivanov VV, Leonov VV (1965) Izv Akad Nauk SSSR FAO 1:803
65. Ivlev LS (1982) Chemical composition and structure of atmospheric aerosols. LGU Press, Leningrad
66. Kondratyev KY, Buznikov AA, Vasilyev OB et al. (1972) A study on natural formations using combined observations from Soyuz-7 space craft and from flying laboratories. Probl Atmos Phys 10:3
67. Isimaru A (1981) Propagation and scattering of waves in randomly inhomogeneous media. Vol 1: Single scattering and the transfer theory. vol. 2. Multiple scattering, turbulence, rough surfaces and remote sounding. Mir Press, Moscow

68. Kondratyev KY et al. (1972) Studies of the environment from manned orbital stations. Gidrometeoizdat, Leningrad
69. Istomina LG, Malkevich MS, Siachinov VI (1970) Izv Akad Nauk SSSR FAO 6:468
70. Kabanov MV (1968) Izv Akad Nauk SSSR FAO 4:385
71. Keiz K, Zveifel P (1972) A linear transfer theory. Mir Press, Moscow
72. Keldysh MV, Marov MYa (1981) Cosmic investigations. Nauka Press, Moscow
73. Klimenko OY, Kozoderov VV (1984) Stud Earth Space 2:67
74. Kozoderov VV (1978) An assessment of the distorting effect of the atmosphere in decoding natural formations from space. In: Zolotukhin VG (ed) Aerospace studies of the Earth. Processing of videoinformation with the use of computers. Nauka Press, Moscow, p 24
75. Kozoderov VV (1979) Tr GosNITsIPR 5:22
76. Kozoderov VV (1980) Tr GosNITsIPR 8:51
77. Kozoderov VV (1983) Stud Earth Space 2:65
78. Kozoderov VV (1983) Stud Earth Space 1:69
79. Kozoderov VV (1987) Thematic interpretation of aerospace images. In: Proc Interdisc Meet "Application of results of air- and space-based studies of the Earth in the interest of agriculture in the USSR", Nauka Press, Leningrad, 8 pp
80. Kozoderov VV, Liubovny ND (1982) A consideration of the atmospheric effect in processing space-derived video information. In: Kondratyev KY (ed) Studies of hydrological cycle using space-borne means. Radio Commun Press, Moscow 90 pp
81. Kozoderov VV, Mishin IV (1980) Tr GosNITsIPR 8:64
82. Kozoderov VV, Mishin IV (1982) On the solution of the problem of radiative correction of remote sounding data. In: Tistschenko AP (ed) Techniques for remote sounding of atmospheric parameters and underlying surface from space. Gidrometeoizdat, Leningrad, p 68
83. Kozoderov VV, Sazhina LM (1983) A consideration on non-orthotropic reflection of surface objects in the problem of radiative correction space-derived images. In: Tistschenko AP (ed) Methods for processing and thematic analysis of multispectral aeroscape videoinformation. Gidrometeoizdat, Leningrad, p 47
84. Kozoderov VV, Liubovny ND, Tischenko AP (1980) Tr GosNITsIPR 8:83
85. Kozoderov VV, Sazhina LM, Shulgina IB (1980) Tr GosNITsIPR 8:35
86. Kolesov AK (1976) Part I: Vestnik LGU 7:136 (1977) Part II: Vestnik LGU 1:127
87. Kolesov AK (1976) Astrophysics 12:83, 12:485
88. Kolesov AK, Smokty OI (1970) Astron J 47:397
89. Kolesov AK, Smokty OI (1971) Astron J 48:1013
90. Kolesov AK, Sobolev VV (1969) Tr Astron Obs LGU 26:3
91. Kondratyev KY (1980) Radiative factors of the present-day golbal climate changes. Gidrometeoizdat, Leningrad
92. Kondratyev KY, Zhvalev VF (1981) First Global GARP Experiment Vol 1: Aerosol and climate. Gidrometeoizdat, Leningrad, p 7
93. Kondratyev KY, Pozdniakov DV (1981) Aerosol models of the atmosphere. Nauka Press, Moscow
94. Kondratyev KY, Smokty OI (1972) Dokl Akad Nauk SSSR 206:1102
95. Kondratyev KY, Smokty OI (1972) Dokl Akad Nauk SSSR 206:1349
96. Kondratyev KY, Smokty OI (1973) Dokl Akad Nauk SSSR 208:77
97. Kondratyev KY, Smokty OI (1973) Tr GGO 295:24
98. Kondratyev KY, Smokty OI (1973) Izv Akad Nauk SSSR FAO 9:1269
99. Kondratyev KY, Smokty OI (1974) Izv Akad Nauk SSSR FAO 10:195
100. Kondratyev KY, Smokty OI (1972) Dokl Akad Nauk SSSR 207:86
101. Kondratyev KY, Smokty OI, Kozoderov VV (1985) The effect of the atmosphere on studies of natural resources from space. Mashinostroenie Press, Moscow
102. Kondratyev KY, Grigoryev AA, Pokrovsky OM (1975) The information content of the data of the remote sounding of the environment and natural resources from space. LGU Press, Leningrad
103. Kondratyev KY, Moskalenko NI, Terzi VF (1980) Dokl Akad Nauk SSSR 253:1354
104. Kondratyev KY, Moskalenko NI, Pozdniakov DV (1983) Atmospheric aerosol. Gidrometeoizdat, Leningrad

105. Kondratyev KY, Kozoderov VV, Korzov VI (1982) Some results of comparisons of the data of air- and space-based measurements. In: Radiative studies in the atmosphere. Gidrometeoizdat, Leningrad, p 29
106. Kondratyev KY, Kozoderov VV (1984) Anomalies of radiation balance and heat content of the oceanic mixed layer as indicators of energy active zones. Progr Sci Technol Ser Atmos Ocean Space, programme "Sections", Vol 4, VINITI, Moscow
107. Kondratyev KY, Kozoderov VV, Fedchenko PP, Somova SA, Karvovskaya LB (1985) Stud Earth Space 2:29
108. Kondratyev KY, Kozoderov VV, Fedchenko PP (1986) Studies of the soils and vegetation from aircraft and from space. Gidrometeoizdat, Leningrad (see also Kondratyev KY, Fedchenko PP (1982) Spectral reflectivity and regognition of vegetation. Gidrometeoizdat, Leningrad)
109. Korzov VI, Krasilshchikov LB (1974) Methods for and some results from aircraft measurements of the angular and spectral distribution of reflected shortwave radiation. In: Kondratyev KY (ed) Radiative processes in the atmosphere and on the ground. Gidrometeoizdat, Leningrad, p 57
110. Kondratyev KY, Grigoryev AA, Pokrovsky AG et al. (1974) Spacebased remote sounding of minor gaseous and aerosol components of the atmosphere. LGU Press, Leningrad
111. Coleman GB, Andrews HS (1979) IEEE 67:82
112. Lazarev AI, Nikolaev AG, Khrunov EV (1979) Optical studies in space. Gidrometeoizdat, Leningrad
113. Loskutov VM (1972) Tr Astron Observ LGU 29:33
114. Malkevich MS (1973) Optical studies of the atmosphere from satellites. Nauka Press, Moscow
115. Malkevich MS (1978) An account of the atmosphere in the problems of studying Earth resources from space. In: Malkevich MS (ed) Studies of Earth resouces from space. Methods and means to measure and to process information. Nauka Press, Moscow, p 110
116. Malkevich MS, Badaev VV (1981) Stud Earth Space 4:45
117. Malkevich MS, Mishin IV (1983) Stud Earth Space 3:105
118. Marchuk GI (1961) Methods to calculate nuclear reactors. Gosatomizdat, Moscow
119. Marchuk GI (1974) Meteorol Hydrol 3:17
120. Marchuk GI (1979) Meteorol Hydrol 7:25
121. Marchuk GI (1982) Mathematical modeling in the environmental problem. Nauka Press, Moscow
122. Marchuk GI (1964) Cosmic Invest 2:462
123. Marchuk GI, Kondratyev KA, Kozoderov VV (1983) Dokl Akad Nauk SSSR 272:1099
124. Marchuk GI, Kondratyev KY, Kozoderov VV (1984) Stud Earth Space 1:3
125. Marchuk GI, Kondratyev KY, Kozoderov VV, Khvorostyanov VI (1986) Clouds and climate. Gidrometeoizdat, Leningrad
126. Maslennikov MV (1968) The Miln anisotropic-scattering problem. Nauka Press, Moscow
127. Matiyasevich LM (1979) Atmospheric interferences in Earth observations using space-based means operating in the optical wavelength region. In: Aerospace studies of the Earth. Nauka Press, Moscow, p 209
128. Marchuk GI et al. (1976) Monte-Carlo simulations in atmospheric optics. Nauka Press, Novosibirsk
129. Minin IN (1969) Vestnik LGU. Math Mech Astron 1:133 and (1963) 3:106
130. Minin IN (1966) Astron J 43:1244
131. Minin IN (1973) Izv Akad Nauk SSSR FAO 9:829
132. Minin IN (1982) On an interpretation of satellite-based measurements of the brightness and light polarization by the cloudless atmosphere and underlying surface. In: Problems in atmospheric physics. Methods for studying the gas and aerosol composition of the atmosphere, No 17, p 11
133. Mishin VP (1982) Izv VUZov Geodes Aerophotosurvey 1:9
134. Mishin IV (1982) Stud Earth Space 4:95
135. Mishin IV, Orlov VM (1979) Izv Akad Nauk SSSR, FAO 15:166
136. Mishin IV, Sushkevich TA (1980) Stud Earth Space 4:69

137. Mishin IV, Tishchenko AP (1981) Stud Earth Space 1:48
138. Mnatsakian MA (1975) Dokl Akad Nauk SSSR 225:1049
139. Kondratyev KY et al. (1981) Modeling the optical characteristics of atmospheric aerosol. In: First Global GARP Experiment. Vol 1: Aerosol and climate. Gidrometeoizdat, Leningrad, p 130
140. Muskhelishvili NI (1968) Singular integral equations: boundary problems of the theory of functions and their applications in mathematical physics. Nauka Press, Moscow
141. Kondratyev KY et al. (1970) Dokl Akad Nauk SSSR 195:1084
142. Kondratyev KY et al. (1971) Dokl Akad Nauk SSSR 197:1066
143. Marchuk GI (1972) Cosmic Invest 10:245
144. Kondratyev KY (1979) Izv Akad Nauk SSSR 15:1280
145. Kondratyev KY (1973) Tr GGO 295:62
146. Zolotukhin VG et al. (1982) Processing of information obtained within the programme "INTERCOSMOS", Moscow, p 257
147. Obukhov AM et al. (1981) Atmospheric physics and climate problem. Nauka press, Moscow, p 216
148. Kondratyev et al. (1972) Optical phenomena in the atmosphere from observations from manned spacecraft. Gidrometeoizdat, Leningrad
149. Panfilov AS (1976) Cosmic Invest 14:316
150. Pivovarova ZI (1977) Radiative characteristics of the climate in the USSR. Gidrometeoizdat, Leningrad
151. Pinsker ISh (1979) Chaotization principle and its application to observational data processing. In: Models, algorithms. Making decisions. Nauka Press, Moscow, p 5
152. Kondratyev KY et al. (1967) Radiation field of the Earth as a planet. Gidrometeoizdat, Leningrad
153. Kondratyev KY et al. (1977) Radiation field of a spherical atmosphere. LGU Press, Leningrad
154. Kondratyev KY, Ter-Markariants NE (eds) (1976) Complete radiative experiment. Gidrometeoizdat, Leningrad
155. Oppenheim E (ed) (1980) Application of the digital processing of signals. Mir Press, Moscow
156. Malkevich MS et al. (1958) Calculation of the light brightness in the atmosphere with anisotropic scattering. Akad Nauk SSSR Press, Moscow, Pt 1, (1962) Pt 2
157. Zuev VE et al. (1974) Izv VUZov Phys 11:30
158. Rachkulik VI, Sitnikova MV (1976) Meteorol Hydrol 4:82
159. Rachkulik VI, Sitnikova MV (1981) Reflectivity and the state of vegetation cover. Gidrometeoizdat, Leningrad
160. Remez EY (1969) Fundamentals of numerical methods of the Chebyshev approximation. Naukova Dumka Press, Kiev
161. Marchuk GI et al. (1968) A solution of direct and some inverse problems in atmospheric optics using Monte-Carlo simulations. Nauka Press, Novosibirsk
162. Rozenberg VI (1972) Scattering and reduction of electromagnetic radiation on atmospheric particles. Gidrometeoizdat, Leningrad
163. Rozenberg GV (1981) Atmospheric aerosol and the optics of scattering media. In: Rozenberg GV (ed) Some problems in present atmospheric physics. Nauka Press, Moscow, p 134
164. Rozenberg GV, Sandomirsky AB (1971) Izv Akad Nauk SSSR FAO 7:737
165. Rozenfeld A (1972) Recognition and processing of images úsing computers. Mir Press, Moscow
166. Romanova LM (1982) Izv Akad Nauk SSSR FAO 5:467
167. Ross YK (1975) Radiative regime and architectonics of vegetation cover. Gidrometeoizdat, Leningrad
168. Rytov SM, Kravtsov YA, Tatarsky VI (1978) Introduction to statistical radiophysics. Pt 2: Random fields. Nauka Press, Moscow
169. Vasilyev MP et al. (eds) (1973) "Salyut" in the orbit. Mashinostroenie Press, Moscow
170. Kondratyev KY et al. (1971) Proc All-Union Conf on meteorological support from supersonic aviation, Leningrad, p 11
171. Seber J (1980) Linear regression analysis. Mir Press, Moscow

172. Sirotenko OD (1981) Mathematical modeling of water-heat regime ecosystems productivity. Gidrometeoizdat, Leningrad
173. Smelov VV (1978) Lectures on the theory of the transfer of neutrons. Atomizdat, Moscow
174. Smokty OI (1976) Proc All-Union Symp on atmospheric optics. Tomsk, Pt 1, p 359
175. Smokty OI (1976) Proc All-Union Symp on atmospheric optics. Tomsk, Pt 2, p 70
176. Smokty OI (1976) Izv Akad Nauk SSSR FAO 12:1053
177. Smokty OI (1976) Izv Akad Nauk SSSR FAO 12:1168
178. Smokty OI (1982) Summaries of papers for the First All-Union Conf Biosphere and climate from studies from space, Baku, 29 Nov–3 Dec 1982. ELM Press, Baku, p 38
179. Smokty OI (1986) Modeling the radiation fields in the problems of space-based spectrophotometry. Nauka Press, Leningrad
180. Smokty OI, Fabrikov VA (1989) Techniques in the theory of systems and transformation in the optics. Nauka Press, Leningrad
181. Snegerev VA (1976) Tr GosNiTsIPR 2:46
182. Snegerev VA, Kozoderov VV (1975) Tr GNTs 165:87
183. Sobolev VV (1948) Dokl Akad Nauk SSSR 61:121
184. Sobolev VV (1956) Radiative energy transfer in the atmospheres of stars and planets. Gostekhteorizdat, Moscow
185. Sobolev VV (1968) Astrophysics 4:325
186. Sobolev VV (1969) Astrophysics 5:5
187. Sobolev VV (1972) Light scattering in the planetary atmospheres. Nauka Press, Moscow
188. Sobolev VV (1974) Astron J 51:50
189. Sobolev VV (1984) Astrophysics 20:123
190. Sobolev VV (1987) Dokl Akad Nauk SSSR 295:60
191. Zuev VE et al. (1978) J Appl Spectrosc 29:1011 (see also (1980) "Soyuz-22" investigates the Earth. Nauka Press, Moscow
192. Sushkevich TA, Mishin IV (1982) Dokl Akad Nauk SSSR 263:60
193. Tikhonov AN, Arsenin VYa (1979) Methods for solving incorrective problems. Nauka Press, Moscow
194. Stepanov BI, Ivanov AP (eds) Theoretical and applied problems of light scattering. Nauka Tech Press, Minsk
195. Tu J, Gonsales R (1978) Principles of image recognition. Mir Press, Moscow
196. Wittacker ET, Watson GN (1963) The course of the current analysis. Fizmatizdat, Moscow
197. Finkelberg VM (1964) J Techn Phys 34:509
198. Fedchenko PP, Kondratyev KY (1981) Spectral reflectivity of some soils. Gidrometeoizdat, Leningrad
199. Feldbaum AA, Butkovsky AG (1971) Methods in the theory of automatic control. Nauka Press, Moscow
200. Shifrin KS et al. (1981) Physical aspects of remote sounding of the ocean-atmosphere system. Nauka Press, Moskow
201. Filippov VL (1976) Izv VUZov Phys 5:34
202. Filippov VL, Ivanov VP (1979) Meteorol Hydrol 4:65
203. Fu K (1977) Structural methods for recognition of images. Mir Press, Moscow
204. Khristov KhY (1961) J Comp Math Math Phys 1:825
205. Chandrasekhar S (1953) Radiative energy transfer. IL, Moscow
206. Shifrin KS (1951) Light scattering in a turbid medium. Gidrometeoizdat, Leningrad
207. Shifrin KS, Kolomiytsov VY (1968) Tr GGO 221:142
208. Yunge X (1966) Chemical composition and radioactivity of the atmosphere. Mir Press, Moscow
209. Yaglom AM (1981) Correlative theory of stationary random functions. Gidrometeoizdat, Leningrad
210. Yanovitsky EG (1972) Izv Akad Nauk SSSR FAO 8:518
211. Yaroslavsky LP (1979) Introduction to the digital processing of images. Sov Radio Press, Moscow
212. Abhyankar KD, Fymat AL (1970) Astrophys J 159:1009
213. Ackerman TP, Baker MB (1977) J Appl Meteorol 16:63

214. Ahern FJ, Goodenough DJ, Jain SC, Rao VR, Rochon G (1977) Use of clear lakes as standard reflectors for atmospheric measurements. Proc 11th Int Symp Remote Sens Environ, p 731
215. Aranuvachapun S (1986) Int J Remote Sens 7:499
216. Arking A, Potter G (1969) Astrophys J 155:565
217. Badhwar GD, Carnes JG, Austin WW (1982) Remote Sens Environ 12:57
218. Bary E de (1975) Beitr Phys Atmos 48:95
219. Bary E de, Eschelbach G (1974) Tellus 26:682
220. Bellman R, Kalaba R (1963) Icarus 1:297
221. Brest CL, Goward SN (1987) Int J Remote Sens 8:351
222. Briegleb BP, Minais P, Ramanathan V, Harrison E (1986) J Clim Appl Meteorol 25:214
223. Brown RJ, Bernier M et al. (1982) Mach Proc Remote Sens Data Spec Emphas Crop Invent and Monit, VIIIth Int Symp, West Lafayette, Ind, July 7–9, 1982 New York, NY, p 374
224. Buriez JC, Bonnel B, Fouquart Y (1986) Contrib Atmos Phys 59:263
225. Burt JE (1985) Phys Geogr 6:240
226. Busbridge IW, Orchard SE (1967) J Astr 149:655
227. Busbridge IW, Orchard SE (1968) Astrophys J 154:729
228. Cadle RD, Gramms GW (1975) Rev Geophys Space Phys 13:475
229. Cano D, Monget JM Albuisson M, Guillard H, Regas N, Wald L (1986) Solar Energy 37:31
230. Canosa J, Wray A (1976) J Quant Spectrasc Radiat Transfer 4:232
231. Charlock TR, Sellers WD (1980) J Atmos Sci 37:1327
232. Chavez PS (1988) Remote Sens Environ 24:459
233. Chen TS, Ohring G (1984) J Atmos Sci 41:150
234. Chu CM, Clark GC, Churchill SM (1957) Tables of angular distribution coefficients for light scattering by spheres. Eng Res Inst Michigan
235. Chylek P, Grams G, Pinnick RC (1976) Science 193:480
236. Coffeen D (1969) Astrophys J 74: N3
237. Coulson KL, Dave JV, Sekera Z (1960) Tables related to radiation emerging from a planetary atmosphere with Rayleigh scattering. Univ Calif Press, Berkeley Los Angeles
238. Coulson KL, Gray EL, Bouricius GM (1965) A study of the reflection and polarization characteristics of selected natural and artificial surface: Rept R-65504. Space Sci Lab Gen Electr Co, Philadelphia
239. Crist EP (1984) Remote Sens Environ 14:3
240. Dana R (1982) Appl Opt 21:4106
241. Darnell WL, Gupta SK, Staylor WF (1983) J Clim Appl Meteorol 22:1956
242. Dave JV (1975) J Atmos Sci Pt 1 / 32:790, Pt 2 32:1463
243. Dave JV (1980) Photogramm Eng Remote Sens 46:1178
244. Dave JV, Canosa JA (1974) J Atmos Sci 31:1089
245. Deshamps P, Herman M, Tanre D et al. (1982) ESA J 6:283
246. Diner DJ, Martonchik JV (1984) J Quant Spectrosc Radiat Transfer 31:97
247. Diner PJ, Martonchik JV (1984) J Quant Spectrosc Radiat Transfer 32:279
248. Diner PJ, Martonchik JV (1985) IEEE Trans Geosci Remote Sens 23:618
249. Diner PJ, Martonchik JV (1985) Appl Opt 24:3503
250. Dugas WA, Heuer ML (1985) J Clim Appl Meteorol 24:751
251. Dunn WL, Maiorino JR (1980) J Quant Spectrosc Radiat Trnasfer 24:203
252. Dunn WL (1981) J Comp Phys 41:154
253. Dunn WL (1983) J Quant Spectrosc Radiat Transfer 29:19
254. Duntley SQ, Gordon YJ, Taylor YH et al. (1964) Appl Opt 3:312
255. Rowell R, Stein R (eds) (1967) Electromagnetic scattering. Gordon & Breach Sci Publ, New York
256. Elterman L (1964) Atmospheric attenuation model 1964 in the ultraviolet, visible and infrared region for altitude to 50 km: Report AFCRL-64-740. Environ Res Pap 46:190
257. Elterman L (1968) UV, visible and IR attenuation for altitudes to 50 km, Report AFCRL-68-0153. Environ Res 285:182
258. Elterman L, Wexler R, Chang DF (1969) Appl Opt 8:693

259. Elterman L (1970) Vertical-attenuation model with eight surface meteorological ranger 2 to 13 kilometers. AFCRL-70-0200, March 1970, Environ Res Pap 318, Air Force Cambridge Res Lab, Bedford, Massachusetts
260. Elterman L, Toolin RB, Essex JD (1973) Appl Opt 12:330
261. Fraser RS, Babethi P, Al-Abbas AH (1975) Proc 2nd Conf Remote Sens, AnnArbor, p 71
262. Fraser RS, Kaufman YJ (1985) IEEE Geosci Remote Sens 23:625
263. Fymat AL, Lenoble J (1979) J Quant Spectrosc Radiat Transfer 21:75
264. Gauthier C, Diak G, Masse S (1984) J Clim Appl Meteorol 23:1380
265. Gordon HR (1978) Appl Opt 17:1631
266. Gordon HR, Clark DK (1981) Appl Opt 20:4175
267. Gordon HR, Clark DK, Brown JW, Brown OB, Evans RH, Broenkow WW (1983) Appl Opt 22:20
268. Gras TL, Laby TE (1978) J Geophys Res 83:1869
269. Grassl H, Newiger M (1982) Sci Total Environ 23:713
270. Greenberg JM, Lind AC, Wang RT (1967) Scattering by nonspherical systems. In: Rowell R, Stein R (eds) Electromagnetic scattering. Gordon & Breach Sci Publ, New York, p 3
271. Gupta SK, Darnell WL, Staylor WF (1983) A parameterization for downward longwave radiation from satellite meteorological data: 5th Conf Atmos Rad, Oct 31 – Nov 4 1983, Baltimore, Maryland. Am Meteorol Soc, Boston, p 436
272. Guzzi R, Rizzi R, Zibordi G (1987) Appl Opt 26:3042
273. Haba Y, Kawata Y, Terashita Y (1978) Atmosphere effects on classification of remotely sensed Earth's imageries. Environ Syst Plann, Des and Cont Proc IFAC Symp Kyoto, 1977, vol 1, Oxford, p 409
274. Halpern P (1984) Remote Sens Environ 15:47
275. Hansen JE (1966) Astrophys J 155:566
276. Hansen JE, Hovenier JW (1974) J Atmos Sci 31:1137
277. Hansen JE, Travis LD (1974) Light scattering in planetary atmospheres. Goddard Inst Space Stud, New York
278. Hatfield JL, Asrar G, Kanemasu ET (1984) Remote Sens Environ 14:65
279. Hoffman DJ, Rosen JM, Pinnick RG (1976) Measurements of stratospheric aeorsol size distribution and optical model calculations: Preprint, Symp Radiation, C Partenkirchen, Aug 19 – 28
280. Holben BN, Fan C (1980) Photogramm Eng Remote Sens 46:651
281. Horak H, Chandrasekhar S (1961) Astrophys J 134:45
282. Horak H, Janousek A (1965) Astrophys J (Suppl) Ser 11:277
283. Horak H, Little S (1965) Astrophys J (Suppl) Ser 11:373
284. Hovenier JW (1978) Astron Astrophys 68:239
285. Hovenier JW (1980) Astron Astrophys 82:61
286. Hovenier JW, Mee C van der (1983) Astron Astrophys 128:1
287. Hovenier JW, Mee C van der, Heer D de (1988) Astron Astrophys 207:194
288. Buznikov AA, Kondratyev KY, Vasilyev OB, Smokty OI (1974) Influence of the atmosphere on spectral radiance and contrast of natural formations measured from space: Proc UCLA Int Conf Univ Calif, Los Angeles, p 309
289. Jackson RD, Slater PH, Pinter PJ (1983) Int J Remote Sens 4:313
290. Justus CG (1984) Atmospheric and surface short-wave energy balance from combined satellite and ground-based data. Conf Satellite Remote Sens Appl, Preprint, Clearwater Beach, Fla, June 25 – 29, 1984, Am Meteorol Soc, Boston, Mass, p 197
291. Justus CG, Tarpley JD (1983) Accuracy and availability of solar radiation data from satellites and from forecast estimates. 5th Conf Atmos Radiat, Baltimore, MD, 31 Oct – 4 Nov 1983, Boston, Mass, p 243
292. Justus CG, Paris MV, Tarpley JD (1986) Remote Sens Environ 20:57
293. Kamiuto K, Seki J (1987) J Quant Spectrosc Radiat Transfer 37:455
294. Kästner M (1980) Ann Meteor Neul Forge 15:181
295. Kästner M, Köpke P, Quenzel H (1983) Adv Space Res 2:119
296. Kaufman YJ (1979) J Geophys Res 84:3165
297. Kaufman YJ (1980) Proc SPIE 475:20

298. Kaufman YJ (1982) J Geophys Res 8:4137
299. Kaufman YJ, Joseph JH (1982) J Geophys Res 87:1287
300. Kaufman YJ, Fraser RS (1984) Remote sensing of aerosol absorption. Proc Int Radiat Symp, Perugia, Italy, 21–28 Aug 1984. Deepak Publ Hampton, USA, p 16
301. Kaufman YJ (1988) IEEE Trans Geosci Remote Sens 26:441
302. Kawata Y, Kusaka T, Ueno S (1987) In: Singh M (ed) Encyclopedia of systems and Control. Pergamon Press, New York p 2652
303. Kawata Y, Ueno S, Kusaka T (1988) J Remote Sens 9:729
304. Kiang RK (1982) IEEE Trans Geosci Remote Sens 20:365
305. King MD (1981) J Atmos Sci 38:2031
306. King MD (1983) J Quant Spectrosc Radiat Transfer 30:143
307. King MD, Harshvardhan (1986) Comparative accuracy of the albedo, transmission and absorption for selected radiative transfer approximations NASA Ref Publ 1160:44
308. King MD, Harshvardhan S (1986) J Atmos Sci 43:784
309. King MD (1987) J Atmos Sci 44:1734
310. Köpke P, Quenzel H (1979) Geophys Res 84:7847
311. Köpke P, Qzenzel H (1983) Adv Space Res 2:29
312. Kondratyev KY, Smokty OI (1977) Applications of the radiation transfer theory to the interpretation of the Earth's brightness measurements from space: Abstr Pap XXVIII IAF Congr, Praha, p 215
313. Kondratyev KY et al. (1979) Tellus 31:13
314. Kondratyev KY, Kozoderov VV, Smokty OI (1983) The atmospheric impact on the spectra and images of natural formations as observed from space: XXXIV Congress of the IAF, Abstr Pap, Budapest, Hungary, Oct 10–13, p 92
315. Kopeika NS, Solomon S, Gencay Y (1981) J OptSoc Am 71:892
316. Kopeika NS (1982) J Opt Soc Am 72:548
317. Kriebel KT, Koepke P, Amann V (1984) Mixed surface reflection functions determined from a priori knowledge and satellite radiance. Adv Space Res 1985 5 N6: Space Observ Clim Stud Proc Symp 4 COSPAR, 25th Plenary Meet, Graz, 25 June–7 July, p 345
318. Kusur J (1956) J Math Phys34:256
319. Liou KN (1980) An introduction to atmospheric radiation. Int Geophys Ser 26, Pergamon Press, New York
320. Kusaka T, Kawata J, Ueno S, Furumoto S (1987) Removal of the atmospheric and topographic effects from the rugged terrain image data remotely sensed by Landsat. IGARSS'87: Int Geosci Remote Sens Symp, Ann Arbor, Mich, May 18–21, Vol 1, New York, NY, p 649
321. Larsen EW (1984) J Math Phys 25:131
322. Lenoble J (ed) (1972) Standard procedures to compute atmospheric radiative transfer in a scattering atmosphere. Int Assoc Meteorol Atmos Phys Rad Comm, Boulder, CO
323. MacFarlane NA (1981) A test of an atmospheric correction of Landsat data over water surfaces. Match Remote Sens Tech Appl Proc 9th Ann Conf, London, 16–18 Dec Reading, p 499
324. Manschhoff R, Jalink A, Hickey J, Swedberg J (1984) J Geophys Res 89:5059
325. McCormick NJ (1979) J Math Phys 20:1504
326. McCormick NJ, Sanches R (1981) J Math Phys 22:199
327. McCormick NJ (1981) Prog Nuclear Energy 8:235
328. McCormick NJ (1984) Transp Theory Stat Phys 13:15
329. McCormick NJ (1986) Transp Theory Stat Phys 15:759
330. McClatchey RA, Fenn RW, Selby JE et al. (1970) Optical properties of the atmosphere. AFCRL-70-0527, p 188
331. McClatchey RA, Bolle HJ, Kondratyev KY (1980) Report of the IAMAP RC WG on a standard radiation atmosphere. WMO/IAMAP, Fort Collins, CO, USA, 10 August, Preprint
332. Meerkötter R, Grassl H (1984) Longwave net flux at the ground from radiances at the top. Proc Int Radiat Symp, Perugia, Italy, 21–28 Aug 1984. Deepak Publ, Hampton, USA, p 220
333. Mekler Y, Kaufman YJ (1980) J Geophys Res 85:4067
334. Mekler Y, Kaufman YJ (1982) Appl Opt 21:310
335. Mekler Y, Kaufman YJ (1984) J Atmos Sci 41:2595

336. Morimoto T, Kumar R et al. (1979) Effect of atmosphere on the classification of Landsat data: Mach Proc Remote Sens Data, 5th Annu Symp IEEE, West Lafayette, Ind, New York, p 43
337. Möser W, Raschke E (1983) Meterol Rdsch 36:33
338. Möser W, Raschke E (1984) J Clim Appl Meteorol 23:166
339. Mullikin TM (1964) Trans Am Math Soc 113:316
340. Mullikin TM (1964) Astron J 139:379
341. Mullikin TM (1964) Ap J 139:1267, (1967) Ap J 147:1858
342. Murtha PA (1978) Photogramm Eng Remote Sens 44:1147
343. Rao NCR (1984) Arch Met Geophys Bioclim B34:353
344. Rao NCR, Takashima T (1986) Quart J R Met Soc 112:1111
345. Rao NCR, Takashima T (1987) Meteorol Atmos Phys 37:205
346. Nilson B (1979) Appl Opt 18:3457
347. Odenweller JB, Johnson KI (1984) Remote Sens Environ 14:39
348. Orchard SE (1967) Astr. 149:665
349. Otterman J, Robinov CJ (1981) Int J Remote Sens 2:351
350. Otterman T, Ungar S, Kaufman Y, Podolak KM (1980) Remote Sens Environ 9:115
351. Otterman J, Tucker CJ (1985) J Clim Appl Meteorol 24:228
352. Paál A (1985) Zb Meteorol Hidrol Radiat 12:216
353. Paál A (1985) Idöjàràs 89:194
354. Pinker RT, Razgaitis JA (1983) J Geophys Res 88:6007
355. Pinker RT, Razgaitis JA (1983) Surface radiation budget from satellites. 5th Conf Atmos Radiat, Oct 31–Nov 4, p 495
356. Pinker RT, Corio LA (1984) Mon Weather Rev 112:209
357. Pinker RT (1985) Determination of surface albedo from satellites. Adv Space Res 5 6:Space Observ Clim Stud Proc Symp 34 COSPAR, 25th Plenary Meet, Graz, 25 June–7 July 1984, p 333
358. Pinty B, Szeiwach G, Stum J (1984) Surface albedo over the Sahel from Meteosat radiances. Conf Satell Remote Sens Appl, Clearwater Beach, Fla, June 25–29, 1984. Boston, Mass, p 73
359. Pinty B, Szeiwach G (1985) J Clim Appl Meteorol 24:741
360. Piotrowski S (1956) Astrophys J 121:325
361. Plass GN, Kattawar CM, Guinn JA (1975) Appl Opt 14:1924
362. Pollak JB, Toon OB, Sagan C et al. (1976) J Geophys Res 81:1071
363. Powell GL, Brazel AJ, Pasqualetti MJ (1984) Prof Geogr 36:227
364. Quenzel H, Kalstner H (1980) Appl Opt 19:1338
365. Quenzel H (1983) Adv Space Res 2:19
366. Raschke E (1985) On the derivation of radiation budget parameters at the surface from satellite measurements. Adv Space Res 5 6: Space Observ Clim Stud Proc Symp 4 COSPAR, 25th Plenar Meet, Graz, 25 June–7 July, p 319
367. Raphael C (1983) Models for estimating solar irradiance from satellite data: an initial assessment. Can Climate Cent Rep 83–1:104
368. Rimóczi-Paál A (1985) Arch Meteorol Geophys Bioklimatol B36:67
369. Rimóczi-Paál A (1983) Idöjárás 87:352
370. Rogers RH (1974) Investigation of techniques for correcting ERTS data for solar and atmosphere effect. Final Rep 207771, NA-5-21863, Goddard Space Flight Center, Greenbelt, MD
371. Sanchez R, McCormick NJ (1981) J Math Phys 22:847
372. Sanchez R, McCormick NJ (1982) J Quant Spectrosc Radiat Transfer 28:169
373. Shaw GE (1976) Physical and climatic properties of the global aerosols. Preprint Symp Radiat, G-Partenkirchen, August 19–28, p 96
374. Shettle EP, Fenn RW (1979) Models for the aerosols of the lower atmosphere and the effects of humidity variations on their optical properties. AFGL-TR-79-0214:94
375. Siewert CE (1978) J Math Phys 19:2619
376. Siewert CE (1979) J Quant Spectrosc Radiat Transfer 22:441

377. Singh SM (1982) A procedure for atmospheric correction of Coastal Zone Color Scanner (CZCS) data. Remote Sens Atmos Proc Annu Tech Conf, Liverpool, 15–17 Dec, Reading, p 169
378. Sjoberg RW, Horn BKP (1983) Appl Opt 22:1702
379. Slater PN (1985) Proc IEEE 73:997
380. Smokty OI (1988) Solving the inverse problems of the atmospheric optics based on the environmental space remote sensing information IRS'88: Abstr Int Radiat Symp, Lille, France, August 1988, p 221
381. Smokty OI (1988) Implementation of the principle of mirror symmetry in the radiation transfer theory. IRS'88: Abstr Int Radiat Symp, Lille, France, August 1988, p 185
382. Stamnes K, Dave JV (1981) J Atmos Sci 78:2696
383. Steven MD, Rollin EM (1986) Int J Remote Sens 7:481
384. Suffles JT, Buckström BR, Timary SN (1981) AIAA Pap 1094:7
385. Switzer P, Kowalik WS, Lyon WS (1981) Photogramm. Eng Remote Sens 47:1469
386. Tanre D, Herman M, Deschamps PY, de Leffe A (1979) Appl Opt 18:3587
387. Tanre D, Herman M, Deschamps PY (1981) Appl Opt 20:3676
388. Tanre D, Herman M, Deschamps PY (1983) Appl Opt 22:733
389. Toon OB, Pollak JB (1976) J Appl Meteorol 15:225
390. Tucker JL, Holben BL, Elgin JH Jr, Memurlrey JE (1980) Photogramm Eng Remote Sens 46:657
391. Turner RE, Spencer MM (1972) Atmospheric model for correction of spacecraft data. Proc 8th Int Symp Remote Sens Environ p 895
392. Turner Re (1977) Signature variations due to atmospheric effects. Proc 10th Int Symp Remote Sens Environ 2:671
393. Ueno S, Kalaba R, Kagiwada H, Bellman R (1970) Publ Astron Soc JPN 22:75
394. Ueno S, Mukai S (1978) Atmospheric effects on remotely sensed data from space. Environ Syst Plann, Des Contr Proc IFAC Symp, Kyoto, 1977, Oxford ea, 1:423
395. Ueno S (1981) Parameter identification of multiple scattering system via quasilinearization. Proc IFAC 8th Triennial World Congr, 24–28 Aug 1981, Kyoto, Pergamon Press, VI:89
396. Ueno S (1982) Real-time solution of linear least-squares estimations problem with semi-degenerate covariance. Proc 1982 Am Control Conf, Arlington, Virginia, 14–16 June, 1982, p 592
397. Ueno S, Kawata Y (1983) Optical assessment of subsurface water parameters using radiance measurements from space. Papers selected for the 16th Int Symp Remote Sens Environ, Buenos Aires, Argentina, June ERIM, p 409
398. Ueno S (1983) Appl Math Comp 13:87
399. Ueno S (1985) J Math Phys 26:85
400. Ueno S, Kawata Y, Kusaka T (1985) Optimum classification of Landsat TM data for ecological study: Proc 19th Int Symp Remote Sens Environ, Ann Arbor, Michigan, 21–25 Oct 1985, ERIM, p 821
401. Ueno S, Kawata Y (1987) Identification of optical parameters in Earth-atmosphere system. In: Singh M (ed) Encyclopedia of systems and control. Pergamon Press, New York, p 1307
402. Ueno S, Kawata Y, Kusaka T (1987) Ground albedo mapping invariant embedding. In: Singh M (ed) Encyclopedia of systems and control. Pergamon Press, New York, p 2046
403. Ueno S, Kawata Y, Kusaka T (1989) Comparative analysis of Landsat 5 and SPOT HRV data in Kanazawa region. Proc 16th Congr Int Soc Photogramm Remote Sens (ISPRS), July 1–11, 1988, Kyoto JPN. Photogramm Eng Remote Sens 55:106
404. Van de Hulst HC (1948) Astrophys J 107:220
405. Van de Hulst HC (1980) Multiple light scattering. Academic Press, Lond New York, Vols 1–2
406. Violier M, Tanre D, Deschamps PJ (1980) Bound Layer Remote Sens Meteorol 18:247
407. Welch RM, Cox SK, Kondratyev KY (1980) Determination of vertical profiles of aerosol size spectra from aircraft radiative flux measurements. Pt 2: The effect of particle nonsphericity. Preprint
408. Wells WC, Gal G, Munn TW (1977) Appl Opt 16:634
409. Werernak CT, Turner RE, Lyzenda DC (1976) Spectral reflectance and radiance characteristics of water pollutions. Rep NASA-CR-2665:218

410. Whitlock CH (1982) Appl Opt 21:4196
411. Wiegand CL, Richardson AJ, Kanemasu ET (1975) Phtogramm Eng Remote Sens 41:1401
412. Yates HM, Tarpley YD (1982) The role of meteorological satellites in agricultural remote sensing: Mach Proc Remote Sens, West Lafayette, Ind, July 7–9, 1982, New York, NY, p 23
413. Zibordi G, Guzzi R, Rizzi R (1986) A model for the atmospheric correction of data measured from flying platforms over the sea. Int Meet Atmos Transparency for Satellite Applications, 15–19 Sept, Capri, Italy, p 169

Subject Index

MIX
Papier aus verantwortungsvollen Quellen
Paper from responsible sources
FSC® C105338

If you have any concerns about our products,
you can contact us on
ProductSafety@springernature.com

In case Publisher is established outside the EU,
the EU authorized representative is:
Springer Nature Customer Service Center GmbH
Europaplatz 3, 69115 Heidelberg, Germany

Printed by Libri Plureos GmbH
in Hamburg, Germany